Pearson New International Edition

Fundamentals of Signals and Systems
Using the Web and MATLAB
Edward W. Kamen Bonnie S Heck
Third Edition

Pearson Education Limited
Edinburgh Gate
Harlow
Essex CM20 2JE
England and Associated Companies throughout the world

Visit us on the World Wide Web at: www.pearsoned.co.uk

© Pearson Education Limited 2014

ISBN 10: 1-292-02598-0
ISBN 13: 978-1-292-02598-8

British Library Cataloguing-in-Publication Data
A catalogue record for this book is available from the British Library

ARP Impression 98
Printed in Great Britain by Clays Ltd, St Ives plc

Table of Contents

Fundamental Concepts

The concepts of signals and systems arise in virtually all areas of technology, including electrical circuits, communication devices, signal processing devices, robotics and automation, automobiles, aircraft and spacecraft, biomedical devices, chemical processes, and heating and cooling devices. The concepts of signals and systems are also of great importance in other areas of human endeavor, such as in science and economics. In this chapter various fundamental aspects of signals and systems are considered. The chapter begins with a brief introduction to continuous-time and discrete-time signals given in Sections 1.1 and 1.2. In Section 1.2 it is shown how discrete-time data can be acquired for analysis by downloading data from the Web. Then in Section 1.3 the concept of a system is introduced, and in Section 1.4 three specific examples of a system are given. In Section 1.5 of the chapter, the basic system properties of causality, linearity, and time invariance are defined. A summary of the chapter is given in Section 1.6.

1.1 CONTINUOUS-TIME SIGNALS

A signal $x(t)$ is a *real-valued,* or *scalar-valued,* function of the time variable t. The term *real valued* means that for any fixed value of the time variable t, the value of the signal at time t is a real number. When the time variable t takes its values from the set of real numbers, t is said to be a *continuous-time variable* and the signal $x(t)$ is said to be a *continuous-time signal* or an *analog signal.* Common examples of continuous-time signals are voltage or current waveforms in an electrical circuit, audio signals such as speech or music waveforms, positions or velocities of moving objects, forces or torques in a mechanical system, bioelectric signals such as an electrocardiogram (ECG) or an electroencephalogram (EEG), flow rates of liquids or gases in a chemical process, and so on.

Given a signal $x(t)$ that is very complicated, it is often not possible to determine a mathematical function that is exactly equal to $x(t)$. An example is a speech signal, such as the 50-millisecond (ms) segment of speech shown in Figure 1.1. The segment of speech shown in Figure 1.1 is the "sh"-to-"u" transition in the utterance of the word "should." Due to their complexity, signals such as speech waveforms are usually not specified in mathematical form. Instead, they may be given by a set of sample values. For example, if $x(t)$ denotes the speech signal in Figure 1.1, the signal can be represented by the set of sample values

$$\{x(t_0), x(t_1), x(t_2), x(t_3), \ldots, x(t_N)\}$$

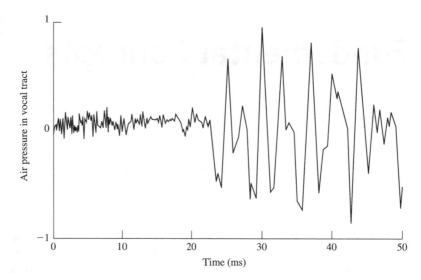

FIGURE 1.1
Segment of speech.

where $x(t_i)$ is the value of the signal at time $t_i, i = 0, 1, 2, \ldots, N$, and $N + 1$ is the number of sample points. This type of signal representation can be generated by sampling the speech signal. Sampling is discussed briefly in Section 1.2 and then is studied in depth in later chapters.

In addition to the representation of a signal in mathematical form or by a set of sample values, signals can also be characterized in terms of their "frequency content" or "frequency spectrum." The representation of signals in terms of the frequency spectrum is accomplished by using the Fourier transform, which is studied in Chapters 3 to 5.

Some simple examples of continuous-time signals that can be expressed in mathematical form are given next.

1.1.1   Step and Ramp Functions

Two simple examples of continuous-time signals are the unit-step function $u(t)$ and the unit-ramp function $r(t)$. These functions are plotted in Figure 1.2.

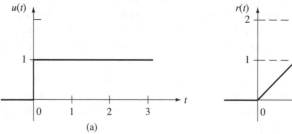
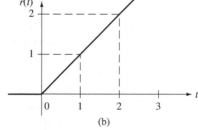

FIGURE 1.2
(a) Unit-step and (b) unit-ramp functions.

The *unit-step function u(t)* is defined mathematically by

$$u(t) = \begin{cases} 1, & t \geq 0 \\ 0, & t < 0 \end{cases}$$

Here *unit step* means that the amplitude of $u(t)$ is equal to 1 for all $t \geq 0$. [Note that $u(0) = 1$; in some textbooks, $u(0)$ is defined to be zero.] If K is an arbitrary nonzero real number, $Ku(t)$ is the step function with amplitude K for $t \geq 0$.

For any continuous-time signal $x(t)$, the product $x(t)u(t)$ is equal to $x(t)$ for $t \geq 0$ and is equal to zero for $t < 0$. Thus multiplication of a signal $x(t)$ with $u(t)$ eliminates any nonzero values of $x(t)$ for $t < 0$.

The *unit-ramp function r(t)* is defined mathematically by

$$r(t) = \begin{cases} t, & t \geq 0 \\ 0, & t < 0 \end{cases}$$

Note that for $t \geq 0$, the slope of $r(t)$ is 1. Thus $r(t)$ has "unit slope," which is the reason $r(t)$ is called the unit-ramp function. If K is an arbitrary nonzero scalar (real number), the ramp function $Kr(t)$ has slope K for $t \geq 0$.

The unit-ramp function $r(t)$ is equal to the integral of the unit-step function $u(t)$; that is,

$$r(t) = \int_{-\infty}^{t} u(\lambda)\, d\lambda$$

Conversely, the first derivative of $r(t)$ with respect to t is equal to $u(t)$, except at $t = 0$, where the derivative of $r(t)$ is not defined.

1.1.2   The Impulse

The *unit impulse* $\delta(t)$, also called the *delta function* or the *Dirac distribution*, is defined by

$$\delta(t) = 0, \quad t \neq 0$$

$$\int_{-\varepsilon}^{\varepsilon} \delta(\lambda)\, d\lambda = 1, \quad \text{for any real number } \varepsilon > 0$$

The first condition states that $\delta(t)$ is zero for all nonzero values of t, while the second condition states that the area under the impulse is 1, so $\delta(t)$ has unit area.

It is important to point out that the value $\delta(0)$ of $\delta(t)$ at $t = 0$ is not defined; in particular, $\delta(0)$ is not equal to infinity. The impulse $\delta(t)$ can be approximated by a pulse centered at the origin with amplitude A and time duration $1/A$, where A is a very large positive number. The pulse interpretation of $\delta(t)$ is displayed in Figure 1.3.

For any real number K, $K\delta(t)$ is the impulse with area K. It is defined by

$$K\delta(t) = 0, \quad t \neq 0$$

$$\int_{-\varepsilon}^{\varepsilon} K\delta(\lambda)\, d\lambda = K, \quad \text{for any real number } \varepsilon > 0$$

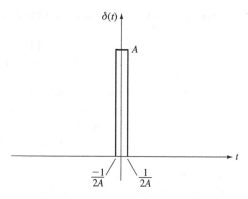

FIGURE 1.3
Pulse interpretation of $\delta(t)$.

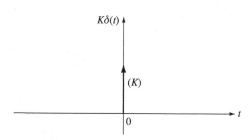

FIGURE 1.4
Graphical representation of the impulse $K\delta(t)$.

The graphical representation of $K\delta(t)$ is shown in Figure 1.4. The notation "(K)" in the figure refers to the area of the impulse $K\delta(t)$.

The unit-step function $u(t)$ is equal to the integral of the unit impulse $\delta(t)$; more precisely,

$$u(t) = \int_{-\infty}^{t} \delta(\lambda)\, d\lambda, \text{ all } t \text{ except } t = 0$$

To verify this relationship, first note that for $t < 0$,

$$\int_{-\infty}^{t} \delta(\lambda)\, d\lambda = 0, \text{ since } \delta(t) = 0 \text{ for all } t < 0$$

For $t > 0$,

$$\int_{-\infty}^{t} \delta(\lambda)\, d\lambda = \int_{-t}^{t} \delta(\lambda)\, d\lambda = 1, \text{ since } \int_{-\varepsilon}^{\varepsilon} \delta(\lambda)\, d\lambda = 1 \text{ for any } \varepsilon > 0$$

1.1.3   Periodic Signals

Let T be a fixed positive real number. A continuous-time signal $x(t)$ is said to be periodic with period T if

$$x(t + T) = x(t) \text{ for all } t, -\infty < t < \infty \tag{1.1}$$

Note that if $x(t)$ is periodic with period T, it is also periodic with period qT, where q is any positive integer. The *fundamental period* is the smallest positive number T for which (1.1) holds.

Signals and Sounds

An example of a periodic signal is the sinusoid

$$x(t) = A \cos(\omega t + \theta), -\infty < t < \infty \tag{1.2}$$

Here A is the amplitude, ω the frequency in radians per second (rad/sec), and θ the phase in radians. The frequency f in hertz (Hz) (or cycles per second) is $f = \omega/2\pi$.

To see that the sinusoid given by (1.2) is periodic, note that for any value of the time variable t,

$$A \cos\left[\omega\left(t + \frac{2\pi}{\omega}\right) + \theta\right] = A \cos(\omega t + 2\pi + \theta) = A \cos(\omega t + \theta)$$

Thus the sinusoid is periodic with period $T = 2\pi/\omega$, and in fact, $2\pi/\omega$ is the fundamental period. The sinusoid $x(t) = A \cos(\omega t + \theta)$ is plotted in Figure 1.5 for the case when $-\pi/2 < \theta < 0$. Note that if $\theta = -\pi/2$, then

$$x(t) = A \cos(\omega t + \theta) = A \sin \omega t$$

An important question for signal analysis is whether or not the sum of two periodic signals is periodic. Suppose that $x_1(t)$ and $x_2(t)$ are periodic signals with fundamental

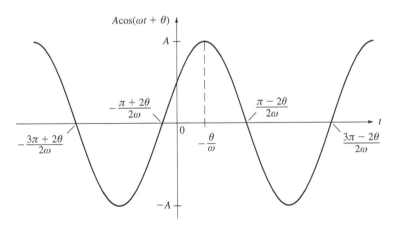

FIGURE 1.5
Sinusoid $x(t) = A \cos(\omega t + \theta)$ with $-\pi/2 < \theta < 0$.

5

periods T_1 and T_2, respectively. Then is the sum $x_1(t) + x_2(t)$ periodic; that is, is there a positive number T such that

$$x_1(t + T) + x_2(t + T) = x_1(t) + x_2(t) \quad \text{for all } t? \tag{1.3}$$

It turns out that (1.3) is satisfied if and only if the ratio T_1/T_2 can be written as the ratio q/r of two integers q and r. This can be shown by noting that if $T_1/T_2 = q/r$, then $rT_1 = qT_2$, and since r and q are integers, $x_1(t)$ and $x_2(t)$ are periodic with period rT_1. Thus the expression (1.3) follows with $T = rT_1$. In addition, if r and q are coprime (i.e., r and q have no common integer factors other than 1), then $T = rT_1$ is the fundamental period of the sum $x_1(t) + x_2(t)$.

Example 1.1   Sum of Periodic Signals

Perio-
dicity of
Sums of
Sinu-
soids

Let $x_1(t) = \cos(\pi t/2)$ and $x_2(t) = \cos(\pi t/3)$. Then $x_1(t)$ and $x_2(t)$ are periodic with fundamental periods $T_1 = 4$ and $T_2 = 6$, respectively. Now

$$\frac{T_1}{T_2} = \frac{4}{6} = \frac{2}{3}$$

Then with $q = 2$ and $r = 3$, it follows that the sum $x_1(t) + x_2(t)$ is periodic with fundamental period $rT_1 = (3)(4) = 12$ seconds.

1.1.4   Time-Shifted Signals

Given a continuous-time signal $x(t)$, it is often necessary to consider a *time-shifted* version of $x(t)$: If t_1 is a positive real number, the signal $x(t - t_1)$ is $x(t)$ shifted to the right by t_1 seconds and $x(t + t_1)$ is $x(t)$ shifted to the left by t_1 seconds. For instance, if $x(t)$ is the unit-step function $u(t)$ and $t_1 = 2$, then $u(t - t_1)$ is the 2-second right shift of $u(t)$ and $u(t + t_1)$ is the 2-second left shift of $u(t)$. These shifted signals are plotted in Figure 1.6. To verify that $u(t - 2)$ is given by the plot in Figure 1.6a, evaluate $u(t - 2)$ for various values of t. For example, $u(t - 2) = u(-2) = 0$ when $t = 0$, $u(t - 2) = u(-1) = 0$ when $t = 1$, $u(t - 2) = u(0) = 1$ when $t = 2$, and so on.

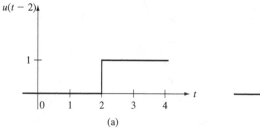
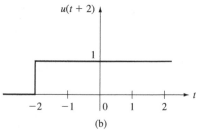

FIGURE 1.6
Two-second shifts of $u(t)$: (a) right shift; (b) left shift.

For any fixed positive or negative real number t_1, the time shift $K\delta(t - t_1)$ of the impulse $K\delta(t)$ is equal to the impulse with area K located at the point $t = t_1$; in other words,

$$K\delta(t - t_1) = 0, \quad t \neq t_1$$

$$\int_{t_1 - \varepsilon}^{t_1 + \varepsilon} K\delta(\lambda - t_1)\, d\lambda = K, \quad \text{any } \varepsilon > 0$$

The time-shifted unit impulse $\delta(t - t_1)$ is useful in defining the *sifting property* of the impulse given by

$$\int_{t_1 - \varepsilon}^{t_1 + \varepsilon} f(\lambda)\delta(\lambda - t_1)\, d\lambda = f(t_1), \quad \text{for any } \varepsilon > 0$$

where $f(t)$ is any real-valued function that is continuous at $t = t_1$. (Continuity of a function is defined subsequently.) To prove the sifting property, first note that since $\delta(\lambda - t_1) = 0$ for all $\lambda \neq t_1$, it follows that

$$f(\lambda)\delta(\lambda - t_1) = f(t_1)\delta(\lambda - t_1)$$

Thus

$$\int_{t_1 - \varepsilon}^{t_1 + \varepsilon} f(\lambda)\delta(\lambda - t_1)\, d\lambda = f(t_1) \int_{t_1 - \varepsilon}^{t_1 + \varepsilon} \delta(\lambda - t_1)\, d\lambda$$

$$= f(t_1)$$

which proves the sifting property.

1.1.5   Continuous and Piecewise-Continuous Signals

A continuous-time signal $x(t)$ is said to be *discontinuous* at a fixed point t_1 if $x(t_1^-) \neq x(t_1^+)$, where $t_1 - t_1^-$ and $t_1^+ - t_1$ are infinitesimal positive numbers. Roughly speaking, a signal $x(t)$ is discontinuous at a point t_1 if the value of $x(t)$ "jumps in value" as t goes through the point t_1.

A signal $x(t)$ is *continuous* at the point t_1 if $x(t_1^-) = x(t_1) = x(t_1^+)$. If a signal $x(t)$ is continuous at all points t, $x(t)$ is said to be a *continuous signal*. The reader should note that the term *continuous* is used in two different ways; that is, there is the notion of a continuous-time signal and there is the notion of a continuous-time signal that is continuous (as a function of t). This dual use of *continuous* should be clear from the context.

Many continuous-time signals of interest in engineering are continuous. Examples are the ramp function $Kr(t)$ and the sinusoid $x(t) = A\cos(\omega t + \theta)$. Another example of a continuous signal is the triangular pulse function displayed in Figure 1.7. As indicated in the figure, the triangular pulse is equal to $(2t/\tau) + 1$ for $-\tau/2 \leq t \leq 0$ and is equal to $(-2t/\tau) + 1$ for $0 \leq t \leq \tau/2$.

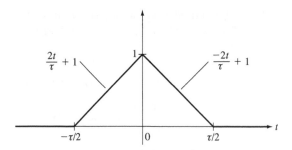

FIGURE 1.7
Triangular pulse function.

There are also many continuous-time signals of interest in engineering that are not continuous at all points t. An example is the step function $Ku(t)$, which is discontinuous at the point $t = 0$ (assuming that $K \neq 0$). Another example of a signal that is not continuous everywhere is the rectangular pulse function $p_\tau(t)$, defined by

$$
p_\tau(t) = \begin{cases} 1, & \dfrac{-\tau}{2} \leq t < \dfrac{\tau}{2} \\[2ex] 0, & t < \dfrac{-\tau}{2}, t \geq \dfrac{\tau}{2} \end{cases}
$$

Here τ is a fixed positive number equal to the time duration of the pulse. The rectangular pulse function $p_\tau(t)$ is displayed in Figure 1.8. It is obvious from Figure 1.8 that $p_\tau(t)$ is continuous at all t except $t = -\tau/2$ and $t = \tau/2$.

Note that $p_\tau(t)$ can be expressed in the form

$$
p_\tau(t) = u\left(t + \frac{\tau}{2}\right) - u\left(t - \frac{\tau}{2}\right)
$$

Note also that the triangular pulse function shown in Figure 1.7 is equal to $(1 - 2|t|/\tau) \times p_\tau(t)$, where $|t|$ is the absolute value of t defined by $|t| = t$ when $t > 0$, $|t| = -t$ when $t < 0$.

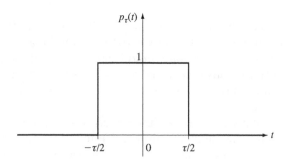

FIGURE 1.8
Rectangular pulse function.

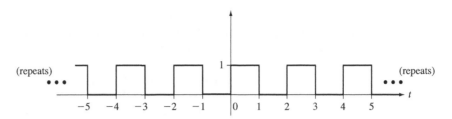

FIGURE 1.9
Signal that is discontinuous at $t = 0, \pm1, \pm2, \ldots$.

A continuous-time signal $x(t)$ is said to be *piecewise continuous* if it is continuous at all t except at a finite or countably infinite collection of points t_i, $i = 1, 2, 3, \ldots$. Examples of piecewise-continuous functions are the step function $Ku(t)$ and the rectangular pulse function $p_\tau(t)$. Another example of a piecewise-continuous signal is the pulse train shown in Figure 1.9. This signal is continuous at all t except at $t = 0, \pm1, \pm2, \ldots$. Note that the pulse train is a periodic signal with fundamental period equal to 2.

1.1.6   Derivative of a Continuous-Time Signal

A continuous-time signal $x(t)$ is said to be *differentiable* at a fixed point t_1 if

$$\frac{x(t_1 + h) - x(t_1)}{h}$$

has a limit as $h \to 0$, independent of whether h approaches zero from above ($h > 0$) or from below ($h < 0$). If the limit exists, $x(t)$ has a *derivative* at the point t_1 defined by

$$\frac{dx(t)}{dt}\bigg|_{t=t_1} = \lim_{h \to 0} \frac{x(t_1 + h) - x(t_1)}{h}$$

This definition of the derivative of $x(t)$ is sometimes called the *ordinary derivative* of $x(t)$.

To be differentiable at a point t_1, it is necessary (but not sufficient, in general) that the signal $x(t)$ be continuous at t_1. Hence continuous-time signals that are not continuous at all points cannot be differentiable at all points. In particular, piecewise-continuous signals are not differentiable at all points. However, piecewise-continuous signals may have a derivative in the generalized sense. Suppose that $x(t)$ is differentiable at all t except $t = t_1$. Then the *generalized derivative* of $x(t)$ is defined to be

$$\frac{dx(t)}{dt} + [x(t_1^+) - x(t_1^-)]\delta(t - t_1)$$

where $dx(t)/dt$ is the ordinary derivative of $x(t)$ at all t except $t = t_1$, and $\delta(t - t_1)$ is the unit impulse concentrated at the point $t = t_1$. Thus the generalized derivative of a signal at a point of discontinuity t_1 is equal to an impulse located at t_1 and with area equal to the amount the function "jumps" at the point t_1.

To illustrate the occurrence of the impulse when taking a generalized derivative, let $x(t)$ be the step function $Ku(t)$. The ordinary derivative of $Ku(t)$ is equal to zero at all t, except at $t = 0$. Therefore, the generalized derivative of $Ku(t)$ is equal to

$$K[u(0^+) - u(0^-)]\delta(t - 0) = K\delta(t)$$

For $K = 1$, it follows that the generalized derivative of the unit-step function $u(t)$ is equal to the unit impulse $\delta(t)$.

1.1.7   Using MATLAB® for Continuous-Time Signals

A continuous-time signal $x(t)$ given by a mathematical expression can be defined and displayed by use of the software MATLAB. Since MATLAB is used throughout this book, the reader should become familiar with the basic commands and is invited to review the short tutorial available from the website that accompanies this text. To illustrate its use, consider the signal $x(t)$ given by

$$x(t) = e^{-0.1t} \sin\tfrac{2}{3}t$$

A plot of $x(t)$ versus t for a range of values of t can be generated by application of the MATLAB software. For example, for t ranging from 0 to 30 seconds with 0.1-second increments, the MATLAB commands for generating $x(t)$ are

```
t = 0:0.1:30;
x = exp(-.1*t).*sin(2/3*t);
plot(t,x)
axis([0 30 -1 1])
grid
xlabel('Time (sec)')
ylabel('x(t)')
```

In this program, the time values for which x is to be plotted are stored as elements in the vector `t`. Each of the expressions `exp(-.1*t)` and `sin(2/3*t)` creates a vector with elements equal to the expression evaluated at the corresponding time values. The resulting vectors must be multiplied element by element to define the vector `x`. As seen from the command `x = exp(-.1*t).*sin(2/3*t)`, element-by-element operations require a dot before the operator. Then by the command `plot(t,x)`, `x` is plotted versus `t`. The `axis` command is used to overwrite the default values. (Usually, the default is acceptable, and this command is not needed.) It should be noted that the use of the `axis` command varies with the version of MATLAB being employed.

The resulting plot of $x(t)$ is shown in Figure 1.10. Note that the MATLAB-generated plot is in box form, the axes labeled as shown. The format of the plot differs from those given previously. In this book, MATLAB-generated plots will always be in box form, whereas plots not generated by MATLAB will be given in the form used previously (as in Figure 1.9).

It is important to note that, in generating a MATLAB plot of a continuous-time signal, the increment in the time step must be chosen to be sufficiently small to yield a smooth

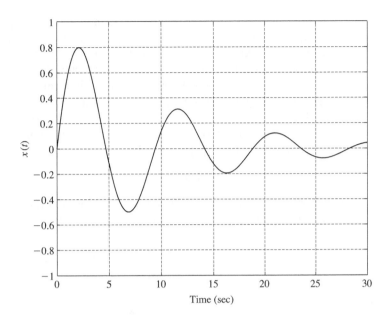

FIGURE 1.10
MATLAB plot of the signal $x(t) = e^{-0.1t} \sin\frac{2}{3}t$.

plot. If the increment is chosen to be too large (for a given signal), then when the values of the signal are connected by straight lines (in the computer generation of the plot), the resulting plot will look jagged. To see this effect, the reader is invited to rerun the preceding program, using a time increment of 1 second to plot $x(t) = e^{-0.1t} \sin\frac{2}{3}t$. For the plots in this book it was found that using 200 to 400 points per plot resulted in a small-enough time increment. See Problem 1.2 for more information on selecting the time increment.

The program given previously is stored as an "M-file" called `fig1_10.m` that is available from the website http://users.ece.gatech.edu/~bonnie/book3/. All MATLAB M-files used in this book are included with a title that matches the figure number or the example number; for example, `ex1_3.m` is the M-file containing the commands for Example 1.3.

1.2   DISCRETE-TIME SIGNALS

The time variable t is said to be a *discrete-time variable* if t takes on only the *discrete values* $t = t_n$ for some range of integer values of n. For example, t could take on the integer values $t = 0, 1, 2, \ldots$; that is, $t = t_n = n$ for $n = 0, 1, 2, \ldots$. A *discrete-time signal* is a signal that is a function of the discrete-time variable t_n; in other words, a discrete-time signal has values (is defined) only at the discrete-time points $t = t_n$, where n takes on only integer values. Discrete-time signals arise in many areas of engineering, science, and economics. In applications to economics, the discrete-time variable t_n may be the day, month, quarter, or year of a specified period of time. In this section an example is given where the discrete-time variable is the day for which the closing price of an index fund is specified.

In this book a discrete-time signal defined at the time points $t = t_n$ will be denoted by $x[n]$. Note that in the notation "$x[n]$," the integer variable n corresponds to the time instants t_n. Also note that brackets are used to denote a discrete-time signal $x[n]$, in contrast to a continuous-time signal $x(t)$, which is denoted by parentheses. A plot of a discrete-time signal $x[n]$ will always be given in terms of the values of $x[n]$ versus the integer time variable n. The values $x[n]$ are often indicated on the plot by filled circles with vertical lines connecting the circles to the time axis. This results in a *stem plot*, which is a common way of displaying a discrete-time signal. For example, suppose that the discrete-time signal $x[n]$ is given by

$$x[0] = 1,\ x[1] = 2,\ x[2] = 1,\ x[3] = 0,\ x[4] = -1$$

with $x[n] = 0$ for all other n. Then the stem plot of $x[n]$ is shown in Figure 1.11. A plot of this signal can be generated by the MATLAB commands

```
n = -2:6;
x = [0 0 1 2 1 0 -1 0 0];
stem (n,x,'filled');
xlabel ('n')
ylabel ('x[n]')
```

The MATLAB-generated plot of $x[n]$ is shown in Figure 1.12. Again note that the MATLAB plot is in box form, in contrast to the format of the plot given in Figure 1.11. As in the continuous-time case, MATLAB plots are always displayed in box form. Plots of discrete-time signals not generated by MATLAB will be given in the form shown in Figure 1.11.

1.2.1   Sampling

One of the most common ways in which discrete-time signals arise is in sampling continuous-time signals: As illustrated in Figure 1.13, suppose that a continuous-time signal $x(t)$ is applied to an electronic switch that is closed briefly every T seconds. If the amount of time during which the switch is closed is much smaller than T, the output of the switch can be viewed as a discrete-time signal that is a function of the discrete-time

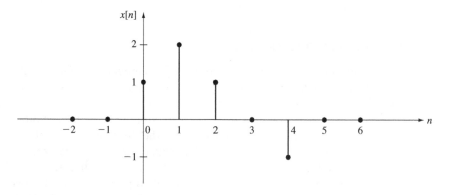

FIGURE 1.11
Stem plot of discrete-time signal.

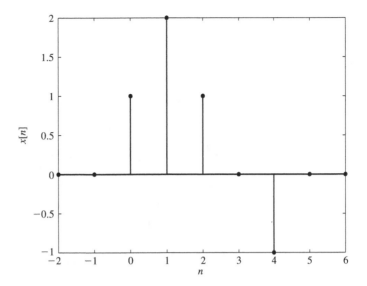

FIGURE 1.12
MATLAB plot of $x[n]$.

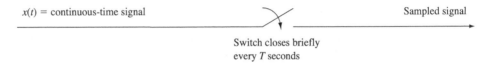

FIGURE 1.13
Sampling process.

points $t_n = nT$, where $n = \ldots, -2, -1, 0, 1, 2, \ldots$. The resulting discrete-time signal is called the *sampled version* of the original continuous-time signal $x(t)$, and T is called the *sampling interval*. Since the time duration T between adjacent sampling instants $t_n = nT$ and $t_{n+1} = (n + 1)T$ is equal to a constant, the sampling process under consideration here is called *uniform sampling*. *Nonuniform sampling* is sometimes utilized in applications, but is not considered in this book.

To be consistent with the notation previously introduced for discrete-time signals, the discrete-time signal resulting from the uniform sampling operation illustrated in Figure 1.13 will be denoted by $x[n]$. Note that in this case, the integer variable n denotes the time instant nT. By definition of the sampling process, the value of $x[n]$ for any integer value of n is given by

$$x[n] = x(t)|_{t=nT} = x(nT)$$

A large class of discrete-time signals can be generated by sampling continuous-time signals. For instance, if the continuous-time signal $x(t)$ displayed in Figure 1.10 is sampled with $T = 1$, the result is the discrete-time signal $x[n]$ plotted in Figure 1.14. This plot can be obtained by running the program that generated Figure 1.10, where the time increment is 1 second and the `plot(t,x)` command is replaced with `stem(t,x,'filled')`.

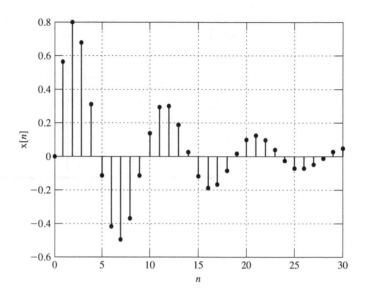

FIGURE 1.14
Sampled continuous-time signal.

1.2.2   Step and Ramp Functions

Two simple examples of discrete-time signals are the discrete-time unit-step function $u[n]$ and the discrete-time unit-ramp function $r[n]$, which are defined by

$$u[n] = \begin{cases} 1, & n = 0, 1, \ldots \\ 0, & n = -1, -2, \ldots \end{cases}$$

$$r[n] = \begin{cases} n, & n = 0, 1, \ldots \\ 0, & n = -1, -2, \ldots \end{cases}$$

These two discrete-time signals are plotted in Figure 1.15.

The discrete-time step function $u[n]$ can be obtained by sampling the continuous-time step function $u(t)$. If the unit-ramp function $r(t) = tu(t)$ is sampled, the result is

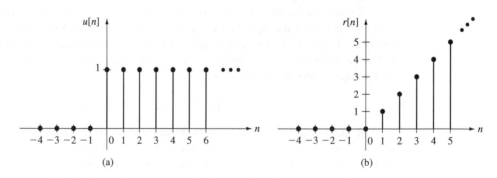

(a)   (b)

FIGURE 1.15
(a) Discrete-time unit-step and (b) discrete-time unit-ramp functions.

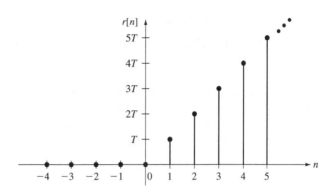

FIGURE 1.16
Discrete-time ramp function.

the discrete-time ramp function $r[n]$, given by

$$r[n] = r(t)|_{t=nT} = r(nT)$$

The discrete-time signal $r[n]$ is plotted in Figure 1.16. Note that although the discrete-time signals in Figures 1.15b and 1.16 are given by the same notation $r[n]$, these two signals are not the same unless the sampling interval T is equal to 1. To distinguish between these two signals, the one plotted in Figure 1.16 could be denoted by $r_T[n]$, but the standard convention (which is followed here) is not to show the dependence on T in the notation for the sampled signal.

1.2.3   Unit Pulse

It should first be noted that there is no sampled version of the unit impulse $\delta(t)$ since $\delta(0)$ is not defined. However, there is a discrete-time signal that is the discrete-time counterpart of the unit impulse. This is the *unit-pulse function* $\delta[n]$, defined by

$$\delta[n] = \begin{cases} 1, & n = 0 \\ 0, & n \neq 0 \end{cases}$$

The unit-pulse function is plotted in Figure 1.17. It should be stressed that $\delta[n]$ is not a sampled version of the unit impulse $\delta(t)$.

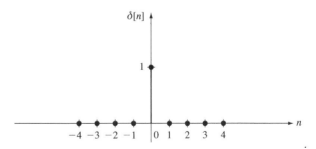

FIGURE 1.17
Unit-pulse function.

1.2.4   Periodic Discrete-Time Signals

A discrete-time signal $x[n]$ is periodic if there exists a positive integer r such that

$$x[n + r] = x[n] \quad \text{for all integers } n$$

Hence $x[n]$ is periodic if and only if there is a positive integer r such that $x[n]$ repeats itself every r time instants, where r is called the *period*. The fundamental period is the smallest value of r for which the signal repeats.

For example, let us examine the periodicity of a discrete-time sinusoid given by

$$x[n] = A \cos(\Omega n + \theta)$$

where Ω is the "discrete-time frequency" in radians per unit time T and θ is the phase in radians. The signal is periodic with period r if

$$A \cos[\Omega(n + r) + \theta] = A \cos(\Omega n + \theta)$$

Recall that the cosine function repeats every 2π radians, so that

$$A \cos(\Omega n + \theta) = A \cos(\Omega n + 2\pi q + \theta)$$

for all integers q. Therefore, the signal $A \cos(\Omega n + \theta)$ is periodic if and only if there exists a positive integer r such that $\Omega r = 2\pi q$ for some integer q, or equivalently, that the discrete-time frequency Ω is such that $\Omega = 2\pi q/r$ for some positive integers q and r. The fundamental period is the smallest integer value of r such that $\Omega = 2\pi q/r$.

The discrete-time sinusoid $x[n] = A \cos(\Omega n + \theta)$ is plotted in Figure 1.18 with $A = 1$ and with two different values of Ω. For the case when $\Omega = \pi/3$ and $\theta = 0$, which is plotted in Figure 1.18a, the signal is periodic since $\Omega = 2\pi q/r$ with $q = 1$ and

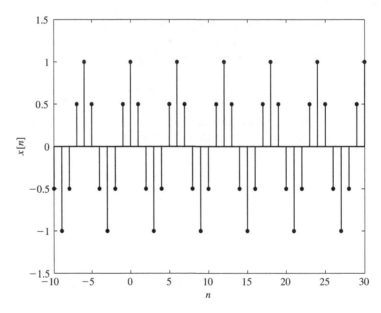

FIGURE 1.18a
Discrete-time sinusoid with $\Omega = \pi/3$ and $\theta = 0$.

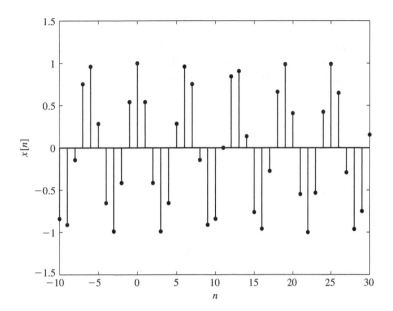

FIGURE 1.18b
Discrete-time sinusoid with $\Omega = 1$ and $\theta = 0$.

$r = 6$, and the fundamental period is equal to 6. The case when $\Omega = 1$ and $\theta = 0$ is plotted in Figure 1.18b. Note that in this case the envelope of the signal is periodic, but the signal itself is not periodic since $1 \neq 2\pi q/r$ for any positive integers q and r.

1.2.5   Discrete-Time Rectangular Pulse

Let L be a positive odd integer. An important example of a discrete-time signal is the *discrete-time rectangular pulse function* $p_L[n]$ of length L defined by

$$p_L[n] = \begin{cases} 1, & n = -(L-1)/2, \ldots, -1, 0, 1, \ldots, (L-1)/2 \\ 0, & \text{all other } n \end{cases}$$

The discrete-time rectangular pulse is displayed in Figure 1.19.

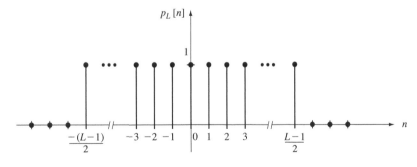

FIGURE 1.19
Discrete-time rectangular pulse.

1.2.6 Digital Signals

Let $\{a_1, a_2, \ldots, a_N\}$ be a set of N real numbers. A *digital signal* $x[n]$ is a discrete-time signal whose values belong to the finite set $\{a_1, a_2, \ldots, a_N\}$; that is, at each time instant $t_n, x(t_n) = x[n] = a_i$ for some i, where $1 \leq i \leq N$. So a digital signal can have only a finite number of different values.

A sampled continuous-time signal is not necessarily a digital signal. For example, the sampled unit-ramp function $r[n]$ shown in Figure 1.16 is not a digital signal since $r[n]$ takes on an infinite range of values when $n = \ldots, -2, -1, 0, 1, 2, \ldots$.

A *binary signal* is a digital signal whose values are equal to 1 or 0; that is, $x[n] = 0$ or 1 for $n = \ldots, -2, -1, 0, 1, 2, \ldots$. The sampled unit-step function and the unit-pulse function are both examples of binary signals.

1.2.7 Time-Shifted Signals

Given a discrete-time signal $x[n]$ and a positive integer q, the discrete-time signal $x[n - q]$ is the q-step right shift of $x[n]$, and $x[n + q]$ is the q-step left shift of $x[n]$. For example, $p_3[n - 2]$ is the two-step right shift of the discrete-time rectangular pulse $p_3[n]$, and $p_3[n + 2]$ is the two-step left shift of $p_3[n]$. The shifted signals are plotted in Figure 1.20.

1.2.8 Downloading Discrete-Time Data from the Web

There exists a large number of websites that contain discrete-time data (often referred to as time series) that arise in the fields of engineering, science, and economics. This is especially the case in economics where numerous sites exist that contain a wide range of economic data such as monthly employment numbers, housing sales, interest rates, daily commodity prices (e.g., oil, gas, gold, silver, wheat, soybeans, etc.), and daily stock prices. The data are usually presented in a table format with the first column corresponding to points in time given day by day, week by week, month by month, etc., and with the other columns in the table containing the data that go with the various time points.

In many cases the time series data contained on websites can be downloaded into one's computer and saved in a file. If spreadsheet software is installed on the computer, the file can then be opened and saved by the spreadsheet software. From many websites, the data can be downloaded directly into a spreadsheet. If the spreadsheet software

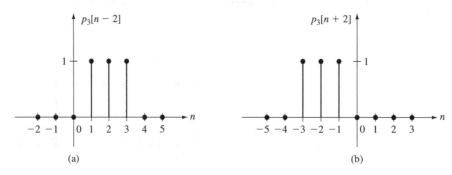

FIGURE 1.20
Two-step shifts of $p_3[n]$: (a) right shift; (b) left shift.

can save the data in a csv (comma-separated value) formatted file, MATLAB will be able to read the file, in which case various analysis techniques can then be applied to the data. The process of downloading time series data into a spreadsheet will be illustrated subsequently. The spreadsheet software that is used for this is Microsoft Excel.

The development given here focuses on downloading historical price data for QQQQ, which is an index fund whose value tracks the stock price of 100 companies having stock traded on the Nasdaq Stock Exchange. For a list of the 100 companies and other information regarding QQQQ, simply type "QQQQ" into a search engine, and a multitude of sites will appear.

The index fund QQQQ can be traded throughout any given business day like an ordinary stock, so it has a daily opening price (the price at 9:30 A.M. EST), a high price for the day, a low price for the day, and a daily closing price (the price at 4:00 P.M. EST). Historical price data for QQQQ and other stocks can be downloaded directly into a spreadsheet from many different websites. Here the Yahoo! site will be used.

The daily price data for QQQQ is an interesting example of a discrete-time signal, or time series, that can be analyzed by various mathematical techniques, including those developed in this book. This is illustrated to some extent in Section 1.5 and in other chapters of the book. Of course, price data for QQQQ is an example of a *financial time series*, not an *engineering time series*. Nevertheless, the authors believe that this type of signal is appropriate as an example for study in a textbook that is intended primarily for the engineering profession. Many individuals in engineering most likely do invest in the stock market, so having some idea as to how technical analysis can be applied to stock price data may be of help in making investment decisions. In addition, some of the methods that can be applied to the analysis of stock price data can also be applied to engineering signals that are characterized by having a good deal of noise in the signal (which is the case for stock price data).

To download price data for QQQQ, first go to the website http://finance.yahoo.com. Near the top of the Web page, enter the symbol QQQQ, click on "GO," and then in the left-hand column of the page that comes up, click on "Historical Prices." The table that appears on your computer screen gives the opening price, high price, low price, closing price, volume, and adjusted close of QQQQ for each day over a time period of several years. The first line of data in the table is the most recent price data for QQQQ, and the last line of the table is the price data for the first date in the time period displayed. To see the historical price data for QQQQ for a different time period, type in the desired start date and end date on the Web page and then click on "Get Prices." To obtain data for a different stock, type the ticker symbol into the Web page at the location to the right of "Get Historical Prices for" and then click on "GO." Once the data for a specific stock has been acquired for some desired time period, the data can be downloaded by clicking on "Download to Spreadsheet" located at the bottom of the page. An example illustrating the procedure is given next.

Example 1.2   Downloading Data from the Web

Suppose that the objective is to download the closing price of QQQQ for the 10-business-day period from March 1, 2004, up to and including March 12, 2004. To accomplish this, carry out the steps just described, and then type in the start date of March 1, 2004, and the end date of

Date	Open	High	Low	Close	Volume	Adj. Close*
12-Mar-04	35.18	35.59	35.15	35.51	1.18E + 08	35.17
11-Mar-04	35.07	35.53	34.8	34.87	1.52E + 08	34.54
10-Mar-04	35.75	36	35.13	35.19	1.34E + 08	34.86
9-Mar-04	35.81	35.98	35.52	35.66	1.26E + 08	35.32
8-Mar-04	36.69	36.82	35.73	35.77	1.13E + 08	35.43
5-Mar-04	36.42	37.15	36.36	36.63	1.18E + 08	36.28
4-Mar-04	36.44	36.83	36.39	36.76	65905600	36.41
3-Mar-04	36.51	36.63	36.21	36.42	83938304	36.07
2-Mar-04	36.98	37.18	36.61	36.62	91536000	36.27
1-Mar-04	36.68	37.07	36.47	37.05	79700704	36.7

FIGURE 1.21
Copy of Excel spreadsheet.

March 12, 2004. Then click on "Get Prices," and the table of data will appear on your computer screen. Click on "Download to Spreadsheet." A copy of the resulting Excel spreadsheet is given in Figure 1.21. The data shown in Figure 1.21 are reproduced with permission of Yahoo! Inc., © 2005 by Yahoo! Inc. YAHOO! and the YAHOO! logo are trademarks of Yahoo! Inc.

Note that the spreadsheet in Figure 1.21 consists of 11 rows and 7 columns. The first row and the first column of the table both contain text (not data), which cannot be read by MATLAB. MATLAB numbers the first row as row 0, and the first column as column 0. Hence, the data in the table in Figure 1.21 are located in rows 1 through 10 and columns 1 through 6. A data value in the table is denoted by the two-tuple (R,C), where R is the row and C is the column containing the data value. For example, (1,1) is the value in the upper left corner of the data table, which has the value 35.18, and (10,6) is the value in the lower right corner, which has the value 36.7.

After the data have been downloaded into Excel, it is necessary to reverse the order of the data so that the earliest date appears first in the table. This is accomplished by first left-clicking on the upper left-hand corner of the table in the Excel spreadsheet so that the entire table is highlighted in blue. Then click on "Sort Ascending" in the tool bar. This will reverse the order of the data so that the first line of data in the table is now the price data for the first day in the range of interest. Then left-click on a point located outside of the table, and the highlighting will be removed. The information can then be saved in a csv-formatted file with any desired file name, such as QQQQdata.csv. Do not forget to add the extension "csv" to the file name. Save the file in a subdirectory under the directory that contains your student version of MATLAB or in a directory in the MATLAB search path.

When MATLAB is opened, all or a portion of the data in the file can be read into MATLAB by use of the command `csvread('filename',R1,C1,[R1 C1 R2 C2])`, where `(R1,C1)` is the upper left corner and `(R2,C2)` is the lower right corner of the portion of the data that is to be read into MATLAB.

Example 1.3   Importing Data into MATLAB

For the spreadsheet created in Example 1.2, reorder the data as instructed previously. Click on "save as" under file, and type in the file name QQQQdata1.csv. Save this file in your MATLAB subdirectory, and answer "yes" to the question regarding keeping this format. The closing prices of QQQQ can then be read into MATLAB by the command `c=csvread('QQQQdata1.csv',1,4, [1 4 10 4])`, which generates a column vector c containing the closing prices of QQQQ. Note that the number 4 appears since we are reading the fourth column of the table containing numerical data.

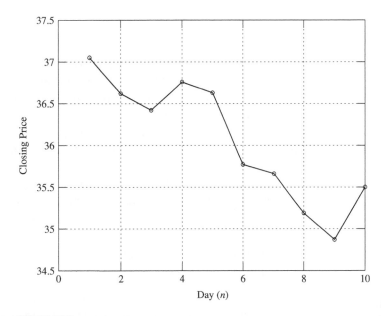

FIGURE 1.22
Closing price of QQQQ from 3/1/04 to 3/12/04.

To verify that this process works, after the file QQQQdata1.csv has been created, run the following MATLAB commands:

```
c=csvread('QQQQdata1.csv',1,4,[1 4 10 4]);
n=1:10;
plot(n,c(n),n,c(n),'o')
grid
xlabel('Day (n)')
ylabel('Closing Price')
```

The resulting MATLAB plot is shown in Figure 1.22, which displays the closing price of QQQQ over the 10-day time period. Note that the value in the plot at $n = 1$ is the closing price on March 1, 2004, and the value at $n = 10$ is the closing price on March 12, 2004. Also note that we are not using the stem command to plot the values. Instead, the data points are plotted as unfilled o's with lines connecting the values. The reason for this is that we are interested in seeing what the trend is in the closing prices, and this is much more clearly revealed by lines connecting the data values than by the stem plot.

1.3   SYSTEMS

A system is a collection of one or more devices, processes, or computer-implemented algorithms that operates on an input signal x to produce an output signal y. When the inputs and outputs are continuous-time signals, $x(t)$ and $y(t)$, the system is said to be a *continuous-time system* or an *analog system*. When the inputs and outputs are discrete-time signals, $x[n]$ and $y[n]$, the system is said to be a *discrete-time system*.

Some common examples of systems that consist of physical devices or processes are listed as follows:

1. An electrical circuit with inputs equal to driving voltages and/or currents and with outputs equal to voltages and/or currents at various points in the circuit.
2. A communications system with inputs equal to the signals to be transmitted and with outputs equal to the received signals.
3. A biological system, such as the human heart with inputs equal to the electrical stimuli applied to the heart muscle and with output equal to the flow rate of blood through the heart.
4. A robotic manipulator with inputs equal to the torques applied to the links of the robot and with output equal to the position of the end effector (hand).
5. An oil refinery with input equal to the flow rate of oil and with output equal to the flow rate of gasoline.
6. A manufacturing system with inputs equal to the flow rate of raw materials and with output equal to the rate of production of the finished product.

In addition to the aforementioned examples of systems, there are numerous types of signal processing systems (referred to as *signal processors*) that operate on an input signal to produce a desired output signal. A common example of a signal processor is a *filter* that is designed to remove sinusoidal components whose frequencies are in some range of frequencies, or to remove the noise that may be present in a signal. To introduce the process of noise removal, suppose that a signal $x(t)$ can be expressed in the form

$$x(t) = s(t) + e(t) \tag{1.4}$$

where $s(t)$ is the *smooth part* of the signal $x(t)$ and $e(t)$ is the *erratic or noisy part* of the signal $x(t)$. Many signals with noise arising in applications can be expressed in the additive form (1.4). For example, suppose that $x(t)$ is the measurement of the distance (e.g., range) from some target to a radar antenna. Since the energy reflected from a target is so small, radar measurements of a target's position are always very noisy and usually are embedded in the "background noise." In this case, the measurement $x(t)$ can be expressed in the form (1.4), where $e(t)$ is the background noise and $s(t)$ is the true distance between the target and the radar antenna.

Given an input signal $x(t)$ having the form (1.4), the objective of a filter is to eliminate $e(t)$ and pass $s(t)$, so that the output $y(t)$ of the filter is equal to $s(t)$. In practice, this is rarely possible to do, although it may be possible to design the filter so that $y(t)$ is "close to" $s(t)$. In the next section a specific type of filter is considered, and then filtering is studied in more depth in other chapters of the book.

Signal processors are often used, not only for filtering, but also to determine the information contained in a signal. In general, this is not a simple problem; in particular, knowing the functional form or sample values of a signal does not directly reveal (in general) the information carried in a signal. An interesting example is the extraction of information carried in a speech signal. For example, it is a nontrivial matter to develop a speech processing scheme that is capable of identifying the person who is speaking, from a segment of speech. Of course, to be able to identify the speaker correctly, the speech processor must have stored in its memory the "speech patterns" of a collection of people, one of whom is the speaker. The question here is, What is an appropriate speech pattern? In other words, exactly what is it about one's voice that distinguishes it from that of others? One way to answer this is to consider the characterization of speech signals in terms of their frequency spectrum. The concept of the frequency spectrum of a signal is studied in Chapters 3 to 5.

The extraction of information from signals is also of great importance in the medical field in the processing of bioelectric signals. For instance, an important problem is determining the health of a person's heart, from the information contained in a collection of ECG signals taken from surface electrodes placed on the person. A specific objective is to be able to detect if there is any heart damage that may be a result of coronary artery disease or from a prolonged state of hypertension. A trained physician may be able to detect heart disease by "reading" ECG signals, but due to the complexity of these signals, it is not likely that a "human processor" will be able to extract all the information contained in these signals. This is a problem area where signal processing techniques can be applied, and in fact, progress has been made on developing automated processing schemes for bioelectric signals.

To undertake an in-depth study of a system, such as one of the examples previously mentioned, it is very useful to have a *mathematical model* of the system. A mathematical model consists of a collection of equations describing the relationships between the signals appearing in the system. In many cases, the equations can be determined from physical principles such as Newton's laws of motion for a mechanical system. This is pursued to some extent in Chapter 2. Another method for modeling, known as system identification, is to devise mathematical relationships that fit a sample set of input data and the corresponding set of output data. System identification based on input/output data is left to a more advanced treatment of signals and systems, and thus is not considered in this text.

A mathematical model of a system is usually an idealized representation of the system. In other words, many actual (physical) systems cannot be described exactly by a mathematical model, since many assumptions must be made in order to obtain the equations for the system. However, a sufficiently accurate mathematical model can often be generated so that system behavior and properties can be studied in terms of the model. Mathematical models are also very useful in the design of new systems having various desirable operating characteristics—for example, in the design of "controllers" whose purpose is to modify system behavior to meet some performance objectives. Thus, mathematical models are used extensively in both system analysis and system design.

If a model of a system is to be useful, it must be tractable, and thus an effort should always be made to construct the simplest possible model of the system under study. But the model must also be sufficiently accurate, which means that all primary characteristics (all first-order effects) must be included in the model. Usually, the more characteristics that are put into a model, the more complicated the model is, and so there is a trade-off between simplicity of the model and accuracy of the model.

There are two basic types of mathematical models: input/output representations describing the relationship between the input and output signals of a system, and the state or internal model describing the relationship among the input, state, and output signals of a system. Input/output representations are studied in the first 10 chapters; the state model is considered in Chapter 11.

Four types of input/output representations are studied in this text:

1. The convolution model
2. The input/output difference equation, or differential equation
3. The Fourier transform representation
4. The transfer function representation

As will be shown, the Fourier transform representation can be viewed as a special case of the transfer function representation. Hence, there are only three fundamentally different types of input/output representations that will be studied in this book.

The first two representations just listed and the state model are referred to as *time-domain models*, since these representations are given in terms of functions of time. The last two of the four representations listed are referred to as *frequency-domain models*, since they are specified in terms of functions of a complex variable that is interpreted as a frequency variable. Both time-domain and frequency-domain models are used in system analysis and design. These different types of models are often used together to maximize understanding of the behavior of the system under study.

1.4.   EXAMPLES OF SYSTEMS

To provide some concreteness to the concept of a system, in this section three specific examples of a system are given. The first two examples are continuous-time systems, and the third example is a discrete-time system.

1.4.1   *RC* Circuit

Consider the *RC* circuit shown in Figure 1.23. The *RC* circuit can be viewed as a continuous-time system with input $x(t)$ equal to the current $i(t)$ into the parallel connection, and with output $y(t)$ equal to the voltage $v_C(t)$ across the capacitor. By Kirchhoff's current law (see Section 2.4),

$$i_C(t) + i_R(t) = i(t) \tag{1.5}$$

where $i_C(t)$ is the current in the capacitor and $i_R(t)$ is the current in the resistor.

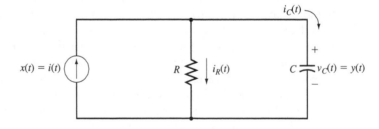

FIGURE 1.23
RC circuit.

Now

$$i_C(t) = C\frac{dv_C(t)}{dt} = C\frac{dy(t)}{dt} \qquad (1.6)$$

and

$$i_R(t) = \frac{1}{R}v_C(t) = \frac{1}{R}y(t) \qquad (1.7)$$

Inserting (1.6) and (1.7) into (1.5) yields the following linear differential equation:

$$C\frac{dy(t)}{dt} + \frac{1}{R}y(t) = i(t) = x(t) \qquad (1.8)$$

The differential equation (1.8) is called the *input/output differential equation* of the circuit. It provides an implicit relationship between the input $x(t)$ and the output $y(t)$. The output $y(t)$ resulting from an input $x(t)$ can be generated by solving the input/output differential equation (1.8). For example, suppose that the input $x(t)$ is equal to the unit-step function $u(t)$ and the initial condition $y(0)$ is equal to zero. Then the response $y(t)$ for $t > 0$ is the solution to the differential equation

$$C\frac{dy(t)}{dt} + \frac{1}{R}y(t) = 1, t > 0 \qquad (1.9)$$

with the initial condition $y(0) = 0$. The solution to (1.9) can be found by the use of the MATLAB symbolic manipulator, which is considered in Chapter 2, or by the use of the Laplace transform, which is studied in Chapter 6. The result is that the output response is given by

$$y(t) = R[1 - e^{-(1/RC)t}], t \geq 0 \qquad (1.10)$$

The output response $y(t)$ given by (1.10) is called the *step response*, since $y(t)$ is the output when the input is the unit-step function $u(t)$ with zero initial condition $[y(0) = 0]$. If at $t = 0$ a constant current source of amplitude 1 is switched on [so that $x(t) = u(t)$], the resulting voltage across the capacitor would be given by (1.10). For the case $R = 1$ and $C = 1$, the step response is as plotted in Figure 1.24. Note that the voltage on the capacitor builds up to a value of 1 as $t \to \infty$ in response to switching on a constant current source of amplitude 1 at time $t = 0$.

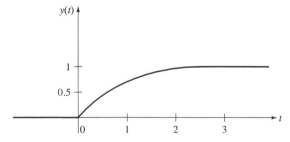

FIGURE 1.24
Step response of *RC* circuit when $R = C = 1$.

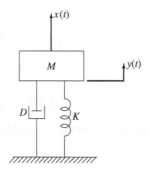

FIGURE 1.25
Schematic diagram of a mass-
spring-damper system.

1.4.2   Mass–Spring–Damper System

The simplest model for many vibratory systems is the mass–spring–damper system, shown schematically in Figure 1.25. The mass–spring–damper system is an accurate representation of many actual structures or devices; examples include an accelerometer (a device for measuring acceleration), a seismometer (a device for measuring the vibration of the earth), and a vibration absorber (a mounting device used to absorb vibration of equipment). Other systems, such as a machine tool or a compressor on a resilient mount, can be modeled as mass–spring–damper systems for simplified analysis. This system, while crude, demonstrates most of the phenomena associated with vibratory systems, and, as such, it is the fundamental building block for the study of vibration.

Physically, the mass M is supported by a spring with stiffness constant K and a damper with damping constant D. An external force $x(t)$ is applied to the mass and causes the mass to move upward or downward with displacement $y(t)$, measured with respect to an equilibrium value. (That is, $y(t) = 0$ when no external force is applied.) When the mass is above its equilibrium value, $y(t) > 0$; and when the mass is below its equilibrium value, $y(t) < 0$. The movement of the mass is resisted by the spring. (If the mass is moving downward, it compresses the spring, which then acts to push upward on the mass.) The damper acts to dissipate energy by converting mechanical energy to thermal energy, which leaves the system in the form of heat. For example, a shock absorber in a car contains a damper.

As shown in Section 2.4, the input/output differential equation for the mass–spring–damper system is given by

$$M\frac{d^2y(t)}{dt^2} + D\frac{dy(t)}{dt} + Ky(t) = x(t)$$

A demonstration of the mass–spring–damper system is available online at the website that accompanies this text. The demo allows the user to select different inputs for $x(t)$, such as a step function or a sinusoid, and view the resulting animated response of the system as the mass moves in response to the input. The user can choose different values of M, D, and K to view their effects on the system response. For many combinations of

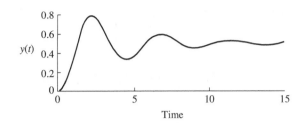

FIGURE 1.26

Response of the mass–spring–damper system to a unit-step input with $M = 1$, $K = 2$, and $D = 0.5$.

Mass–
Spring–
Damper
System

values for *M, D,* and *K*, the response $y(t)$ to a step input, $x(t) = u(t)$, is a decaying oscillation that settles to a constant (or steady-state) value, as seen in Figure 1.26. The oscillation is due to the transfer of energy between kinetic energy (proportional to the velocity squared of the mass) and the potential energy (energy stored in the spring as it compresses or stretches). The decay of the oscillation is due to the dissipation of energy that occurs in the damper.

A detailed discussion of vibrations is not the objective of this example or the online demo. However, the mass–spring–damper is a system whose response can be visualized readily via animation. A series *RLC* circuit is governed by the same general equation and responds in the same manner as this system, but the response cannot be visualized easily via animation. Therefore, the mass–spring–damper system and the accompanying online demo will be used throughout this text to demonstrate basic system input/output concepts.

1.4.3   Moving Average Filter

Given a positive integer *N*, the *N-point moving average (MA) filter* is a discrete-time system given by the input/output relationship

$$y[n] = \frac{1}{N}[x[n] + x[n-1] + x[n-2] + \cdots + x[n-N+1]] \qquad (1.11)$$

where $x[n]$ is the input applied to the filter and $y[n]$ is the resulting output response. For example, if $N = 3$, the 3-point MA filter is given by the input/output relationship

$$y[n] = \frac{1}{3}[x[n] + x[n-1] + x[n-2]]$$

From (1.11), it is seen that the output $y[n]$ at time *n* of the *N*-point MA filter is the average of the *N* input values $x[n], x[n-1], x[n-2], \ldots, x[n-N+1]$. Hence, the term "*N*-point" refers to the number of input values used in the computation of the filter's output. The filter is referred to as a "moving average filter," since we compute the next value $y[n+1]$ of the output by moving the range of time points over which the filter output is computed. In particular, $y[n + 1]$ is the average of $x[n + 1]$, $x[n]$, $x[n - 1], \ldots, x[n - N + 2]$, so that

$$y[n + 1] = \frac{1}{N}[x[n + 1] + x[n] + x[n - 1] + \cdots + x[n - N + 2]] \quad (1.12)$$

Note that (1.12) follows from (1.11) by the setting of $n = n + 1$ in (1.11). Some authors refer to the MA filter as the *running average filter*.

MA filters are often used to reduce the magnitude of the noise that may be present in a signal. To see how this is possible, suppose that the input $x[n]$ is given in the form $x[n] = s[n] + e[n]$, where $s[n]$ is the smooth part of $x[n]$ and $e[n]$ is the erratic or noisy part of $x[n]$. Then the output $y[n]$ of the N-point MA filter is given by

$$y[n] = \frac{1}{N}[s[n] + s[n - 1] + \cdots + s[n - N + 1]]$$

$$+ \frac{1}{N}[e[n] + e[n - 1] + \cdots + e[n - N + 1]] \quad (1.13)$$

The noisy part of the MA filter output $y[n]$ given by (1.13) is the average of the noise values $e[n], e[n - 1], \ldots, e[n - N + 1]$, which is equal to

$$\frac{1}{N}[e[n] + e[n - 1] + \cdots + e[n - N + 1]] \quad (1.14)$$

If $e[n]$ varies randomly about zero, the noisy term given by the average (1.14) can be made as small as desired (in theory) by taking the value of N to be sufficiently large. This explains why MA filters can work well in reducing the magnitude of the erratic or noisy part of a signal.

If the value of N is sufficiently large, the output of the MA filter is approximately equal to a time delay of the smooth part $s[n]$ of $x[n]$ with the amount of the delay equal to $(N - 1)/2$ time units. The occurrence of the time delay will be verified mathematically in Chapter 5 with the discrete Fourier transform. An illustration of the N-point MA filter and the time delay that can occur is given in the next example, where the filter is applied to price data for the stock fund QQQQ. First, it should be pointed out that the output $y[n]$ of the filter given by (1.11) is easily computed by the MATLAB command `sum`. In particular, if the input signal $x[n]$ is written as the column vector

$$x = \begin{bmatrix} x[n - N + 1] \\ x[n - N + 2] \\ \vdots \\ x[n] \end{bmatrix}$$

then the output $y[n]$ at time n is equal to the MATLAB command `(1/N)*sum(x)`.

Example 1.4   Application to Stock Price Data

The closing price of QQQQ for the 50-business-day period from March 1, 2004, up to and including May 10, 2004, will be considered. To acquire the data, follow the procedure described in Section 1.2, and then save the data in the file QQQQdata2.csv. Setting $N = 11$ days, the 11-day MA filter can then be applied to the closing prices of QQQQ by the MATLAB commands

```
c=csvread('QQQQdata2.csv',1,4,[1 4 50 4]);
for i=11:50;
  y(i)=(1/11)*sum(c(i-10:i));
end;
n=11:50;
plot(n,c(n),n,c(n),'o',n,y(n),n,y(n),'*')
grid
xlabel('Day(n)')
ylabel('c[n] and y[n]')
```

Note that in this case the input $x[n]$ to the filter is the closing price $c[n]$ of QQQQ. Also note that the first value of the filter output $y[n]$ is the output $y[11]$ at day 11, since the 11-day MA filter requires the input values $c[1], c[2], c[3], \ldots, c[11]$ in order to compute $y[11]$. The resulting MATLAB plot for the filter input $c[n]$ and the filter output $y[n]$ is given in Figure 1.27. In the plot, the values of $c[n]$ are plotted with o's, and the values of $y[n]$ are plotted with *'s. Note that even though the closing prices are quite erratic, the filter does a good job of smoothing out the data. However, it's clear from the plot that the filter delays the input signal by several days. As noted previously, the delay is approximately equal to $(N - 1)/2$, which is equal to five days in this example. If $y[n]$ is shifted to the left by five days (which corresponds to the left-shifted signal

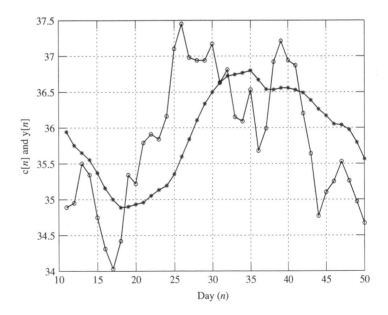

FIGURE 1.27

MATLAB plot of filter input and output.

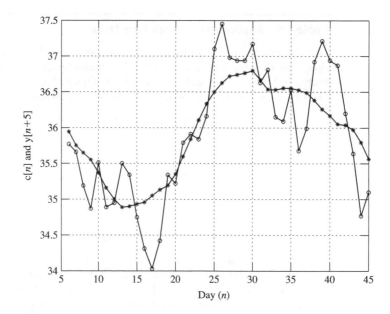

FIGURE 1.28
MATLAB plot of filter input and left-shifted output.

$y[n + 5]$) and then plotted with $c[n]$, the result is as shown in Figure 1.28. Note that the left-shifted output $y[n + 5]$ fits the input data very nicely. Unfortunately, the shifted output cannot be generated in real time, since its computation requires future values of $c[n]$. To see this, let $\rho[n] = y[n + 5]$. Then, replacing n by $n + 5$ and setting $N = 11$ in (1.11) give

$$\rho[n] = \frac{1}{11}[c[n + 5] + c[n + 4] + c[n + 3] + c[n + 2] + c[n + 1]$$

$$+ c[n] + c[n - 1] + \cdots + c[n - 5]] \tag{1.15}$$

Hence, the computation of $\rho[n]$ at time n requires the future values $c[n + 5]$, $c[n + 4]$, $c[n + 3]$, $c[n + 2]$, and $c[n + 1]$ of the input $c[n]$.

Application to trading QQQQ.   Individuals who trade stocks sometimes use MA filters to determine when to buy and when to sell a particular stock (such as QQQQ). In the application to trading, MA filters are often referred to as SMA filters, where SMA stands for "simple moving average." An even more common type of filter used in trading is the EWMA filter, also called the EMA filter, where EWMA stands for "exponentially-weighted moving average." The N-point EWMA filter is defined in Section 2.1; in Section 7.5, a "recursive version" of the EWMA filter is defined. In Section 7.5, an approach to buying and selling QQQQ is given in terms of the difference in the responses to two EWMA filters having different parameter values. For details on the use of moving average filters in trading, type "moving average crossover" into a search engine.

1.5   BASIC SYSTEM PROPERTIES

The extent to which a system can be studied by the use of analytical techniques depends on the properties of the system. Two of the most fundamental properties are linearity and time invariance. It will be seen in this book that there exists an extensive analytical theory for the study of systems possessing the properties of linearity and time invariance. These two properties and the property of causality are defined in this section.

In the definitions that follow, it is assumed that $y(t)$ is the output response of a system resulting from input $x(t)$. The systems considered in this section are limited to those for which an input of $x(t) = 0$ for all t, where $-\infty < t < \infty$, yields an output of $y(t) = 0$ for all t. In the following development, the time variable t may take on real values or only the discrete values $t = nT$; that is, the system may be continuous time or discrete time.

1.5.1   Causality

A system is said to be *causal* or *nonanticipatory* if for any time t_1, the output response $y(t_1)$ at time t_1 resulting from input $x(t)$ does not depend on values of the input $x(t)$ for $t > t_1$. In a causal system, if $y(t)$ is the response due to input $x(t)$ and $x(t) = 0$ for all $t < t_2$, for some t_2, then $y(t) = 0$ for all $t < t_2$. A system is said to be *noncausal* or *anticipatory* if it is not causal. Although all systems that arise in nature are causal (or appear to be causal), there are applications in engineering where noncausal systems arise. An example is the off-line processing (or batch processing) of data. This will be discussed in a later chapter.

Example 1.5   Ideal Predictor

Consider the continuous-time system given by the input/output relationship

$$y(t) = x(t + 1)$$

This system is noncausal, since the value $y(t)$ of the output at time t depends on the value $x(t + 1)$ of the input at time $t + 1$. Noncausality can be seen also by considering the response of the system to a 1-second input pulse shown in Figure 1.29a. From the relationship $y(t) = x(t + 1)$, it can be seen that the output $y(t)$ resulting from the input pulse is the pulse shown in Figure 1.29b. Since the output pulse appears before the input pulse is applied, the system is noncausal. The system with the input/output relationship $y(t) = x(t + 1)$ is called an *ideal* predictor.

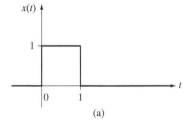
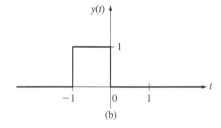

FIGURE 1.29
(a) Input and (b) output pulse in Example 1.5.

Example 1.6   Ideal Time Delay

Consider the system with input/output relationship

$$y(t) = x(t - 1)$$

This system is causal, since the value of the output at time t depends only on the value of the input at time $t - 1$. If the pulse shown in Figure 1.30a is applied to this system, the output is the pulse shown in Figure 1.30b. From Figure 1.30 it is clear that the system delays the input pulse by 1 second. In fact, the system delays all inputs by 1 second; in other words, the system is an *ideal time delay*.

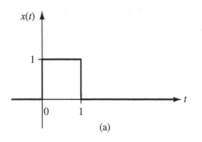
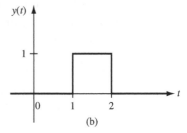

FIGURE 1.30
(a) Input and (b) output pulse of system in Example 1.6.

Example 1.7   MA Filter

Consider the N-point MA filter given by the input/output relationship

$$y[n] = \frac{1}{N}[x[n] + x[n - 1] + x[n - 2] + \cdots + x[n - N + 1]] \tag{1.16}$$

The filter is causal, since the output $y[n]$ at time n depends only on the input values $x[i]$ for $i = n, n - 1, n - 2, \ldots, n - N + 1$. However, the MA filter given by the input/output relationship (1.15) is noncausal, since the filter output $\rho[n]$ at time n requires the future values $c[n + 5], c[n + 4], c[n + 3], c[n + 2]$, and $c[n + 1]$ of the input $c[n]$.

Memoryless systems and systems with memory.   A causal system is memoryless, or static, if for any time t_1, the value of the output at time t_1 depends only on the value of the input at time t_1.

Example 1.8   Ideal Amplifier/Attenuator

Suppose that $y(t) = Kx(t)$, where K is a fixed real number. At any time t_1, $y(t_1) = Kx(t_1)$, and thus $y(t_1)$ depends only on the value of the input at time t_1. Hence, the system is memoryless. Since an ideal amplifier or attenuator can be represented by the input/output relationship $y(t) = Kx(t)$, it is obvious that these devices are memoryless.

A causal system that is not memoryless is said to have *memory*. A system has memory if the output at time t_1 depends in general on the past values of the input $x(t)$ for some range of values of t up to $t = t_1$.

Example 1.9   MA filter

Again consider the N-point MA filter with the input/output relationship (1.16), and suppose that $N \geq 2$. Since the filter output $y[n]$ at time n depends on the input values $x[i]$ for $i = n, n - 1, n - 2, \ldots, n - N + 1$, the MA filter does have memory. When $N = 1$, the one-point MA filter has the input/output relationship $y[n] = x[n]$, and thus is memoryless in this case.

1.5.2   Linearity

A system is said to be *additive* if, for any two inputs $x_1(t)$ and $x_2(t)$, the response to the sum of inputs $x_1(t) + x_2(t)$ is equal to the sum of the responses to the inputs. More precisely, if $y_1(t)$ is the response to input $x_1(t)$ and $y_2(t)$ is the response to input $x_2(t)$, the response to $x_1(t) + x_2(t)$ is equal to $y_1(t) + y_2(t)$. A system is said to be *homogeneous* if, for any input $x(t)$ and any real scalar a, the response to the input $ax(t)$ is equal to a times the response to $x(t)$.

A system is linear if it is both additive and homogeneous; that is, for any inputs $x_1(t)$, $x_2(t)$, and any scalars a_1, a_2, the response to the input $a_1x_1(t) + a_2x_2(t)$ is equal to a_1 times the response to input $x_1(t)$ plus a_2 times the response to input $x_2(t)$. So, if $y_1(t)$ is the response to $x_1(t)$ and $y_2(t)$ is the response to $x_2(t)$, the response to $a_1x_1(t) + a_2x_2(t)$ is equal to $a_1y_1(t) + a_2y_2(t)$. A system that is not linear is said to be *nonlinear*.

Linearity is an extremely important property. If a system is linear, it is possible to apply the vast collection of existing results on linear operations in the study of system behavior and structure. In contrast, the analytical theory of nonlinear systems is very limited in scope. In practice, a given nonlinear system is often approximated by a linear system so that analytical techniques for linear systems can then be applied.

A very common type of nonlinear system is a circuit containing diodes, as shown in the following example.

Example 1.10   Circuit with Diode

Consider the circuit with the ideal diode shown in Figure 1.31. Here the output $y(t)$ is the voltage across the resistor with resistance R_2. The ideal diode is a short circuit when the voltage $x(t)$ is

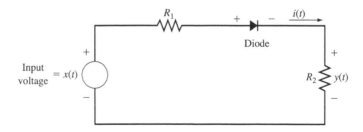

FIGURE 1.31
Resistive circuit with ideal diode.

positive, and it is an open circuit when $x(t)$ is negative. Thus the input/output relationship of the circuit is given by

$$y(t) = \begin{cases} \dfrac{R_2}{R_1 + R_2}x(t) & \text{when } x(t) \geq 0 \\ 0 & \text{when } x(t) \leq 0 \end{cases} \qquad (1.17)$$

Now suppose that the input $x(t)$ is the unit-step function $u(t)$. Then, from (1.17), the resulting response is

$$y(t) = \frac{R_2}{R_1 + R_2}u(t) \qquad (1.18)$$

If the unit-step input is multiplied by the scalar -1, so that the input is $-u(t)$, by (1.17) the resulting response is zero for all $t \geq 0$. But this is not equal to -1 times the response to $u(t)$ given by (1.18). Hence the system is not homogeneous, and thus it is not linear. It is also easy to see that the circuit is not additive.

Nonlinearity may be a result also of the presence of signal multipliers. This is illustrated by the following example.

Example 1.11   Square-Law Device

Consider the continuous-time system with the input/output relationship

$$y(t) = x^2(t) \qquad (1.19)$$

This system can be realized by the use of a signal multiplier, as shown in Figure 1.32. The signal multiplier in Figure 1.32 can be built (approximately) with operational amplifiers and diodes. The system defined by (1.19) is sometimes called a square-law device. Note that the system is memoryless. Given a scalar a and an input $x(t)$, by (1.19) the response to $ax(t)$ is $a^2x^2(t)$. But a times the response to $x(t)$ is equal to $ax^2(t)$, which is not equal to $a^2x^2(t)$ in general. Thus the system is not homogeneous, and the system is not linear.

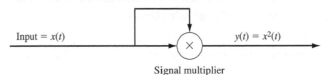

Signal multiplier

FIGURE 1.32
Realization of $y(t) = x^2(t)$.

Another way in which nonlinearity arises is in systems containing devices that go into "saturation" when signal levels become too large, as in the following example.

Example 1.12   Amplifier

Consider an ideal amplifier with the input/output relationship $y(t) = Kx(t)$, where K is a fixed positive real number. A plot of the output $y(t)$ versus the input $x(t)$ is given in Figure 1.33. The ideal

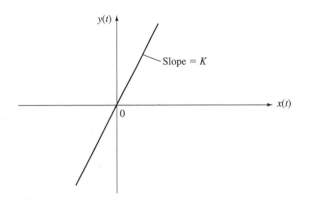

FIGURE 1.33
Output versus input in an ideal amplifier.

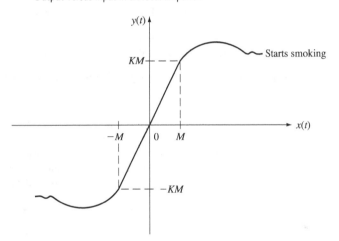

FIGURE 1.34
Output versus input in a nonideal amplifier.

amplifier is clearly linear, but this is not the case for an actual (nonideal) amplifier, since the output $y(t)$ will not equal $Kx(t)$ for arbitrarily large input signals. In a nonideal amplifier, the output versus input characteristics may be as shown in Figure 1.34. From the figure it is clear that $y(t) = Kx(t)$ only when the magnitude $|x(t)|$ of the input is less than M. The nonideal amplifier is not homogeneous, since the response to $ax(t)$ is not equal to a times the response to $x(t)$ unless $|x(t)| < M$ and $|ax(t)| < M$. The nonideal amplifier can be viewed as a linear system only if it can be guaranteed that the magnitude of the input applied to the amplifier will never exceed M.

Although nonlinear systems are very common, many systems arising in practice are linear systems or can be approximated by a linear system. For example, the MA filter is a linear system. Linearity of the MA filter follows easily from the fact that the filter output is a linear combination of the values of the input [see (1.16)]. In addition, both the *RC* circuit and the mass–spring–damper system considered in Section 1.4 are linear systems. In these two examples, linearity follows from the fact that each system is defined by a linear input/output differential equation.

Except for the brief introduction to nonlinear systems given previously, this book deals only with linear systems. A general class of linear systems is defined in the next chapter.

1.5.3   Time Invariance

Given a real number t_1 and a signal $x(t)$, recall that $x(t - t_1)$ is equal to $x(t)$ shifted to the right by t_1 seconds if $t_1 > 0$, and that $x(t - t_1)$ is equal to $x(t)$ shifted to the left by t_1 seconds if $t_1 < 0$. Now consider a system with input $x(t)$ and output $y(t)$. The system is said to be *time invariant* or *constant* if for any input $x(t)$ and any t_1, the response to the shifted input $x(t - t_1)$ is equal to $y(t - t_1)$, where $y(t)$ is the response to $x(t)$. Therefore, in a time-invariant system the response to a left or right shift of the input $x(t)$ is equal to a corresponding shift in the response $y(t)$ to $x(t)$. In a time-invariant system, there are no changes in the system structure as a function of time t. A system is *time varying* or *time variant* if it is not time invariant.

Example 1.13   Amplifier with Time-Varying Gain

Suppose that $y(t) = tx(t)$. It is easy to see that this system is memoryless and linear. Now for any t_1,

$$y(t - t_1) = (t - t_1)x(t - t_1)$$

But the response to input $x(t - t_1)$ is $tx(t - t_1)$, which does not equal $(t - t_1)x(t - t_1)$, in general. Hence $y(t - t_1)$ is not equal to the t_1-second shift of the response to $x(t)$, and thus the system is time varying. Note that this system can be viewed as an ideal amplifier with time-varying gain t.

Example 1.14   MA Filter

Again consider the N-point MA filter with the input/output relationship (1.16). Given a positive or negative integer q, consider the shifted version $x[n - q]$ of the input $x[n]$ to the filter. Then, replacing $x[n]$ by $x[n - q]$ in (1.16) reveals that the filter response $\rho[n]$ to $x[n - q]$ is given by

$$\rho[n] = \frac{1}{N}[x[n - q] + x[n - q - 1] + x[n - q - 2] + \cdots + x[n - q + N - 1]] \qquad (1.20)$$

In addition, if n is replaced by $n - q$ in (1.16), the resulting expression for $y[n - q]$ is equal to the right-hand side of (1.20), and thus, $y[n - q] = \rho[n]$. Hence, the MA filter is time invariant.

In addition to the MA filter, the *RC* circuit and the mass–spring–damper system defined in Section 1.4 are also time invariant, since the input/output differential equation for each of these systems is a constant-coefficient differential equation.

In this book the focus is on systems that are both linear and time invariant. The study of a general class of such systems begins in the next chapter with time-domain representations.

1.6   CHAPTER SUMMARY

This chapter introduces the concepts of signals and systems. Both signals and systems are broken into the categories of continuous time and discrete time. Continuous-time signals are signals $x(t)$ in which time t assumes values from the set of real numbers, while discrete-time signals $x[n]$ are defined only at integer values of the index n. Continuous-time signals arise naturally, with common examples being voltages and currents in circuits, velocities and positions of moving objects, flow rates and pressures in chemical processes, and speech and ECGs in humans. Discrete-time signals are often obtained by sampling a continuous-time signal, that is, defining the signal only at discrete points in time $t = nT$, where T is the sampling time. Discrete-time signals also arise "naturally," as shown in the example given in this chapter involving the closing price of a particular stock at the end of each trading day.

There are several common signals defined in this chapter that will be used throughout the text. The names are the same for the continuous-time and discrete-time versions of these signals: step function, ramp function, and sinusoidal function. In addition, there is the unit-impulse function for continuous-time signals and the unit-pulse function for discrete-time signals. Impulse and pulse functions are very important for analyzing engineering systems. It will be shown in Chapter 2 that the response of the system to the unit impulse or unit pulse is a means of characterizing the system. The discrete-time step function, ramp function, and sinusoidal function can be viewed as sampled versions of their continuous-time versions, but a scaled version of the sampled ramp function is more often used, $r[n] = nu[n]$. It is interesting to note that a continuous-time sinusoid $\cos(\omega t + \theta)$ is periodic, but a discrete-time sinusoid $\cos(\Omega n + \theta)$ is periodic if and only if there exists a positive integer r such that $\Omega = 2\pi q/r$ for some integer q.

A system is a collection of one or more devices or processes that operates on an input signal to produce an output signal. Examples of systems include electrical circuits, communication systems, signal processors, biological systems, robotic manipulators, chemical processes, and manufacturing systems. There are several representations of systems; some of them are considered as time domain, and others are considered as frequency domain models. Time domain representations include the convolution model and the input/output differential, or difference equation, to be studied in Chapter 2. The frequency domain models include the Fourier transform, to be studied in Chapters 3–5, and the transform models to be studied in Chapters 6–8.

There are several important properties of continuous-time and discrete-time systems, including linearity, time invariance, causality, and memory. Several powerful analysis methods exist, such as the Fourier transform and transfer function approaches, that can be used to study the behavior of systems which have these properties. The remaining chapters in the text focus on these analysis methods.

PROBLEMS

1.1. Consider the continuous-time signals displayed in Figure P1.1.
   (i) Show that each of these signals is equal to a sum of rectangular pulses $p_\tau(t)$ and/or triangular pulses $(1 - 2|t|/\tau)p_\tau(t)$.
   (ii) Use MATLAB to plot the signals in Figure P1.1.

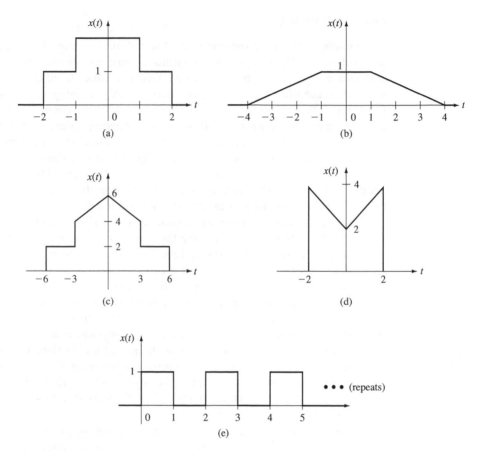

FIGURE P1.1

1.2. Obtaining a computer-generated plot of a continuous-time signal requires some care in choosing the time increment Δt (the spacing between points). As mentioned in Section 1.1, too large of an increment will cause a jagged plot. Moreover, a very large time increment may introduce a phenomenon known as aliasing, which distorts the information given in the signal. (Aliasing is covered in greater detail in Chapter 5.) To avoid aliasing when defining a computer-generated sinusoid such as $x(t) = \cos(\omega t + b)$, choose $\Delta t \leq \pi/\omega$. A rule of thumb in the case of a decaying sinusoid such as $x(t) = e^{-at} \cos(\omega t + b)$ is to choose $\Delta t \leq \pi/\left(4\sqrt{a^2 + \omega^2}\right)$. (Choosing even smaller values of Δt creates smoother plots.)

(a) Compute the maximum time increment for plotting $x(t) = \sin \pi t$ in MATLAB. Verify your result by plotting $x(t)$ for $t = 0$ to $t = 20$ sec with the following time increments: $\Delta t = 0.1$ sec, $\Delta t = 0.5$ sec, $\Delta t = 0.9$ sec, $\Delta t = 1.5$ sec. Note the apparent change in frequency in the plot due to the aliasing effect for $\Delta t = 1.5$. How do you expect your plot to appear when $\Delta t = 2$ sec? Verify your result.

(b) Compute the maximum time increment for plotting $x(t) = e^{-0.1t} \cos \pi t$. Verify your result by plotting $x(t)$ for $t = 0$ to $t = 20$ sec with $\Delta t = 0.1, 0.5, 1.5,$ and 2 sec.

(c) Compute the maximum time increment for plotting $x(t) = e^{-t}\cos(\pi t/4)$. Verify your result by plotting $x(t)$ for $t = 0$ to $t = 10$ sec with $\Delta t = 0.1, 1, 2,$ and 3 sec.

The problem with aliasing exists not only with plotting, but with all digital processing of continuous-time signals. A computer program that emulates a continuous-time system needs to have an input signal defined so that the signal has very little aliasing.

1.3. Use MATLAB to plot the functions (a)–(h) for $-1 \le t \le 5$ sec. Label your axes appropriately.

(a) the unit-step function $u(t)$

(b) the unit-ramp function $r(t)$

(c) $x(t) = \begin{cases} 1, & 0 \le t \le 2 \\ 0, & \text{otherwise} \end{cases}$

(d) $x(t) = 10e^{-3t}\, u(t)$

(e) $x(t) = 3e^{-t}\cos 2t\, u(t)$

(f) $x(t) = 3e^{t}\cos 2t\, u(t)$

(g) $x(t) = 2\sin(3t - \pi/2) - \cos 2t$

(h) $x(t) = \sin 5t + \sin \pi t$

1.4. Sketch the continuous-time signals in parts (a) to (c).

(a) $x(t) = u(t + 1) - 2u(t - 1) + u(t - 3)$

(b) $x(t) = (t + 1)u(t - 1) - tu(t) - u(t - 2)$

(c) $x(t) = 2(t - 1)u(t - 1) - 2(t - 2)u(t - 2) + 2(t - 3)u(t - 3)$

(d) Use MATLAB to plot the signals defined in parts (a) to (c).

1.5. Use MATLAB to plot the following signals over the range $-1 \le t \le 5$:

(a) $x(t) = e^{-t}u(t) + e^{-t}[\exp(2t - 4) - 1]u(t - 2) - e^{t-4}u(t - 4)$

(b) $x(t) = \cos t\left[u\left(t + \dfrac{\pi}{2}\right) - 2u(t - \pi)\right] + (\cos t)u\left(t - \dfrac{3\pi}{2}\right)$

1.6. Given a continuous-time signal $x(t)$ and a constant c, consider the signal $x(t)u(t - c)$.

(a) Show that there exists a signal $v(t)$ such that

$$x(t)u(t - c) = v(t - c)u(t - c)$$

Express $v(t)$ in terms of $x(t)$.

(b) Determine the simplest possible analytical form for $v(t)$ when

(i) $x(t) = e^{-2t}$ and $c = 3$

(ii) $x(t) = t^2 - t + 1$ and $c = 2$

(iii) $x(t) = \sin 2t$ and $c = \pi/4$

1.7. Plot the following discrete-time signals:

(a) $x[n] = $ discrete-time unit-step function $u[n]$

(b) $x[n] = $ discrete-time unit-ramp function $r[n]$

(c) $x[n] = (0.5)^n\, u[n]$

(d) $x[n] = (-0.5)^n\, u[n]$

(e) $x[n] = 2^n\, u[n]$

(f) $x[n] = \sin(\pi n/4)$

(g) $x[n] = \sin(\pi n/2)$

(h) $x[n] = (0.9)^n[\sin(\pi n/4) + \cos(\pi n/4)]$

(i) $x[n] = \begin{cases} 1, & -4 \le n \le 4 \\ 0, & \text{otherwise} \end{cases}$

(j) Verify the plots for parts (a)–(i) by using MATLAB with the `stem` command `stem(n,x,'filled')`. Label your axes appropriately.

1.8. Sketch the following discrete-time signals:

(a) $x[n] = u[n] - 2u[n - 1] + u[n - 4]$

(b) $x[n] = (n + 2)u[n + 2] - 2u[n] - nu[n - 4]$

(c) $x[n] = \delta[n + 1] - \delta[n] + u[n + 1] - u[n - 2]$

(d) $x[n] = e^{0.8n}u[n + 1] + u[n]$

(e) Use MATLAB to plot the signals defined in parts (a) to (d).

1.9. Use an analytical method to determine if the signals (a)–(f) are periodic; if so, find the fundamental period. Use MATLAB to plot each signal, and verify your prediction of periodicity. Use a small enough time increment for the continuous-time signals to make your plot smooth (see Problem 1.2).

(a) $x(t) = \cos \pi t + \cos(4\pi t/5)$

(b) $x(t) = \cos(2\pi(t - 4)) + \sin 5\pi t$

(c) $x(t) = \cos 2\pi t + \sin 10t$

(d) $x[n] = \sin 10n$

(e) $x[n] = \sin(10 \pi n/3)$

(f) $x[n] = \cos(\pi n^2)$ (*Hint:* To get n^2 when n is stored in a vector, type n.^2.)

1.10. For the N-point MA filter given by the input/output relationship (1.16), where N is any positive integer, derive a mathematical expression for the filter's output response $y[n]$ when the input $x[n]$ is

(a) the unit-pulse function $\delta[n]$.

(b) the discrete-time unit-step function $u[n]$.

(c) the discrete-time unit-ramp function $r[n]$.

1.11. In Problem 1.10c, determine the time delay in the N-point MA filter response $y[n]$ when $n > N$.

1.12. Write a MATLAB M-file to compute the output response $y[n]$ of the 6-point MA filter to the input $x[n] = 5 \sin(\pi n/10 + \pi/4)$, for $0 \le n \le 69$. Express your answer by giving the M-file and the MATLAB plot.

1.13. For your solution in Problem 1.12, determine the time delay in the filter response.

1.14. Consider the system given by the input/output relationship

$$y[n] = \frac{32}{63}\left[x[n] + \frac{1}{2}x[n - 1] + \frac{1}{4}x[n - 2] + \frac{1}{8}x[n - 3] + \frac{1}{16}x[n - 4] + \frac{1}{32}x[n - 5] \right]$$

(a) When $x[n]$ is equal to a constant c for all $n \ge 0$, show that $y[n]$ is equal to c for all $n \ge 5$.

(b) Write a MATLAB M-file to compute the output response $y[n]$ resulting from the input $x[n] = 5 \sin(\pi n/10 + \pi/4)$, for $0 \le n \le 69$. Express your answer by giving the M-file and the MATLAB plot.

(c) Compare your result in part (b) with the MA filter response obtained in Problem 1.12.

1.15. The continuous-time counterpart to the N-point discrete-time MA filter is the I-interval continuous-time MA filter given by the input/output relationship

$$y(t) = \frac{1}{I} \int_{t-I}^{t} x(\lambda) \, d\lambda$$

where I is a positive number (the time interval over which the input is integrated). Derive a mathematical expression for the filter's output response $y(t)$ when the input $x(t)$ is

(a) the unit impulse $\delta(t)$.

(b) the continuous-time unit-step function $u(t)$.

(c) the continuous-time unit-ramp function $r(t)$.

(d) For your result in part c, determine the delay in the filter's response when $t > I$.

1.16. Using the procedure given in Section 1.2, download from the Web the closing prices of QQQQ for the 60-business-day period April 27, 2004, through July 22, 2004. Give the MATLAB plot for $9 \le n \le 60$ for the closing prices and the nine-day MA filter response to the data.

1.17. Determine whether these continuous-time systems are causal or noncausal, have memory or are memoryless. Justify your answers. In the following parts, $x(t)$ is an arbitrary input and $y(t)$ is the output response to $x(t)$:

(a) $y(t) = |x(t)| = \begin{cases} x(t) & \text{when } x(t) \ge 0 \\ -x(t) & \text{when } x(t) < 0 \end{cases}$

(b) $y(t) = e^{x(t)}$

(c) $y(t) = (\sin t)x(t)$

(d) $y(t) = \begin{cases} x(t) & \text{when } |x(t)| \le 10 \\ 10 & \text{when } |x(t)| > 10 \end{cases}$

(e) $y(t) = \int_{0}^{t} (t - \lambda)x(\lambda) \, d\lambda$

(f) $y(t) = \int_{0}^{t} \lambda x(\lambda) \, d\lambda$

1.18. Prove that the integral system and the differential system given next are both linear:

(i) $y(t) = \int_{0}^{t} x(\lambda) \, d\lambda$

(ii) $y(t) = \dfrac{dx(t)}{dt}$

1.19. For each of the systems in Problem 1.17, determine whether the system is linear or nonlinear. Justify your answers.

1.20. For each of the systems in Problem 1.17, determine whether the system is time invariant or time varying. Justify your answers.

1.21. A continuous-time system is said to have a *dead zone* if the output response $y(t)$ is zero for any input $x(t)$ with $|x(t)| < A$, where A is a constant called the *threshold*. An example is a dc motor that is unable to supply any torque until the input voltage exceeds a threshold value. Show that any system with a dead zone is nonlinear.

1.22. Determine whether the circuit with the ideal diode in Figure P1.22 is causal or noncausal, linear or nonlinear, and time invariant or time varying. Justify your answers.

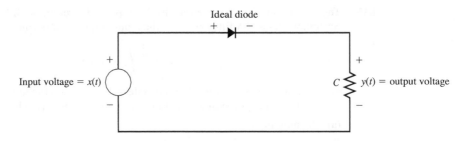

FIGURE P1.22

1.23. Suppose that $x(t)$ is the input to a linear time-invariant system with corresponding output $y(t)$. Prove that the system has the following properties:

(a) An input of $x_1(t) = \displaystyle\int_0^t x(\lambda)\,d\lambda$ to the same system yields an output of $y_1(t) = \displaystyle\int_0^t y(\lambda)\,d\lambda$.

(b) An input of $x_1(t) = \dfrac{dx(t)}{dt}$ to the same system yields an output of $y_1(t) = \dfrac{dy(t)}{dt}$.

1.24. A linear time-invariant continuous-time system responds to the following inputs with the corresponding outputs: If $x(t) = u(t)$, then $y(t) = 2(1 - e^{-t})\,u(t)$ and if $x(t) = \cos t$, then $y(t) = 1.414\cos(t - \pi/4)$. Find $y(t)$ for the following inputs (Hint: Use Problem 1.23 for part (d)):

(a) $x(t) = 2u(t) - 2u(t - 1)$

(b) $x(t) = 4\cos(2(t - 2))$

(c) $x(t) = 5u(t) + 10\cos(2t)$

(d) $x(t) = tu(t)$

1.25. To understand better the concept of linearity in discrete-time systems, write a MATLAB M-file that generates the output response $y[n]$ for $0 \le n \le 30$ of the five-day MA filter to the following various inputs:

(a) Compute and give the MATLAB plot of the response $y_1[n]$ to the input $x[n] = u[n]$.

(b) Compute and give the MATLAB plot of the response $y_2[n]$ to the input $x[n] = 2u[n]$. Compare this response with that obtained in part (a).

(c) Compute and give the MATLAB plot of the response $y_3[n]$ to the input $x[n] = \sin(\pi n/4)\,u[n]$.

(d) Compute and give the MATLAB plot of the response $y_4[n]$ to the input $x[n] = 2u[n] + \sin(\pi n/4)\,u[n]$. Compare this response with $2y_1[n] + y_3[n]$.

1.26. To understand better the concept of time-invariance in discrete-time systems, write a MATLAB M-file that generates the output response $y[n]$ for $0 \le n \le 30$ to the following various inputs for the system defined in Problem 1.14:

(a) Compute and give the MATLAB plot of the response $y_1[n]$ to the input $x[n] = u[n]$.

(b) Compute and give the MATLAB plot of the response $y_2[n]$ to the input $x[n] = u[n - 2]$. Compare this response with that obtained in part (a).

(c) Compute and give the MATLAB plot of the response $y_3[n]$ to the input $x[n] = \sin(\pi n/4)\, u[n]$.

(d) Compute and give the MATLAB plot of the response $y_4[n]$ to the input $x[n] = \sin(\pi[n-4]/4)u[n-4]$. Compare this response with $y_3[n-4]$.

1.27. Prove that the following system is linear:

$$y[n] = \sum_{i=0}^{n} a_i x[n-i]$$

where the coefficients a_i are constants.

1.28. Determine whether the system defined in Problem 1.14 is causal or noncausal, linear or nonlinear, and time invariant or time varying. Justify your answers.

1.29. Determine whether the following discrete-time systems are causal or noncausal, have memory or are memoryless, are linear or nonlinear, are time invariant or time varying. Justify your answers. In the following parts, $x[n]$ is an arbitrary input and $y[n]$ is the response to $x[n]$.

(a) $y[n] = x[n] + 2x[n-2]$

(b) $y[n] = x[n] + 2x[n+1]$

(c) $y[n] = nx[n]$

(d) $y[n] = u[n]x[n]$

(e) $y[n] = |x[n]|$

(f) $y[n] = \sin x[n]$

(g) $y[n] = \sum_{i=0}^{n} (0.5)^n x[i], \; n \geq 0$

Time-Domain Models of Systems

This chapter deals with the study of linear time-invariant discrete-time and continuous-time systems given by time-domain models that describe the relationship between the system input and the resulting system output. The development begins in Section 2.1 with the convolution representation of linear time-invariant discrete-time systems, and then in Section 2.2 the evaluation of the convolution operation for discrete-time signals is considered. In Section 2.3 the presentation focuses on discrete-time systems specified by a linear constant-coefficient input/output difference equation that can be solved very easily by the use of recursion. The recursion process is implemented with a MATLAB program that yields a software realization of the discrete-time system under consideration. Then in Section 2.4 the study of causal linear time-invariant continuous-time systems is given in terms of input/output differential equations. It is shown how such equations arise in the modeling of electrical circuits and mechanical systems. In Section 2.5, it is shown how an input/output differential equation can be solved by numerical methods and the MATLAB symbolic manipulator. The convolution of continuous-time signals and the convolution representation of linear time-invariant continuous-time systems are studied in Section 2.6. In Section 2.7, a summary of the chapter is given.

2.1 INPUT/OUTPUT REPRESENTATION OF DISCRETE-TIME SYSTEMS

In Section 1.4 of Chapter 1, the N-point moving average (MA) filter was defined by the input/output relationship

$$y[n] = \frac{1}{N}\Big[x[n] + x[n-1] + x[n-2] + \ldots + x[n-N+1]\Big] \tag{2.1}$$

where N is a positive integer. Note that (2.1) can be expressed in the form

$$y[n] = \sum_{i=0}^{N-1} \frac{1}{N} x[n-i] \tag{2.2}$$

The relationship given by (2.2) or (2.1) is often referred to as the *input/output representation* of the system. It is an example of a time-domain model, since it is expressed in terms of the system input $x[n]$ and the system output $y[n]$, both of which are time signals.

It turns out that we can generalize (2.2) to be a large class of causal linear time-invariant discrete-time systems by taking the input/output representation to be

$$y[n] = \sum_{i=0}^{N-1} w_i x[n - i] \tag{2.3}$$

where the w_i (that is, $w_0, w_1, w_2, \ldots, w_{N-1}$) are real numbers, and are called the *weights* of the linear combination given by (2.3). Note that if all the weights are equal to $1/N$, that is, $w_i = 1/N$ for $i = 0, 1, 2, \ldots, N - 1$, then (2.3) reduces to (2.2). Hence, the N-point MA filter is an example of a system whose input/output representation can be written in the form (2.3).

2.1.1   Exponentially Weighted Moving Average Filter

Another example of a system that can be expressed in the form (2.3) is the *N-point exponentially weighted moving average filter* defined by

$$y[n] = \sum_{i=0}^{N-1} a(b^i x[n - i]) \tag{2.4}$$

where b is a real number with $0 < b < 1$ and a is a positive constant given by

$$a = \frac{1 - b}{1 - b^N} \tag{2.5}$$

Note that if $b = 0$, then $a = 1$, and (2.4) reduces to $y[n] = x[n]$, in which case there is no filtering of the input signal.

From (2.4) it is seen that the weights of the N-point exponentially weighted moving average (EWMA) filter are given by

$$w_i = ab^i, i = 0, 1, 2, \ldots, N - 1 \tag{2.6}$$

Here the term "exponentially weighted" refers to the exponent of b in the weights given by (2.6). Since $0 < b < 1$, the weights given by (2.6) decrease in value as i increases in value. For example, if $N = 3$ and $b = 0.5$, then from (2.5) it follows that

$$a = \frac{1 - 0.5}{1 - 0.5^3} = \frac{0.5}{0.875} = 0.571$$

and using (2.6) gives $w_0 = a = 0.571$, $w_1 = ab = 0.286$, and $w_2 = ab^2 = 0.143$. Hence, the 3-point EWMA filter with $b = 0.5$ is given by the input/output relationship

$$y[n] = 0.571x[n] + 0.286x[n - 1] + 0.143x[n - 2] \tag{2.7}$$

From (2.7) it is seen that in the computation of the filter output $y[n]$, a larger "weight" is given to the more recent values of the input $x[n]$, whereas in the 3-point MA filter, the same weight is given to all the input values; that is, in the 3-point MA filter the output is

$$y[n] = 0.333x[n] + 0.333x[n - 1] + 0.333x[n - 2]$$

Due to the exponential weighting, the N-point EWMA filter given by (2.4) has a quicker response to time variations in the filter input $x[n]$ in comparison with the N-point MA filter. In other words, for a fixed value of N, the time delay through the EWMA filter is less than that of the MA filter; however, in general, the MA filter does a better job of removing noise than the EWMA filter does. For a given value of N, the time delay through the EWMA filter depends on the choice of b: the smaller b is, the smaller the time delay will be. In particular, if $b = 0$, then as previously noted, the input/output relationship is $y[n] = x[n]$, and thus there is no delay through the system in this case. In the following example, the EWMA is applied to filtering of price data for the index fund QQQQ.

Example 2.1   EWMA Filtering of QQQQ Price Data

As in Example 1.4 in Chapter 1, the closing price $c[n]$ of QQQQ for the 50-business-day period from March 1, 2004, up to May 10, 2004, will be considered. The time series $c[n]$ will be applied to an 11-day EWMA filter with $b = 0.7$. The output $y[n]$ of the filter is then computed by evaluating (2.4). To accomplish this, the weights $w_i = ab^i$ are computed and then are arranged in a row vector w having the form

$$w = [w_{N-1} \quad w_{N-2} \quad w_{N-3} \quad \cdots \quad w_0]$$

Then $y[n]$ is given by the MATLAB multiplication $w*c(n-10:n)$. The MATLAB commands for computing the filter output $y[n]$ for $11 \leq n \leq 50$ are as follows:

```
c=csvread('QQQQdata2.csv',1,4,[1 4 50 4]);
b=0.7;
a=(1-b)/(1-b^11);
i=1:11;
w=a*(b.^(11-i));
for n=11:50;
  y(n)=w*c(n-10:n);
end;
n=11:50;
plot(n,c(n),n,c(n),'o',n,y(n),n,y(n),'*')
grid
xlabel('Day(n)')
ylabel('c[n] and y[n]')
```

The resulting MATLAB plot for the filter input $c[n]$ and the filter output $y[n]$ is given in Figure 2.1, with the values of $c[n]$ plotted using o's and the values of $y[n]$ plotted using *'s. Comparing Figure 1.27 in Chapter 1 with Figure 2.1 shows that there is less time delay through the 11-day EWMA filter with $b = 0.7$ than there is through the 11-day MA filter. To see this more clearly, in Figure 2.2 the output of the 11-day MA filter is combined with the plot of $c[n]$ and the output of the 11-day EWMA filter. The values of the MA filter response are plotted with +'s.

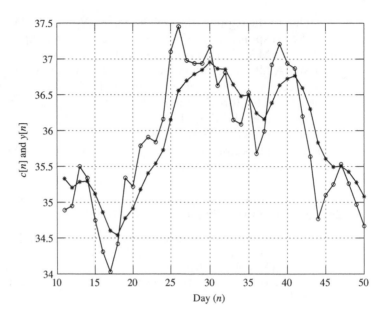

FIGURE 2.1
MATLAB plot of EWMA filter input $c[n]$ and output $y[n]$.

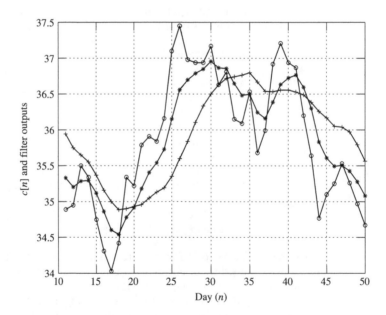

FIGURE 2.2
MATLAB plot of $c[n]$ and filter outputs.

2.1.2   A General Class of Systems

The class of systems given by the input/output representation (2.3) can be generalized further by replacing the upper index $N-1$ in the summation with n; that is, the input/output relationship is now given by

$$y[n] = \sum_{i=0}^{n} w_i x[n-i], n \geq 0 \tag{2.8}$$

where, in general, there are an infinite number of nonzero weights $w_0, w_1, w_2, \ldots$. It turns out that any causal linear time-invariant discrete-time system with the input $x[n]$ equal to zero for all $n < 0$ can be expressed in the form (2.8).

A system with the input/output representation (2.8) is usually expressed in terms of its unit-pulse response, which is defined as follows: The *unit-pulse response*, denoted by $h[n]$, is the output response of the system resulting from the application of the unit pulse $\delta[n]$; that is, $x[n] = \delta[n]$. (Recall that $\delta[0] = 1$ and $\delta[n] = 0$ for all $n \neq 0$.) Note that, since $\delta[n] = 0$ for $n = -1, -2, \ldots$, by causality the unit-pulse response $h[n]$ must be zero for all integers $n < 0$ (see Section 1.5 in Chapter 1).

To compute the unit-pulse response for a system given by (2.8), simply insert $x[n] = \delta[n]$ into (2.8), which gives

$$h[n] = \sum_{i=0}^{n} w_i \delta[n-i], n \geq 0 \tag{2.9}$$

Now, since $\delta[n-i] = 0$ for all $i \neq n$, and $\delta[n-i] = 1$ when $i = n$, (2.9) reduces to

$$h[n] = w_n, n \geq 0$$

Hence, the value $h[n]$ of the unit-pulse response at time n is equal to the weight w_n.

Example 2.2   Unit-Pulse Responses of the MA and EWMA Filters

From the preceding development, the weights of the 11-day MA filter are given by $w_i = 1/11$ for $i = 0, 1, \ldots, 10$, and by (2.5) and (2.6), the weights of the 11-day EWMA with $b = 0.7$ are given by $w_i = ab^i = (0.3061)(0.7)^i, i = 0, 1, \ldots, 10$. For both filters, $w_i = 0$ for $i \geq 11$. Then setting $h[n] = w_n, n = 0, 1, \ldots, 10$, and $h[n] = 0$ for $n \geq 11$ and $n \leq -1$ yields the stem plots of the unit-pulse responses for the 11-day MA and EWMA filters shown in Figure 2.3. In the figure, $h1[n]$ is the unit-pulse response of the 11-day MA filter and $h2[n]$ is the unit-pulse response of the 11-day EWMA filter with $b = 0.7$. Note that the unit-pulse responses of the 11-day MA and EWMA filters are finite-duration signals; that is, $h[n]$ is nonzero for only a finite number of values of n.

Rewriting (2.8) in terms of the unit-pulse response $h[n]$ gives

$$y[n] = \sum_{i=0}^{n} h[i] x[n-i], n \geq 0 \tag{2.10}$$

The operation defined by the expression

$$\sum_{i=0}^{n} h[i] x[n-i]$$

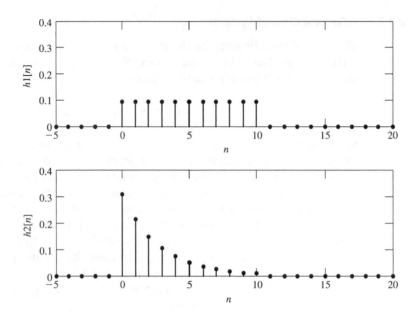

FIGURE 2.3
Unit-pulse responses of 11-day MA filter, $h1[n]$, and EWMA filter, $h2[n]$.

is called the *convolution* of $h[n]$ and $x[n]$, and is denoted by the symbol "*"; that is,

$$h[n] * x[n] = \sum_{i=0}^{n} h[i]x[n - i]$$

Rewriting (2.10) in terms of the convolution symbol gives

$$y[n] = h[n] * x[n], n \geq 0 \tag{2.11}$$

By (2.11), the output response $y[n]$ resulting from input $x[n]$ with $x[n] = 0$ for all $n < 0$ is equal to the convolution of the unit-pulse response $h[n]$ with the input $x[n]$. Equation (2.11) [or (2.10)] is called the *convolution representation* of the system. This is a time domain model, since the components of (2.11) are functions of the discrete-time index n.

An interesting consequence of the convolution representation (2.10) is the result that the system is determined completely by the unit-pulse response $h[n]$. In particular, if $h[n]$ is known, the output response resulting from any input $x[n]$ can be computed by evaluating (2.10). The evaluation of the convolution operation is studied in the next section.

2.2   CONVOLUTION OF DISCRETE-TIME SIGNALS

In the previous section, the convolution of an input $x[n]$ and the unit-pulse response $h[n]$ were defined, with both $x[n]$ and $h[n]$ equal to zero for all $n < 0$. In this section the convolution operation is defined for arbitrary discrete-time signals $x[n]$ and $v[n]$ that are not necessarily zero for $n < 0$.

Given two discrete-time signals $x[n]$ and $v[n]$, the convolution of $x[n]$ and $v[n]$ is defined by

$$x[n] * v[n] = \sum_{i=-\infty}^{\infty} x[i]v[n-i] \qquad (2.12)$$

The summation on the right-hand side of (2.12) is called the *convolution sum*. It is important to note that the convolution operation $x[n] * v[n]$ is commutative; that is, $x[n] * v[n] = v[n] * x[n]$, and therefore,

$$\sum_{i=-\infty}^{\infty} x[i]v[n-i] = \sum_{i=-\infty}^{\infty} v[i]x[n-i]$$

If $x[n]$ and $v[n]$ are zero for all integers $n < 0$, then $x[i] = 0$ for all integers $i < 0$ and $v[n-i] = 0$ for all integers $n - i < 0$ (or $n < i$). Thus the summation on i in (2.12) may be taken from $i = 0$ to $i = n$, and the convolution operation is given by

$$x[n] * v[n] = \begin{cases} 0, & n = -1, -2, \ldots \\ \sum_{i=0}^{n} x[i]v[n-i], & n = 0, 1, 2, \ldots \end{cases} \qquad (2.13)$$

or by commutativity,

$$x[n] * v[n] = \begin{cases} 0, & n = -1, -2, \ldots \\ \sum_{i=0}^{n} v[i]x[n-i], & n = 0, 1, 2, \ldots \end{cases} \qquad (2.14)$$

If the signals $x[n]$ and $v[n]$ are given by simple mathematical expressions, the convolution $x[n] * v[n]$ can be computed analytically by inserting $x[n]$ and $v[n]$ into (2.13) or (2.14). This is illustrated by the following example:

Example 2.3   Use of Analytical Form

Suppose that $x[n] = a^n u[n]$ and $v[n] = b^n u[n]$, where $u[n]$ is the discrete-time unit-step function and a and b are fixed nonzero real numbers. Inserting $x[i] = a^i u[i]$ and $v[n-i] = b^{n-i} u[n-i]$ into (2.13) gives

$$x[n] * v[n] = \sum_{i=0}^{n} a^i u[i] b^{n-i} u[n-i], \quad n = 0, 1, 2, \ldots \qquad (2.15)$$

Now $u[i] = 1$ and $u[n - i] = 1$ for all integer values of i ranging from $i = 0$ to $i = n$, and thus (2.15) reduces to

$$x[n] * v[n] = \sum_{i=0}^{n} a^i b^{n-i} = b^n \sum_{i=0}^{n} \left(\frac{a}{b}\right)^i, \quad n = 0, 1, 2, \ldots \tag{2.16}$$

If $a = b$,

$$\sum_{i=0}^{n} \left(\frac{a}{b}\right)^i = n + 1$$

and

$$x[n] * v[n] = b^n(n + 1) = a^n(n + 1), \quad n = 0, 1, 2, \ldots$$

If $a \neq b$,

$$\sum_{i=0}^{n} \left(\frac{a}{b}\right)^i = \frac{1 - (a/b)^{n+1}}{1 - a/b} \tag{2.17}$$

The relationship (2.17) can be verified by multiplying both sides of (2.17) by $1 - (a/b)$. Inserting (2.17) into (2.16) yields (assuming that $a \neq b$)

$$x[n] * v[n] = b^n \frac{1 - (a/b)^{n+1}}{1 - a/b}$$

$$= \frac{b^{n+1} - a^{n+1}}{b - a}, \quad n = 0, 1, 2, \ldots$$

It is easy to generalize (2.13) or (2.14) to the case when $x[n]$ and $v[n]$ are not necessarily zero for all integers $n < 0$. In particular, suppose that $x[n] = 0$ for all $n < Q$ and $v[n] = 0$ for all $n < P$, where P and Q are positive or negative integers. In this case the convolution operation (2.12) can be written in the form

$$x[n] * v[n] = \begin{cases} 0, & n < P + Q \\ \sum_{i=Q}^{n-P} x[i]v[n - i], & n \geq P + Q \end{cases} \tag{2.18}$$

Note that the convolution sum in (2.18) is still finite, and thus the convolution $x[n] * v[n]$ exists.

The convolution operation (2.18) can be evaluated with an array, as follows:

	$x[Q]$	$x[Q + 1]$	$x[Q + 2]$	$x[Q + 3]$
$v[P]$	$v[P]x[Q]$	$v[P]x[Q + 1]$	$v[P]x[Q + 2]$	$v[P]x[Q + 3]$
$v[P + 1]$	$v[P + 1]x[Q]$	$v[P + 1]x[Q + 1]$	$v[P + 1]x[Q + 2]$	$v[P + 1]x[Q + 3]$
$v[P + 2]$	$v[P + 2]x[Q]$	$v[P + 2]x[Q + 1]$	$v[P + 2]x[Q + 2]$	$v[P + 2]x[Q + 3]$
$v[P + 3]$	$v[P + 3]x[Q]$	$v[P + 3]x[Q + 1]$	$v[P + 3]x[Q + 2]$	$v[P + 3]x[Q + 3]$

The top of the array is labeled with the values $x[Q], x[Q + 1], \ldots$, and the left side is labeled with the values $v[P], v[P + 1], \ldots$. The elements of the array are filled in by multiplication of the corresponding column and row labels. The values of $y[n]$ of the convolution $x[n] * v[n]$ are then determined by the sum of the elements in the backwards diagonals, where the diagonal that begins at $x[Q + i]$ and ends at $v[P + i]$ is summed to yield $y[Q + P + i]$.

The process is illustrated by the following example.

Example 2.4   Array Method

Suppose that $x[n] = 0$ for $n < -1$, $x[-1] = 1$, $x[0] = 2$, $x[1] = 3$, $x[2] = 4$, $x[3] = 5, \ldots$, and $v[n] = 0$ for $n < -2$, $v[-2] = -1$, $v[-1] = 5$, $v[0] = 3$, $v[1] = -2$, $v[2] = 1, \ldots$. In this case, $Q = -1$, $P = -2$, and the array is as follows:

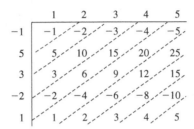

The values $y[n]$ of the convolution $x[n] * v[n]$ are computed by the sum of the elements in the diagonals that are indicated by the dashed lines, starting with the element in the upper left corner of the array. Summing the indices of the first elements of $x[n]$ and $v[n]$, which are $n = -1$ and $n = -2$, determines the index for the first element in the sequence for y. Since $(-1) + (-2) = -3$, the index for the first element in y is -3. Thus, the first nonzero value of $y[n]$ is the value $y[-3]$ that is equal to the diagonal element -1. The next value $y[-2]$ is equal to the sum $-2 + 5 = 3$, $y[-1]$ is equal to the sum $-3 + 10 + 3 = 10$, $y[0]$ is equal to the sum $-4 + 15 + 6 - 2 = 15$, $y[1]$ is equal to the sum $-5 + 20 + 9 - 4 + 1 = 21$, and so on. In this example, $y[n] = 0$ for $n < -3$.

Convolution of Discrete-Time Signals

The convolution of two discrete-time signals can be carried out with the MATLAB M-file conv. To illustrate this, consider the convolution of the pulse $p[n]$ with itself, where $p[n]$ is defined by $p[n] = 1$ for $0 \leq n \leq 10$, $p[n] = 0$ for all other n. The MATLAB commands for computing the convolution in this case are

```
p = [0 ones(1,10) zeros(1,5)]; % corresponds to n=-1 to n=14
x = p; v = p;
y = conv(x,v);
n = -2:25;
stem(n,y(1:length(n)),'filled')
```

The command $y = \text{conv}(x,v)$ in this example results in a vector y that has length 32. As in the case with the array method, the index n corresponding to the first element in the vector y is determined from the sum of the indices of the first elements for the

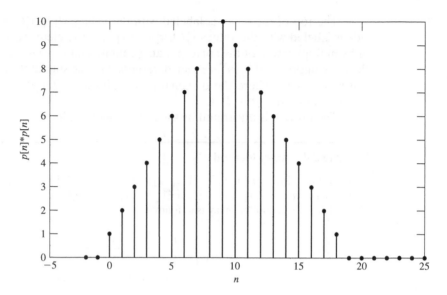

FIGURE 2.4
Convolution of the pulse with itself.

vectors x and v, resulting in the first element of y corresponding to $y[-2]$. The values corresponding to $n = -2$ to $n = 25$ are then plotted, which results in the stem plot shown in Figure 2.4. The reader is encouraged to read the comments in the MATLAB tutorial on the website.

2.2.1 Computation of System Output

Consider a causal linear time-invariant discrete-time system with $x[n] = 0$ for $n < 0$. Since the output $y[n]$ is equal to the discrete-time convolution $h[n] * x[n]$ given by (2.10), the response $y[n]$ for any finite range of values of n can be computed with the MATLAB M-file conv. The procedure is illustrated by the following example.

Example 2.5   Computation of Output Response by Use of MATLAB

Suppose that the unit-pulse response $h[n]$ is equal to $\sin(0.5n)$ for $n \geq 0$, and the input $x[n]$ is equal to $\sin(0.2n)$ for $n \geq 0$. Plots of $h[n]$ and $x[n]$ are given in Figure 2.5. Now, to compute the response $y[n]$ for $n = 0, 1, \ldots, 40$, use the commands

```
n=0:40;
x = sin(.2*n);
h = sin(.5*n);
y = conv(x,h);
stem(n,y(1:length(n)),'filled')
```

A MATLAB-generated plot of the response values is given in Figure 2.6.

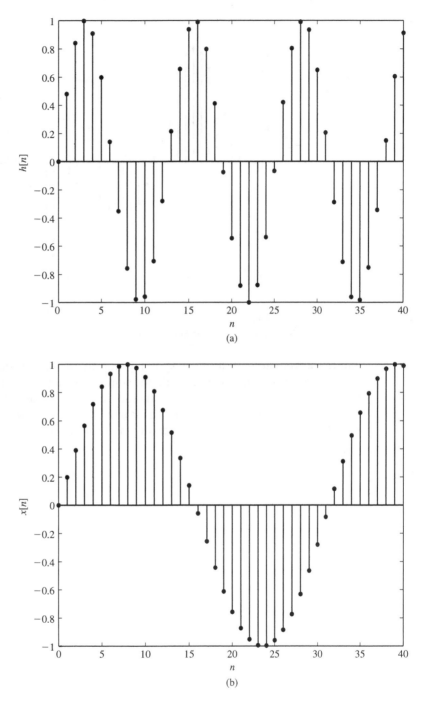

FIGURE 2.5
(a) Plot of $h[n]$ in Example 2.5. (b) Plot of $x[n]$ in Example 2.5.

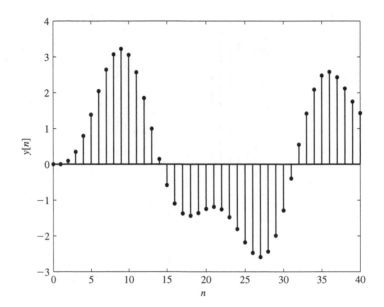

FIGURE 2.6
Plot of output response in Example 2.5.

Noncausal systems.   If the given system is noncausal, the unit-pulse response $h[n]$ will not be zero for $n < 0$, and as a result, the summation in (2.10) for computing $y[n]$ must run from $i = 0$ to $i = \infty$ (not $i = n$). In addition, if the input $x[n]$ is nonzero for values of n ranging from 0 to $-\infty$, the summation in (2.10) must start at $i = -\infty$. Hence the input/output convolution expression for a noncausal system (with $x[n] \neq 0$ for $n < 0$) is given by

$$y[n] = h[n] * x[n] = \sum_{i=-\infty}^{\infty} h[i]x[n - i] \tag{2.19}$$

It is interesting to note that, although (2.19) is the input/output relationship for the system, in general (2.19) cannot be computed since it is a bi-infinite sum; that is, an infinite summation cannot be evaluated in a finite number of (computational) steps.

2.3   DIFFERENCE EQUATION MODELS

In many applications, a causal linear time-invariant discrete-time system is given by an input/output difference equation instead of an input/output convolution model. For example, the repayment of a bank loan can be modeled by a difference equation representation as follows: With $n = 1, 2, \ldots$, the input $x[n]$ is the amount of the loan payment in the nth month, and the output $y[n]$ is the balance of the loan after the nth month. Here n is the time index that denotes the month, and the input $x[n]$ and output $y[n]$ are discrete-time signals that are functions of n. The initial condition $y[0]$ is the amount of the loan. Usually, the loan payments $x[n]$ are constant; that is, $x[n] = c, n = 1, 2, 3, \ldots$, where c is a constant. In this example, $x[n]$ is allowed to vary from month to month (i.e., the loan payments may not be equal).

The repayment of the loan is described by the input/output difference equation

$$y[n] - \left(1 + \frac{I}{12}\right)y[n-1] = -x[n], \quad n = 1, 2, \ldots \tag{2.20}$$

where I is the yearly interest rate in decimal form. For example, if the yearly interest rate were 10 percent, I would be equal to 0.1. The term $(I/12)y[n-1]$ in (2.20) is the interest on the loan in the nth month; and thus in the model given by (2.20), the interest is compounded monthly. Equation (2.20) is a *first-order linear difference equation*. It is the input/output difference equation of the system consisting of the loan-repayment process. It is important to note that the output response $y[n]$ in (2.20) is the response resulting from the application of the input $x[n]$ for $n \geq 1$ and the initial condition $y[0]$. This formulation differs from the input/output convolution model in that in the latter model there are no initial conditions.

You can compute the output $y[n]$ by solving (2.20) recursively as follows. First, rewrite (2.20) in the form

$$y[n] = \left(1 + \frac{I}{12}\right)y[n-1] - x[n] \tag{2.21}$$

Now, inserting $n = 1$ in (2.21) yields

$$y[1] = \left(1 + \frac{I}{12}\right)y[0] - x[1] \tag{2.22}$$

Inserting $n = 2$ into (2.21) gives

$$y[2] = \left(1 + \frac{I}{12}\right)y[1] - x[2] \tag{2.23}$$

Taking $n = 3$ in (2.21) gives

$$y[3] = \left(1 + \frac{I}{12}\right)y[2] - x[3] \tag{2.24}$$

By continuing in this manner, $y[n]$ can be computed for any finite range of integer values of n. From (2.22) through (2.24), it is seen that the next value of the output is computed from the present value of the output plus an input term. This is why the process is called a *recursion*. In this example, the recursion is a first-order recursion.

A MATLAB program for carrying out the recursion defined by (2.21) is given in Figure 2.7. The inputs to the program are the loan amount, the interest rate, and the monthly payment. The statement "$y = [\ ];$" is used to initialize y as a vector with

```
% Loan Balance program
% Program computes loan balance y[n]
y0 = input ('Amount of loan ');
I = input ('Yearly Interest rate, in decimal ');
c = input ('Monthly loan payment '); % x[n] = c
y = [ ];              % defines y as an empty vector
y(1) = (1 + (I/12))*y0 - c;
for n = 2:360,
    y(n) = (1 + (I/12))*y(n-1) - c;
    if y(n) < 0, break, end
end
```

FIGURE 2.7
MATLAB program for computing loan balance.

no elements. The elements of y are then computed recursively to be the loan balance at the end of the nth month where the index of the vector corresponds to month n. Note that elements in vectors are denoted in MATLAB with parentheses. The program continues in a loop until the loan balance is negative, which means that the loan is paid off.

As an example, the MATLAB program was run with $y[0] = \$6000$, interest rate equal to 12 percent, and monthly payment equal to $\$200$ (so that $I = 0.12$ and $c = 200$). The resulting loan balance $y[n]$ is shown in Table 2.1. When the monthly payment is $\$300$, the loan balance $y[n]$ is as displayed in Table 2.2. Note that in the first case, it takes 36 months to pay off the loan, whereas in the latter case, the loan is paid off in 23 months. When a loan is taken out, the number of months in the payoff period is usually specified and then the monthly payment is determined. It is possible to solve for the monthly payment by the use of the representation (2.20) [or (2.21)], but this is not pursued here. (See Problem 2.14.)

2.3.1   Nth-Order Input/Output Difference Equation

The first-order input/output difference equation (2.20) is easily generalized to the Nth-order case, where N is any positive integer. This results in a class of causal linear time-invariant discrete-time systems given by the input/output representation

$$y[n] + \sum_{i=1}^{N} a_i y[n-i] = \sum_{i=0}^{M} b_i x[n-i] \tag{2.25}$$

where n is the integer-valued discrete-time index, $x[n]$ is the input, and $y[n]$ is the output. Here it is assumed that the coefficients $a_1, a_2, \ldots, a_N$ and $b_0, b_1, \ldots, b_M$ are constants. It should be noted that the integer N in (2.25) is not related to the value of N in N-point MA and EWMA filters defined in Section 1.4 of Chapter 1 and in Section 2.1 of this chapter. This dual use of the notation "N" is very common in the signals and systems field and should not result in any confusion.

TABLE 2.1   Loan Balance with $200 Monthly Payments

n	$y[n]$	n	$y[n]$
1	$5859.99	19	$3086.47
2	5718.59	20	2917.33
3	5575.78	21	2746.51
4	5431.54	22	2573.97
5	5285.85	23	2399.71
6	5138.71	24	2223.71
7	4990.1	25	2045.95
8	4840	26	1866.41
9	4688.4	27	1685.07
10	4535.29	28	1501.92
11	4380.64	29	1316.94
12	4224.44	30	1130.11
13	4066.69	31	941.41
14	3907.36	32	750.83
15	3746.43	33	558.33
16	3583.89	34	363.92
17	3419.73	35	167.56
18	3253.93	36	−30.77

TABLE 2.2   Loan Balance with $300 Monthly Payments

n	$y[n]$	n	$y[n]$
1	$5759.99	13	$2685.76
2	5517.59	14	2412.61
3	5272.77	15	2136.74
4	5025.5	16	1858.11
5	4775.75	17	1576.69
6	4523.51	18	1292.46
7	4268.75	19	1005.38
8	4011.43	20	715.43
9	3751.55	21	422.59
10	3489.06	22	126.81
11	3223.95	23	−171.92
12	2956.19		

Linear input/output difference equations of the form (2.25) can be solved by a direct numerical procedure. More precisely, the output $y[n]$ for some finite range of integer values of n can be computed recursively as follows. First, rewrite (2.25) in the form

$$y[n] = -\sum_{i=1}^{N} a_i y[n-i] + \sum_{i=0}^{M} b_i x[n-i] \qquad (2.26)$$

Then, setting $n = 0$ in (2.26) gives

$$y[0] = -a_1 y[-1] - a_2 y[-2] - \cdots - a_N y[-N] + b_0 x[0] + b_1 x[-1]$$
$$+ \cdots + b_M x[-M]$$

Thus the output $y[0]$ at time 0 is a linear combination of $y[-1], y[-2], \ldots, y[-N]$ and $x[0], x[-1], \ldots, x[-M]$.

Setting $n = 1$ in (2.26) gives

$$y[1] = -a_1 y[0] - a_2 y[-1] - \cdots - a_N y[-N + 1] + b_0 x[1]$$
$$+ b_1 x[0] + \cdots + b_M x[-M + 1]$$

So, $y[1]$ is a linear combination of $y[0], y[1], \ldots, y[-N + 1]$ and $x[1], x[0], \ldots, x[-M + 1]$.

If this process is continued, it is clear that the next value of the output is a linear combination of the N past values of the output and $M + 1$ values of the input. At each step of the computation, it is necessary to store only the N past values of the output (plus, of course, the input values). This process is called an Nth-*order recursion*. Here the term *recursion* refers to the property that the next value of the output is computed from the N previous values of the output (plus the input values). The discrete-time system defined by (2.25) [or (2.26)] is sometimes called a *recursive discrete-time system* or a *recursive digital filter*, since its output can be computed recursively. Here it is assumed that at least one of the coefficients a_i in (2.25) is nonzero. If all the a_i are zero, the input/output difference equation (2.25) reduces to

$$y[n] = \sum_{i=0}^{M} b_i x[n - i]$$

In this case, the output at any fixed time point depends only on values of the input $x[n]$, and thus the output is not computed recursively. Such systems are said to be *nonrecursive*. Examples of nonrecursive systems include MA and EWMA filters.

From the expression (2.26) for $y[n]$, it follows that if $M = N$, the computation of $y[n]$ for each integer value of n requires (in general) $2N$ additions and $2N + 1$ multiplications. So the "computational complexity" of the Nth-order recursion is directly proportional to the order N of the recursion. In particular, note that the number of computations required to compute $y[n]$ does not depend on n.

Finally, from (2.25) or (2.26) it is clear that the computation of the output response $y[n]$ for $n \geq 0$ requires that the N initial conditions $y[-N], y[-N + 1], \ldots, y[-1]$ must be specified. In addition, if the input $x[n]$ is not zero for $n < 0$, the evaluation of (2.25) or (2.26) also requires the M initial input values $x[-M], x[-M + 1], \ldots, x[-1]$. Hence, the output response $y[n]$ for $n \geq 0$ given in (2.25) or (2.26) is the response to the input $x[n]$ applied for $n \geq -M$ and the initial conditions $y[-N], y[-N + 1], \ldots, y[-1]$. Note again how this differs from the convolution model where there are no initial conditions given in terms of values of the output.

It should also be pointed out that the unit-pulse response $h[n]$ of a system given by (2.25) or (2.26) can be computed by setting $x[n] = \delta[n]$, with zero initial conditions, that is, $y[n]$ is zero for $n \leq -1$. The evaluation of (2.26) is illustrated via the following example:

Example 2.6   Second-Order System

Consider the discrete-time system given by the second-order input/output difference equation

$$y[n] - 1.5y[n-1] + y[n-2] = 2x[n-2] \qquad (2.27)$$

Writing (2.27) in the form (2.26) results in the input/output equation

$$y[n] = 1.5y[n-1] - y[n-2] + 2x[n-2] \qquad (2.28)$$

To compute the unit-pulse response $h[n]$ of the system, set $x[n] = \delta[n]$ in (2.28) with the initial conditions $y[-1]$ and $y[-2]$ both equal to zero. This gives $h[0] = 0$, $h[1] = 0$, $h[2] = 2\delta[0] = 2$, $h[3] = (1.5)h[2] = 3$, $h[4] = (1.5)h[3] - h[2] = 2.5$, and so on.

Now suppose that the input $x[n]$ is the discrete-time unit-step function $u[n]$ and that the initial output values are $y[-2] = 2$ and $y[-1] = 1$. Then, setting $n = 0$ in (2.28) gives

$$y[0] = 1.5y[-1] - y[-2] + 2x[-2]$$
$$y[0] = (1.5)(1) - 2 + (2)(0) = -0.5$$

Setting $n = 1$ in (2.28) gives

$$y[1] = 1.5y[0] - y[-1] + 2x[-1]$$
$$y[1] = (1.5)(-0.5) - 1 + 2(0) = -1.75$$

Continuing the process yields

$$y[2] = (1.5)y[1] - y[0] + 2x[0]$$
$$= (1.5)(-1.75) + 0.5 + (2)(1) = -0.125$$
$$y[3] = (1.5)y[2] - y[1] + 2x[1]$$
$$= (1.5)(-0.125) + 1.75 + (2)(1) = 3.5625$$

and so on.

In solving (2.25) or (2.26) recursively, we see that the process of computing the output $y[n]$ can begin at any time point desired. In the preceding development, the first value of the output that was computed was $y[0]$. If the first desired value is the output $y[q]$ at time q, we should start the recursion process by setting $n = q$ in (2.26). In this case, the initial values of the output that are required are $y[q - N]$, $y[q - N + 1], \ldots,$ $y[q - 1]$.

The Nth-order difference equation (2.25) can be solved with the MATLAB M-file recur, available from the website. An abbreviated version of the program that contains all of the important steps is given in Figure 2.8. To use the recur M-file, the user must input the system coefficients (i.e., the a_i and b_i), the initial values of $y[n]$ and $x[n]$, the desired range of n, and the input $x[n]$. The program first initializes the solution vector y by

```
N = length(a);

M = length(b)-1;
y = [y0 zeros(1,length(n))];
x = [x0 x];
a1 = a(length(a):-1:1);      % reverses the elements in a
b1 = b(length(b):-1:1);
for i=N+1:N+length(n),
    y(i) = -a1*y(i-N:i-1)' + b1*x(i-N:i-N+M)';
end
y = y(N+1:N+length(n));
```

FIGURE 2.8
MATLAB program recur.

augmenting the initial conditions with zeros. This predefines the length of y, which makes the loop more efficient. The input vector x is also augmented by the initial conditions of the input. The summations given in (2.25) are computed by the multiplication of vectors; for example, the summation on the left-hand side of (2.25) can be written as

$$\sum_{i=1}^{N} a_i y[n-i] = [a_N\ a_{N-1} \cdots a_1] \begin{bmatrix} y[n-N] \\ y[n-N+1] \\ \vdots \\ y[n-1] \end{bmatrix}$$

Each summation could also be evaluated by the use of an inner loop in the program; however, MATLAB does not process loops very efficiently, so loops are avoided whenever possible. In the last line of the program, the initial conditions are removed from the vector y, with the result being a vector y that contains the values of $y[n]$ for the time indices defined in the vector n. It should be noted that there are other programs in the MATLAB toolboxes that solve the equation in (2.25); however, these programs utilize concepts that have not yet been introduced.

The following commands demonstrate how recur is used to compute the output response when $x[n] = u[n]$ for the system in Example 2.6:

```
a = [-1.5 1]; b = [0 0 2];
y0 = [2 1]; x0 = [0 0];
n = 0:20;
x = ones(1, length(n));
y = recur (a, b, n, x, x0, y0);
stem(n,y,'filled')        % produces a "stem plot"
xlabel ('n')
ylabel ('y[n]')
```

The M-file computes the response y for $n = 0, 1, \dots, 20$ and then plots y versus n with labels on the axes. The resulting output response is given in Figure 2.9. In future examples,

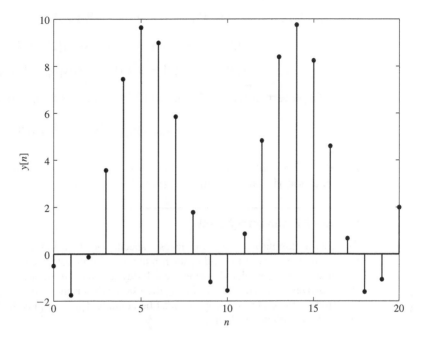

FIGURE 2.9
Plot of output response resulting from $x[n] = u[n]$ in Example 2.6.

the plotting commands will not be shown, except when the variables to be plotted are not obvious.

2.3.2   Complete Solution

By solving (2.25) or (2.26) recursively, it is possible to generate an expression for the complete solution $y[n]$ resulting from initial conditions and the application of the input $x[n]$. We illustrate the process by considering the first-order linear difference equation

$$y[n] = -ay[n-1] + bx[n], n = 1, 2, \ldots \qquad (2.29)$$

with the initial condition $y[0]$. First, setting $n = 1$, $n = 2$, and $n = 3$ in (2.29) gives

$$y[1] = -ay[0] + bx[1] \qquad (2.30)$$

$$y[2] = -ay[1] + bx[2] \qquad (2.31)$$

$$y[3] = -ay[2] + bx[3] \qquad (2.32)$$

Inserting the expression (2.30) for $y[1]$ into (2.31) gives

$$y[2] = -a(-ay[0] + bx[1]) + bx[2]$$

$$y[2] = a^2 y[0] - abx[1] + bx[2] \qquad (2.33)$$

Inserting the expression (2.33) for $y[2]$ into (2.32) yields

$$y[3] = -a(a^2 y[0] - abx[1] + bx[2]) + bx[3]$$
$$y[3] = -a^3 y[0] + a^2 bx[1] - abx[2] + bx[3] \tag{2.34}$$

From the pattern in (2.30), (2.33), and (2.34), it can be seen that, for $n \geq 1$,

$$y[n] = (-a)^n y[0] + \sum_{i=1}^{n} (-a)^{n-i} bx[i] \tag{2.35}$$

Equation (2.35) gives the complete output response $y[n]$ for $n \geq 1$ resulting from initial condition $y[0]$ and the input $x[n]$ applied for $n \geq 1$.

Example 2.7   Inventory Level

Consider a manufacturer that produces a specific product. Let $y[n]$ denote the number of the product in inventory at the end of the nth day, let $p[n]$ denote the number of the product whose manufacturing is completed during the nth day, and let $d[n]$ denote the number of the product that is delivered (to customers) during the nth day. Then the number $y[n]$ of the product in the inventory at the end of the nth day must be equal to $y[n-1]$ plus the difference between $p[n]$ and $d[n]$. In mathematical terms,

$$y[n] = y[n-1] + p[n] - d[n] \quad \text{for } n = 1, 2, \ldots \tag{2.36}$$

Here $y[0]$ is the initial number of the product in inventory. Now with $x[n]$ defined to be the difference $x[n] = p[n] - d[n]$, (2.36) is in the form (2.29) with $a = -1$ and $b = 1$. Hence, from (2.35), the solution is

$$y[n] = y[0] + \sum_{i=1}^{n} x[i], n = 1, 2, \ldots$$

$$y[n] = y[0] + \sum_{i=1}^{n} (p[i] - d[i]), n = 1, 2, \ldots \tag{2.37}$$

One of the objectives in manufacturing is to keep the level of the inventory fairly constant; in particular, depletion of the inventory should obviously be avoided; otherwise, there will be a delay in delivery of the product to customers. From the previous expression for $y[n]$, it is seen that $y[n]$ can be kept constant by the setting of $p[n] = d[n]$. In other words, the number of the product whose manufacturing is completed during the nth day should be equal to the number of the product delivered during the nth day. However, it is not possible to set $p[n] = d[n]$, since a product cannot be manufactured "instantaneously" and $d[n]$ depends on customer orders and is not known in advance. If the manufacture of the product requires less than one day, it is possible to set

$$p[n] = d[n-1], n = 2, 3, \ldots \tag{2.38}$$

That is, the number $p[n]$ of the product whose manufacturing is to be completed during the nth day is set equal to the number $d[n-1]$ of deliveries during the preceding day.

In order to use (2.38) in (2.37), it is first necessary to write (2.37) in the form

$$y[n] = y[0] + p[1] - d[1] + \sum_{i=2}^{n} (p[i] - d[i]), n = 2, 3, \ldots \tag{2.39}$$

Then, inserting $p[i] = d[i-1]$ into (2.39) yields

$$y[n] = y[0] + p[1] - d[1] + \sum_{i=2}^{n}(d[i-1] - d[i]), n = 2, 3, \dots$$

$$y[n] = y[0] + p[1] - d[1] + d[1] - d[n], n = 2, 3, \dots$$

$$y[n] = y[0] + p[1] - d[n], n = 2, 3, \dots$$

From this result, it is clear that the inventory will never be depleted if the initial inventory is sufficiently large to handle the variations in the number of deliveries from day to day. More precisely, depletion of the inventory will not occur if

$$y[0] > d[n] - p[1]$$

This is an interesting result, for it tells the manufacturer how much of the product should be kept in stock to avoid delays in delivery due to inventory depletion.

Closed-form expressions (i.e., expressions that do not contain summations) for the solution of linear constant-coefficient difference equations can be computed by application of the z-transform. This is pursued in Chapter 7.

2.4   DIFFERENTIAL EQUATION MODELS

Continuous-time systems are often specified by an input/output differential equation that can be generated by application of the laws of physics. This section begins with the process of determining the input/output differential equation for a class of electrical circuits and mechanical systems. Then, solution methods are given for solving differential equations.

2.4.1   Electrical Circuits

Schematic representations of a resistor, capacitor, and inductor are shown in Figure 2.10. With respect to the voltage $v(t)$ and the current $i(t)$ defined in Figure 2.10, the voltage–current relationship for the resistor is

$$v(t) = Ri(t) \tag{2.40}$$

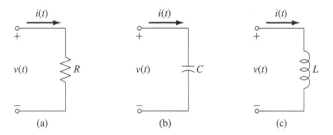

FIGURE 2.10
Basic circuit elements: (a) resistor; (b) capacitor; (c) inductor.

For the capacitor it is

$$\frac{dv(t)}{dt} = \frac{1}{C}i(t) \qquad \text{or} \qquad v(t) = \frac{1}{C}\int_{-\infty}^{t} i(\lambda)\,d\lambda \qquad (2.41)$$

and for the inductor it is

$$v(t) = L\frac{di(t)}{dt} \qquad \text{or} \qquad i(t) = \frac{1}{L}\int_{-\infty}^{t} v(\lambda)\,d\lambda \qquad (2.42)$$

Now consider the process of determining the input/output differential equation of an electrical circuit consisting of an interconnection of resistors, capacitors, and inductors. The input $x(t)$ to the circuit is a voltage or current driving source, and the output $y(t)$ is a voltage or current at some point in the circuit. The input/output differential equation of the circuit can be determined by the use of Kirchhoff's voltage and current laws. The voltage law states that at any fixed time the sum of the voltages around a closed loop in the circuit must be equal to zero. The current law states that at any fixed time the sum of the currents entering a node (a junction of circuit elements) must equal the sum of the currents leaving the node.

Via the voltage–current relationships (2.40)–(2.42) and Kirchhoff's voltage and current laws, node and/or loop equations can be written for the circuit, which can then be combined to yield the input/output differential equation. An illustration of the use of Kirchhoff's current law was given for the parallel connection of a resistor and capacitor that was considered in Section 1.4 of Chapter 1. The use of Kirchhoff's voltage law is illustrated in the next example for a series connection of a resistor and capacitor. For more complicated circuits, the system model is usually generated on the basis of the Laplace transform representation of a circuit, which is discussed in Section 6.6. For an in-depth treatment of the writing of circuit equations, see Hayt et al. [2002] or Nilsson and Riedel [2004].

Example 2.8   Series RC Circuit

Consider the series RC circuit shown in Figure 2.11. As indicated in the figure, the input $x(t)$ is the voltage $v(t)$ applied to the series connection and the output $y(t)$ is equal to the voltage $v_C(t)$

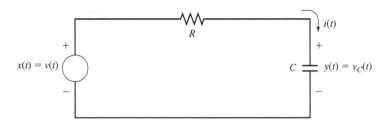

FIGURE 2.11
Series RC circuit.

across the capacitor. By Kirchhoff's voltage law, the sum of the voltages around the loop is equal to zero, and thus

$$Ri(t) + y(t) - x(t) = 0 \tag{2.43}$$

where $i(t)$ is the current in the loop. (See Figure 2.11.) Using (2.41) gives

$$i(t) = C\frac{dv_C(t)}{dt} = C\frac{dy(t)}{dt} \tag{2.44}$$

and inserting (2.44) into (2.43) yields

$$RC\frac{dy(t)}{dt} + y(t) - x(t) = 0 \tag{2.45}$$

Then, dividing both sides of (2.45) by RC and rearranging terms results in the following linear input/output differential equation:

$$\frac{dy(t)}{dt} + \frac{1}{RC}y(t) = \frac{1}{RC}x(t) \tag{2.46}$$

It follows from (2.46) that the series RC circuit is a causal linear time-invariant continuous-time system. In the next section methods are given for solving (2.46).

2.4.2   Mechanical Systems

Motion in a mechanical system can always be resolved into translational and rotational components. Translational motion is considered first.

In linear translational systems there are three fundamental types of forces that resist motion. They are inertia force of a moving body, damping force due to viscous friction, and spring force. By Newton's second law of motion, the inertia force $x(t)$ of a moving body is equal to its mass M times its acceleration; that is,

$$x(t) = M\frac{d^2y(t)}{dt^2} \tag{2.47}$$

where $y(t)$ is the position of the body at time t.

The damping force $x(t)$ due to viscous friction is proportional to velocity, so that

$$x(t) = k_d\frac{dy(t)}{dt} \tag{2.48}$$

where k_d is the damping constant. Viscous friction is often represented by a dashpot consisting of an oil-filled cylinder and piston. A schematic representation of the dashpot is shown in Figure 2.12.

The restoring force $x(t)$ of a spring is proportional to the amount $y(t)$ it is stretched; that is,

$$x(t) = k_s y(t) \tag{2.49}$$

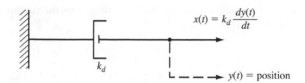

$$x(t) = k_d \frac{dy(t)}{dt}$$

$y(t)$ = position

FIGURE 2.12
A dashpot.

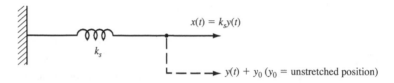

$$x(t) = k_s y(t)$$

$y(t) + y_0$ (y_0 = unstretched position)

FIGURE 2.13
A spring.

where k_s is a constant representing the stiffness of the spring. The schematic representation of a spring is shown in Figure 2.13.

The input/output differential equation of a translational mechanical system can be determined by applying D'Alembert's principle, which is a slight variation of Newton's second law of motion. By D'Alembert's principle, at any fixed time the sum of all the external forces applied to a body in a given direction and all the forces resisting motion in that direction must be equal to zero. D'Alembert's principle is the mechanical analog of Kirchhoff's laws in circuit analysis. The application of D'Alembert's principle is illustrated in the following example:

Example 2.9   Mass–Spring–Damper System

Consider the mass–spring–damper system that was defined in Section 1.4. The schematic diagram of the system is reproduced in Figure 2.14. As discussed in Section 1.4., the input $x(t)$ to

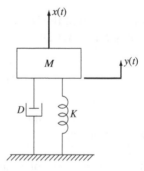

FIGURE 2.14
Schematic diagram of the
mass–spring–damper system.

the system is the external force applied to the mass, which causes the mass to move up or down. The output of the system is the displacement $y(t)$ of the mass, which is measured with respect to an equilibrium position. When the mass is moved upward by the external force $x(t)$ from its equilibrium position, the displacement $y(t)$ will be positive. In this case, the spring is expanded and thus will resist the upward motion, resulting in a negative force applied to the mass. In addition, the inertia force and damping force will also resist the upward motion and will therefore result in negative forces applied to the mass. By D'Alembert's principle, the sum of the external force $x(t)$ and the forces resisting motion in the upward direction must be zero. Hence,

$$x(t) - k_s y(t) - M \frac{d^2 y(t)}{dt^2} - k_d \frac{dy(t)}{dt} = 0 \tag{2.50}$$

Equation (2.50) is also valid in the case when the external force $x(t)$ moves the mass downward from its equilibrium position. Then, setting $k_d = D$ and $k_s = K$ and rearranging terms in (2.50) yield the following second-order input/output differential equation for the mass–spring–damper system:

$$M \frac{d^2 y(t)}{dt^2} + D \frac{dy(t)}{dt} + K y(t) = x(t)$$

As noted in Section 1.5, this result shows that the mass–spring–damper system is a linear time-invariant continuous-time system.

Rotational Mechanical Systems.   In analogy with the three types of forces resisting translational motion, there are three types of forces resisting rotational motion. They are inertia torque given by

$$x(t) = I \frac{d^2 \theta(t)}{dt^2} \tag{2.51}$$

damping torque given by

$$x(t) = k_d \frac{d\theta(t)}{dt} \tag{2.52}$$

and spring torque given by

$$x(t) = k_s \theta(t) \tag{2.53}$$

In (2.51)–(2.53), $\theta(t)$ is the angular position at time t, I is the moment of inertia, and k_d and k_s are the rotational damping and viscous friction (or stiffness) constants, respectively.

For rotational systems, D'Alembert's principle states that at any fixed time the sum of all external torques applied to a body about any axis and all torques resisting motion about that axis must be equal to zero. The input/output differential equation of a rotational system can be derived using (2.51)–(2.53) and D'Alembert's principle. The process is very similar to the steps carried out for the mass–spring–damper system in Example 2.9.

Example 2.10   Motor with Load

Consider a motor with load as shown schematically in Figure 2.15. The load indicated in the figure is some structure, such as a valve or plate, to which the motor shaft is connected. The input in this example is the torque $T(t)$ applied to the motor shaft that is generated by the motor. The torque $T(t)$ is generated by the motor via a process that depends on the type of motor being used. Later in the text, a field-controlled dc motor will be considered, at which point the generation of $T(t)$ will be specified. The output of the motor with load is the angular position $\theta(t)$ of the motor shaft relative to a reference position. The torque $T(t)$ is resisted by the inertia torque and the damping torque, and thus by D'Alembert's principle, the equation for the motor with load is given by

$$T(t) - I\frac{d^2\theta(t)}{dt^2} - k_d\frac{d\theta(t)}{dt} = 0 \tag{2.54}$$

where I is the moment of inertia of the motor and load and k_d is the viscous friction coefficient of the motor and load. Rearranging terms in (2.54) results in the following second-order input/output differential equation for the motor with load

$$I\frac{d^2\theta(t)}{dt^2} + k_d\frac{d\theta(t)}{dt} = T(t) \tag{2.55}$$

The motor with load given by (2.55) will be used in Chapter 9, which deals with control applications.

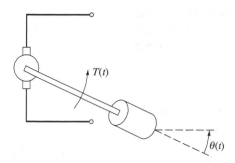

FIGURE 2.15
Motor with load.

2.5   SOLUTION OF DIFFERENTIAL EQUATIONS

There are several methods for solving linear constant-coefficient input/output differential equations, some of which are described in this text. The most familiar method to students is a classical mathematical technique that solves for the homogeneous solution and the nonhomogeneous solution separately. Since this classical method is treated in depth in mathematical texts on differential equations, it will not be covered in this text. Another common method for solving differential equations uses the Laplace transform, which is introduced in Chapter 6. In this section, two additional solution methods are described: a numerical method that discretizes the differential equation and a method that uses the Symbolic Math Toolbox available with MATLAB.

2.5.1   Numerical Solution Method

A differential equation can be discretized in time, resulting in a difference equation that can then be solved by recursion. This discretization in time actually yields an approximate discrete-time representation of the continuous-time system defined by the given input/output differential equation.

Consider the linear time-invariant continuous-time system with the first-order input/output differential equation

$$\frac{dy(t)}{dt} = -ay(t) + bx(t) \tag{2.56}$$

where a and b are constants. Equation (2.56) can be discretized in time by setting $t = nT$, where T is a fixed positive number and n takes on integer values only. This results in the equation

$$\left.\frac{dy(t)}{dt}\right|_{t=nT} = -ay(nT) + bx(nT) \tag{2.57}$$

Now the derivative in (2.57) can be approximated by

$$\left.\frac{dy(t)}{dt}\right|_{t=nT} = \frac{y(nT + T) - y(nT)}{T} \tag{2.58}$$

If T is suitably small and $y(t)$ is continuous, the approximation (2.58) to the derivative $dy(t)/dt$ will be accurate. This approximation is called the *Euler approximation* of the derivative.

Inserting the approximation (2.58) into (2.57) gives

$$\frac{y(nT + T) - y(nT)}{T} = -ay(nT) + bx(nT) \tag{2.59}$$

To be consistent with the notation that is being used for discrete-time signals, the input signal $x(nT)$ and the output signal $y(nT)$ will be denoted by $x[n]$ and $y[n]$, respectively; that is,

$$x[n] = x(t)|_{t=nT} \quad \text{and} \quad y[n] = y(t)|_{t=nT}$$

In terms of this notation, (2.59) becomes

$$\frac{y[n + 1] - y[n]}{T} = -ay[n] + bx[n] \tag{2.60}$$

Finally, multiplying both sides of (2.60) by T and replacing n by $n - 1$ results in a discrete-time approximation to (2.56) given by the first-order input/output difference equation

$$y[n] - y[n - 1] = -aTy[n - 1] + bTx[n - 1]$$

or

$$y[n] = (1 - aT)y[n - 1] + bTx[n - 1] \tag{2.61}$$

The difference equation (2.61) is called the *Euler approximation* of the given input/ output differential equation (2.56), since it is based on the Euler approximation of the derivative.

The discrete values $y[n] = y(nT)$ of the solution $y(t)$ to (2.56) can be computed by solution of the difference equation (2.61). The solution of (2.61) with initial condition $y[0]$ and with $x[n] = 0$ for all n is given by

$$y[n] = (1 - aT)^n y[0], \quad n = 0, 1, 2, \ldots \tag{2.62}$$

To verify that (2.62) is the solution, insert the expression (2.62) for $y[n]$ into (2.61) with $x[n] = 0$. This gives

$$(1 - aT)^n y[0] = (1 - aT)(1 - aT)^{n-1} y[0]$$
$$= (1 - aT)^n y[0]$$

Hence, (2.61) is satisfied, which shows that (2.62) is the solution.

The expression (2.62) for $y[n]$ gives approximate values of the solution $y(t)$ to (2.56) at the times $t = nT$ with arbitrary initial condition $y[0]$ and with zero input. To compare (2.62) with the exact values of $y(t)$ for $t = nT$, first note that the exact solution $y(t)$ to (2.56) with initial condition $y(0)$ and with zero input is given by

$$y(t) = e^{-at} y(0), t \geq 0 \tag{2.63}$$

The solution given by (2.63) can be generated by the Laplace transform, as shown in Chapter 6. Setting $t = nT$ in (2.63) gives the following exact expression for $y[n]$:

$$y[n] = e^{-anT} y[0], \quad n = 0, 1, 2, \ldots \tag{2.64}$$

Now, since

$$e^{ab} = (e^a)^b \text{ for any real numbers } a \text{ and } b$$

(2.64) can be written in the form

$$y[n] = (e^{-aT})^n y[0], \quad n = 0, 1, 2, \ldots \tag{2.65}$$

Further, inserting the expansion

$$e^{-aT} = 1 - aT + \frac{a^2 T^2}{2} - \frac{a^3 T^3}{6} + \cdots$$

for the exponential into (2.65) results in the following exact expression for the values of $y(t)$ at the times $t = nT$:

$$y[n] = \left(1 - aT + \frac{a^2 T^2}{2} - \frac{a^3 T^3}{6} + \cdots \right)^n y[0], \quad n = 0, 1, 2, \ldots \qquad (2.66)$$

Comparing (2.62) with (2.66) shows that (2.62) is an accurate approximation if $1 - aT$ is a good approximation to the exponential $\exp(-aT)$. This will be the case if the magnitude of aT is much less than 1, in which case the magnitudes of the powers of aT will be much smaller than the quantity $1 - aT$.

Example 2.11   Series RC Circuit

Consider the series RC circuit given in Figure 2.11. As shown in Example 2.8 [see (2.46)], the circuit has the input/output differential equation

$$\frac{dy(t)}{dt} + \frac{1}{RC} y(t) = \frac{1}{RC} x(t) \qquad (2.67)$$

where $x(t)$ is the input voltage applied to the circuit and $y(t)$ is the voltage across the capacitor. Writing (2.67) in the form (2.56) reveals that in this case, $a = 1/RC$ and $b = 1/RC$. Hence the discrete-time representation (2.61) for the RC circuit is given by

$$y[n] = \left(1 - \frac{T}{RC} \right) y[n - 1] + \frac{T}{RC} x[n - 1] \qquad (2.68)$$

The difference equation (2.68) can be solved recursively to yield approximate values $y[n]$ of the voltage on the capacitor resulting from initial voltage $y[0]$ and input voltage $x(t)$ applied for $t \geq 0$. The recursion can be carried out with the MATLAB M-file `recur`, where the coefficients are identified by comparing (2.68) with (2.26). This yields $a_1 = -(1 - T/RC)$, $b_0 = 0$, and $b_1 = T/RC$. The commands for the case when $R = C = 1$, $y[0] = 0$, $x(t) = 1, t \geq 0$, and $T = 0.2$ are found to be

```
R = 1; C = 1; T = .2;
a = -(1-T/R/C); b = [0 T/R/C];
y0 = 0; x0 = 1;
n = 1:40;
x = ones(1, length(n));
y1 = recur(a, b, n, x, x0, y0);     % approximate solution
```

The computation of the exact solution can be carried out with the Laplace transform, which yields the result

$$y(t) = 1 - e^{-t}, t \geq 0$$

The MATLAB commands used to compute the exact y for $t = 0$ to $t = 8$ and the commands used to plot both solutions are

```
t = 0:0.04:8;
y2 = 1 - exp(-t);  % exact solution
% augment the initial condition onto the vector
y1 = [y0 y1];
n = 0:40;    % redefines n accordingly
plot(n*T,y1,'o',t,y2,'-');
```

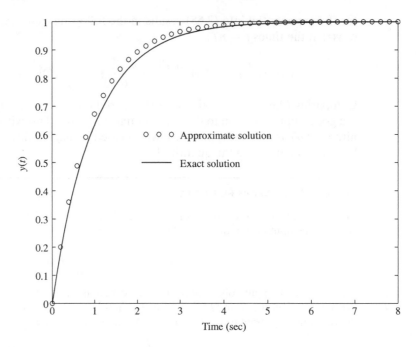

FIGURE 2.16
Exact and approximate step response in Example 2.11.

A plot of the resulting output (the step response) for the approximation is displayed in Figure 2.16 along with the exact step response. Since `y1` is an approximation to a continuous-time signal, the plot is not displayed by the use of the stem form (which is used for discrete-time signals).

Note that, since the value of $aT = T/RC = 0.2$ is small compared with 1, the approximate step response is close to that of the exact response. We can obtain a better approximation by taking a smaller value for T and then using the preceding MATLAB program. The reader is invited to try this.

The discretization technique for first-order differential equations previously described can be generalized to second- and higher-order differential equations. In the second-order case the following approximations can be used:

$$\frac{dy(t)}{dt}\bigg|_{t=nT} = \frac{y(nT + T) - y(nT)}{T} \tag{2.69}$$

$$\frac{d^2y(t)}{dt^2}\bigg|_{t=nT} = \frac{dy(t)/dt|_{t=nT+T} - dy(t)/dt|_{t=nT}}{T} \tag{2.70}$$

Combining (2.69) and (2.70) yields the following approximation to the second derivative:

$$\frac{d^2y(t)}{dt^2}\bigg|_{t=nT} = \frac{y(nT + 2T) - 2y(nT + T) + y(nT)}{T^2} \tag{2.71}$$

The approximation (2.71) is the Euler approximation to the second derivative. Details of the Euler approximation for a second-order system and an example are given in the extra notes available on the textbook website.

For continuous-time systems given by an input/output differential equation, there are a number of numerical solution techniques that are much more accurate (for a given value of T) than the preceding technique based on the Euler approximation of derivatives. Additional discretization methods are described in Chapter 10. Also, MATLAB provides an ODE solver that contains several approximation methods for solving first-order ordinary differential equations numerically. An example method is Runge–Kutta, which is accessible with the command ode45. The ODE solver requires a MATLAB function to be defined that contains the numerical computations needed to define the first derivative.

Example 2.12   RC Circuit Using MATLAB ODE Solver

Consider the RC circuit given in Example 2.11. The command ode45 returns a vector t and the corresponding solution vector y when given a function that computes dy/dt, the initial and final times, and the initial condition for $y(t)$. The specific MATLAB commands for the RC circuit are

```
tspan = [0 8];    %vector of initial and final times
y0 = 0;           %initial value for y(t)
[t,y] = ode45(@ex2_12_func,tspan,y0);
```

where ex2_12_func is a user-defined MATLAB function that computes the first derivative of $y(t)$. The differential equation for the RC circuit given in equation (2.67) is used to obtain an

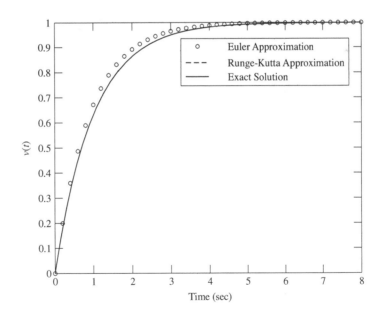

FIGURE 2.17
Comparison of Euler and Runge–Kutta approximate step responses to the exact step response in Example 2.12.

expression for *dy/dt*. The corresponding MATLAB function is stored in the file ex2_12_func.m and contains the following commands:

```
function dy = ex2_12_func(t,y);
R = 1; C = 1; x = 1;
dy = x/R/C - y/R/C;
```

Figure 2.17 shows a comparison of the numerical solutions for the step response of the *RC* circuit using Runge–Kutta and Euler approximations. From the plots, the Runge–Kutta is indistinguishable from the exact solution, and, therefore, is a much better approximation than is the Euler method.

2.5.2   MATLAB Symbolic Math Solution

A symbolic manipulator is useful for finding closed form solutions to simple algebraic and differential equations. The MATLAB Symbolic Math Toolbox differential equation solver is dsolve, which is invoked with the expression dsolve('*expression*') where *expression* is the differential equation to be solved.

Example 2.13   RC Circuit Solved Using the Symbolic Math Toolbox

The series *RC* circuit in Example 2.11 is described by the differential equation given in Equation (2.67). For the case when $R = 1$ and $C = 1$ and $x(t) = 1$ for $t \geq 0$, dsolve can be used as follows:

```
y = dsolve('Dy = 1 - y','y(0) = 0')
```

The resulting expression from MATLAB is

```
y =
1-exp(-t)
```

Higher-order differential equations can be solved as easily. For example, the differential equation

$$\frac{d^2y}{dt^2} + 4\frac{dy}{dt} + 13y(t) = 0; \quad \left.\frac{dy}{dt}\right|_{t=0} = 0, \quad y(0) = 1$$

is solved by the command

```
y = dsolve('D2y = -4*Dy - 13*y ','Dy(0) = 0','y(0) = 1')
```

which yields the solution

```
y =
2/3*exp(-2*t)*sin(3*t)+exp(-2*t)*cos(3*t)
```

2.6   CONVOLUTION REPRESENTATION OF CONTINUOUS-TIME SYSTEMS

Given two continuous-time signals $x(t)$ and $v(t)$, the convolution of $x(t)$ and $v(t)$ is defined by

$$x(t) * v(t) = \int_{-\infty}^{\infty} x(\lambda)v(t - \lambda)\, d\lambda \tag{2.72}$$

The integral in the right-hand side of (2.72) is called the *convolution integral*. As is the case for discrete-time signals, the convolution operation $x(t) * v(t)$ is commutative, so that $x(t) * v(t) = v(t) * x(t)$, and thus

$$\int_{-\infty}^{\infty} x(\lambda)v(t - \lambda)\, d\lambda = \int_{-\infty}^{\infty} v(\lambda)x(t - \lambda)\, d\lambda$$

If $x(t)$ and $v(t)$ are both zero for all $t < 0$, then $x(\lambda) = 0$ for all $\lambda < 0$ and $v(t - \lambda) = 0$ for all $t - \lambda < 0$ (or $t < \lambda$). In this case the integration in (2.72) may be taken from $\lambda = 0$ to $\lambda = t$, and the convolution operation is given by

$$x(t) * v(t) = \begin{cases} 0, & t < 0 \\ \displaystyle\int_{0}^{t} x(\lambda)v(t - \lambda)\, d\lambda, & t \geq 0 \end{cases}$$

Now consider a causal linear time-invariant continuous-time system with input $x(t)$ and output $y(t)$, where $y(t)$ is the response resulting from $x(t)$ with $x(t) = 0$ for $t < 0$. Then the output $y(t)$ is given by the convolution model

$$y(t) = h(t) * x(t) = \int_{0}^{t} h(\lambda)x(t - \lambda)\, d\lambda, t \geq 0 \tag{2.73}$$

or by commutativity,

$$y(t) = x(t) * h(t) = \int_{0}^{t} x(\lambda)h(t - \lambda)\, d\lambda, t \geq 0 \tag{2.74}$$

where $h(t)$ is the impulse response of the system. The *impulse response* $h(t)$ of a causal linear time-invariant continuous-time system is the output response when the input $x(t)$ is the unit impulse $\delta(t)$. Since the system is assumed to be causal and $\delta(t) = 0$ for all $t < 0$, the impulse response $h(t)$ is zero for all $t < 0$.

Note that (2.73) is a natural continuous-time counterpart of the convolution representation (2.10) in the discrete-time case. Also note that the major difference between (2.73) and (2.10) is that the convolution sum in the discrete-time case becomes a convolution integral in the continuous-time case.

By the previous results, the input/output relationship of a linear time-invariant continuous-time system is a convolution operation between the input $x(t)$ and the impulse response $h(t)$. One consequence of this relationship is that the system is completely determined by $h(t)$ in the sense that if $h(t)$ is known, the response to any input can be computed. Again, this corresponds to the situation in the discrete-time case where knowledge of the unit-pulse response $h[n]$ determines the system uniquely.

If the input and the impulse response are given by simple mathematical expressions, we can compute the convolution $h(t) * x(t)$ by inserting the expressions for $h(t)$ and $x(t)$ into (2.73) or (2.74). This is illustrated by the following example.

Example 2.14   Output Response of *RC* Circuit

Again consider the *RC* circuit shown in Figure 2.11. Recall that the input $x(t)$ is the voltage applied to the series connection, and the output $y(t)$ is the voltage $v_C(t)$ across the capacitor. In Example 2.8 it was shown that the input/output differential equation of the *RC* circuit is given by

$$\frac{dy(t)}{dt} + \frac{1}{RC}y(t) = \frac{1}{RC}x(t) \tag{2.75}$$

To determine the convolution model for the *RC* circuit, it is first necessary to compute the impulse response $h(t)$ of the circuit. As will be shown in Chapter 6, this is easy to accomplish by taking the Laplace transform of (2.75) with $x(t) = \delta(t)$. The result is

$$h(t) = \frac{1}{RC}e^{-(1/RC)t}, t \geq 0, \text{ and } h(t) = 0, t < 0 \tag{2.76}$$

We can verify the expression for $h(t)$ given by (2.76) by showing that (2.75) is satisfied with $h(t)$ given by (2.76) and with $x(t) = \delta(t)$. The reader is invited to check this.

In the case when $R = C = 1$, the response $y(t)$ of the *RC* circuit will be computed when the input $x(t)$ is equal to the pulse $p(t) = 1$ for $0 \leq t \leq 1$, $p(t) = 0$ for all other t. Due to the

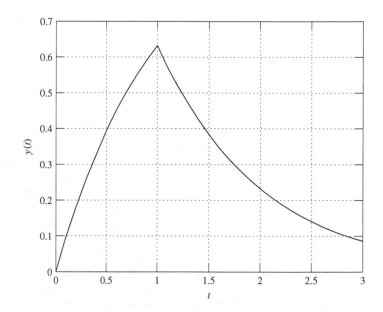

FIGURE 2.18
Output response of *RC* circuit in Example 2.14.

form of $p(t)$, it is easier to compute the response $y(t)$ by (2.74) rather than by (2.73), and thus (2.74) will be used: Inserting $h(t - \lambda)$ and $x(\lambda)$ into (2.74) gives

$$y(t) = \int_0^t (1)e^{-(t-\lambda)} \, d\lambda = e^{-t} \int_0^t e^{\lambda} = e^{-t}(e^t - 1) = 1 - e^{-t}, 0 \le t \le 1 \qquad (2.77)$$

$$y(t) = \int_0^1 (1)e^{-(t-\lambda)} \, d\lambda = e^{-t} \int_0^1 e^{\lambda} \, d\lambda = e^{-t}(e - 1), t \ge 1 \qquad (2.78)$$

Hence, the response $y(t)$ is specified by (2.77) and (2.78) for the two time intervals $0 \le t \le 1$ and $t \ge 1$. A plot of $y(t)$ is given in Figure 2.18. From the plot it is seen that the application of the input voltage at time $t = 0$ causes the capacitor to build up voltage, and then when the input voltage is switched off (at $t = 1$), the capacitor starts to discharge.

2.6.1 Graphical Approach to Convolution

To compute the convolution $x(t) * v(t)$, it is often useful to graph the functions in the integrand of the convolution integral. This can help to determine the integrand and integration limits of the convolution integral, especially in cases where either $x(t)$ or $v(t)$ is defined piecewise. The main procedure is to plot $x(\lambda)$ and $v(t - \lambda)$ as functions of λ, determine where they overlap and what is the analytical form of $x(\lambda)v(t - \lambda)$, and integrate the product. When $x(t)$ or $v(t)$ is defined piecewise, the analytical form of the product changes, depending on the interval of time t. To determine the appropriate functional form of the product and the corresponding limits of integration, slide the plot of $v(t - \lambda)$ from left to right to see how the overlap between $v(t - \lambda)$ and $x(\lambda)$ changes. The steps of this graphical aid to computing the convolution integral are listed subsequently. Here it is assumed that both $x(t)$ and $v(t)$ are zero for all $t < 0$. If $x(t)$ and $v(t)$ are not zero for all $t < 0$, the shift property can be used to reduce the problem to the case when $x(t)$ and $v(t)$ are zero for all $t < 0$.

The steps for carrying out the graphical approach to convolution are as follows:

Step 1. Graph $x(\lambda)$ and $v(-\lambda)$ as functions of λ. The function $v(-\lambda)$ is equal to the function $v(\lambda)$ reflected about the vertical axis.

Step 2. Graph $v(t - \lambda)$ for an arbitrary value of t such that $t < 0$. Note that $v(t - \lambda)$ is equal to $v(-\lambda)$ shifted such that the origin of the plot of $v(-\lambda)$ is at $\lambda = t$. Since $x(t)$ and $v(t)$ are zero for all $t < 0$, there is no nonzero overlap between $v(t - \lambda)$ and $x(\lambda)$. The product $x(\lambda)v(t - \lambda)$ is determined by multiplication of $x(\lambda)$ and $v(t - \lambda)$ point by point with respect to λ; therefore, when $t < 0$ the product $x(\lambda)v(t - \lambda) = 0$ for all λ.

Step 3. Slide $v(t - \lambda)$ to the right until there is a nonzero overlap between $v(t - \lambda)$ and $x(\lambda)$. Suppose that the first value of t for which this occurs is $t = a$. Then $x(t) * v(t) = 0$ for $t < a$.

Step 4. Continue sliding $v(t - \lambda)$ to the right past $t = a$. Determine the interval of time $a \le t < b$ for which the product $x(\lambda)v(t - \lambda)$ has the same analytical form. Integrate the product $x(\lambda)v(t - \lambda)$ as a function of λ with the limits of integration from $\lambda = a$ to $\lambda = t$. The result is the expression for $x(t) * v(t)$ for $a \le t < b$.

Step 5. Slide $v(t - \lambda)$ to the right past $t = b$. Determine the next interval of time $b \leq t < c$ for which the product $x(\lambda)v(t - \lambda)$ has the same analytical form. Integrate the product $x(\lambda)v(t - \lambda)$ as a function of λ with the limits of integration from $\lambda = a$ to $\lambda = t$, where $b \leq t < c$. The integral is computed piecewise.

$$x(t) * v(t) = \int_a^b x(\lambda)v(t - \lambda)\, d\lambda + \int_b^t x(\lambda)v(t - \lambda)\, d\lambda, \text{ for } b \leq t < c$$

Repeat the pattern set forth in Steps 4 and 5 as many times as necessary until $x(t) * v(t)$ is computed for all $t > 0$.

This procedure is illustrated by the following two examples:

Example 2.15   Convolution of Pulses

Suppose that $x(t) = u(t) - 2u(t - 1) + u(t - 2)$, and that $v(t)$ is the pulse $v(t) = u(t) - u(t - 1)$. The convolution of $x(t)$ and $v(t)$ is carried out via the steps previously given.

Convolution of Continuous-Time Signals

Step 1.   The functions $x(\lambda)$ and $v(-\lambda)$ are plotted in Figure 2.19.

Steps 2-3. There is no overlap of $x(\lambda)$ and $v(t - \lambda)$ for $t < 0$, so $a = 0$ and $x(t) * v(t) = 0$ for $t < 0$.

Step 4.   For $0 \leq t < 1$, the plots of $x(\lambda)$, $v(t - \lambda)$ and the product $x(\lambda)v(t - \lambda)$ are given in Figure 2.20. For $1 \leq t < 2$, the form of the product $x(\lambda)v(t - \lambda)$ changes, as shown in Figure 2.21c. Thus the value of b is 1. Integrating the product $x(\lambda)v(t - \lambda)$ displayed in Figure 2.20c for $0 \leq t \leq 1$ gives

$$x(t) * v(t) = \int_0^t 1\, d\lambda = t$$

Step 5.   For $2 \leq t < 3$, the product $x(\lambda)v(t - \lambda)$ is plotted in Figure 2.22c. From Figures 2.21c and 2.22c, it is seen that the form of the product $x(\lambda)v(t - \lambda)$ changes from the

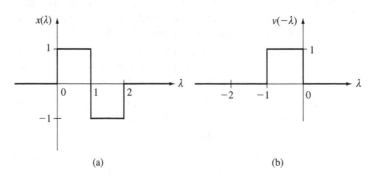

(a)   (b)

FIGURE 2.19
Plots of (a) $x(\lambda)$ and (b) $v(-\lambda)$.

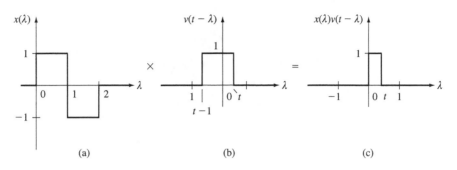

FIGURE 2.20
Plots of (a) $x(\lambda)$, (b) $v(t - \lambda)$, and (c) $x(\lambda)v(t - \lambda)$ for $0 \le t < 1$.

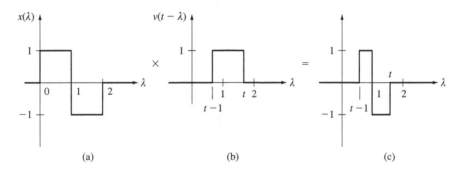

FIGURE 2.21
Plots of (a) $x(\lambda)$, (b) $v(t - \lambda)$, and (c) $x(\lambda)v(t - \lambda)$ for $1 \le t < 2$.

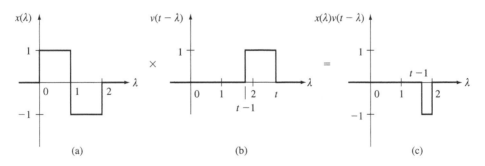

FIGURE 2.22
Plots of (a) $x(\lambda)$, (b) $v(t - \lambda)$, and (c) $x(\lambda)v(t - \lambda)$ for $2 \le t < 3$.

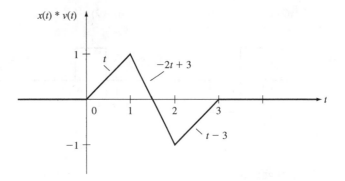

FIGURE 2.23
Sketch of $x(t) * v(t)$.

interval $1 \le t < 2$ to the interval $2 \le t < 3$. Thus the value of c is 2. Integrating the product plotted in Figure 2.21c for $1 \le t \le 2$ yields

$$x(t) * v(t) = \int_{t-1}^{1} (1)\, d\lambda + \int_{1}^{t} (-1)\, d\lambda$$

$$= 1 - (t-1) + (-1)(t-1) = -2t + 3, \text{ for } 1 \le t < 2$$

Repeating Step 5 for the interval $2 \le t < 3$, from Figure 2.22c,

$$x(t) * v(t) = \int_{t-1}^{2} (-1)\, d\lambda$$

$$= (-1)[2 - (t-1)] = t - 3, \text{ for } 2 \le t \le 3$$

Finally, for $t \ge 3$, the product $x(\lambda)v(t - \lambda)$ is zero, since there is no overlap between $x(\lambda)$ and $v(t - \lambda)$. Hence,

$$x(t) * v(t) = 0 \text{ for } t \ge 3$$

A sketch of the convolution $x(t) * v(t)$ is shown in Figure 2.23.

Example 2.16   Convolution of Exponential Segments

Consider the signals $x(t)$ and $v(t)$ defined by

$$x(t) = \begin{cases} e^t, & 0 \le t < 1 \\ e^{2-t}, & 1 \le t < 2 \\ 0, & \text{all other } t \end{cases} \qquad v(t) = \begin{cases} e^{-t}, & 0 \le t \le 4 \\ 0, & \text{all other } t \end{cases}$$

The signals $x(t)$ and $v(t)$ are plotted in Figure 2.24. The functions $x(\lambda)$ and $v(-\lambda)$ are displayed in Figure 2.25. For $0 \le t < 1$, the functions $v(t - \lambda)$ and $x(\lambda)v(t - \lambda)$ are plotted in Figure 2.26,

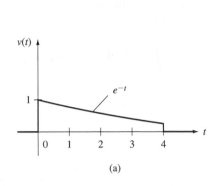

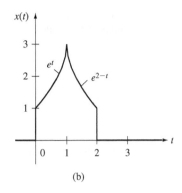

FIGURE 2.24
Plots of (a) $v(t)$ and (b) $x(t)$.

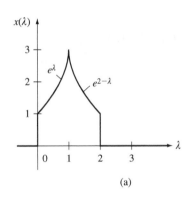

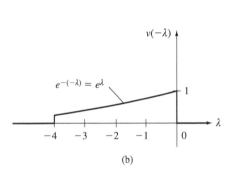

FIGURE 2.25
Functions (a) $x(\lambda)$ and (b) $v(-\lambda)$.

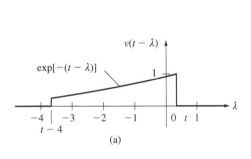

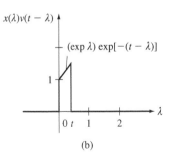

FIGURE 2.26
Functions (a) $v(t - \lambda)$ and (b) $x(\lambda)v(t - \lambda)$ for $0 \leq t < 1$.

and for $1 \leq t < 2$ these functions are plotted in Figure 2.27. Integrating the product $x(\lambda)v(t - \lambda)$ displayed in Figure 2.26b for $0 \leq t < 1$ yields

$$x(t) * v(t) = \int_0^t e^\lambda e^{-(t-\lambda)}\, d\lambda = e^{-t} \int_0^t e^{2\lambda}\, d\lambda$$

$$= \frac{1}{2}(e^t - e^{-t})$$

Integrating the product displayed in Figure 2.27b for $1 \leq t < 2$ gives

$$x(t) * v(t) = \int_0^1 e^\lambda e^{-(t-\lambda)}\, d\lambda + \int_1^t e^{2-\lambda} e^{-(t-\lambda)}\, d\lambda$$

$$= \left[\frac{-e^2 - 1}{2} + e^2 t \right] e^{-t}$$

Continuing with the steps previously described, for $2 \leq t < 4$,

$$x(t) * v(t) = \int_0^1 e^\lambda e^{-(t-\lambda)}\, d\lambda + \int_1^2 e^{2-\lambda} e^{-(t-\lambda)}\, d\lambda$$

$$= (3e^2 - 1) \frac{e^{-t}}{2}$$

For $4 \leq t < 5$,

$$x(t) * v(t) = \int_{t-4}^1 e^\lambda e^{-(t-\lambda)}\, d\lambda + \int_1^2 e^{2-\lambda} e^{-(t-\lambda)}\, d\lambda$$

$$= \frac{1}{2}[3e^2 - e^{2(t-4)}]e^{-t}$$

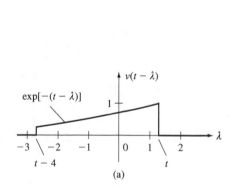
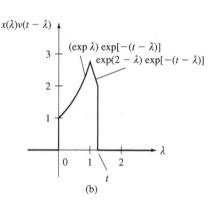

FIGURE 2.27
Functions (a) $v(t - \lambda)$ and (b) $x(\lambda)v(t - \lambda)$ for $1 \leq t < 2$.

For $5 \le t < 6$,

$$x(t) * v(t) = \int_{t-4}^{2} e^{2-\lambda} e^{-(t-\lambda)} \, d\lambda$$

$$= e^2 (6 - t) e^{-t}$$

Finally, for $t \ge 6$, $x(t) * v(t) = 0$, since the functions $v(\lambda)$ and $x(t - \lambda)$ do not overlap when $t \ge 6$.

While all of these integrals can be computed by hand, the MATLAB Symbolic Math Toolbox computes the integrals easily with the following commands:

```
syms t lambda y
y = int(exp(2*lambda-t),lambda,0,t);      % for 0<t<1
simplify(y)
y = int(exp(2*lambda-t),lambda,0,1)+int(exp(2-t),lambda,1,t); %for 1<t<2
simplify(y)
y = int(exp(2*lambda-t),lambda,0,1)+int(exp(2-t),lambda,1,2); %for 2<t<4
simplify(y)
y = int(exp(2*lambda-t),lambda,t-4,1)+int(exp(2-t),lambda,1,2); %for 4<t<5
simplify(y)
int(exp(2-t),lambda,t-4,2)  %for 5<t<6
```

The command `syms t lambda y` constructs symbolic objects `t`, `lambda`, and `y`. The command `int` integrates the expression defined in the first argument with respect to the second argument `lambda`, with limits of integration given by the third and fourth arguments. The `simplify` command groups terms into a simpler form.

Noncausal systems.   If the given system is noncausal, $h(t)$ will not be zero for $t < 0$, and in this case the upper limit of the integral in the convolution expression (2.73) must be taken to be ∞ (not t). In addition, if $x(t)$ is nonzero for $t < 0$, the lower limit of the integral in (2.73) must be taken to be $-\infty$. Thus, for a noncausal system with input $x(t) \ne 0$ for $t < 0$, the input/output convolution relationship is given by

$$y(t) = h(t) * x(t) = \int_{-\infty}^{\infty} h(\lambda) x(t - \lambda) \, d\lambda$$

2.7   CHAPTER SUMMARY

This chapter examines time-domain modeling and analysis of linear time-invariant systems. Both discrete-time and continuous-time systems are considered. A discrete-time system can be represented as a difference equation, which can be solved easily by recursion. Alternatively, convolution can be used to find the response to an arbitrary input, given the unit-pulse response $h[n]$. The unit-pulse response is the response of the system to a unit-pulse input when there is no energy stored in the system, or equivalently, when there are zero initial conditions. Common examples of discrete-time systems are digital filters such as moving average filters.

A continuous-time system can be represented as a differential equation, which in many cases we can find by using the laws of physics to determine equations that govern the behavior of a system. Examples of physical systems that have differential equation models are circuits, translational and rotational mechanical systems, and electromechanical systems such as motors. Classical mathematical solution methods for differential equations are not considered in depth in this text. Instead, numerical solutions are presented, as well as a symbolic solution method. Chapter 6 will cover a Laplace transform solution method that is often easier to apply than the classical solution methods.

As in the case of discrete-time systems, convolution is a time-domain solution method used to determine the response of a continuous-time system to an arbitrary input. The system is characterized by its impulse response $h(t)$, which is the response of the system with zero initial conditions to an input of a unit impulse. The impulse response is convolved with the input to obtain the corresponding output. Moreover, the convolution integral can be viewed as a representation of a system.

PROBLEMS

2.1. Consider the N-point EWMA filter given by (2.4) and (2.5), where $0 < b < 1$. When the input $x[n]$ is equal to a constant c for $n \geq 0$, show that the resulting response $y[n] = c$ for all $n \geq N$.

2.2. Generate a MATLAB plot of the output response $y[n]$ for $5 \leq n \leq 40$ of the 5-point EWMA filter with $b = 0.7$ when the input $x[n]$ is
  (a) $x[n] = 1, 0 \leq n \leq 40$
  (b) $x[n] = n, 0 \leq n \leq 40$
  (c) $x[n] = \sin(\pi n/10), 0 \leq n \leq 40$
  (d) $x[n] = \sin(.9\pi n), 0 \leq n \leq 40$
  (e) Explain the differences in the responses obtained in parts (c) and (d).

2.3. Show that the input/output relationship (2.4) of the N-point EWMA filter approaches that of the N-point MA filter as b approaches 1.

2.4. Consider the closing prices of QQQQ for the 50-business-day period from March 1, 2004, up to May 10, 2004.
  (a) For the 40-day time period $11 \leq n \leq 50$, compute the difference $D[n] = y1[n] - y2[n]$, where $y1[n]$ is the response of the 11-day EWMA filter with $b = 0.3$ and $y2[n]$ is the response of the 11-day EWMA filter with $b = 0.7$. In the literature on trading, $D[n]$ is sometimes referred to as the "MACD signal." (See Section 7.5.)
  (b) Suppose that the stock is bought at the close on Day n when the value of $D[n]$ rises above zero, and then the stock is sold at the close on Day n when $D[n]$ falls below zero. Determine the days when there is a buy and the days when there is a sell.
  (c) For each buy/sell signal found in part (b), determine the loss or gain per share.
  (d) Using your result in part (c), determine the net gain or loss per share over the 40-day period.

2.5. Compute the unit-pulse response $h[n]$ for $n = 0, 1, 2, 3$ for each of the following discrete-time systems:
  (a) $y[n + 1] + y[n] = 2x[n]$
  (b) $y[n + 1] + 1/2y[n] = x[n]$
  (c) $y[n + 2] + 1.5y[n + 1] + 0.5y[n] = x[n]$
  (d) $y[n + 2] + 1/2y[n + 1] + 1/4y[n] = x[n + 1] - x[n]$
  (e) $y[n + 2] + 1/4y[n + 1] - 3/8y[n] = 2x[n + 2] - 3x[n]$

2.6. Compute the unit-pulse response $h[n]$ for all integers $n \geq 0$ for each of the following discrete-time systems:

    (a) $y[n+1] + y[n] = x[n]$
    (b) $y[n+1] + 1/2y[n] = x[n+1]$
    (c) $y[n+1] + 2y[n] = 2x[n+1] - 2x[n]$
    (d) $y[n+1] - 1/2y[n] = x[n+1] + 1/2x[n]$
    (e) $y[n+2] + 1/2y[n] = 2x[n+2] - x[n]$

2.7. For the discrete-time signals $x[n]$ and $v[n]$ shown in Figure P2.7, do the following:

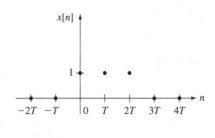
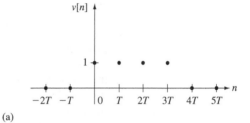

(a)

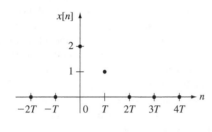
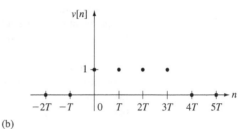

(b)

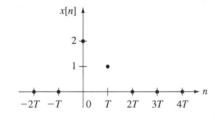
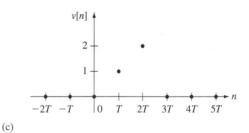

(c)

FIGURE P2.7

    (a) Compute the convolution $x[n] * v[n]$ for all $n \geq 0$. Sketch the results.
    (b) Repeat part (a), using the M-file conv.

2.8. For the discrete-time signals $x[n]$ and $v[n]$ given in each of the following parts, compute the convolution $x[n] * v[n]$ for $n \geq 0$ [compute $x[n] * v[n]$ for $n \geq -2$ in part (b)].

    (a) $x[0] = 4$, $x[1] = 1$, $x[3] = -1$, $x[n] = 0$ for all other integers n; $v[0] = 1$, $v[1] = -2$, $v[2] = 3$, $v[3] = -4$, $v[n] = 0$ for all other integers n.
    (b) $x[n] = 1$ for $-1 \leq n \leq 2$ and $x[n] = 0$ for all other integers n; $v[-1] = -1$; $v[0] = 0$, $v[1] = 8$, $v[2] = 2$, $v[3] = 3$ and $v[n] = 0$ for all other integers n.
    (c) $x[n] = 2^n$ for $n \leq 3$ and $x[n] = 0$ for $n \geq 4$; $v[0] = 2$, $v[1] = -3$, $v[2] = 0$, $v[3] = 6$, $v[n] = 0$ for all other integers n.

   (d) $x[n] = 1/n$ for $2 \leq n \leq 5$ and $x[n] = 0$ for all other integers n; $v[2] = -2$, $v[3] = -5$, $v[n] = 0$ for all other integers n.

   (e) $x[n] = u[n]$, $v[n] = u[n]$, where $u[n]$ is the discrete-time step function.

   (f) $x[n] = u[n]$, $v[n] = \ln(n)$ for all integers $n \geq 1$ and $v[n] = 0$ for all integers $n < 1$.

   (g) $x[n] = \delta[n] - \delta[n - 2]$, where $\delta[n]$ is the unit pulse concentrated at $n = 0$; $v[n] = \cos(\pi n/3)$ for all integers $n \geq 0$, $v[n] = 0$ for all integers $n < 0$.

2.9. Convolve $v[n]$ with $x[n]$ for each of the cases (a)–(c) that follows. Express your answer in closed form.

   (a) $v[n] = 2^n u[n]$ and $x[n] = u[n]$

   (b) $v[n] = (0.25)^n u[n]$ and $x[n] = u[n]$

   (c) $v[n] = 2^n u[n]$ and $x[n] = (0.5)^n u[n]$

   (d) Use the M-file conv to compute the convolution in parts (a) to (c) for $0 \leq n \leq 20$; that is, define x and v for this range of n, compute the convolution, and then save the values of $x[n] * v[n]$ only for this range of n. Plot the results, using a stem plot. (See the comments in Problem 2.10 for more information regarding numerical convolution of infinite-duration signals.)

2.10. You must take care when using a computer to perform convolution on infinite-duration signals (i.e., signals that have nonzero values for an infinite number of points). Since you can store only a finite number of values for the signal, the numerical convolution returns an answer that is equivalent to the signal being zero outside the range of n defined for the stored points. In MATLAB, if $x[n]$ is defined for the range $0 \leq n \leq q$ and $v[n]$ is defined for the range $0 \leq n \leq r$, the result $y[n] = x[n] * v[n]$ will be defined over the range $0 \leq n \leq q + r$. However, the answer will be correct only for the range $0 \leq n \leq \min\{q, r\}$. As an example, consider the convolution of two step functions, $u[n] * u[n]$.

   (a) Compute a closed-form expression for the actual convolution. [See Problem 2.8(e).]

   (b) Define a signal that is the truncated version of a step, $x[n] = u[n]$ for $n \leq q$ and $x[n] = 0$ for all other integers n. Compute $x[n] * x[n]$ for $q = 5$. Compare this result with that found in part (a) to see the effect of the truncation.

   (c) Now, define a vector in MATLAB that is the truncated version of the signal; that is, x contains only the elements of $u[n]$ for $n \leq q$. Take $q = 5$. Compute the numerical convolution $x[n] * x[n]$ and plot the result for $0 \leq n \leq 2q$. Compare this result with the answers found in parts (a) and (b). For what range of n does the result accurately represent the convolution of the two step functions?

   (d) Repeat parts (b) and (c) for $q = 10$.

2.11. Use the M-file conv to convolve the signals defined in Problem 2.8, and compare your answers with those found in Problem 2.8. Use the comments in Problems 2.9 and 2.10 when computing the convolutions for infinite-duration signals.

2.12. A discrete-time system has the following unit-pulse response:

$$h[n] = 0.3(0.7)^n u[n]$$

   (a) Use conv to calculate the response of this system to $x[n] = u[n]$, and plot the response.

   (b) Use conv to calculate the response of this system to $x[n] = \sin(n\pi/8)u[n]$, and plot the response.

   (c) Use conv to calculate the response of this system to $x[n] = u[n] + \sin(n\pi/8)u[n]$, and plot the response.

   (d) Find the first-order difference equation that describes this system where x is the input and y is the output.

   (e) Using the result in part (d) and the M-file recur, calculate the response of the system to $x[n] = u[n]$, and compare with the answer obtained in part (a).

2.13. A discrete-time system has the following unit-pulse response:

$$h[n] = ((0.5)^n - (0.25)^n)u[n]$$

(a) Use conv to calculate the response of this system to $x[n] = u[n]$, and plot the response.

(b) Use conv to calculate the response of this system to $x[n] = \sin(n\pi/4)u[n]$, and plot the response.

(c) Use conv to calculate the response of this system to $x[n] = u[n] + \sin(n\pi/4)u[n]$, and plot the response.

(d) Show that the following difference equation has the unit-pulse response given in this problem:

$$y[n + 2] - 0.75y[n + 1] + 0.125y[n] = 0.25x[n + 1]$$

(e) Using the difference equation in part (d) and the M-file recur, calculate the response of the system to $x[n] = u[n]$, and compare with the answer obtained in part (a).

2.14. Again consider the loan-balance system with the input/output difference equation

$$y[n] - \left(1 + \frac{I}{12}\right)y[n - 1] = -x[n], \, n = 1, 2, \ldots$$

Recall that $y[0]$ is the amount of the loan, $y[n]$ is the loan balance at the end of the nth month, $x[n]$ is the loan payment in the nth month, and I is the yearly interest rate in decimal form. It is assumed that the monthly payments $x[n]$ for $n \geq 1$ are equal to a constant c. Suppose that the number of months in the repayment period is N. Derive an expression for the monthly payments c in terms of $y[0]$, N, and I.

2.15. A savings account in a bank with interest accruing quarterly can be modeled by the input/output difference equation

$$y[n] - \left(1 + \frac{I}{4}\right)y[n - 1] = x[n], \, n = 1, 2, \ldots$$

where $y[n]$ is the amount in the account at the end of the nth quarter, $x[n]$ is the amount deposited in the nth quarter, and I is the yearly interest rate in decimal form.

(a) Suppose that $I = 10\%$. Compute $y[n]$ for $n = 1, 2, 3, 4$ when $y[0] = 1000$ and $x[n] = 1000$ for $n \geq 1$.

(b) Suppose that $x[n] = c$ for $n \geq 1$ and $y[0] = 0$. Given an integer N, suppose that it is desired to have an amount $y[N]$ in the savings account at the end of the Nth quarter. Derive an expression for N in terms of $y[N]$, c, and I.

(c) Suppose that an IRA (individual retirement account) is set up with $y[0] = 2000$, $I = 5\%$, and $x[n] = \$5,000$, $n \geq 1$ ($n =$ quarter). How many years will it take to amass $\$500,000$ in the account?

(d) Modify the loan balance program given in Figure 2.7 to compute the savings amount. Repeat parts (a) and (c) using your new MATLAB program.

2.16. For each of the following difference equations

(i) $y[n + 1] + 1.5y[n] = x[n]$

(ii) $y[n + 1] + 0.8y[n] = x[n]$

(iii) $y[n + 1] - 0.8y[n] = x[n]$

use the method of recursion to solve the following problems:

(a) Compute $y[n]$ for $n = 0, 1, 2$, when $x[n] = 0$ for all n and $y[-1] = 2$.

(b) Compute $y[n]$ for $n = 0, 1, 2$, when $x[n] = u[n]$ and $y[-1] = 0$.

(c) Compute $y[n]$ for $n = 0, 1, 2$, when $x[n] = u[n]$ and $y[-1] = 2$.

2.17. For the difference equations given in Problem 2.16:

(a) Find a closed-form solution for $y[n]$ when $x[n] = 0$ for all n and $y[0] = 2$.

(b) Find a closed-form solution for $y[n]$ when $x[n] = u[n]$ and $y[0] = 0$.

(c) Find a closed-form solution for $y[n]$ when $x[n] = u[n]$ and $y[0] = 2$.

(d) Use the M-file `recur` to solve the difference equations for the cases defined in parts (a) to (c). Plot the corresponding answers from parts (a) to (c) along with those found from MATLAB.

2.18. For the difference equations given next, solve for the sequence $y[n]$, using the program `recur` for $0 \le n \le 10$, and plot y versus n on a stem plot.

(a) $y[n] = y[n - 1] + u[n - 1];\ y[-1] = 0$

(b) $y[n] = 0.5y[n - 1];\ y[-1] = 1$

(c) $y[n] = 0.5y[n - 1] + 0.1y[n - 2] + u[n - 1];\ y[-2] = 1,\ y[-1] = 0$

(d) $y[n] = 0.5y[n - 2] + 0.1y[n - 1] + (0.5)^n u[n];\ y[-1] = y[-2] = 0$

2.19. A discrete-time system is given by the following input/output difference equation:

$$y[n + 2] + 0.75y[n + 1] + 0.125y[n] = x[n]$$

(a) Compute $y[n]$ for $n = 0, 1, 2, 3$ when $y[-2] = -1$, $y[-1] = 2$, and $x[n] = 0$ for all n.

(b) Compute $y[n]$ for $n = 0, 1, 2, 3$ when $y[-2] = y[-1] = 0$, and $x[n] = 1$ for $n \ge -2$.

(c) Compute $y[n]$ for $n = 0, 1, 2, 3$ when $y[-2] = -1$, $y[-1] = 2$, and $x[n] = 1$ for $n \ge -2$.

(d) Compute $y[n]$ for $n = 0, 1, 2, 3$ when $y[-2] = 2$, $y[-1] = 3$, and $x[n] = \sin(\pi n/2)$ for $n \ge 0$.

(e) Compute $y[n]$ for $n = 0, 1, 2, 3$ when $y[-2] = -2$, $y[-1] = 4$, and $x[n] = (0.5)^{n-1} u[n - 1]$ for all n.

2.20. For the RLC circuit in Figure P2.20, find the input/output differential equation when the following conditions are met:

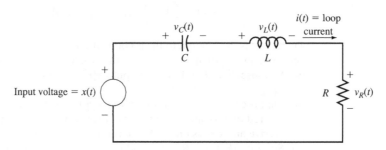

FIGURE P2.20

(a) The output is the voltage $v_C(t)$ across the capacitor.

(b) The output is the current $i(t)$ in the loop.

2.21. Find the input/output differential equations for the RC circuits in Figure P2.21.

2.22. A mass M sits on top of a vibration absorber, as illustrated in Figure P2.22. As shown in Figure P2.22, a force $x(t)$ (e.g., a vibrational force) is applied to the mass M, whose base is located at position $y(t)$. Derive the input/output differential equation of the system.

2.23. Consider the system consisting of two masses and three springs shown in Figure P2.23. The masses are on wheels that are assumed to be frictionless. The input $x(t)$ to the system is the force $x(t)$ applied to the first mass. The position of the first mass is $q(t)$ and the position of

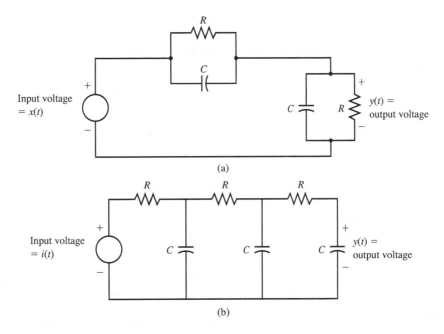

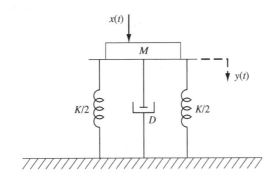

FIGURE P2.21

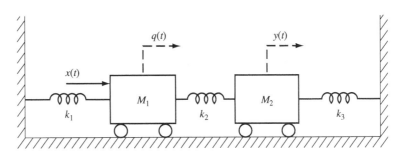

FIGURE P2.22

FIGURE P2.23

the second mass is the output $y(t)$, where both $q(t)$ and $y(t)$ are defined with respect to some equilibrium position. Determine the input/output differential equation of the system.

2.24. Consider the RL circuit shown in Figure P2.24.

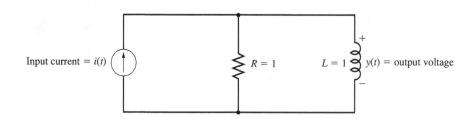

Input current = $i(t)$ $R = 1$ $L = 1$ $y(t)$ = output voltage

FIGURE P2.24

    (a) Write a differential equation for this circuit with the output $y(t)$ and the input $i(t)$.

    (b) Use the MATLAB Symbolic Math Toolbox to solve for an analytical expression for the output voltage $y(t)$ for all $t > 0$ when $y(0) = 0$ and $i(t) = u(t) - u(t - 1)$, where $u(t)$ is the step function.

    (c) Using Euler's approximation of derivatives with T arbitrary and input $x(t)$ arbitrary, derive a difference equation model for the RL circuit.

    (d) Using your answer to part (c) and the M-file recur with $T = 0.1$, and $i(t) = u(t) - u(t - 1)$, plot the approximation to $y(t)$ for $t = 0$ to $t = 2$ seconds. Take $y(-T) = 0$. Compare your results with the exact solution plotted from the answer obtained in part (a).

    (e) Use the MATLAB ODE solver ode45 to solve for the output when the input is as given in part (d). Compare this solution with the exact solution.

2.25. Consider the following differential equation:

$$\frac{d^2y(t)}{dt^2} + \frac{dy(t)}{dt} + 4.25y(t) = 0, \quad y(0) = 2, \quad \dot{y}(0) = 1$$

    (a) Show that the solution is given by $y(t) = e^{-0.5t}(\sin 2t + 2\cos 2t)$.

    (b) Using Euler's approximation of derivatives with T arbitrary and input $x(t)$ arbitrary, derive a difference equation model.

    (c) Using the answer in part (b) and the M-file recur with $T = 0.1$, compute the approximation to $y(t)$.

    (d) Repeat part (c) for $T = 0.05$.

    (e) Plot the responses obtained in parts (a), (c), and (d) for $0 \leq t \leq 10$, and compare the results.

2.26. Consider the following differential equation:

$$\frac{d^2y(t)}{dt^2} + 3\frac{dy(t)}{dt} + 2y(t) = 0, \quad y(0) = 1, \quad \dot{y}(0) = 0$$

    (a) Solve for $y(t)$, using the MATLAB Symbolic Math Toolbox.

    (b) Using Euler's approximation of derivatives with T arbitrary and input $x(t)$ arbitrary, derive a difference equation model. Using the M-file recur with $T = 0.4$, compute the approximation to $y(t)$.

(c) Repeat the numerical approximation in part (b) for $T = 0.1$.

(d) Find a numerical solution to this problem, using the Runge–Kutta approximation with the command ode45.

(e) Plot the responses obtained in parts (a), (b), (c), and (d) for $0 \le t \le 10$, and compare the results.

2.27. Consider the following differential equation:

$$\frac{d^2y(t)}{dt^2} + 2\frac{dy(t)}{dt} + y(t) = 0, \quad y(0) = 2, \quad \dot{y}(0) = -1$$

(a) Show that the solution is given by $y(t) = 2e^{-t} + te^{-t}$.

(b) Using Euler's approximation of derivatives with T arbitrary and input $x(t)$ arbitrary, derive a difference equation model.

(c) Using the answer in part (b) and the M-file recur with $T = 0.4$, compute the approximation to $y(t)$.

(d) Repeat part (c) for $T = 0.1$ seconds.

(e) Plot the responses obtained in parts (a), (c), and (d) for $0 \le t \le 10$, and compare the results.

2.28. Consider the mass–spring–damper system described in Example 2.9 and in the online demo. The differential equation for the system is given by

Mass–
Spring–
Damper
System

$$M\frac{d^2y(t)}{dt^2} + D\frac{dy(t)}{dt} + Ky(t) = x(t)$$

(a) Use the Runge–Kutta approximation with the command ode45 to simulate the unit-step response (that is, the response $y(t)$ when $x(t) = u(t)$) for $M = 10$, $D = 1$, and $K = 1$. Simulate the response long enough so that $y(t)$ appears to reach a constant steady-state value.

(b) Plot your approximation for $y(t)$ versus time.

(c) Compute the response for an input of $x(t) = 10\sin(0.2\pi t)$, and plot $y(t)$. Determine the amplitude of the resulting sinusoid.

(d) Use the online demo to check your results in parts (b) and (c). The "Show Input/Output Summary" button can be used to view the results.

2.29. For the continuous-time signals $x(t)$ and $v(t)$ shown in Figure P2.29, compute the convolution $x(t) * v(t)$ for all $t \ge 0$, and plot your resulting signal.

2.30. Compute the convolution $x(t) * v(t)$ for $-\infty < t < \infty$, where $x(t) = u(t) + u(t-1) - 2u(t-2)$ and $v(t) = 2u(t+1) - u(t) - u(t-1)$.

2.31. A continuous-time system has the input/output relationship

$$y(t) = \int_{-\infty}^{t} (t - \lambda + 2)x(\lambda)\, d\lambda$$

(a) Determine the impulse response $h(t)$ of the system.

(b) Compute the output response $y(t)$ for $1 \le t \le 2$ when $x(t) = u(t) - u(t-1)$.

2.32. A causal linear time-invariant continuous-time system has impulse response

$$h(t) = e^{-t} + \sin t, t \ge 0$$

(a) Compute the output response for all $t \ge 0$ when the input is the unit-step function $u(t)$.

(b) Compute the output response $y(t)$ for all $t \ge 0$ resulting from the input $u(t) - u(t-2)$.

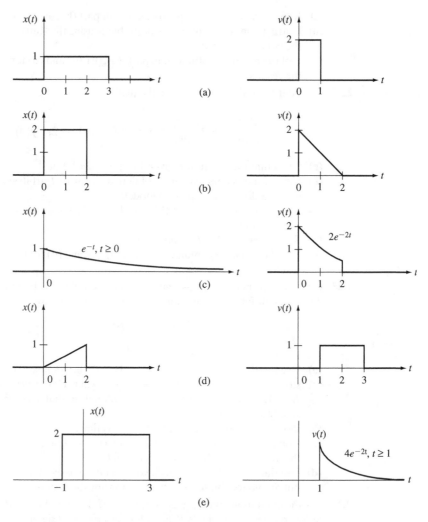

FIGURE P2.29

2.33. A causal linear time-invariant continuous-time system has impulse response $h(t) = (\sin t)u(t - 2)$. Compute the output response $y(t)$ for all $t \geq 0$ when $x(t) = u(t) - u(t - 1)$.

2.34. Consider the series RLC circuit shown in Figure P2.34a. The circuit is equivalent to the cascade connection shown in Figure P2.34b, that is, the system in Figure P2.34b has the same input/output differential equation as the RLC circuit.

   (a) Find the impulse responses of each of the subsystems in Figure P2.34b.
   (b) Using your results in part (a), compute the impulse response of the RLC circuit.
   (c) Use the Runge–Kutta approximation with command ode45 to determine the output response when $x(t) = \sin(t)u(t)$ and $y(0) = \dot{y}(0) = 0$.
   (d) Again suppose that $x(t) = \sin(t)u(t)$. Use the Euler approximation and the M-file recur to compute $y[n]$ for $0 \leq n \leq 100$ with $T = 0.1$. Take $y(0) = 0$, $\dot{y}(0) = 0$. Compare your results with those obtained in part (c).
   (e) Which approximation scheme gives the better results?

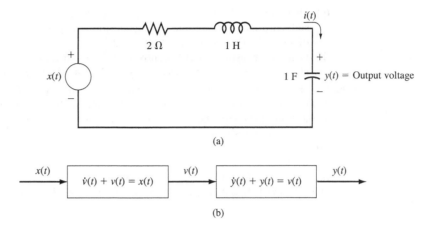

(a)

(b)

FIGURE P2.34

2.35. Consider the top view of the single human eye shown in Figure P2.35. The input $x(t)$ is the angular position $\theta_T(t)$ of the target, and the output $y(t)$ is the angular position $\theta_E(t)$ of the eye, with both angular positions defined relative to the resting position. An idealized model for eye movement is given by the equations

$$T_e \frac{d\theta_E(t)}{dt} + \theta_E(t) = R(t)$$

$$R(t) = b\theta_T(t - d) - b\theta_T(t - d - c) + \theta_T(t - d)$$

where $R(t)$ is the firing rate of action potentials in the nerve to the eye muscle, d is the time delay through the central nervous system, and T_e, b, and c are positive constants.

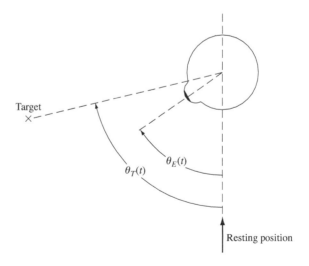

FIGURE P2.35

    (a) Using the symbolic manipulator, derive an expression for $\theta_E(t)$ when the target suddenly moves from the resting position to position A at time $t = 0$; that is, $\theta_T(t) = Au(t)$. Assume that $\theta_E(0) = 0$.

    (b) Using your result in part (a), show that there is a value of b for which $\theta_E(t) = A$ for all $t \geq d + c$; in other words, the eye locks onto the target at time $d + c$.

2.36. For the single-eye system in Problem 2.35, assume that $T_e = c = 0.1$, $d = 0.2$, and $\theta_T(t) = u(t)$. For the values of b given next, simulate the system by using a Runge–Kutta method with the command `ode45` to compute the approximate values $\theta_E(t)$ of the resulting output (eye position) for $0 \leq t \leq 2$.

    (i) $b = 1$

    (ii) $b = 0.2$

    (iii) $b =$ value found in part (b) of Problem 2.35

Does the eye lock onto the target for these values of b? Discuss your results.

CHAPTER 3

The Fourier Series and Fourier Transform

The fundamental notion of the frequency spectrum of a continuous-time signal is introduced in this chapter. As will be seen, the frequency spectrum displays the various sinusoidal components that make up a given continuous-time signal. In general, the frequency spectrum is a complex-valued function of the frequency variable, and thus it is usually specified in terms of an amplitude spectrum and a phase spectrum.

The chapter begins with a study of signals that can be expressed as a sum of sinusoids, which includes periodic signals if an infinite number of terms is allowed in the sum. In the case of a periodic signal, the frequency spectrum can be generated by computation of the *Fourier series*. The Fourier series is named after the French physicist Jean Baptiste Fourier (1768–1830), who was the first one to propose that periodic waveforms could be represented by a sum of sinusoids (or complex exponentials). It is interesting to note that, in addition to his contributions to science and mathematics, Fourier was also very active in the politics of his time. For example, he played an important role in Napoleon's expeditions to Egypt during the late 1790s.

In Section 3.1, a frequency-domain analysis is given for continuous-time signals that can be expressed as a finite sum of sinusoids. This then leads to the trigonometric Fourier series representation of periodic signals presented in Section 3.2. In Section 3.3 the complex exponential form of the Fourier series is considered. In this section, the frequency spectrum of a periodic signal is defined in terms of the magnitudes and angles of the coefficients of the complex exponential terms comprising the Fourier series.

In Section 3.4 the Fourier transform of a nonperiodic signal is defined. In contrast to a periodic signal, the amplitude and phase spectra of a nonperiodic signal consist of a continuum of frequencies. In Chapter 5 it will be seen that the characterization of input signals in terms of their frequency spectrum is very useful in determining how a linear time-invariant system processes inputs.

In Section 3.5 the frequency spectrum is generated for specific signals by the MATLAB symbolic manipulator, and in Section 3.6 the properties of the Fourier transform are given. A brief treatment of the generalized Fourier transform is given in Section 3.7, and then in Section 3.8 the application of the Fourier transform to modulation and demodulation is studied. A summary of the chapter is presented in Section 3.9.

3.1 REPRESENTATION OF SIGNALS IN TERMS OF FREQUENCY COMPONENTS

A fundamental concept in the study of signals is the notion of the *frequency content* of a signal. For a large class of signals, we can generate the frequency content by decomposing the signal into frequency components given by sinusoids. For example, consider the continuous-time signal $x(t)$ defined by the finite sum of sinusoids

$$x(t) = \sum_{k=1}^{N} A_k \cos(\omega_k t + \theta_k), \quad -\infty < t < \infty \qquad (3.1)$$

In (3.1), N is a positive integer, the A_k (which are assumed to be nonnegative) are the amplitudes of the sinusoids, the ω_k are the frequencies (in rad/sec) of the sinusoids, and the θ_k are the phases of the sinusoids.

In the case of the signal given by (3.1), the frequencies "present in the signal" are the frequencies $\omega_1, \omega_2, \ldots, \omega_N$ of the sinusoids constituting the signal, and the frequency components of the signal are the sinusoids $A_k \cos(\omega_k t + \theta_k)$ constituting the signal. It is important to observe that the signal given by (3.1) is characterized completely by the frequencies $\omega_1, \omega_2, \ldots, \omega_N$, the amplitudes $A_1, A_2, \ldots, A_N$, and the phases $\theta_1, \theta_2, \ldots, \theta_N$ in the representation given by (3.1).

The characteristics or "features" of a signal given by (3.1) can be studied in terms of the frequencies, amplitudes, and phases of the sinusoidal terms composing the signal. In particular, the amplitudes $A_1, A_2, \ldots, A_N$ specify the relative weights of the frequency components composing the signal, and these weights are a major factor in determining the "shape" of the signal. This is illustrated by the following example.

Example 3.1   Sum of Sinusoids

Consider the continuous-time signal given by

$$x(t) = A_1 \cos t + A_2 \cos(4t + \pi/3) + A_3 \cos(8t + \pi/2), \, -\infty < t < \infty \qquad (3.2)$$

This signal obviously has three frequency components with frequencies 1, 4, 8 rad/sec, amplitudes A_1, A_2, A_3, and phases $0, \pi/3, \pi/2$ rad. The goal here is to show that the shape of the signal depends on the relative magnitudes of the frequency components making up the signal, which are specified in terms of the amplitudes A_1, A_2, A_3. For this purpose, the following MATLAB commands were used to generate $x(t)$ for arbitrary values of A_1, A_2, and A_3:

```
t = 0:20/400:20;
w1 = 1; w2 = 4; w3 = 8;
A1 = input('Input the amplitude A1 for w1 = 1: ');
A2 = input('Input the amplitude A2 for w2 = 4: ');
A3 = input('Input the amplitude A3 for w3 = 8: ');
x = A1*cos(w1*t)+A2*cos(w2*t+pi/3)+A3*cos(w3*t+pi/2);
```

By the preceding commands, MATLAB plots of $x(t)$ were generated for the three cases $A_1 = 0.5, A_2 = 1, A_3 = 0$; $A_1 = 1, A_2 = 0.5, A_3 = 0$; and $A_1 = 1, A_2 = 1, A_3 = 0$. The resulting plots are given in Figure 3.1. In all three of these cases, only the 1- and 4-rad/sec frequency components are present. In the first case, the 4-rad/sec component is twice as large as the 1-rad/sec component. The dominance of the 4-rad/sec component is obvious from Figure 3.1a. In the second case, the 1-rad/sec component dominates, which results in the signal shape shown in Figure 3.1b. In the third case, both frequency components have the same amplitude, which results in the waveform shown in Figure 3.1c.

The MATLAB program was then run again for the cases $A_1 = 0.5, A_2 = 1, A_3 = 0.5$; $A_1 = 1, A_2 = 0.5, A_3 = 0.5$; and $A_1 = 1, A_2 = 1, A_3 = 1$. In these three cases, all three frequency

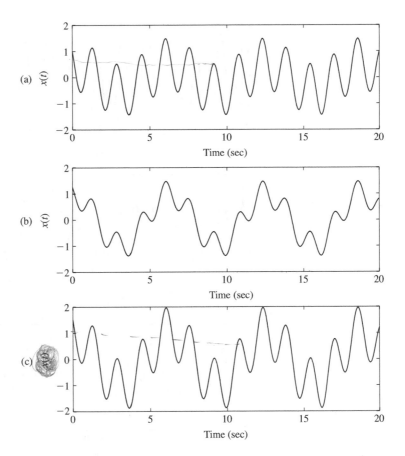

FIGURE 3.1
Plots of $x(t)$ for (a) $A_1 = 0.5, A_2 = 1, A_3 = 0$; (b) $A_1 = 1, A_2 = 0.5, A_3 = 0$; and (c) $A_1 = 1, A_2 = 1, A_3 = 0$.

components are present, with the 4-rad/sec component dominating in the first case, the 1-rad/sec component dominating in the second case, and with all three components having the same amplitude in the third case. Figure 3.2 shows the plots of $x(t)$ for these three cases. For each of the plots in Figure 3.2, the reader should be able to distinguish all three of the frequency components making up the signal.

Again consider the signal given by (3.1). With ω equal to the *frequency variable* (a real variable), the amplitudes A_k can be plotted versus ω. Since there are only a finite number of frequencies present in $x(t)$, the plot of A_k versus ω will consist of a finite number of points plotted at the frequencies ω_k present in $x(t)$. Usually, vertical lines are drawn connecting the values of the A_k with the points ω_k. The resulting plot is an example of a *line spectrum* and is called the *amplitude spectrum* of the signal $x(t)$. The amplitude spectrum shows the relative magnitudes of the various frequency components that make up the signal. For instance, consider the signal in Example 3.1 given by (3.2). For the various versions of the signal plotted in Figure 3.2, the amplitude spectrum is shown

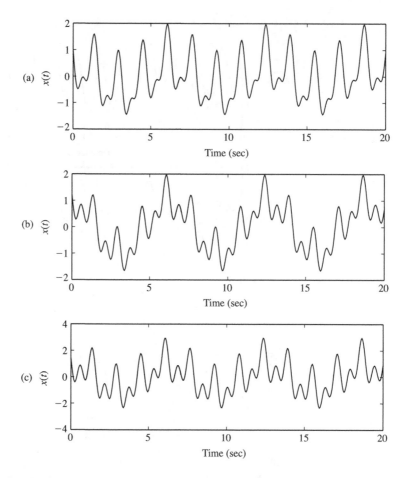

FIGURE 3.2
Plots of $x(t)$ for (a) $A_1 = 0.5, A_2 = 1, A_3 = 0.5$; (b) $A_1 = 1, A_2 = 0.5, A_3 = 0.5$; and (c) $A_1 = 1, A_2 = 1, A_3 = 1$.

in Figure 3.3. Note the direct correspondence between the magnitudes of the spectral components shown in Figure 3.3 and the shape of the signals in Figure 3.2.

In addition to the amplitude spectrum, the signal defined by (3.1) also has a *phase spectrum*, which is a plot of the phase θ_k in degrees (or radians) versus the frequency variable ω. Again, in generating this plot, vertical lines are drawn connecting the values of the θ_k with the frequency points ω_k, so the phase spectrum is also a line spectrum. For example, the phase spectrum of the signal given by (3.2) is plotted in Figure 3.4. As discussed in Section 3.3, the amplitude and phase spectra of a sum of sinusoids are usually generated with respect to the complex exponential form of the sinusoids. Hence, the spectra shown in Figures 3.3 and 3.4 are not in the standard form, but the format considered in this section provides a simple introduction to the concept of line spectra, which is pursued in Section 3.3.

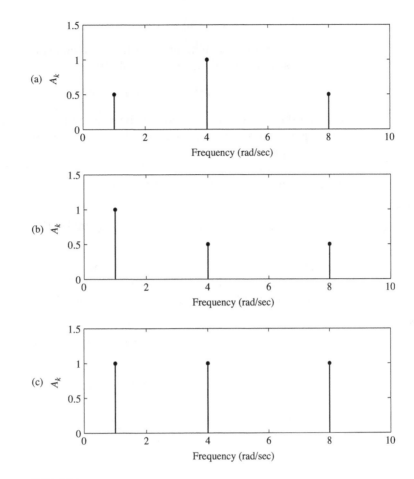

FIGURE 3.3
Amplitude spectra of the versions of $x(t)$ plotted in Figure 3.2.

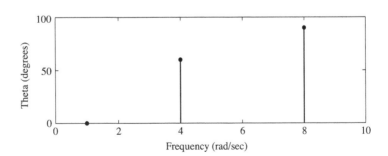

FIGURE 3.4
Phase spectrum of signal $x(t)$ defined by (3.2).

3.2   TRIGONOMETRIC FOURIER SERIES

Let T be a fixed positive real number. As first defined in Section 1.1, a continuous-time signal $x(t)$ is said to be periodic with period T if

$$x(t + T) = x(t), \quad -\infty < t < \infty \tag{3.3}$$

Recall that the fundamental period T is the smallest positive number for which (3.3) is satisfied. For example, the rectangular pulse train shown in Figure 3.5 is periodic with fundamental period $T = 2$.

Let $x(t)$ be a periodic signal with fundamental period T. Then $x(t)$ can be expressed as a (in general, infinite) sum of sinusoids

$$x(t) = a_0 + \sum_{k=1}^{\infty} [a_k \cos(k\omega_0 t) + b_k \sin(k\omega_0 t)], \quad -\infty < t < \infty \tag{3.4}$$

In the representation (3.4), a_0, the a_k, and the b_k are real numbers, and ω_0 is the *fundamental frequency* (in rad/sec) given by $\omega_0 = 2\pi/T$, where T is the fundamental period. The coefficients a_k and b_k are computed with the formulas

$$a_k = \frac{2}{T} \int_0^T x(t) \cos(k\omega_0 t) \, dt, \quad k = 1, 2, \ldots \tag{3.5}$$

$$b_k = \frac{2}{T} \int_0^T x(t) \sin(k\omega_0 t) \, dt, \quad k = 1, 2, \ldots \tag{3.6}$$

It should be noted that the a_k and b_k given by (3.5) and (3.6) can be computed by integration over any full period. For instance,

$$a_k = \frac{2}{T} \int_{-T/2}^{T/2} x(t) \cos(k\omega_0 t) \, dt, \quad k = 1, 2, \ldots$$

The term a_0 in (3.4) is the constant or dc component of $x(t)$ given by

$$a_0 = \frac{1}{T} \int_0^T x(t) \, dt \tag{3.7}$$

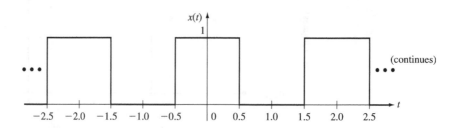

(continues)

FIGURE 3.5
Periodic signal with fundamental period $T = 2$.

The representation (3.4) is called the *trigonometric Fourier series* of the periodic signal $x(t)$. The first harmonic of $x(t)$ is the term $a_1 \cos(\omega_0 t) + b_1 \sin(\omega_0 t)$, the second harmonic is the term $a_2 \cos(2\omega_0 t) + b_2 \sin(2\omega_0 t)$, and the kth harmonic is the term $a_k \cos(k\omega_0 t) + b_k \sin(k\omega_0 t)$. Note that the frequencies of the harmonics that make up $x(t)$ are integer multiples $k\omega_0$ of the fundamental frequency ω_0. This is a key property of periodic signals.

The trigonometric Fourier series given by (3.4) can be written in the *cosine-with-phase form*

$$x(t) = a_0 + \sum_{k=1}^{\infty} A_k \cos(k\omega_0 t + \theta_k), \quad -\infty < t < \infty \tag{3.8}$$

where

$$A_k = \sqrt{a_k^2 + b_k^2}, k = 1, 2, \ldots \tag{3.9}$$

and

$$\theta_k = \begin{cases} \tan^{-1}\left(-\dfrac{b_k}{a_k}\right), & k = 1, 2, \ldots, \text{when } a_k \geq 0 \\[4mm] \pi + \tan^{-1}\left(-\dfrac{b_k}{a_k}\right), & k = 1, 2, \ldots, \text{when } a_k < 0 \end{cases} \tag{3.10}$$

It is worth noting that if a constant term a_0 is added to the sum in (3.1) and the sum is allowed to be infinite (i.e., $N = \infty$), the Fourier series (3.8) is a special case of (3.1) in that all the frequencies present in the signal are integer multiples of the fundamental frequency ω_0. To say this another way, with $N = \infty$ in (3.1) and with the addition of a constant term a_0, the class of signals that can be expressed in the form (3.1) includes the class of periodic signals given by (3.8).

The Fourier series representation of a periodic signal is a remarkable result. In particular, it shows that a periodic signal such as the waveform with "corners" in Figure 3.5 can be expressed as a sum of sinusoids. Since sinusoids are infinitely smooth functions (i.e., they have ordinary derivatives of arbitrarily high order), it is difficult to believe that signals with corners can be expressed as a sum of sinusoids. Of course, the key here is that the sum is an infinite sum. It is not surprising that Fourier had a difficult time convincing his peers (in this case, the members of the French Academy of Science) that his theorem was true.

Fourier believed that any periodic signal could be expressed as a sum of sinusoids. However, this turned out not to be the case, although virtually all periodic signals arising in engineering do have a Fourier series representation. In particular, a periodic signal $x(t)$ has a Fourier series if it satisfies the following *Dirichlet conditions*:

1. $x(t)$ is absolutely integrable over any period; that is,

$$\int_a^{a+T} |x(t)| \, dt < \infty \quad \text{for any } a$$

2. $x(t)$ has only a finite number of maxima and minima over any period.

3. $x(t)$ has only a finite number of discontinuities over any period.

Example 3.2   Rectangular Pulse Train

Consider the rectangular pulse train shown in Figure 3.5. This signal is periodic with fundamental period $T = 2$, and thus the fundamental frequency is $\omega_0 = 2\pi/2 = \pi$ rad/sec. The signal obviously satisfies the Dirichlet conditions, and thus it has a Fourier series representation. From (3.7), the constant component of $x(t)$ is

$$a_0 = \frac{1}{2}\int_0^2 x(t)\,dt = \frac{1}{2}\int_0^{0.5}(1)\,dt + \frac{1}{2}\int_{1.5}^2(1)\,dt = \frac{1}{4} + \frac{1}{4} = \frac{1}{2}$$

Evaluating (3.5) gives

$$a_k = \frac{2}{2}\int_0^2 x(t)\cos(k\pi t)\,dt$$

$$= \int_0^{0.5}\cos(k\pi t)\,dt + \int_{1.5}^2\cos(k\pi t)\,dt$$

$$= \frac{1}{\pi k}\sin(k\pi t)\Big|_{t=0}^{t=0.5} + \frac{1}{\pi k}\sin(k\pi t)\Big|_{t=1.5}^{t=2}$$

$$= \frac{1}{\pi k}\left[\sin\left(\frac{k\pi}{2}\right) - \sin\left(\frac{3k\pi}{2}\right)\right]$$

$$= \frac{1}{\pi k}\left[\sin\left(\frac{k\pi}{2}\right) - \sin\left(\frac{k\pi}{2} + k\pi\right)\right]$$

$$= \frac{1}{\pi k}\left[2\sin\left(\frac{k\pi}{2}\right)\right], \quad k = 1, 2, \ldots$$

Evaluating (3.6) gives

$$b_k = \frac{2}{2}\int_0^2 x(t)\sin(k\pi t)\,dt$$

$$= \int_0^{0.5}\sin(k\pi t)\,dt + \int_{1.5}^2\sin(k\pi t)\,dt$$

$$= -\frac{1}{\pi k}\cos(k\pi t)\Big|_{t=0}^{t=0.5} - \frac{1}{\pi k}\cos(k\pi t)\Big|_{t=1.5}^{t=2}$$

$$= -\frac{1}{\pi k}\left[\frac{\pi}{2} - \frac{\pi}{2}\right]$$

$$= 0, \quad k = 1, 2, \ldots$$

Then inserting the values for a_0, a_k, and b_k into (3.4) results in the following Fourier series representation of the pulse train shown in Figure 3.5:

$$x(t) = \frac{1}{2} + \frac{2}{\pi}\sum_{k=1}^{\infty}\frac{1}{k}\sin\left(\frac{k\pi}{2}\right)\cos(k\pi t), \quad -\infty < t < \infty \qquad (3.11)$$

Note that, since $\sin\left(\dfrac{k\pi}{2}\right) = 0$ for $k = 2, 4, 6, \ldots$, the signal $x(t)$ contains only odd harmonics, and thus (3.11) can be written in the form

$$x(t) = \frac{1}{2} + \frac{2}{\pi} \sum_{\substack{k=1 \\ k \text{ odd}}}^{\infty} \frac{1}{k} \sin\left(\frac{k\pi}{2}\right) \cos(k\pi t), \quad -\infty < t < \infty \tag{3.12}$$

3.2.1   Even or Odd Symmetry

A signal $x(t)$ is said to be an *even function* of t if $x(t) = x(-t)$ for $-\infty < t < \infty$, and $x(t)$ is an *odd function* of t if $x(t) = -x(-t)$ for $-\infty < t < \infty$. Examples of signals that are even are $A\cos(\rho t)$ for any real numbers A and ρ, and the pulse train shown in Figure 3.5. For any real numbers A and ρ, the signal $A\sin(\rho t)$ is an odd function of t.

If $x(t)$ and $v(t)$ are any two even (or odd) functions of t, then for any constant $\eta > 0$,

$$\int_{-\eta}^{\eta} x(t)v(t)\, dt = 2\int_{0}^{\eta} x(t)v(t)\, dt \tag{3.13}$$

If $x(t)$ is even and $v(t)$ is odd, then

$$\int_{-\eta}^{\eta} x(t)v(t)\, dt = 0 \tag{3.14}$$

The reader is asked to prove these results in Problem 3.5.

Then, since $\cos(k\pi t)$ is an even function of t and $\sin(k\pi t)$ is an odd function of t, if $x(t)$ is an even periodic signal with period T, (3.5) and (3.6) reduce to

$$a_k = \frac{4}{T} \int_{0}^{T/2} x(t) \cos(k\omega_0 t)\, dt, \quad k = 1, 2, \ldots \tag{3.15}$$

$$b_k = 0, k = 1, 2, \ldots \tag{3.16}$$

If $x(t)$ is an odd periodic signal with period T, (3.5) and (3.6) reduce to

$$a_k = 0, k = 1, 2, \ldots \tag{3.17}$$

$$b_k = \frac{4}{T} \int_{0}^{T/2} x(t) \sin(k\omega_0 t)\, dt, \quad k = 1, 2, \ldots \tag{3.18}$$

The expressions given by (3.15)-(3.16) and (3.17)-(3.18) greatly simplify the computation of the Fourier series coefficients in the case when $x(t)$ is even or odd. To illustrate this, the coefficients for the Fourier series of the pulse train in Figure 3.5 are recomputed by the use of symmetry.

Example 3.3   Use of Symmetry

Again consider the pulse train shown in Figure 3.5. As noted, this signal is an even function of t, and thus (3.15) and (3.16) can be used to compute the Fourier series coefficients a_k and b_k as follows. First, since $x(t)$ is even, by (3.16), $b_k = 0$, $k = 1, 2, \ldots$. Then, using (3.15) gives

$$a_k = \frac{4}{2} \int_0^1 x(t) \cos(k\pi t)\, dt$$

$$= 2 \int_0^{.5} \cos(k\pi t)\, dt$$

$$= \frac{2}{k\pi} \sin(k\pi t) \Big|_{t\,=\,0}^{t\,=\,.5}$$

$$= \frac{2}{k\pi} \sin\left(\frac{k\pi}{2}\right),\, k = 1, 2, \ldots$$

These values for a_k and b_k are the same as the values found in Example 3.2, so the use of symmetry does yield the same result.

3.2.2   Gibbs Phenomenon

Again, consider the pulse train $x(t)$ with the trigonometric Fourier series representation (3.12). Given an odd positive integer N, let $x_N(t)$ denote the finite sum

$$x_N(t) = \frac{1}{2} + \frac{2}{\pi} \sum_{\substack{k=1 \\ k\,\text{odd}}}^{N} \frac{1}{k} \sin\left(\frac{k\pi}{2}\right) \cos(k\pi t),\, \infty < t < \infty$$

Convergence of Fourier Series

By Fourier's theorem, $x_N(t)$ should converge to $x(t)$ as $N \to \infty$. In other words $|x_N(t) - x(t)|$ should be getting close to zero for all t as N is increased. Thus, for a suitably large value of N, $x_N(t)$ should be a close approximation to $x(t)$. To see if this is the case, $x_N(t)$ can simply be plotted for various values of N. The MATLAB commands for generating $x_N(t)$ are

```
t = -3:6/1000:3;
N = input('Number of harmonics =  ');
a0 = 0.5;
w0 = pi;
xN = a0*ones(1, length(t)); % dc component
for k = 1:2:N, % even harmonics are zero
xN = xN + 2/k/pi*sin(k*pi/2)*cos(k*w0*t);
end
```

For the signal $x_N(t)$ given previously, the even harmonics are zero, and thus these terms are excluded in the loop to make the MATLAB program more efficient.

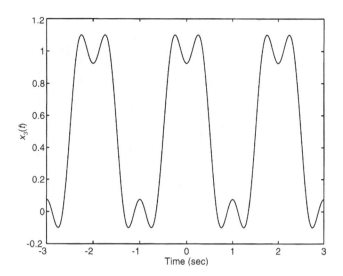

FIGURE 3.6
Plot of $x_N(t)$ when $N = 3$.

Now with $N = 3$, $x_N(t)$ becomes

$$x_3(t) = \frac{1}{2} + \frac{2}{\pi}\cos(\pi t) - \frac{2}{3\pi}\cos(3\pi t), \quad -\infty < t < \infty$$

Setting $N = 3$ in the preceding program results in the plot of $x_3(t)$ shown in Figure 3.6. Note that, even though $x_3(t)$ consists of the constant component and only two harmonics (the first and third), $x_3(t)$ does resemble the pulse train in Figure 3.5.

Increasing N to 9 produces the result shown in Figure 3.7. Comparing Figures 3.6 and 3.7 reveals that $x_9(t)$ is a much closer approximation to the pulse train $x(t)$ than $x_3(t)$. Of course, $x_9(t)$ contains the constant component and the first, third, fifth, seventh, and ninth harmonics of $x(t)$, and thus it would be expected to be a much closer approximation than $x_3(t)$.

Setting $N = 21$ produces the result in Figure 3.8. Except for the overshoot at the corners of the pulse, the waveform in Figure 3.8 is a much better approximation to $x(t)$ than $x_9(t)$. From a careful examination of the plot in Figure 3.8, it can be seen that the magnitude of the overshoot is approximately equal to 9%.

Taking $N = 45$ yields the result displayed in Figure 3.9. Note that the 9% overshoot at the corners is still present. In fact, the 9% overshoot is present even in the limit as N approaches ∞. This characteristic was first discovered by Josiah Willard Gibbs (1839–1903), and thus the overshoot is referred to as the *Gibbs phenomenon*. Gibbs demonstrated the existence of the overshoot from mathematical properties rather than by direct computation.

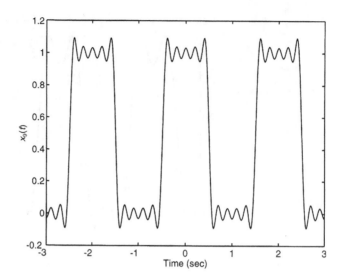

FIGURE 3.7
Approximation $x_9(t)$.

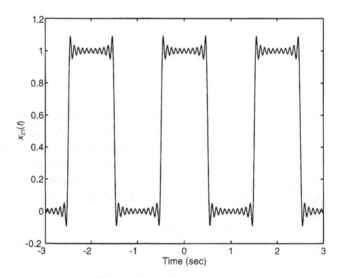

FIGURE 3.8
Approximation $x_{21}(t)$.

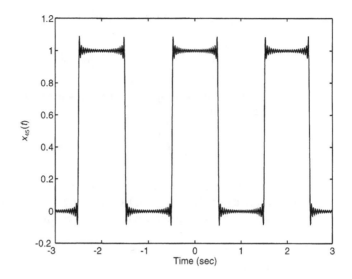

FIGURE 3.9
The signal $x_{45}(t)$.

Now let $x(t)$ be an arbitrary periodic signal. As a consequence of the Gibbs phenomenon, the Fourier series representation of $x(t)$ is not actually equal to the true value of $x(t)$ at any points where $x(t)$ is discontinuous. If $x(t)$ is discontinuous at $t = t_1$, the Fourier series representation is off by approximately 9% at t_1^- and t_1^+.

3.3   COMPLEX EXPONENTIAL SERIES

The trigonometric Fourier series given by (3.4) or (3.8) can be expressed in a complex exponential form given by

$$x(t) = \sum_{k=-\infty}^{\infty} c_k e^{jk\omega_0 t}, \quad -\infty < t < \infty \qquad (3.19)$$

In the representation (3.19), c_0 is a real number and the c_k for $k \neq 0$ are, in general, complex numbers. As in (3.4) or (3.8), ω_0 is the fundamental frequency (in rad/sec) given by $\omega_0 = 2\pi/T$, where T is the fundamental period. Note that, in contrast to the trigonometric Fourier series given by (3.4) or (3.8), the complex exponential form is a bi-infinite sum, in general; that is, $k = 0, \pm 1, \pm 2, \dots$.

The coefficients c_k of the complex exponentials in (3.19) can be computed from the coefficients of the trigonometric Fourier series given by (3.4) by the formulas

$$c_0 = a_0, \text{ and } c_k = \frac{1}{2}(a_k - jb_k), c_{-k} = \frac{1}{2}(a_k + jb_k), k = 1, 2, \dots \qquad (3.20)$$

In addition, the c_k can be computed directly from the signal $x(t)$ by the formula

$$c_k = \frac{1}{T}\int_0^T x(t)e^{-jk\omega_0 t}\, dt, \quad k = 0, \pm 1, \pm 2, \ldots \tag{3.21}$$

It should be noted that the c_k given by (3.21) can be computed by integration over any full period. For instance,

$$c_k = \frac{1}{T}\int_{-T/2}^{T/2} x(t)e^{-jk\omega_0 t}\, dt, \quad k = 0, \pm 1, \pm 2, \ldots$$

In Problem 3.8, the reader is asked to derive the relationships (3.20) and (3.21).

If a periodic signal $x(t)$ is given by the complex exponential form (3.19), it is possible to express $x(t)$ in the trigonometric form (3.4) by the relationships

$$a_0 = c_0, \text{ and } a_k = c_k + c_{-k} = 2\text{Re}(c_k),\, b_k = j(c_k - c_{-k}) = -2\,\text{Im}(c_k), \tag{3.22}$$
$$k = 1, 2, \ldots$$

The relationships in (3.22) follow easily from (3.20). The verification is left to the reader.

Note that if $x(t)$ is an even function of t, the coefficients of the complex exponential form are real numbers given by

$$c_0 = a_0 \text{ and } c_k = \frac{1}{2}a_k,\, c_{-k} = \frac{1}{2}a_k,\, k = 1, 2, \ldots \tag{3.23}$$

If $x(t)$ is an odd function of t, the coefficients of the complex exponential form are purely imaginary numbers (except for the value c_0) given by

$$c_0 = a_0 \text{ and } c_k = -j\frac{1}{2}b_k,\, c_{-k} = j\frac{1}{2}b_k,\, k = 1, 2, \ldots \tag{3.24}$$

Example 3.4   Rectangular Pulse Train

Again consider the rectangular pulse train shown in Figure 3.5. From the result in Example 3.2 and (3.23), the coefficients of the complex Fourier series are

$$c_0 = \frac{1}{2}, \text{ and } c_k = \frac{1}{\pi k}\sin\left(\frac{k\pi}{2}\right),\, k = \pm 1, \pm 3, \pm 5, \ldots$$

Hence, the complex exponential form of the Fourier series for the pulse train in Figure 3.5 is given by

$$x(t) = \frac{1}{2} + \frac{1}{\pi}\sum_{\substack{k=-\infty \\ k \text{ odd}}}^{\infty} \frac{1}{k}\sin\left(\frac{k\pi}{2}\right)e^{jk\pi t}, \quad -\infty < t < \infty \tag{3.25}$$

Solving for Fourier coefficients can be very tedious for all but the simplest forms of $x(t)$, such as a rectangular pulse train. The Symbolic Math Toolbox in MATLAB can be used to perform the integrations and simplify the expressions. To illustrate this, the

Fourier coefficients c_k for the rectangular pulse train in Figure 3.5 can be computed for $k = 1{:}5$ by the following commands:

```
k = 1:5;
syms ck t
ck = 0.5*int(exp(-j*k*pi*t),t,-0.5,0.5)
```

The command `syms ck t` constructs symbolic objects `ck` and `t`. The `int` command integrates the expression defined in the first argument with respect to the second argument `t` from -0.5 to 0.5. The result is a vector of `ck` corresponding to the values of `k` defined.

```
ck =
[    1/pi,       0, -1/3/pi,       0,  1/5/pi]
```

3.3.1   Line Spectra

Given a periodic signal $x(t)$ with period T, consider the trigonometric Fourier series given by the cosine-with-phase form

$$x(t) = a_0 + \sum_{k=1}^{\infty} A_k \cos(k\omega_0 t + \theta_k), \quad -\infty < t < \infty \tag{3.26}$$

As noted in Section 3.1, the frequency components constituting this signal may be displayed in terms of the amplitude and phase spectra specified by plots of A_k (with $A_0 = a_0$) and θ_k (with $\theta_0 = 0$) versus $\omega = k\omega_0, k = 0, 1, 2, \ldots$. This results in line spectra defined for nonnegative frequencies only. However, the line spectra for a signal $x(t)$ consisting of a sum of sinusoids are usually defined with respect to the complex exponential form (3.19). In this case, the amplitude spectrum is the plot of the magnitudes $|c_k|$ versus $\omega = k\omega_0, k = 0, \pm 1, \pm 2, \ldots$, and the phase spectrum is a plot of the angles $\angle c_k$ versus $\omega = k\omega_0, k = 0, \pm 1, \pm 2, \ldots$. This results in line spectra that are defined for both positive and negative frequencies. It should be stressed that the negative frequencies are a result of the complex exponential form (consisting of a positive and a negative frequency component) and have no physical meaning.

From (3.20) it can be see that

$$|c_k| = |c_{-k}| \text{ for } k = 1, 2, \ldots$$

and thus the amplitude spectrum is symmetrical about $\omega = 0$. That is, the values of the amplitude spectrum for positive frequencies are equal to the values of the amplitude spectrum for the corresponding negative frequencies. In other words, the amplitude spectrum is an even function of the frequency variable ω. It also follows from (3.20) that

$$\angle(c_{-k}) = -\angle c_k \quad k = 1, 2, \ldots$$

which implies that the phase spectrum is an odd function of the frequency variable ω.

To determine the line spectra for a periodic signal $x(t)$ given by (3.26), it is first necessary to determine the magnitudes $|c_k|$ and the angles $\angle c_k$ for $k = 1, 2, \ldots$. From (3.20),

$$|c_k| = \sqrt{a_k^2 + b_k^2}, k = 1, 2, \ldots$$

and

$$
\angle c_k = \begin{cases} \tan^{-1}\left(-\dfrac{b_k}{a_k}\right), & k = 1, 2, \ldots, \text{when } a_k \geq 0 \\[3mm] \pi + \tan^{-1}\left(-\dfrac{b_k}{a_k}\right), & k = 1, 2, \ldots, \text{when } a_k < 0 \end{cases}
$$

Thus, using (3.9) and (3.10) gives

$$
|c_k| = \frac{1}{2} A_k, k = 1, 2, \ldots \text{ and } \angle c_k = \theta_k, k = 1, 2, \ldots \tag{3.27}
$$

Example 3.5   Line Spectra

Consider the signal

$$
x(t) = \cos t + 0.5 \cos(4t + \pi/3) + \cos(8t + \pi/2)
$$

Using (3.27) gives

$$
c_1 = \frac{1}{2} = 0.5, c_4 = \frac{0.5}{2} e^{j\pi/3} = 0.25 \angle 60°, c_8 = \frac{1}{2} e^{j\pi/2} = 0.5 \angle 90°
$$

$$
c_{-1} = \frac{1}{2} = 0.5, c_{-4} = \frac{0.5}{2} e^{-j\pi/3} = 0.25 \angle -60°, c_{-8} = \frac{1}{2} e^{-j\pi/2} = 0.5 \angle -90°
$$

The amplitude and phase spectra are plotted in Figure 3.10.

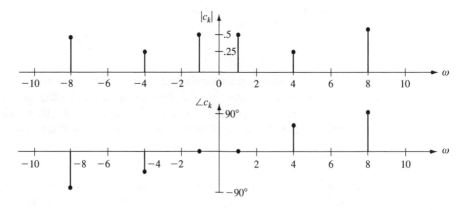

FIGURE 3.10
Line spectra for the signal in Example 3.5.

Example 3.6   Line Spectra of Rectangular Pulse Train

Recall that the complex exponential form of the Fourier series for the pulse train shown in Figure 3.5 is given by

$$x(t) = \frac{1}{2} + \frac{1}{\pi} \sum_{\substack{k=-\infty \\ k \text{ odd}}}^{\infty} \frac{1}{k} \sin\left(\frac{k\pi}{2}\right) e^{jk\pi t}, \quad -\infty < t < \infty$$

To compute the amplitude and phase spectra for the rectangular pulse train, first note that

$$|c_k| = \begin{cases} 0, & k = 2, 4, \ldots \\ \dfrac{1}{k\pi}, & k = 1, 3, \ldots \end{cases}$$

$$\angle c_k = \begin{cases} 0, & k = 2, 4, \ldots \\ [(-1)^{(k-1)/2} - 1]\dfrac{\pi}{2}, & k = 1, 3, \ldots \end{cases}$$

The frequency spectra (amplitude and phase) are plotted in Figure 3.11.

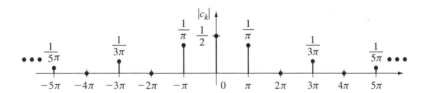

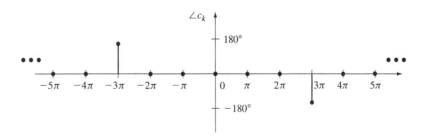

FIGURE 3.11
Line spectra for the rectangular pulse train.

3.3.2   Truncated Complex Fourier Series

It should be noted that the finite sum $x_N(t)$ resulting from the truncation of the trigonometric Fourier series of the rectangular pulse train in Figure 3.5 can also be calculated by truncating the exponential form of the Fourier series as follows:

$$x_N(t) = \sum_{k=-N}^{N} c_k e^{jk\omega_0 t}$$

The MATLAB commands for computing the truncated exponential Fourier series for the pulse train are

```
t = -3:6/1000:3;
N = input ('Number of harmonics = ');
c0 = 0.5;
w0 = pi;
xN = c0*ones (1, length(t));   % dc component
for k = 1:N,
  ck = 1/k/pi*sin(k*pi/2);
  c_k = ck;
  xN = xN + ck*exp(j*k*w0*t) + c_k*exp(-j*k*w0*t);
end
```

The expression for c_k is given in Example 3.4, where it is seen by inspection that $c_{-k} = c_k$. Running the preceding program for $N = 3, 9, 21$, and 45 yields the same plots as those in Figures 3.6 to 3.9.

3.3.3   Parseval's Theorem

Let $x(t)$ be a periodic signal with period T. The average power P of the signal is defined by

$$P = \frac{1}{T} \int_{-T/2}^{T/2} x^2(t)\, dt \tag{3.28}$$

If $x(t)$ is the voltage across a 1-ohm resistor or the current in a 1-ohm resistor, the average power is given by (3.28). So, the expression (3.28) is a generalization of the notion of average power to arbitrary signals.

Again let $x(t)$ be an arbitrary periodic signal with period T, and consider the Fourier series of $x(t)$ given by (3.19). By Parseval's theorem, the average power P of the signal $x(t)$ is given by

$$P = \sum_{k=-\infty}^{\infty} |c_k|^2 \tag{3.29}$$

The relationship (3.29) is useful, since it relates the average power of a periodic signal to the coefficients of the Fourier series of the signal. The proof of Parseval's theorem is beyond the scope of this book.

3.4   FOURIER TRANSFORM

A key feature of the Fourier series representation of periodic signals is the description of such signals in terms of the frequency content given by sinusoidal components. The question then arises as to whether or not nonperiodic signals, also called *aperiodic signals,* can be described in terms of frequency content. The answer is yes, and the analytical construct for doing this is the Fourier transform. As will be seen, the frequency components of nonperiodic signals are defined for all real values of the frequency variable ω, not just for discrete values of ω as in the case of a periodic signal. In other words, the spectra for a nonperiodic signal are not line spectra (unless the signal is equal to a sum of sinusoids).

Given a signal $x(t)$, the Fourier transform $X(\omega)$ of $x(t)$ is defined to be the frequency function

$$X(\omega) = \int_{-\infty}^{\infty} x(t)e^{-j\omega t}\, dt, \quad -\infty < \omega < \infty \tag{3.30}$$

where ω is the continuous frequency variable. In this book, the Fourier transform will always be denoted by an uppercase letter or symbol, whereas signals will usually be denoted by lowercase letters or symbols.

Note that, due to the presence of the complex exponential $\exp(-j\omega t)$ in the integrand of the integral in (3.30), the values of $X(\omega)$ may be complex. Hence, in general, the Fourier transform $X(\omega)$ is a complex-valued function of the frequency variable ω, and thus in order to specify $X(\omega)$, in general, it is necessary to display the magnitude function $|X(\omega)|$ and the angle function $\angle X(\omega)$. The amplitude spectrum of a signal $x(t)$ is defined to be the magnitude function $|X(\omega)|$ of the Fourier transform $X(\omega)$, and the phase spectrum of $x(t)$ is defined to be the angle function $\angle X(\omega)$. The amplitude and phase spectra of a signal $x(t)$ are natural generalizations of the line spectra of periodic signals.

A signal $x(t)$ is said to have a Fourier transform in the ordinary sense if the integral in (3.30) converges (i.e., exists). The integral does converge if $x(t)$ is "well behaved" and if $x(t)$ is absolutely integrable, where the latter condition means that

$$\int_{-\infty}^{\infty} |x(t)|\, dt < \infty \tag{3.31}$$

Well behaved means that the signal has a finite number of discontinuities, maxima, and minima within any finite interval of time. All actual signals (i.e., signals that can be physically generated) are well behaved and satisfy (3.31).

Since any well-behaved signal of finite duration in time is absolutely integrable, any such signal has a Fourier transform in the ordinary sense. An example of a signal that does not have a Fourier transform in the ordinary sense follows.

Example 3.7   Constant Signal

Consider the dc or constant signal

$$x(t) = 1, \quad -\infty < t < \infty$$

Clearly, the constant signal is not an actual signal, since no signal can be generated physically that is nonzero for all time. Nevertheless, the constant signal plays a very important role in the theory of signals and systems. The Fourier transform of the constant signal is

$$X(\omega) = \int_{-\infty}^{\infty} (1)e^{-j\omega t}\, dt \qquad (3.32)$$

$$= \lim_{T \to \infty} \int_{-T/2}^{T/2} e^{-j\omega t}\, dt$$

$$= \lim_{T \to \infty} -\frac{1}{j\omega}[e^{-j\omega t}]_{t=-T/2}^{t=T/2}$$

$$X(\omega) = \lim_{T \to \infty} -\frac{1}{j\omega}\left[\exp\left(-\frac{j\omega T}{2}\right) - \exp\left(\frac{j\omega T}{2}\right)\right]$$

But $\exp(j\omega T/2)$ does not have a limit as $T \to \infty$, and thus the integral in (3.32) does not converge. Hence, a constant signal does not have a Fourier transform in the ordinary sense. This can be seen by checking (3.31): The area under the constant signal is infinite, so the integral in (3.31) is not finite. In Section 3.7, it will be shown that a constant signal has a Fourier transform in a generalized sense.

Example 3.8   Exponential Signal

Now consider the signal

Fourier Transform of Exponential Signal

$$x(t) = e^{-bt}u(t)$$

where b is a real constant and $u(t)$ is the unit-step function. Note that $x(t)$ is equal to $u(t)$ when $b = 0$. For an arbitrary value of b, the Fourier transform $X(\omega)$ of $x(t)$ is given by

$$X(\omega) = \int_{-\infty}^{\infty} e^{-bt}u(t)e^{-j\omega t}\, dt$$

and since $u(t) = 0$ for $t < 0, u(t) = 1$ for $t \geq 1$,

$$X(\omega) = \int_{0}^{\infty} e^{-bt}e^{-j\omega t}\, dt$$

$$= \int_{0}^{\infty} e^{-(b+j\omega)t}\, dt$$

Evaluating the integral gives

$$X(\omega) = -\frac{1}{b + j\omega}\left[e^{-(b+j\omega)t}\, \Big|_{t=0}^{t=\infty}\right]$$

The upper limit $t = \infty$ cannot be evaluated when $b \leq 0$, and thus for this range of values of b, $x(t)$ does not have an ordinary Fourier transform. Since $x(t) = u(t)$ when $b = 0$, it is seen that the unit-step function $u(t)$ does not have a Fourier transform in the ordinary sense. (But as shown in Section 3.7, $u(t)$ does have a generalized Fourier transform.)

When $b > 0$, $\exp(-bt) \to 0$ as $t \to \infty$, and thus

$$\lim_{t \to \infty} e^{-(b+j\omega)t} = \lim_{t \to \infty} e^{-bt} e^{-j\omega t} = 0$$

Hence for $b > 0$, $x(t)$ has a Fourier transform given by

$$X(\omega) = -\frac{1}{b + j\omega}(0 - 1) = \frac{1}{b + j\omega}$$

and the amplitude and phase spectra are given by

$$|X(\omega)| = \frac{1}{\sqrt{b^2 + \omega^2}}$$

$$\angle X(\omega) = -\tan^{-1}\frac{\omega}{b}$$

Plots of the amplitude spectrum $|X(\omega)|$ and the phase spectrum $\angle X(\omega)$ can be generated for the case $b = 10$ by the following MATLAB commands:

```
w = 0:0.2:50;
b = 10;
X = 1./(b+j*w);
subplot(211), plot(w, abs (X));     % plot magnitude of X
subplot(212), plot(w, angle (X));   % plot angle of X
```

Note that the explicit expressions for $|X(\omega)|$ and $\angle X(\omega)$ in the preceding expressions are not needed to generate MATLAB plots of the amplitude and phase spectra. For the case $b = 10$, the preceding MATLAB program was run with the results displayed in Figure 3.12. From Figure 3.12a it is seen that most of the spectral content of the signal is concentrated in the low-frequency range with the amplitude spectrum decaying to zero as $\omega \to \infty$.

3.4.1   Rectangular and Polar Form of the Fourier Transform

Consider the signal $x(t)$ with Fourier transform

$$X(\omega) = \int_{-\infty}^{\infty} x(t) e^{-j\omega t}\, dt$$

As noted previously, $X(\omega)$ is a complex-valued function of the real variable ω. In other words, if a particular value of ω is inserted into $X(\omega)$, then in general, the result will be a complex number. Since complex numbers can be expressed in either rectangular or polar form, the Fourier transform $X(\omega)$ can be expressed in either rectangular or polar form.

These forms are defined next.

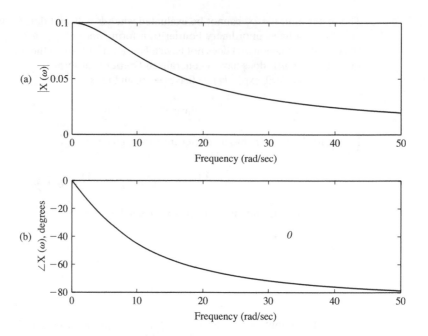

FIGURE 3.12
Plots of the (a) amplitude and (b) phase spectra of $x(t) = \exp(-10t)u(t)$.

By Euler's formula, $X(\omega)$ can be written in the form

$$X(\omega) = \int_{-\infty}^{\infty} x(t) \cos \omega t \, dt - j \int_{-\infty}^{\infty} x(t) \sin \omega t \, dt$$

Now let $R(\omega)$ and $I(\omega)$ denote the real-valued functions of ω defined by

$$R(\omega) = \int_{-\infty}^{\infty} x(t) \cos \omega t \, dt$$

$$I(\omega) = - \int_{-\infty}^{\infty} x(t) \sin \omega t \, dt$$

Then the rectangular form of $X(\omega)$ is

$$X(\omega) = R(\omega) + jI(\omega) \tag{3.33}$$

The function $R(\omega)$ is the real part of $X(\omega)$, and the function $I(\omega)$ is the imaginary part of $X(\omega)$. Note that $R(\omega)$ and $I(\omega)$ could be computed first, and then $X(\omega)$ can be found by the use of (3.33).

Now the polar form of the Fourier transform $X(\omega)$ is given by

$$X(\omega) = |X(\omega)| \exp[j\angle X(\omega)] \tag{3.34}$$

where $|X(\omega)|$ is the magnitude of $X(\omega)$ and $\angle X(\omega)$ is the angle of $X(\omega)$. It is possible to go from the rectangular form to the polar form by use of the relationships

$$|X(\omega)| = \sqrt{R^2(\omega) + I^2(\omega)}$$

$$\angle X(\omega) = \begin{cases} \tan^{-1}\dfrac{I(\omega)}{R(\omega)}, & R(\omega) \geq 0 \\[2mm] \pi + \tan^{-1}\dfrac{I(\omega)}{R(\omega)}, & R(\omega) < 0 \end{cases}$$

Note that if $x(t)$ is real valued, by (3.30)

$$X(-\omega) = \overline{X(\omega)} = \text{complex conjugate of } X(\omega)$$

Then taking the complex conjugate of the polar form (3.34) gives

$$\overline{X(\omega)} = |X(\omega)| \exp[-j\angle X(\omega)]$$

Thus

$$X(-\omega) = |X(\omega)| \exp[-j\angle X(\omega)]$$

which implies that

$$|X(-\omega)| = |X(\omega)|$$

$$\angle X(-\omega) = -\angle X(\omega)$$

This result shows that $|X(\omega)|$ is an even function of ω and therefore is symmetrical about $\omega = 0$, and $\angle X(\omega)$ is an odd function of ω.

3.4.2 Signals with Even or Odd Symmetry

Again, suppose that $x(t)$ has Fourier transform $X(\omega)$ with $X(\omega)$ given in the rectangular form (3.33). As noted in Section 3.2, a signal $x(t)$ is said to be even if $x(t) = x(-t)$, and the signal is said to be odd if $x(-t) = -x(t)$. If the signal $x(t)$ is even, it follows that the imaginary part $I(\omega)$ of the Fourier transform is zero and the real part $R(\omega)$ can be rewritten as

$$R(\omega) = 2\int_0^\infty x(t) \cos \omega t \, dt$$

Hence the Fourier transform of an even signal $x(t)$ is a real-valued function of ω given by

$$X(\omega) = 2 \int_0^\infty x(t) \cos \omega t \, dt \tag{3.35}$$

If the signal $x(t)$ is odd, that is, $x(t) = -x(-t)$ for all $t > 0$, the Fourier transform of $x(t)$ is a purely imaginary function of ω given by

$$X(\omega) = -j2 \int_0^\infty x(t) \sin \omega t \, dt \tag{3.36}$$

The expression (3.35) may be used to compute the Fourier transform of an even signal, and the expression (3.36) may be used to compute the Fourier transform of an odd signal.

Example 3.9   Rectangular Pulse

Given a fixed positive number τ, let $p_\tau(t)$ denote the rectangular pulse of duration τ seconds defined by

$$p_\tau(t) = \begin{cases} 1, & -\dfrac{\tau}{2} \leq t \leq \dfrac{\tau}{2} \\ 0, & \text{all other } t \end{cases}$$

The rectangular pulse $p_\tau(t)$, which is plotted in Figure 3.13, is clearly an even signal, and thus (3.35) can be used to compute the Fourier transform. Setting $x(t) = p_\tau(t)$ in (3.35) yields

$$X(\omega) = 2 \int_0^{\tau/2} (1) \cos \omega t \, dt$$

$$= \frac{2}{\omega} \left[\sin(\omega t) \Big|_{t=0}^{t=\tau/2} \right]$$

$$= \frac{2}{\omega} \sin \frac{\omega \tau}{2}$$

The Fourier transform $X(\omega)$ can be expressed in terms of the sinc function defined by

$$\text{sinc}(a\omega) = \frac{\sin(a\pi\omega)}{a\pi\omega} \quad \text{for any real number } a \tag{3.37}$$

Setting $a = \dfrac{\tau}{2\pi}$ in (3.37) gives

$$\text{sinc}\left(\frac{\tau\omega}{2\pi}\right) = \frac{2}{\tau\omega} \sin\left(\frac{\tau\omega}{2}\right)$$

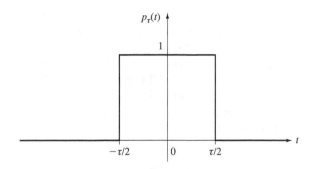

FIGURE 3.13
Rectangular pulse of τ seconds.

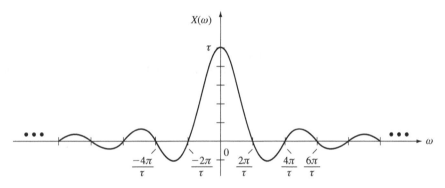

FIGURE 3.14
Fourier transform of the τ-second rectangular pulse.

and thus

$$X(\omega) = \tau \, \text{sinc}\left(\frac{\tau\omega}{2\pi}\right)$$

Note that, since the Fourier transform $X(\omega)$ in this example is real valued, $X(\omega)$ can be plotted versus ω. The result is displayed in Figure 3.14.

3.4.3   Bandlimited Signals

A signal $x(t)$ is said to be *bandlimited* if its Fourier transform $X(\omega)$ is zero for all $\omega > B$, where B is some positive number, called the *bandwidth* of the signal. If a signal $x(t)$ is bandlimited with bandwidth B, the signal does not contain any spectral components with frequency higher than B, which justifies the use of the term *bandlimited*. It turns out that any bandlimited signal must be of infinite duration in time; that is, bandlimited signals cannot be time limited. (A signal $x(t)$ is time limited if there exists a positive number T such that $x(t) = 0$ for all $t < -T$ and $t > T$.)

If a signal $x(t)$ is not bandlimited, it is said to have *infinite bandwidth* or an *infinite spectrum*. Since bandlimited signals cannot be time limited, time-limited signals cannot be bandlimited, and thus all time-limited signals have infinite bandwidth. In addition, since all (physical) signals are time limited, any such signal must have an infinite bandwidth. However, for any well-behaved time-limited signal $x(t)$, it can be proved that the Fourier transform $X(\omega)$ converges to zero as $\omega \to \infty$. Therefore, for any time-limited signal arising in practice, it is always possible to assume that $|X(\omega)| \approx 0$ for all $\omega > B$, where B is chosen to be suitably large.

Example 3.10   Frequency Spectrum

Again consider the rectangular pulse function $x(t) = p_\tau(t)$. In Example 3.9 it was shown that the Fourier transform $X(\omega)$ is

$$X(\omega) = \tau \operatorname{sinc}\left(\frac{\tau\omega}{2\pi}\right)$$

The plots of the amplitude and phase spectra for this example are given in Figure 3.15. From Figure 3.15a, it is clear that the spectrum of the rectangular pulse is infinite; however, since the

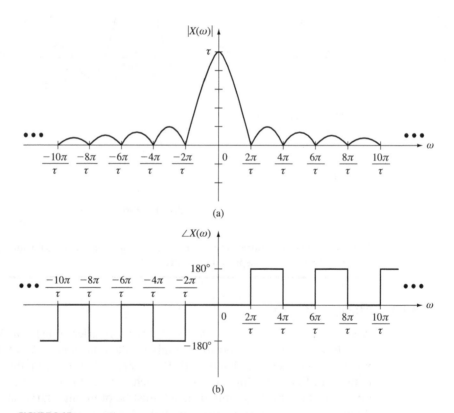

(a)

(b)

FIGURE 3.15
(a) Amplitude and (b) phase spectra of the rectangular pulse.

sidelobes shown in Figure 3.15a decrease in magnitude as the frequency ω is increased, it is clear that for any $c > 0$, there is a B (in general, depending on c) such that $|X(\omega)| < c$ for all $\omega > B$. So, if B is chosen to be sufficiently large, the rectangular pulse can be viewed as being "approximately bandlimited" with bandwidth B. Note also that if the time duration τ of the rectangular pulse is made smaller, the amplitude spectrum "spreads out." This result shows that shorter-time-duration signals (e.g., a pulse with smaller time duration) have more spectral content at higher frequencies than longer-time-duration signals.

3.4.4   Inverse Fourier Transform

Given a signal $x(t)$ with Fourier transform $X(\omega)$, $x(t)$ can be recomputed from $X(\omega)$ by application of the inverse Fourier transform given by

$$x(t) = \frac{1}{2\pi} \int_{-\infty}^{\infty} X(\omega) e^{j\omega t} \, d\omega \tag{3.38}$$

To denote the fact that $X(\omega)$ is the Fourier transform of $x(t)$, or that $X(\omega)$ is the inverse Fourier transform of $x(t)$, the transform pair notation

$$x(t) \leftrightarrow X(\omega)$$

will sometimes be used. One of the most fundamental transform pairs in the Fourier theory is the pair

$$p_\tau(t) \leftrightarrow \tau \operatorname{sinc} \frac{\tau \omega}{2\pi} \tag{3.39}$$

The transform pair (3.39) follows from the results of Example 3.9. Note that by (3.39), a rectangular function in time corresponds to a sinc function in frequency, and conversely, a sinc function in frequency corresponds to a rectangular function in time.

It is sometimes possible to compute the Fourier transform or the inverse Fourier transform without having to evaluate the integrals in (3.30) and (3.38). In particular, it is possible to derive new transform pairs from a given transform pair [such as (3.39)] by use of the properties of the Fourier transform. These properties are given in Section 3.6.

3.5   SPECTRAL CONTENT OF COMMON SIGNALS

As mentioned in Example 3.10, the Fourier transform can be used to determine the spectral content of a signal. However, the computation of the Fourier transform is often tedious for all but the simplest functions. This section uses the MATLAB Symbolic Math Toolbox to compute the Fourier transform of several common signals so that their spectral content can be compared. There are MATLAB commands for computing both the Fourier transform and the inverse Fourier transform, `fourier(f)` and `ifourier(F)`, where both `f` and `F` are symbolic objects that have been defined. These commands actually use the command `int`. For example, `fourier` uses the command `int(f*exp(-i*w*x),-inf,inf)`, where the limits of integration are $-\infty$ and ∞, and the Fourier transform of f is defined with respect to the independent variable x. For the functions in this text, it is generally easier to use int directly.

Example 3.11   Triangular Pulse

Consider a triangular pulse function shown in Figure 3.16a and given by the expression

$$x(t) = \begin{cases} 1 - \dfrac{2|t|}{\tau} & -\dfrac{\tau}{2} \le t \le \dfrac{\tau}{2} \\[3mm] 0 & otherwise \end{cases}$$

where $\tau = 1$. The Fourier transform is computed by the following commands:

```
syms x t w X
tau = 1;
X = int((1-2*abs(t)/tau)*exp(-i*w*t),t,-tau/2,tau/2);
X = simplify(X)
```

This results in the answer

```
X = -4*(cos(1/2*w)-1)/w^2
```

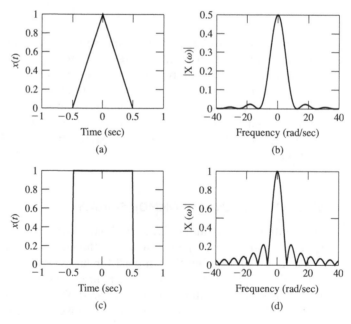

FIGURE 3.16
(a) Triangular pulse and corresponding magnitude of the Fourier transform in (b);
(c) rectangular pulse and magnitude of corresponding Fourier transform in (d).

The magnitude of the Fourier transform for the triangular pulse is shown in Figure 3.16b. For a comparison, the rectangular pulse of duration 1 sec. is shown in Figure 3.16c, with its corresponding spectrum shown in Figure 3.16d. In the time domain, high frequency content is characterized by the signal changing quickly, such as the discontinuity in the rectangular pulse. While there are sidelobes in the spectrums of both the triangular pulse and the rectangular pulse, the sidelobes in the rectangular pulse have higher amplitude, indicating higher frequency content. Also, the main lobe of the triangular pulse is wider, indicating more energy is concentrated in the low-frequency part of the spectrum.

From Example 3.11, it is shown that faster transitions in the time domain indicate higher-frequency content. This same concept was discussed in Example 3.10, where it was noted that if the time duration τ of the rectangular pulse is made smaller, the amplitude spectrum "spreads out." This is consistent with the notion of a quicker transition between the up and the down discontinuities, indicating higher frequency content. In the limit as the time duration of the pulse becomes infinitesimal, the amplitude spectrum becomes a constant over all frequencies, indicating as much high frequency content in the signal as low frequency.

The exponential decay signal examined in Example 3.8 can be viewed as well, from a frequency content perspective. The constant $b > 0$ in the expression for $x(t)$, $x(t) = e^{-bt}u(t)$, is the rate of decay. The larger the value of b is, the faster will be the decay in the signal, indicating fast transitions in the time domain. The amplitude spectrum is the plot of

$$|X(\omega)| = \frac{1}{\sqrt{b^2 + \omega^2}}$$

As b gets larger, the spectrum spreads out, indicating higher frequency content in the signal.

Example 3.12   Decaying Sinusoid

Consider a decaying sinusoid of the form

$$x(t) = e^{-at}\sin(b\pi t)u(t)$$

Figure 3.17a displays $x(t)$ for the case $a = 2$ and $b = 2$. The following commands are used to determine the Fourier transform of $x(t)$ for this case:

```
syms x t X omega
a = 2; b = 2;
x = exp(-a*t)*sin(b*pi*t);
X = simplify(int(x*exp(-j*omega*t),t,0,inf));
```

This yields the result

```
X =
2*pi/(4-w^2+4*pi^2+4*i*w)
```

The equation form of $X(\omega)$ is

$$X(\omega) = \frac{2\pi}{4 - \omega^2 + 4\pi^2 + 4j\omega}$$

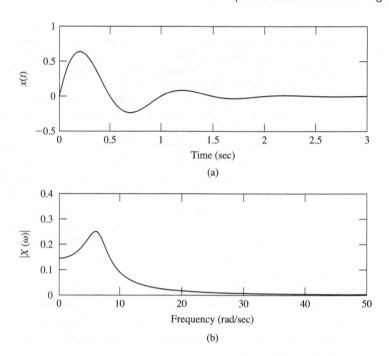

FIGURE 3.17
(a) Plot of $x(t) = e^{-2t} \sin(2\pi t)u(t)$, and (b) corresponding amplitude spectrum.

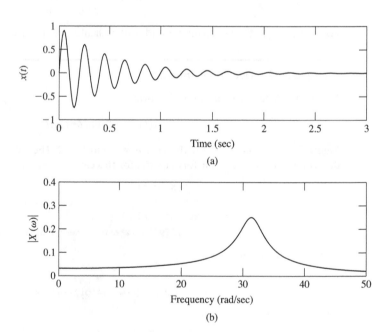

FIGURE 3.18
(a) Plot of $x(t) = e^{-2t} \sin(10\pi t)u(t)$, and (b) corresponding amplitude spectrum.

The corresponding amplitude spectrum is shown in Figure 3.17b. The dominant frequency components in $x(t)$ are in the frequency range surrounding the peak in the spectrum. This peak corresponds to the frequency of the sinusoid, which is equal to 2π.

The program is rerun with values $a = 2$ and $b = 10$, yielding an expression for $X(\omega)$ given by

$$X(\omega) = \frac{10\pi}{4 - \omega^2 + 100\pi^2 + 4j\omega}$$

The time domain signal $x(t)$ is shown in Figure 3.18a, and the corresponding amplitude spectrum is shown in Figure 3.18b. Comparing the plots of Figure 3.17a and Figure 3.18a, the higher-frequency content when $b = 10$ is evident in the faster transitions in the time domain signal. The spectrum for the case of $b = 10$ also shows that the peak has shifted to a higher frequency range, centered near the sinusoidal frequency of 10π.

3.6   PROPERTIES OF THE FOURIER TRANSFORM

The Fourier transform satisfies a number of properties that are useful in a wide range of applications. These properties are given in this section. In Section 3.8, some of these properties are applied to the study of modulation.

3.6.1   Linearity

The Fourier transform is a linear operation; that is, if $x(t) \leftrightarrow X(\omega)$ and $v(t) \leftrightarrow V(\omega)$, then for any real or complex scalars a, b,

$$ax(t) + bv(t) \leftrightarrow aX(\omega) + bV(\omega) \tag{3.40}$$

The property of linearity can be proved by computing the Fourier transform of $ax(t) + bv(t)$: By definition of the Fourier transform,

$$ax(t) + bv(t) \leftrightarrow \int_{-\infty}^{\infty} [ax(t) + bv(t)]e^{-j\omega t}\, dt$$

By linearity of integration,

$$\int_{-\infty}^{\infty} [ax(t) + bv(t)]e^{-j\omega t}\, dt = a\int_{-\infty}^{\infty} x(t)e^{-j\omega t}\, dt + b\int_{-\infty}^{\infty} v(t)e^{-j\omega t}\, dt$$

and thus

$$ax(t) + bv(t) \leftrightarrow aX(\omega) + bV(\omega)$$

Example 3.13   Sum of Rectangular Pulses

Consider the signal shown in Figure 3.19. As illustrated in the figure, this signal is equal to a sum of two rectangular pulse functions. More precisely,

$$x(t) = p_4(t) + p_2(t)$$

Then, by using linearity and the transform pair (3.39), it follows that the Fourier transform of $x(t)$ is

$$X(\omega) = 4 \operatorname{sinc} \frac{2\omega}{\pi} + 2 \operatorname{sinc} \frac{\omega}{\pi}$$

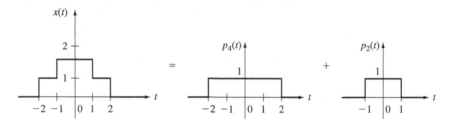

FIGURE 3.19
Signal in Example 3.13.

3.6.2   Left or Right Shift in Time

If $x(t) \leftrightarrow X(\omega)$, then for any positive or negative real number c,

$$x(t - c) \leftrightarrow X(\omega)e^{-j\omega c} \tag{3.41}$$

Note that if $c > 0$, then $x(t - c)$ is a c-second right shift of $x(t)$; and if $c < 0$, then $x(t - c)$ is a $(-c)$-second left shift of $x(t)$. Thus the transform pair (3.41) is valid for both left and right shifts of $x(t)$.

To verify the validity of the transform pair (3.41), first apply the definition of the Fourier transform to the shifted signal $x(t - c)$, which gives

$$x(t - c) \leftrightarrow \int_{-\infty}^{\infty} x(t - c)e^{-j\omega t} \, dt \tag{3.42}$$

In the integral in (3.42), consider the change of variable $\bar{t} = t - c$. Then $t = \bar{t} + c$, $dt = d\bar{t}$, and (3.42) becomes

$$x(t - c) \leftrightarrow \int_{-\infty}^{\infty} x(\bar{t})e^{-j\omega(\bar{t} + c)} \, d\bar{t}$$

$$\leftrightarrow \left[\int_{-\infty}^{\infty} x(\bar{t})e^{-j\omega\bar{t}} \, d\bar{t} \right] e^{-j\omega c}$$

$$\leftrightarrow X(\omega)e^{-j\omega c}$$

Hence, (3.41) is obtained.

Example 3.14   Right Shift of Pulse

The signal $x(t)$ shown in Figure 3.20 is equal to a 1-second right shift of the rectangular pulse function $p_2(t)$; that is,

$$x(t) = p_2(t - 1)$$

The Fourier transform $X(\omega)$ of $x(t)$ can be computed with the time-shift property (3.41) and the transform pair (3.39). The result is

$$X(\omega) = 2\left(\operatorname{sinc}\frac{\omega}{\pi} \right)e^{-j\omega}$$

Note that, since

$$|e^{-j\omega}| = 1 \quad \text{for all } \omega$$

the amplitude spectrum $|X(\omega)|$ of $x(t) = p_2(t - 1)$ is the same as the amplitude spectrum of $p_2(t)$.

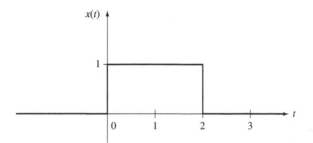

FIGURE 3.20
Signal in Example 3.14.

3.6.3   Time Scaling

If $x(t) \leftrightarrow X(\omega)$, for any positive real number a,

$$x(at) \leftrightarrow \frac{1}{a}X\left(\frac{\omega}{a}\right) \tag{3.43}$$

To prove (3.43), first apply the definition of the Fourier transform to $x(at)$, which gives

$$x(at) \leftrightarrow \int_{-\infty}^{\infty} x(at)e^{-j\omega t}\, dt \tag{3.44}$$

In the integral in (3.44), consider the change of variable $\bar{t} = at$. Then $t = \bar{t}/a$, $d\bar{t} = a(dt)$, and (3.44) becomes

$$x(at) \leftrightarrow \int_{-\infty}^{\infty} x(\bar{t}) \exp\left[-j\left(\frac{\omega}{a}\right)\bar{t}\right]\frac{1}{a}d\bar{t}$$

$$\leftrightarrow \frac{1}{a}\int_{-\infty}^{\infty} x(\bar{t}) \exp\left[-j\left(\frac{\omega}{a}\right)\bar{t}\right] d\bar{t}$$

$$\leftrightarrow \frac{1}{a}X\left(\frac{\omega}{a}\right)$$

Hence (3.43) is verified.

Given an arbitrary signal $x(t)$, if $a > 1$, $x(at)$ is a *time compression* of $x(t)$. For example, suppose that $x(t)$ is the 2-second rectangular pulse $p_2(t)$ and $a = 2$. The signals $p_2(t)$ and $p_2(2t)$ are displayed in Figure 3.21. Clearly, $p_2(2t)$ is a time compression of $p_2(t)$.

Now by (3.39), the Fourier transform of $p_2(t)$ is equal to $2 \text{ sinc}(\omega/\pi)$, and by (3.43) the Fourier transform of $p_2(2t)$ is equal to $\text{sinc}(\omega/2\pi)$. These transforms are displayed in Figure 3.22. As seen from this figure, the Fourier transform of $p_2(2t)$ is a

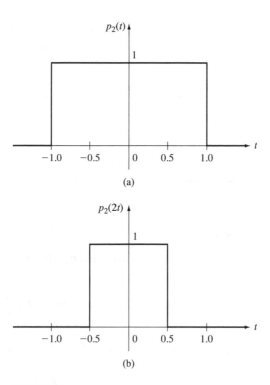

FIGURE 3.21
Signals (a) $p_2(t)$, and (b) $p_2(2t)$.

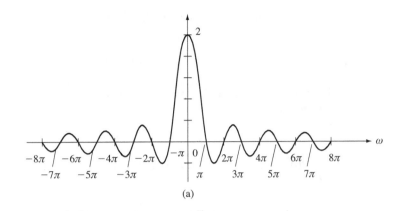

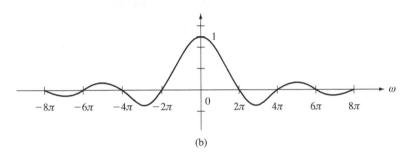

FIGURE 3.22
Fourier transforms of (a) $p_2(t)$, and (b) $p_2(2t)$.

frequency expansion of the Fourier transform of $p_2(t)$. Hence, the shorter-duration pulse $p_2(2t)$ has a wider bandwidth than the longer-duration pulse $p_2(t)$.

For an arbitrary signal $x(t)$ with Fourier transform $X(\omega)$, if $a > 1$, $X(\omega/a)$ is a frequency expansion of $X(\omega)$. Thus, by (3.43) it is seen that a time compression of a signal $x(t)$ corresponds to a frequency expansion of the Fourier transform $X(\omega)$ of the signal. This again shows that shorter-time-duration signals have wider bandwidths than those of longer-time-duration signals.

Again, let $x(t)$ be an arbitrary signal with Fourier transform $X(\omega)$. If $0 < a < 1$, $x(at)$ is a *time expansion* of $x(t)$ and $X(\omega/a)$ is a *frequency compression* of $X(\omega)$. In this case, it follows from (3.43) that a time expansion of $x(t)$ corresponds to a frequency compression of $X(\omega)$. Thus, longer-duration signals have smaller bandwidths.

3.6.4 Time Reversal

Given a signal $x(t)$, consider the time-reversed signal $x(-t)$. The signal $x(-t)$ is equal to $x(t)$ folded about the vertical axis. Now, if $x(t) \leftrightarrow X(\omega)$, then

$$x(-t) \leftrightarrow X(-\omega) \tag{3.45}$$

To prove (3.45), simply replace t by $-t$ in the definition of the Fourier transform of $x(t)$.

If the signal $x(t)$ is real valued, from the definition (3.30) of the Fourier transform it follows that

$$X(-\omega) = \overline{X(\omega)}$$

where $\overline{X(\omega)}$ is the complex conjugate of $X(\omega)$. Hence, the transform pair (3.45) can be rewritten as

$$x(-t) \leftrightarrow \overline{X(\omega)} \qquad (3.46)$$

By (3.46), time reversal in the time domain corresponds to conjugation in the frequency domain.

Example 3.15   Time-Reversed Exponential Signal

Given a real number $b > 0$, consider the signal

$$x(t) = \begin{cases} 0, & t > 0 \\ e^{bt}, & t \leq 0 \end{cases}$$

Note that

$$x(-t) = e^{-bt}u(t)$$

and from the result in Example 3.8, the Fourier transform of $x(-t)$ is $1/(b + j\omega)$. Hence, the Fourier transform of $x(t)$ is

$$X(\omega) = \overline{\frac{1}{b + j\omega}} = \frac{1}{b - j\omega}$$

3.6.5   Multiplication by a Power of t

If $x(t) \leftrightarrow X(\omega)$, for any positive integer n,

$$t^n x(t) \leftrightarrow (j)^n \frac{d^n}{d\omega^n} X(\omega) \qquad (3.47)$$

Setting $n = 1$ in (3.47) yields the result that multiplication by t in the time domain corresponds to differentiation with respect to ω in the frequency domain (plus multiplication by j).

To prove (3.47) for the case $n = 1$, start with the following definition of the Fourier transform:

$$X(\omega) = \int_{-\infty}^{\infty} x(t)e^{-j\omega t}\, dt \qquad (3.48)$$

Differentiating both sides of (3.48) with respect to ω and multiplying by j yield

$$j\frac{dX(\omega)}{d\omega} = j\int_{-\infty}^{\infty}(-jt)x(t)e^{-j\omega t}\,dt$$

$$j\frac{dX(\omega)}{d\omega} = \int_{-\infty}^{\infty}tx(t)e^{-j\omega t}\,dt \tag{3.49}$$

The right-hand side of (3.49) is equal to the Fourier transform of $tx(t)$, and thus (3.47) is verified for the case $n = 1$. The proof for $n \geq 2$ follows by taking second- and higher-order derivatives of $X(\omega)$ with respect to ω. The details are omitted.

Example 3.16   Product of *t* and a Pulse

Let $x(t) = tp_2(t)$, which is plotted in Figure 3.23. The Fourier transform $X(\omega)$ of $x(t)$ can be computed by use of the property (3.47) and the transform pair (3.39). This yields

$$X(\omega) = j\frac{d}{d\omega}\left(2\operatorname{sinc}\frac{\omega}{\pi}\right)$$

$$= j2\frac{d}{d\omega}\left(\frac{\sin \omega}{\omega}\right)$$

$$= j2\frac{\omega\cos \omega - \sin \omega}{\omega^2}$$

The amplitude spectrum $|X(\omega)|$ is plotted in Figure 3.24.

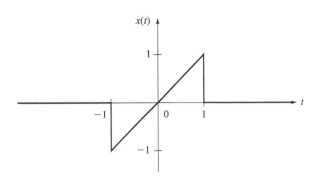

FIGURE 3.23
The signal $x(t) = tp_2(t)$.

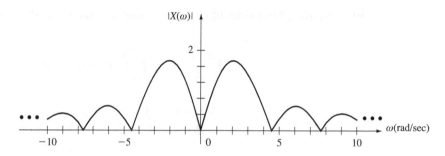

FIGURE 3.24
Amplitude spectrum of the signal in Example 3.16.

3.6.6   Multiplication by a Complex Exponential

If $x(t) \leftrightarrow X(\omega)$, then for any real number ω_0,

$$x(t)e^{j\omega_0 t} \leftrightarrow X(\omega - \omega_0) \tag{3.50}$$

So, multiplication by a complex exponential in the time domain corresponds to a frequency shift in the frequency domain. The proof of (3.50) follows directly from the definition of the Fourier transform. The verification is left to the reader.

3.6.7   Multiplication by a Sinusoid

If $x(t) \leftrightarrow X(\omega)$, then for any real number ω_0,

$$x(t)\sin(\omega_0 t) \leftrightarrow \frac{j}{2}[X(\omega + \omega_0) - X(\omega - \omega_0)] \tag{3.51}$$

$$x(t)\cos(\omega_0 t) \leftrightarrow \frac{1}{2}[X(\omega + \omega_0) + X(\omega - \omega_0)] \tag{3.52}$$

The proof of (3.51) and (3.52) follow from (3.50) and Euler's identity. The details are omitted.

As discussed in Section 3.8, the signals $x(t)\sin(\omega_0 t)$ and $x(t)\cos(\omega_0 t)$ can be viewed as amplitude-modulated signals. More precisely, in forming the signal $x(t)\sin(\omega_0 t)$, the carrier $\sin \omega_0 t$ is modulated by the signal $x(t)$. As a result of this characterization of $x(t)\sin(\omega_0 t)$ [and $x(t)\cos(\omega_0 t)$], the relationships (3.51) and (3.52) are called the *modulation theorems* of the Fourier transform. The relationships (3.51) and (3.52) show that modulation of a carrier by a signal $x(t)$ results in the frequency translations $X(\omega + \omega_0)$, $X(\omega - \omega_0)$ of the Fourier transform $X(\omega)$.

Example 3.17   Sinusoidal Burst

Consider the signal $x(t) = p_\tau(t)\cos(\omega_0 t)$, which can be interpreted as a sinusoidal burst. For the case when $\tau = 0.5$ and $\omega_0 = 60$ rad/sec, the signal is plotted in Figure 3.25. By the modulation

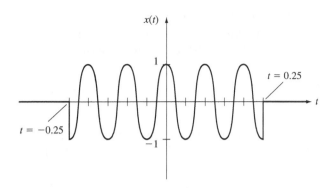

FIGURE 3.25
Sinusoidal burst.

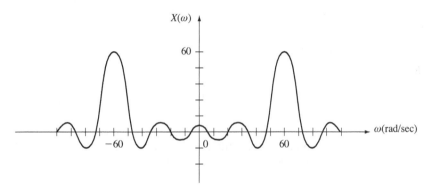

FIGURE 3.26
Fourier transform of the sinusoidal burst $x(t) = p_{0.5}(t)\cos(60t)$.

property (3.52) and the transform pair (3.39), the Fourier transform of the sinusoidal burst is equal to

$$\frac{1}{2}\left[\tau\,\text{sinc}\left(\frac{\tau(\omega + \omega_0)}{2\pi}\right) + \tau\,\text{sinc}\left(\frac{\tau(\omega - \omega_0)}{2\pi}\right)\right]$$

For the case $\tau = 0.5$ and $\omega_0 = 60$ rad/sec, the transform of the sinusoidal burst is plotted in Figure 3.26.

3.6.8   Differentiation in the Time Domain

If $x(t) \leftrightarrow X(\omega)$, then for any positive integer n,

$$\frac{d^n}{dt^n}x(t) \leftrightarrow (j\omega)^n X(\omega) \tag{3.53}$$

For the case $n = 1$, it follows from (3.53) that differentiation in the time domain corresponds to multiplication by $j\omega$ in the frequency domain. To prove (3.53) for the case $n = 1$, first observe that the Fourier transform of $dx(t)/dt$ is

$$\int_{-\infty}^{\infty} \frac{dx(t)}{dt} e^{-j\omega t}\, dt \tag{3.54}$$

The integral in (3.54) can be computed "by parts" as follows: With $v = e^{-j\omega t}$ and $w = x(t)$, $dv = -j\omega e^{-j\omega t}\, dt$ and $dw = [dx(t)/dt]\, dt$. Then,

$$\int_{-\infty}^{\infty} \frac{dx(t)}{dt} e^{-j\omega t}\, dt = vw\big|_{t=-\infty}^{t=\infty} - \int_{-\infty}^{\infty} w\, dv$$

$$= e^{-j\omega t} x(t)\big|_{t=-\infty}^{t=\infty} - \int_{-\infty}^{\infty} x(t)(-j\omega) e^{-j\omega t}\, dt$$

Then, if $x(t) \to 0$ as $t \to \pm\infty$,

$$\int_{-\infty}^{\infty} \frac{dx(t)}{dt} e^{-j\omega t}\, dt = (j\omega) X(\omega)$$

and thus (3.53) is valid for the case $n = 1$. The proof of (3.53) for $n \geq 2$ follows by repeated application of integration by parts.

3.6.9   Integration in the Time Domain

Given a signal $x(t)$, the integral of $x(t)$ is the function

$$\int_{-\infty}^{t} x(\lambda)\, d\lambda$$

Suppose that $x(t)$ has Fourier transform $X(\omega)$. In general, the integral of $x(t)$ does not have a Fourier transform in the ordinary sense, but it does have the generalized transform

$$\frac{1}{j\omega} X(\omega) + \pi X(0)\delta(\omega)$$

where $\delta(\omega)$ is the impulse function in the frequency domain. This results in the transform pair

$$\int_{-\infty}^{t} x(\lambda)\, d\lambda \leftrightarrow \frac{1}{j\omega} X(\omega) + \pi X(0)\delta(\omega) \tag{3.55}$$

Note that if the signal $x(t)$ has no dc component (e.g., $X(0) = 0$), then (3.55) reduces to

$$\int_{-\infty}^{t} x(\lambda)\, d\lambda \leftrightarrow \frac{1}{j\omega} X(\omega)$$

Hence, the second term on the right-hand side of the transform pair (3.55) is due to a possible dc component in $x(t)$.

Example 3.18   Transform of a Triangular Pulse

Consider the triangular pulse function $v(t)$ displayed in Figure 3.27. As first noted in Chapter 1, the triangular pulse can be expressed mathematically by

$$v(t) = \left(1 - \frac{2|t|}{\tau}\right) p_\tau(t)$$

where again $p_\tau(t)$ is the rectangular pulse of duration τ seconds. To compute the Fourier transform $V(\omega)$ of $v(t)$, the Fourier transform of the derivative of $v(t)$ will be computed first. Then, by the integration property (3.55), it will be possible to determine $V(\omega)$.

The derivative of $v(t)$, which is denoted by $x(t)$, is shown in Figure 3.28. From the plot, it is clear that the derivative can be expressed mathematically as

$$x(t) = \frac{2}{\tau} p_{\tau/2}\left(t + \frac{\tau}{4}\right) - \frac{2}{\tau} p_{\tau/2}\left(t - \frac{\tau}{4}\right)$$

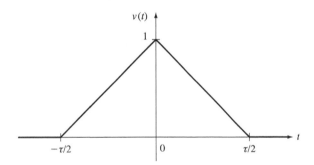

FIGURE 3.27
Triangular pulse.

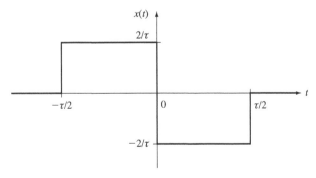

FIGURE 3.28
Derivative of the triangular pulse.

The Fourier transform $X(\omega)$ of $x(t)$ can be computed by the transform pair (3.39) and the shift property (3.41). This yields

$$X(\omega) = \left(\operatorname{sinc} \frac{\tau\omega}{4\pi} \right)\left[\exp\left(\frac{j\tau\omega}{4} \right) - \exp\left(-\frac{j\tau\omega}{4} \right) \right]$$

$$= \left(\operatorname{sinc} \frac{\tau\omega}{4\pi} \right)\left(j2 \sin \frac{\tau\omega}{4} \right)$$

Now, since $v(t)$ is the integral of $x(t)$, by the integration property (3.55), the Fourier transform $V(\omega)$ of $v(t)$ is

$$V(\omega) = \frac{1}{j\omega}\left(\operatorname{sinc} \frac{\tau\omega}{4\pi} \right)\left(j2 \sin \frac{\tau\omega}{4} \right) + \pi X(0)\delta(\omega)$$

$$= \frac{2}{\omega}\left[\frac{\sin(\tau\omega/4)}{\tau\omega/4} \right] \sin \frac{\tau\omega}{4}$$

$$= \frac{\tau}{2} \frac{\sin^2(\tau\omega/4)}{(\tau\omega/4)^2}$$

$$= \frac{\tau}{2} \operatorname{sinc}^2 \frac{\tau\omega}{4\pi}$$

Hence, the end result is the transform pair

$$\left(1 - \frac{2|t|}{\tau} \right) p_\tau(t) \leftrightarrow \frac{\tau}{2} \operatorname{sinc}^2 \frac{\tau\omega}{4\pi} \tag{3.56}$$

By (3.56), it is seen that the triangular pulse in the time domain corresponds to a sinc-squared function in the Fourier transform domain. In the case $\tau = 1$, the Fourier transform of the triangular pulse is plotted in Figure 3.29.

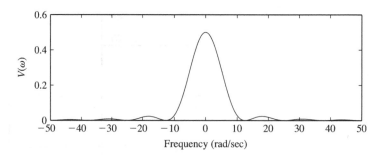

FIGURE 3.29
Fourier transform of the 1-second triangular pulse.

3.6.10 Convolution in the Time Domain

Given two signals $x(t)$ and $v(t)$ with Fourier transforms $X(\omega)$ and $V(\omega)$, the Fourier transform of the convolution $x(t)*v(t)$ is equal to the product $X(\omega)V(\omega)$, which results in the transform pair

$$x(t)*v(t) \leftrightarrow X(\omega)V(\omega) \tag{3.57}$$

Note that by (3.57), convolution in the time domain corresponds to multiplication in the frequency domain. In Chapter 5 it will be seen that this property is very useful in determining the relationship between the input and the output of a linear time-invariant continuous-time system.

To prove (3.57), first recall that by definition of convolution,

$$x(t)*v(t) = \int_{-\infty}^{\infty} x(\lambda)v(t-\lambda)\,d\lambda$$

Hence, the Fourier transform of $x(t)*v(t)$ is given by

$$\int_{-\infty}^{\infty}\left[\int_{-\infty}^{\infty} x(\lambda)v(t-\lambda)\,d\lambda\right]e^{-j\omega t}\,dt$$

This can be rewritten in the form

$$\int_{-\infty}^{\infty} x(\lambda)\left[\int_{-\infty}^{\infty} v(t-\lambda)e^{-j\omega t}\,dt\right]d\lambda$$

Using the change of variable $\bar{t} = t - \lambda$ in the second integral gives

$$\int_{-\infty}^{\infty} x(\lambda)\left[\int_{-\infty}^{\infty} v(\bar{t})e^{-j\omega(\bar{t}+\lambda)}\,d\bar{t}\right]d\lambda$$

This expression can be rewritten in the form

$$\left[\int_{-\infty}^{\infty} x(\lambda)e^{-j\omega\lambda}\,d\lambda\right]\left[\int_{-\infty}^{\infty} v(\bar{t})e^{-j\omega\bar{t}}\,d\bar{t}\right]$$

Clearly, the preceding expression is equal to $X(\omega)V(\omega)$, and thus (3.57) is verified.

3.6.11 Multiplication in the Time Domain

If $x(t) \leftrightarrow X(\omega)$ and $v(t) \leftrightarrow V(\omega)$, then

$$x(t)v(t) \leftrightarrow \frac{1}{2\pi}[X(\omega)*V(\omega)] = \frac{1}{2\pi}\int_{-\infty}^{\infty} X(\lambda)V(\omega-\lambda)\,d\lambda \tag{3.58}$$

From (3.58) it is seen that multiplication in the time domain corresponds to convolution in the Fourier transform domain. The proof of (3.58) follows from the definition of the Fourier transform and the manipulation of integrals. The details are omitted.

3.6.12 Parseval's Theorem

Again, suppose that $x(t) \leftrightarrow X(\omega)$ and $v(t) \leftrightarrow V(\omega)$. Then,

$$\int_{-\infty}^{\infty} x(t)v(t)\, dt = \frac{1}{2\pi} \int_{-\infty}^{\infty} \overline{X(\omega)} V(\omega)\, d\omega \tag{3.59}$$

where $\overline{X(\omega)}$ is the complex conjugate of $X(\omega)$. The relationship (3.59), which is called *Parseval's theorem,* follows directly from the transform pair (3.58). To see this, first note that the Fourier transform of the product $x(t)v(t)$ is equal to

$$\int_{-\infty}^{\infty} x(t)v(t)e^{-j\omega t}\, dt$$

But, by the transform pair (3.58), the Fourier transform of $x(t)v(t)$ is equal to

$$\frac{1}{2\pi} \int_{-\infty}^{\infty} X(\omega - \lambda)V(\lambda)\, d\lambda$$

Thus,

$$\int_{-\infty}^{\infty} x(t)v(t)e^{-j\omega t}\, dt = \frac{1}{2\pi} \int_{-\infty}^{\infty} X(\omega - \lambda)V(\lambda)\, d\lambda \tag{3.60}$$

The relationship (3.60) must hold for all real values of ω. Taking $\omega = 0$ gives

$$\int_{-\infty}^{\infty} x(t)v(t)\, dt = \frac{1}{2\pi} \int_{-\infty}^{\infty} X(-\lambda)V(\lambda)\, d\lambda \tag{3.61}$$

If $x(t)$ is real valued, $X(-\omega) = \overline{X(\omega)}$, and thus changing the variable of integration from λ to ω on the right-hand side of (3.61) results in (3.59).

Note that if $v(t) = x(t)$, Parseval's theorem becomes

$$\int_{-\infty}^{\infty} x^2(t)\, dt = \frac{1}{2\pi} \int_{-\infty}^{\infty} \overline{X(\omega)} X(\omega)\, d\omega \tag{3.62}$$

From the properties of complex numbers,

$$\overline{X(\omega)} X(\omega) = |X(\omega)|^2$$

and thus (3.62) can be written in the form

$$\int_{-\infty}^{\infty} x^2(t)\, dt = \frac{1}{2\pi} \int_{-\infty}^{\infty} |X(\omega)|^2\, d\omega \tag{3.63}$$

The left-hand side of (3.63) can be interpreted as the energy of the signal $x(t)$. Thus, (3.63) relates the energy of the signal and the integral of the square of the magnitude of the Fourier transform of the signal.

3.6.13 Duality

Suppose that $x(t) \leftrightarrow X(\omega)$. A new continuous-time signal can be defined by setting $\omega = t$ in $X(\omega)$. This results in the continuous-time signal $X(t)$. The duality property states that the Fourier transform of $X(t)$ is equal to $2\pi x(-\omega)$; that is,

$$X(t) \leftrightarrow 2\pi x(-\omega) \tag{3.64}$$

In (3.64), $x(-\omega)$ is the frequency function constructed by setting $t = -\omega$ in the expression for $x(t)$.

For any given transform pair $x(t) \leftrightarrow X(\omega)$, by the use of duality, the new transform pair (3.64) can be constructed. For example, applying the duality property to the pair (3.39) yields the transform pair

$$\tau \operatorname{sinc}\frac{\tau t}{2\pi} \leftrightarrow 2\pi p_\tau(-\omega) \tag{3.65}$$

Since $p_\tau(\omega)$ is an even function of ω, $p_\tau(-\omega) = p_\tau(\omega)$, and (3.65) can be rewritten as

$$\tau \operatorname{sinc}\frac{\tau t}{2\pi} \leftrightarrow 2\pi p_\tau(\omega) \tag{3.66}$$

From (3.66) it is seen that a sinc function in time corresponds to a rectangular pulse function in frequency.

Applying the duality property to the transform pair (3.56) gives

$$\frac{\tau}{2}\operatorname{sinc}^2\frac{\tau t}{4\pi} \leftrightarrow 2\pi\left(1 - \frac{2|\omega|}{\tau}\right)p_\tau(\omega) \tag{3.67}$$

Thus, the sinc-squared time function has a Fourier transform equal to the triangular pulse function in frequency.

The duality property is easy to prove: First, by definition of the Fourier transform,

$$X(\omega) = \int_{-\infty}^{\infty} x(t)e^{-j\omega t}\,dt \tag{3.68}$$

Setting $\omega = t$ and $t = -\omega$ in (3.68) gives

$$X(t) = \int_{-\infty}^{\infty} x(-\omega)e^{j\omega t}\,d\omega$$

$$= \frac{1}{2\pi}\int_{-\infty}^{\infty} 2\pi x(-\omega)e^{j\omega t}\,d\omega$$

Thus, $X(t)$ is the inverse Fourier transform of the frequency function $2\pi x(-\omega)$, which proves (3.64).

For the convenience of the reader, the properties of the Fourier transform are summarized in Table 3.1.

TABLE 3.1   Properties of the Fourier Transform

Property	Transform Pair/Property
Linearity	$ax(t) + bv(t) \leftrightarrow aX(\omega) + bV(\omega)$
Right or left shift in time	$x(t - c) \leftrightarrow X(\omega)e^{-j\omega c}$
Time scaling	$x(at) \leftrightarrow \dfrac{1}{a}X\left(\dfrac{\omega}{a}\right) a > 0$
Time reversal	$x(-t) \leftrightarrow X(-\omega) = \overline{X(\omega)}$
Multiplication by a power of t	$t^n x(t) \leftrightarrow j^n \dfrac{d^n}{d\omega^n}X(\omega)\ n = 1, 2, \ldots$
Multiplication by a complex exponential	$x(t)e^{j\omega_0 t} \leftrightarrow X(\omega - \omega_0)\ \ \omega_0$ real
Multiplication by $\sin(\omega_0 t)$	$x(t)\sin(\omega_0 t) \leftrightarrow \dfrac{j}{2}[X(\omega + \omega_0) - X(\omega - \omega_0)]$
Multiplication by $\cos(\omega_0 t)$	$x(t)\cos(\omega_0 t) \leftrightarrow \frac{1}{2}[X(\omega + \omega_0) + X(\omega - \omega_0)]$
Differentiation in the time domain	$\dfrac{d^n}{dt^n}x(t) \leftrightarrow (j\omega)^n X(\omega)\ n = 1, 2, \ldots$
Integration in the time domain	$\displaystyle\int_{-\infty}^{t} x(\lambda)\,d\lambda \leftrightarrow \dfrac{1}{j\omega}X(\omega) + \pi X(0)\delta(\omega)$
Convolution in the time domain	$x(t) * v(t) \leftrightarrow X(\omega)V(\omega)$
Multiplication in the time domain	$x(t)v(t) \leftrightarrow \dfrac{1}{2\pi}X(\omega) * V(\omega)$
Parseval's theorem	$\displaystyle\int_{-\infty}^{\infty} x(t)v(t)\,dt = \dfrac{1}{2\pi}\int_{-\infty}^{\infty}\overline{X(\omega)}V(\omega)\,d\omega$
Special case of Parseval's theorem	$\displaystyle\int_{-\infty}^{\infty} x^2(t)\,dt = \dfrac{1}{2\pi}\int_{-\infty}^{\infty}|X(\omega)|^2\,d\omega$
Duality	$X(t) \leftrightarrow 2\pi x(-\omega)$

3.7   GENERALIZED FOURIER TRANSFORM

In Example 3.8 it was shown that the unit-step function $u(t)$ does not have a Fourier transform in the ordinary sense. It is also easy to see that $\cos(\omega_0 t)$ and $\sin(\omega_0 t)$ do not have a Fourier transform in the ordinary sense. Since the step function and sinusoidal functions often arise in the study of signals and systems, it is very desirable to be able to define the Fourier transform of these signals. We can do this by defining the notion of the generalized Fourier transform, which is considered in this section.

First, the Fourier transform of the unit impulse $\delta(t)$ will be computed. Recall that $\delta(t)$ is defined by

$$\delta(t) = 0, t \neq 0$$

$$\int_{-\varepsilon}^{\varepsilon} \delta(\lambda)\,d\lambda = 1, \text{all } \varepsilon > 0 \tag{3.69}$$

The Fourier transform of $\delta(t)$ is given by

$$\int_{-\infty}^{\infty} \delta(t) e^{-j\omega t} \, dt$$

Since $\delta(t) = 0$ for all $t \neq 0$,

$$\delta(t) e^{-j\omega t} = \delta(t)$$

and the Fourier transform integral reduces to

$$\int_{-\infty}^{\infty} \delta(t) \, dt$$

By (3.69) this integral is equal to 1, which results in the transform pair

$$\delta(t) \leftrightarrow 1 \tag{3.70}$$

This result shows that the frequency spectrum of $\delta(t)$ contains all frequencies with amplitude 1.

Now applying the duality property to (3.70) yields the transform pair

$$x(t) = 1, \ -\infty < t < \infty \leftrightarrow 2\pi\delta(\omega) \tag{3.71}$$

Hence, the Fourier transform of a constant signal of amplitude 1 is equal to an impulse in frequency with area 2π. But, from the results in Example 3.7, it was seen that the constant signal does not have a Fourier transform in the ordinary sense. The frequency function $2\pi\delta(\omega)$ is called the *generalized Fourier transform* of the constant signal $x(t) = 1, \ -\infty < t < \infty$.

Now consider the signal $x(t) = \cos(\omega_0 t), \ -\infty < t < \infty$, where ω_0 is a fixed, but arbitrary real number. Using the transform pair (3.71) and the modulation property reveals that $x(t)$ has the generalized Fourier transform

$$\pi[\delta(\omega + \omega_0) + \delta(\omega - \omega_0)]$$

Hence,

$$\cos(\omega_0 t) \leftrightarrow \pi[\delta(\omega + \omega_0) + \delta(\omega - \omega_0)] \tag{3.72}$$

In a similar manner, it can be shown that $\sin(\omega_0 t)$ has the generalized transform

$$j\pi[\delta(\omega + \omega_0) - \delta(\omega - \omega_0)]$$

and thus

$$\sin \omega_0 t \leftrightarrow j\pi[\delta(\omega + \omega_0) - \delta(\omega - \omega_0)] \tag{3.73}$$

The plot of the Fourier transform of $\cos(\omega_0 t)$ is given in Figure 3.30. Note that the spectrum consists of two impulses located at $\pm\omega_0$, with each impulse having area π.

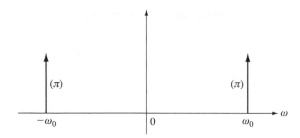

FIGURE 3.30
Fourier transform of $\cos(\omega_0 t)$.

3.7.1   Fourier Transform of a Periodic Signal

Using the transform pair (3.71) and the property (3.50) results in the transform pair

$$e^{j\omega_0 t} \leftrightarrow 2\pi\delta(\omega - \omega_0) \tag{3.74}$$

The transform pair (3.74) can be used to compute the generalized Fourier transform of a periodic signal: Let $x(t)$ be periodic for $-\infty < t < \infty$ with period T. Then $x(t)$ has the complex exponential Fourier series

$$x(t) = \sum_{k=-\infty}^{\infty} c_k e^{jk\omega_0 t} \tag{3.75}$$

where $\omega_0 = 2\pi/T$. The Fourier transform of the right-hand side of (3.75) can be taken by the use of linearity and the transform pair (3.74). This gives

$$X(\omega) = \sum_{k=-\infty}^{\infty} 2\pi c_k \delta(\omega - k\omega_0)$$

So, the Fourier transform of a periodic signal is a train of impulse functions located at $\omega = k\omega_0, k = 0, \pm 1, \pm 2, \ldots$.

3.7.2   Transform of the Unit-Step Function

The (generalized) Fourier transform of the unit step $u(t)$ can be computed by the integration property given by the transform pair (3.55). Since $u(t)$ is equal to the integral of the impulse $\delta(t)$ and the Fourier transform of $\delta(t)$ is the constant unit function, from (3.55) we see that the Fourier transform of $u(t)$ is given by

$$\frac{1}{j\omega}(1) + \pi(1)\delta(\omega) = \frac{1}{j\omega} + \pi\delta(\omega)$$

thus resulting in the transform pair

$$u(t) \leftrightarrow \frac{1}{j\omega} + \pi\delta(\omega) \tag{3.76}$$

TABLE 3.2   Common Fourier Transform Pairs

$$1, -\infty < t < \infty \leftrightarrow 2\pi\delta(\omega)$$

$$-0.5 + u(t) \leftrightarrow \frac{1}{j\omega}$$

$$u(t) \leftrightarrow \pi\delta(\omega) + \frac{1}{j\omega}$$

$$\delta(t) \leftrightarrow 1$$

$$\delta(t - c) \leftrightarrow e^{-j\omega c}, c \text{ any real number}$$

$$e^{-bt}u(t) \leftrightarrow \frac{1}{j\omega + b}, b > 0$$

$$e^{j\omega_0 t} \leftrightarrow 2\pi\delta(\omega - \omega_0), \omega_0 \text{ any real number}$$

$$p_\tau(t) \leftrightarrow \tau \operatorname{sinc} \frac{\tau\omega}{2\pi}$$

$$\tau \operatorname{sinc} \frac{\tau t}{2\pi} \leftrightarrow 2\pi p_\tau(\omega)$$

$$\left(1 - \frac{2|t|}{\tau}\right)p_\tau(t) \leftrightarrow \frac{\tau}{2}\operatorname{sinc}^2\left(\frac{\tau\omega}{4\pi}\right)$$

$$\frac{\tau}{2}\operatorname{sinc}^2\left(\frac{\tau t}{4\pi}\right) \leftrightarrow 2\pi\left(1 - \frac{2|\omega|}{\tau}\right)p_\tau(\omega)$$

$$\cos(\omega_0 t) \leftrightarrow \pi[\delta(\omega + \omega_0) + \delta(\omega - \omega_0)]$$

$$\cos(\omega_0 t + \theta) \leftrightarrow \pi[e^{-j\theta}\delta(\omega + \omega_0) + e^{j\theta}\delta(\omega - \omega_0)]$$

$$\sin(\omega_0 t) \leftrightarrow j\pi[\delta(\omega + \omega_0) - \delta(\omega - \omega_0)]$$

$$\sin(\omega_0 t + \theta) \leftrightarrow j\pi[e^{-j\theta}\delta(\omega + \omega_0) - e^{j\theta}\delta(\omega - \omega_0)]$$

In Table 3.2 a list of common Fourier transform pairs is given, which includes the pairs that were derived in this chapter.

3.8   APPLICATION TO SIGNAL MODULATION AND DEMODULATION

To illustrate the use of the Fourier transform, in this section an introduction is given to the transmission of information (in the form of a signal generated by a source) over a channel and the reception of the information by a user. The channel may consist of free space or a cable. A key component of the transmission process is the use of modulation to convert the signal source into an appropriate form for transmission over the channel. In the modulation process, some parameter of a carrier signal is varied, based on the signal that is being transmitted. There are two basic types of modulation: analog and digital. In analog modulation the parameter being varied can take on a continuous range of values, whereas in digital modulation the parameter takes on only a finite number of different possible values. After transmission over a channel,

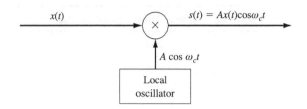

FIGURE 3.31
Amplitude modulation.

the transmitted signal is reconstructed by a receiver that uses a demodulation process to extract the original signal. Analog modulation and demodulation are considered in this section. Digital modulation and demodulation are not covered in this text. (See Proakis [2000].)

3.8.1   Analog Modulation

Let $x(t)$ be a continuous-time signal, such as an audio signal that is to be transmitted over a channel consisting of free space or a cable. As noted previously, the signal is transmitted by modulation of a carrier. The most common type of carrier is a sinusoid, given by $A \cos(\omega_c t)$, where A is the amplitude and ω_c is the frequency in rad/sec. In *amplitude modulation* (AM), the amplitude of the sinusoidal carrier is modulated by the signal $x(t)$. In one form of AM transmission, the signal $x(t)$ and carrier $A \cos(\omega_c t)$ are simply multiplied together to produce the modulated carrier $s(t) = Ax(t) \cos(\omega_c t)$. The process is illustrated in Figure 3.31. The *local oscillator* in Figure 3.31 is a device that produces the sinusoidal signal $A \cos(\omega_c t)$. The signal multiplier may be realized by the use of a nonlinear element, such as a diode.

Example 3.19   Amplitude Modulation

Suppose that $x(t)$ is the signal shown in Figure 3.32a and that the carrier is equal to $\cos(5\pi t)$. The modulated carrier $s(t) = x(t) \cos(5\pi t)$ is plotted in Figure 3.32b.

The frequency spectrum of the modulated carrier $s(t) = Ax(t) \cos(\omega_c t)$ can be determined by the use of the modulation property of the Fourier transform. First, we assume that the signal $x(t)$ is bandlimited with bandwidth B, that is,

$$|X(\omega)| = 0, \text{ for all } \omega > B$$

where $X(\omega)$ is the Fourier transform of $x(t)$. It is also assumed that $\omega_c > B$; that is, the frequency ω_c of the carrier is greater than the bandwidth B of the signal. If $x(t)$ is an audio signal, such as a music waveform, the bandwidth B can be taken to be 20 kHz, since an audio signal is not likely to contain any significant frequency components above 20 kHz.

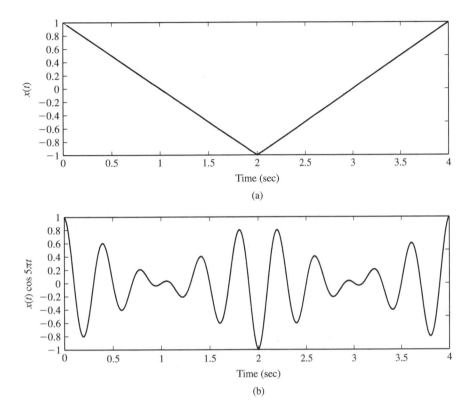

FIGURE 3.32
(a) Signal and (b) modulated carrier $Ax(t) \cos(\omega_c t)$ in Example 3.19.

Now, by the modulation property, the Fourier transform $S(\omega)$ of the modulated carrier $s(t) = Ax(t) \cos(\omega_c t)$ is given by

$$S(\omega) = \frac{A}{2}[X(\omega + \omega_c) + X(\omega - \omega_c)]$$

This result shows that the modulation process translates the Fourier transform $X(\omega)$ of $x(t)$ up to the frequency range from $\omega_c - B$ to $\omega_c + B$ (and to the negative frequency range from $-\omega_c - B$ to $-\omega_c + B$). For example, if the transform $X(\omega)$ has the shape shown in Figure 3.33a, then the transform of the modulated carrier has the form shown in Figure 3.33b. As illustrated, the portion of $X(\omega - \omega_c)$ from $\omega_c - B$ to ω_c is called the lower *sideband*, and the portion of $X(\omega - \omega_c)$ from ω_c to $\omega_c + B$ is called the upper *sideband*. Each sideband contains all the spectral components of the signal $x(t)$.

A key property of amplitude modulation in the transmission of a signal $x(t)$ is the up-conversion of the spectrum of $x(t)$. The higher-frequency range of the modulated carrier makes it possible to achieve good propagation properties in transmission through cable or free space. For example, in optical communications, a beam of light is

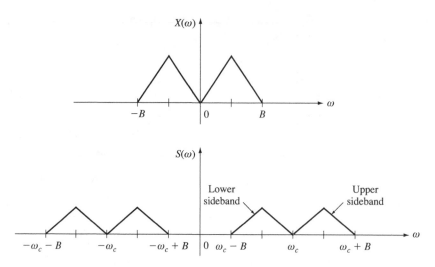

FIGURE 3.33
Fourier transform of (a) signal $x(t)$, and (b) modulated carrier $s(t) = Ax(t)\cos(\omega_c t)$.

modulated, with the result that the spectrum of the signal $x(t)$ is up-converted to an optical frequency range. The up-converted signal is often referred to as the *passband signal*, since it consists of the frequencies in the up-converted spectrum. The source signal $x(t)$ is referred to as the *baseband signal*.

3.8.2   Alternative Form of AM

In some types of AM transmission, such as AM radio, the modulated carrier $s(t)$ is given by

$$s(t) = A[1 + kx(t)]\cos(\omega_c t) \tag{3.77}$$

where k is a positive constant called the *amplitude sensitivity*, which is chosen so that $1 + kx(t) > 0$ for all t. This condition ensures that the envelope of the modulated carrier $s(t)$ is a replica of the signal $x(t)$. In this form of AM transmission, it is also assumed that the carrier frequency ω_c is much larger than the bandwidth B of $x(t)$.

Example 3.20   Alternative Form of AM

Again consider the signal in Figure 3.32a, and let the carrier be $\cos(5\pi t)$. Then, with $k = 0.8$, the modulated signal $s(t) = [1 + kx(t)]\cos(5\pi t)$ is shown in Figure 3.34. Note that the envelope is a replica of the signal $x(t)$, whereas such is not the case for the modulated signal in Figure 3.32b.

We can determine the frequency spectrum $S(\omega)$ of the modulated carrier $s(t) = A[1 + kx(t)]\cos(\omega_c t)$ also by taking the Fourier transform. This gives

$$S(\omega) = \pi A[\delta(\omega + \omega_c) + \delta(\omega - \omega_c)] + \frac{Ak}{2}[X(\omega + \omega_c) + X(\omega - \omega_c)]$$

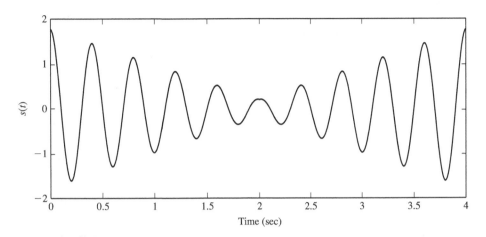

FIGURE 3.34
Modulated carrier in Example 3.20.

Note the frequency components at $\omega = \pm\omega_c$, which are due to the presence of the carrier $A \cos(\omega_c t)$ in the modulated signal $A[1 + kx(t)] \cos(\omega_c t)$. Thus, in this alternative form of AM, the spectrum of the transmitted signal $s(t)$ contains the carrier and the upper and lower sidebands. In contrast, the spectrum of the modulated carrier $Ax(t) \cos(\omega_c t)$ contains only the upper and lower sidebands; the carrier is suppressed. Hence, when the modulated carrier is of the form $Ax(t) \cos(\omega_c t)$, it is referred to as *double-sideband–suppressed carrier* (DSB–SC) *transmission*. When the modulated carrier is of the form $s(t) = A[1 + kx(t)] \cos(\omega_c t)$, it is referred to as *double-sideband* (DSB) *transmission*. A major advantage of DSB–SC over DSB is that, since DSB–SC does not require that the carrier be transmitted, it uses much less power than DSB to transmit the source signal $x(t)$. On the other hand, as will be seen in this section, DSB signals can be demodulated with a simple envelope detector, whereas demodulation of DSB–SC signals requires synchronization between the transmitter and the receiver.

3.8.3   Angle Modulation

In addition to amplitude modulation, a signal $x(t)$ can be "put on" a sinusoidal carrier by modulation of the angle of the carrier. In this form of transmission, called *angle modulation*, the modulated carrier is given by $s(t) = A \cos[\theta(t)]$, where the angle $\theta(t)$ is a function of the baseband signal $x(t)$. There are two basic types of angle modulation: *phase modulation* and *frequency modulation*. In phase modulation (PM), the angle is given by

$$\theta(t) = \omega_c t + k_p x(t)$$

where k_p is the *phase sensitivity* of the modulator. In frequency modulation (FM), the angle is given by

$$\theta(t) = \omega_c t + 2\pi k_f \int_0^t x(\tau)\, d\tau$$

where k_f is the *frequency sensitivity* of the modulator. Thus, the modulated carrier in PM transmission is equal to

$$s(t) = A \cos[\omega_c t + k_p x(t)]$$

and the modulated carrier in FM transmission is equal to

$$s(t) = A \cos\left[\omega_c t + 2\pi k_f \int_0^t x(\tau)\, d\tau \right] \tag{3.78}$$

Note that if $x(t)$ is the sinusoid $x(t) = \alpha \cos(\omega_x t)$, the FM signal (3.78) becomes

$$s(t) = A \cos\left[\omega_c t + \frac{2\pi k_f \alpha}{\omega_x} \sin \omega_x t \right]$$

Example 3.21   PM and FM Modulation

Suppose that $x(t) = \cos(\pi t)$, which is plotted in Figure 3.35a. Then, with $\omega_c = 10\pi$, $A = 1$, $k_p = 5$, and $k_f = 5/2$, the PM and FM signals are plotted in Figure 3.35b and 3.35c.

3.8.4   Pulse-Amplitude Modulation

Instead of modulating a sinusoid, information in the form of a signal $x(t)$ can be transmitted by modulation of other types of waveforms, such as the pulse train $p(t)$ shown in Figure 3.36. The amplitude of $p(t)$ can be modulated by multiplication of $x(t)$ and $p(t)$, as illustrated in Figure 3.37. This process is called *pulse-amplitude modulation* (PAM).

Example 3.22   PAM

Consider the signal displayed in Figure 3.38a. With $T = 0.2$ and $\varepsilon \ll 0.2$, the PAM signal is shown in Figure 3.38b.

A PAM signal can be generated by the application of $x(t)$ to a switch that is closed for ε seconds every T seconds. In the limit as $\varepsilon \to 0$, the modulated signal $s(t) = x(t)p(t)$ is actually a sampled version of $x(t)$, where T is the sampling interval. Thus, sampling is closely related to PAM. In fact, the Fourier transform of a PAM signal is approximately equal to that of the idealized sampled signal considered in Section 5.4.

3.8.5   Demodulation of DSB Signals

In the case of DSB transmission where the carrier is not suppressed, we can reconstruct $x(t)$ from the modulated carrier $s(t) = A[1 + kx(t)] \cos(\omega_c t)$ by applying $s(t)$ to an *envelope detector* given by the circuit in Figure 3.39. As seen from the figure, the circuit consists of a source resistance R_s, diode, capacitor with capacitance C, and load resistance R_L. When there is no voltage on the capacitor and the modulated carrier $s(t)$ increases from 0 to some peak value, current flows through the diode, and the capacitor charges to a voltage equal to the peak value of $s(t)$. When $s(t)$ decreases in

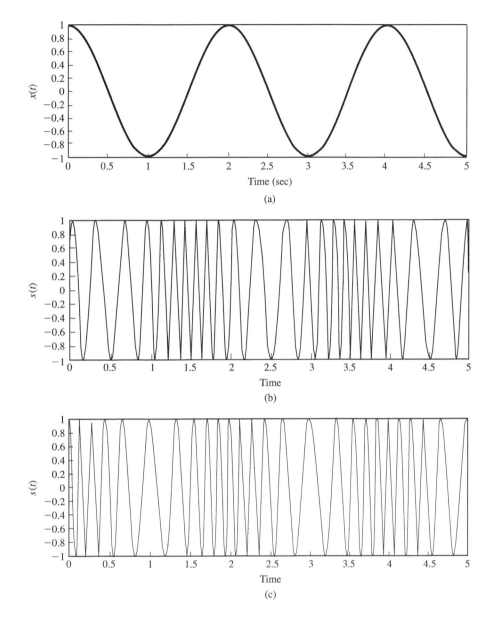

FIGURE 3.35
(a) Signal, (b) PM signal, and (c) FM signal in Example 3.21.

value from the peak value, the diode becomes an open circuit, and the voltage on the capacitor slowly discharges through the load resistance R_L. The discharging continues until $s(t)$ reaches a value that exceeds the value of the voltage across the capacitor, at which time the capacitor again charges up to the peak value of $s(t)$, and then the process repeats. To ensure that the charging of the capacitor is sufficiently fast so that the capacitor voltage reaches the peak value of $s(t)$ on every cycle, the charging time

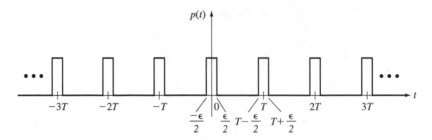

FIGURE 3.36
Pulse train with period T.

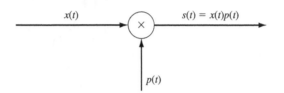

FIGURE 3.37
Pulse-amplitude modulation.

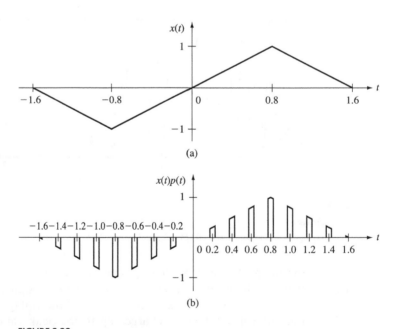

FIGURE 3.38
(a) Signal and (b) PAM signal.

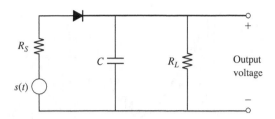

FIGURE 3.39
Envelope detector.

constant must be very small in comparison with the period $2\pi/\omega_c$ of the carrier $A\cos(\omega_c t)$. Assuming that the diode has zero resistance in the forward-biased region, the charging time constant of the envelope detector is equal to $R_s C$, and thus it is required that $R_s C \ll 2\pi/\omega_c$. In addition, the discharging time constant of the envelope detector must be large enough to ensure that the capacitor discharge between positive peaks of $s(t)$ is sufficiently slow. The discharging time constant is equal to $R_L C$, and thus it must be true that $R_L C \gg 2\pi/\omega_c$. It also must be true that the discharging time constant is small in comparison with the maximum rate of change of $x(t)$. If $x(t)$ has bandwidth B, the maximum rate of change of $x(t)$ can be taken to be $2\pi/B$, and thus it is also required that $R_L C \ll 2\pi/B$.

Example 3.23

For the case when $x(t) = \cos(\pi t)$, $\omega_c = 20\pi$, $k = 0.5$, $R_s = 100$ ohms, $C = 10$ microfarads, and $R_L = 40{,}000$ ohms, the modulated carrier and the output of the envelope detector are shown in Figure 3.40.

3.8.6   Demodulation of Other Signal Types

The demodulation of PM and FM signals is left to a more advanced treatment of communication systems. The demodulation of a PAM signal can be carried out by lowpass filtering the PAM signal. The analysis is very similar to the reconstruction of a signal from samples of the signal, which is considered in Section 5.4.

3.9   CHAPTER SUMMARY

This chapter explores the frequency domain analysis of continuous-time signals. As discussed in the chapter, periodic signals can be represented as a Fourier series, while the Fourier transform is used for aperiodic signals.

The Fourier series can be in the form of a trigonometric series or a complex exponential series. The trigonometric series can be expressed either as a series of sines and cosines,

$$x(t) = a_0 + \sum_{k=1}^{\infty}[a_k\cos(k\omega_0 t) + b_k\sin(k\omega_0 t)], \quad -\infty < t < \infty$$

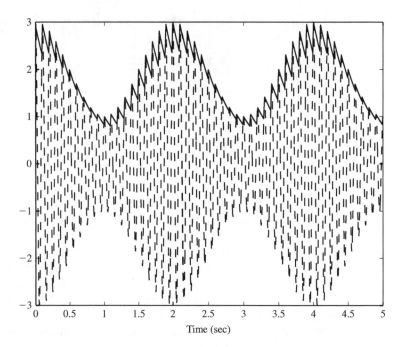

FIGURE 3.40
Modulated carrier and output of envelope detector in Example 3.23.

or as a series of cosine-with-phase,

$$x(t) = a_0 + \sum_{k=1}^{\infty} A_k \cos(k\omega_0 t + \theta_k), \quad -\infty < t < \infty$$

The complex exponential Fourier series has the form

$$x(t) = \sum_{k=-\infty}^{\infty} c_k e^{jk\omega_0 t}, \quad -\infty < t < \infty$$

In each of these forms of the Fourier series, ω_0 represents the fundamental frequency of the periodic signal $x(t)$. Each series contains a constant, or dc, term that represents the average value of the signal. The remaining terms in the series are the harmonics, that is, terms with frequencies that are integer multiples of the fundamental frequency.

The frequency content of the signal can be determined from the magnitudes and the angles of the coefficients in the trigonometric cosine-with-phase series or in the complex exponential series. The line spectra are plots of the magnitudes and angles from the complex exponential series versus the corresponding frequencies. A signal with high-frequency content, that is, large magnitude for a high-frequency term in the series, is seen in the time domain as having fast transitions. Thus, a square wave signal has more high-frequency content than does a triangular wave signal with the same fundamental frequency, due to the discontinuities in the square wave.

The Fourier transform gives the frequency content of an aperiodic signal and is defined by

$$X(\omega) = \int_{-\infty}^{\infty} x(t)e^{-j\omega t}\, dt, \quad -\infty < \omega < \infty$$

The frequency spectrum of an aperiodic signal contains a continuum of frequencies, unlike the Fourier series, which contains frequency components only at the harmonic frequencies. Not every aperiodic signal, such as the unit-step function, has a Fourier transform. To handle these cases, a generalized Fourier transform is introduced that yields a Fourier transform containing an impulse function $\delta(\omega)$. This generalized transform is useful for sinusoids as well, where the impulses are displaced to $\delta(\omega + \omega_0)$ and $\delta(\omega - \omega_0)$, where ω_0 is the frequency of the sinusoid.

Several important properties exist for the Fourier transform, including linearity, shifts in time, multiplication by a sinusoid, convolution in the time domain, and duality. Multiplication by a sinusoid, also known as amplitude modulation, results in a frequency spectrum that has scaled duplicates of the spectrum of the original signal located at $\pm\omega_c$ where ω_c is the carrier frequency. Amplitude modulation, frequency modulation, phase modulation, and phase amplitude modulation are all important concepts in the field of communication systems.

PROBLEMS

3.1. Each of the signals in Figure P3.1 is generated from a sum of sinusoids. Find the frequencies and the amplitudes of the sinusoids, and draw the line spectrum (amplitude only) for each signal.

3.2. Using complex notation, combine the expressions to form a single sinusoid for each of the cases (a)–(d). (See Appendix A.)
   (a) $2\cos(3t) - \cos(3t - \pi/4)$
   (b) $\sin(2t - \pi/4) + 2\cos(2t - \pi/3)$
   (c) $\cos(t) - \sin(t)$
   (d) $10\cos(\pi t + \pi/3) + 8\cos(\pi t - \pi/3)$

3.3. Compute the (sine/cosine) trigonometric Fourier series for each of the periodic signals shown in Figure P3.3. Use even or odd symmetry whenever possible.

3.4. Express each of the trigonometric Fourier series found in Problem 3.3 in cosine-with-phase form.

3.5. Prove the formulas (3.13) and (3.14) for functions $x(t)$ and $v(t)$ that are either even or odd functions of t.

3.6. Express the following terms in polar notation:
   (a) $e^{j\pi/4} + e^{-j\pi/8}$
   (b) $(2 + 5j)e^{j10}$
   (c) $e^{j2} + 1 + j$
   (d) $1 + e^{j4}$
   (e) $e^{j(\omega t + \pi/2)} + e^{j(\omega t - \pi/3)}$

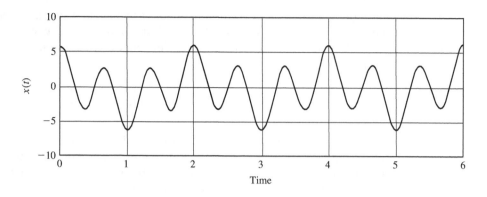

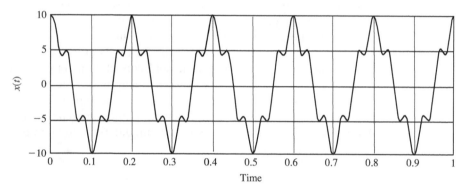

FIGURE P3.1

3.7. Use MATLAB to plot the signals given in Problem 3.2, and verify the expression derived in Problem 3.2.

3.8. Starting with the trigonometric Fourier series given by (3.4)–(3.6), derive the formulas (3.20) and (3.21) for the complex exponential form of the Fourier series.

3.9. For each signal shown in Figure P3.9, do the following:
   (a) Compute the trigonometric and complex exponential Fourier series. You may want to use the MATLAB Symbolic Math Toolbox to solve for the coefficients.
   (b) Compute and plot the truncated exponential series for N = 3, 10, and 30, using MAT-LAB when T = 2 and a = 0.5.
   (c) Repeat part (b), using the truncated trigonometric series, and compare your answer with part (b).

3.10. For each of the periodic signals shown in Figure P3.3, do the following:
   (i) Compute the complex exponential Fourier series. You may want to use the MATLAB Symbolic Math Toolbox to solve for the coefficients.
   (ii) Sketch the amplitude and phase spectra for k = ±1, ±2, ±3, ±4, ±5.
   (iii) Plot the truncated complex exponential series for N = 1, N = 5, and N = 30.

3.11. For each of the following signals, compute the complex exponential Fourier series by using trigonometric identities, and then sketch the amplitude and phase spectra for all values of k.
   (a) $x(t) = \cos(5t - \pi/4)$
   (b) $x(t) = \sin t + \cos t$

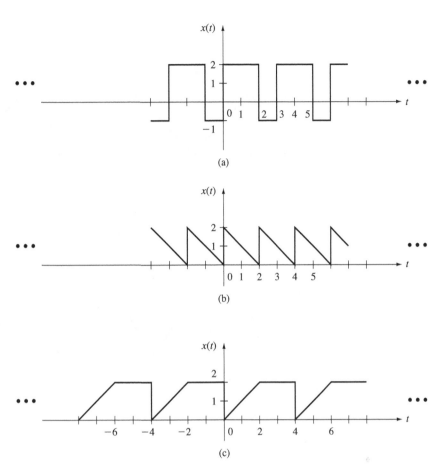

(a)

(b)

(c)

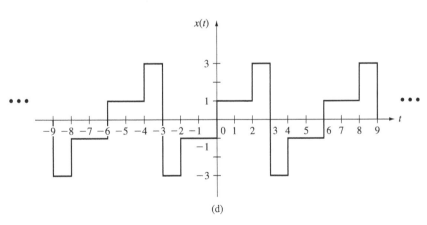

(d)

FIGURE P3.3

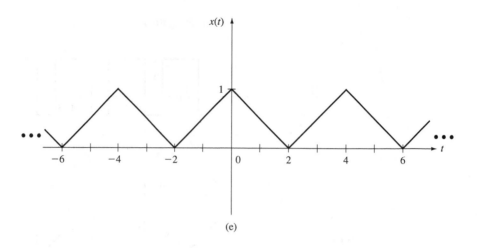

(e)

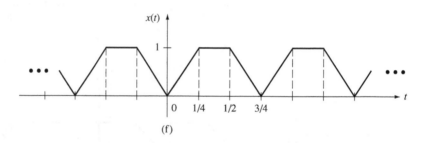

(f)

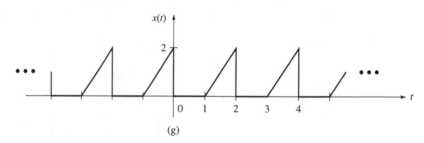

(g)

FIGURE P3.3 *(Continued)*

(c) $x(t) = \cos(t - 1) + \sin\left(t - \tfrac{1}{2}\right)$
(d) $x(t) = \cos 2t \sin 3t$
(e) $x(t) = \cos^2 5t$
(f) $x(t) = \cos 3t + \cos 5t$

3.12. Determine the exponential Fourier series for the following periodic signals:

(a) $x(t) = \dfrac{\sin 2t + \sin 3t}{2 \sin t}$

(b) $x(t) = \displaystyle\sum_{k=-\infty}^{\infty} \delta(t - kT)$

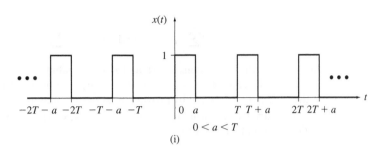

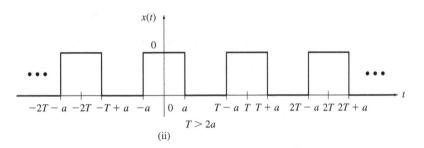

FIGURE P3.9

3.13. A periodic signal with period T has Fourier coefficients c_k^x; that is,

$$x(t) = \sum_{k=-\infty}^{\infty} c_k^x \exp(jk\omega_0 t), \quad \omega_0 = \frac{2\pi}{T}, \quad -\infty < t < \infty$$

Compute the Fourier coefficients c_k^v for the periodic signal $v(t)$, where

(a) $v(t) = x(t-1)$

(b) $v(t) = \dfrac{dx(t)}{dt}$

(c) $v(t) = x(t) \exp[j(2\pi/T)t]$

(d) $v(t) = x(t) \cos\left(\dfrac{2\pi}{T}t\right)$

3.14. The derivation of the complex exponential Fourier series uses *orthogonal basis functions*, which are a set of functions of time, $\phi_k(t)$, such that the following holds over some specified time interval T:

$$\int_T \overline{\phi_k(t)}\phi_m(t)\, dt = 0$$

for all k and m such that $k \neq m$.

(a) Prove that $\phi_k(t) = e^{jk\omega_0 t}$ for $k = 0, \pm 1, \pm 2, \pm 3, \dots$ are orthogonal basis functions over the time interval $T = 2\pi/\omega_0$.

(b) Suppose that $x(t)$ is periodic with period $T = 2\pi/\omega_0$. Approximate $x(t)$ by its Fourier series:

$$x(t) = \sum_{k=-\infty}^{\infty} c_k e^{jk\omega_0 t}$$

Using the notion of orthogonal basis functions, derive the expressions for c_k given in (3.21). [*Hint:* To derive (3.21), multiply both sides of the Fourier series by $e^{jk\omega_0 t}$, and integrate over T.]

3.15. Let

$$f(t) = \sum_{k=-\infty}^{\infty} f_k e^{jk\omega_0 t} \quad \text{and} \quad g(t) = \sum_{k=-\infty}^{\infty} g_k e^{jk\omega_0 t}$$

be the Fourier series expansions for $f(t)$ and $g(t)$. State whether the following is true or false: If $f_k = g_k$ for all k, then $f(t) = g(t)$ for all t. Justify your answer by giving a counterexample if the statement is false, or by proving that the statement is true.

3.16. Using the Fourier transform, determine the complex exponential Fourier series of the periodic signals in Figure P3.16.

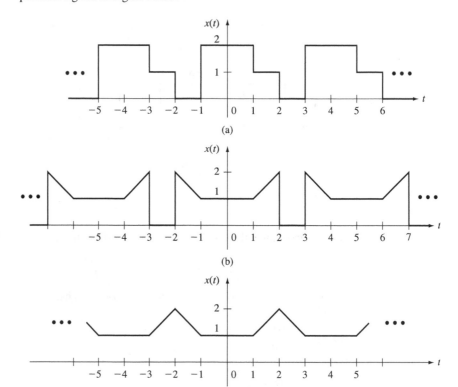

(a)

(b)

(c)

FIGURE P3.16

3.17. Compute the Fourier transform of the following signals, using the symbolic manipulator to perform the integrations. In each case, plot the signal $x(t)$ and the magnitude $|X(\omega)|$ of the Fourier transform.
   (a) $x(t) = 2e^{-4t} \cos(10t)u(t)$
   (b) $x(t) = 2e^{-4t} \sin(10t)u(t)$
   (c) $x(t) = 2te^{-2t}u(t)$
   (d) $x(t) = e^{-t}(\cos 5t + \cos 30t)u(t)$

3.18. Match the signals in Figure P3.18 Part 1 to their appropriate amplitude spectrums in Figure P3.18 Part 2.

3.19. By first expressing $x(t)$ in terms of rectangular pulse functions and triangular pulse functions, compute the Fourier transform of the signals in Figure P3.19. Plot the magnitude and phase of the Fourier transform.

160

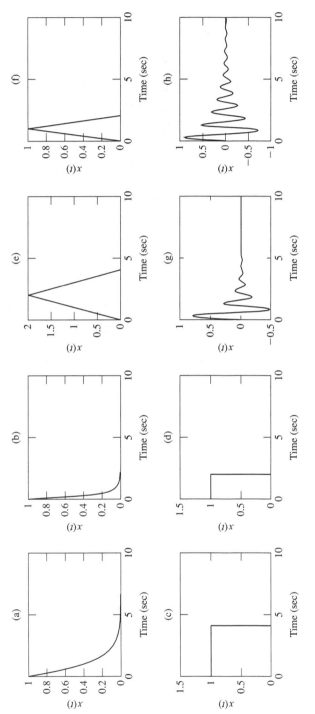

FIGURE P3.18   Part 1

161

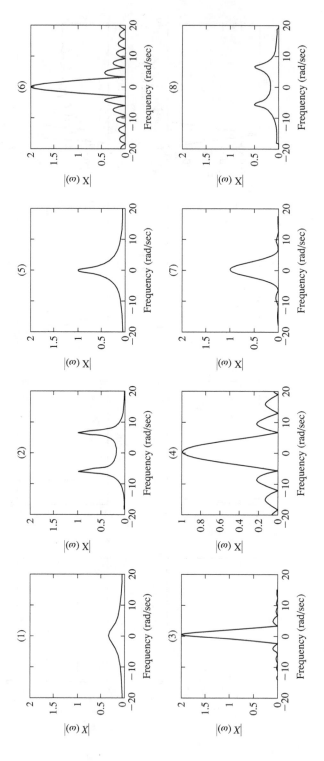

FIGURE P3.18   Part 2

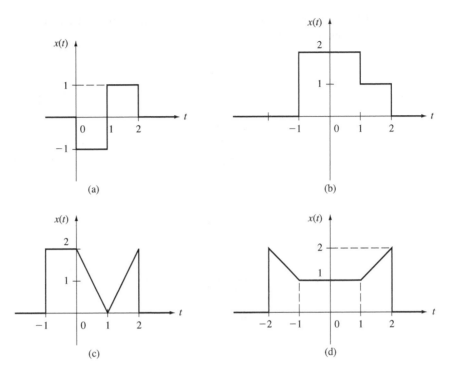

FIGURE P3.19

3.20. Compute the Fourier transform of the signals in Figure P3.20. Plot the magnitude and phase of the Fourier transform. Use the symbolic manipulator to perform the integrations or to check the answers.

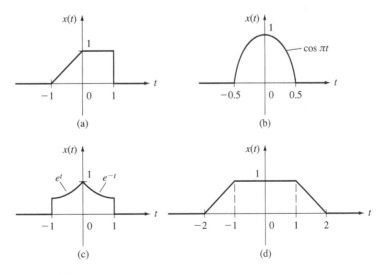

FIGURE P3.20

3.21. Compute the inverse Fourier transforms of the frequency functions $X(\omega)$ shown in Figure P3.21.

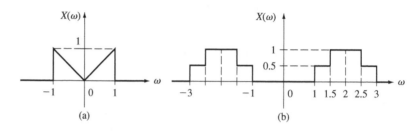

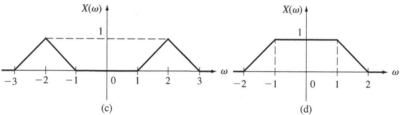

FIGURE P3.21

3.22. Compute the inverse Fourier transform of the following frequency functions:
   (a) $X(\omega) = \cos(4\omega)$, $-\infty < \omega < \infty$
   (b) $X(\omega) = \sin^2(3\omega)$, $-\infty < \omega < \infty$
   (c) $X(\omega) = p_4(\omega)\cos\dfrac{\pi\omega}{2}$
   (d) $X(\omega) = \dfrac{\sin(\omega/2)}{j\omega + 2}e^{-j\omega 2}$, $-\infty < \omega < \infty$

3.23. A signal $x(t)$ has Fourier transform

$$X(\omega) = \frac{1}{j}\left[\operatorname{sinc}\left(\frac{2\omega}{\pi} - \frac{1}{2}\right) - \operatorname{sinc}\left(\frac{2\omega}{\pi} + \frac{1}{2}\right)\right]$$

   (a) Compute $x(t)$.
   (b) Let $x_p(t)$ denote the periodic signal defined by

$$x_p(t) = \sum_{k=-\infty}^{\infty} x(t - 16k)$$

   Compute the Fourier transform $X_p(\omega)$ of $x_p(t)$.

3.24. A continuous-time signal $x(t)$ has Fourier transform

$$X(\omega) = \frac{1}{j\omega + b}$$

   where b is a constant. Determine the Fourier transform $V(\omega)$ of the following signals:
   (a) $v(t) = x(5t - 4)$
   (b) $v(t) = t^2 x(t)$

(c) $v(t) = x(t)e^{j2t}$

(d) $v(t) = x(t)\cos 4t$

(e) $v(t) = \dfrac{d^2 x(t)}{dt^2}$

(f) $v(t) = x(t) * x(t)$

(g) $v(t) = x^2(t)$

(h) $v(t) = \dfrac{1}{jt - b}$

3.25. Using the properties of Fourier transforms, compute the Fourier transform of the following signals:

(a) $x(t) = (e^{-t}\cos 4t)u(t)$

(b) $x(t) = te^{-t}u(t)$

(c) $x(t) = (\cos 4t)u(t)$

(d) $x(t) = e^{-|t|},\ -\infty < t < \infty$

(e) $x(t) = e^{-t^2},\ -\infty < t < \infty$

3.26. For the Fourier transforms $X(\omega)$ given in Figure P3.26, what characteristics does $x(t)$ have (i.e., real-valued, complex-valued, even, odd)? Calculate $x(0)$.

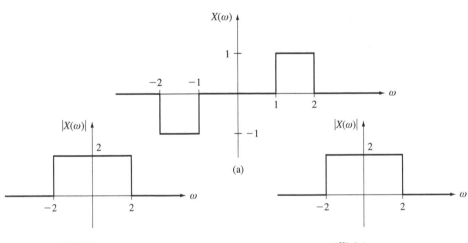

(a)

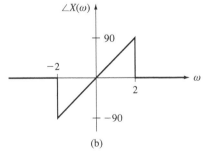

(b)

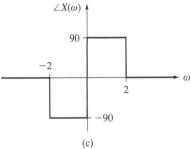

(c)

FIGURE P3.26

3.27. The Fourier transforms of $x(t)$ and $v(t)$ are defined as follows:

$$X(\omega) = \begin{cases} 2, & |\omega| < \pi \\ 0, & \text{otherwise} \end{cases}$$

$$V(\omega) = X(\omega - \omega_0) + X(\omega + \omega_0)$$

    (a) Find $x(t)$ in closed form.
    (b) Find $v(t)$ in closed form.

3.28. Compute the generalized Fourier transform of the following signals:
    (a) $x(t) = 1/t, -\infty < t < \infty$

    (b) $x(t) = 1 + 2e^{-j\frac{2t}{\pi}} + 2e^{j2\pi t}, -\infty < t < \infty$

    (c) $x(t) = 3\cos t + 2\sin 2t, -\infty < t < \infty$

    (d) $x(t) = (2 + 3\cos(\pi t - \pi/4))u(t)$

    (e) $x(t)$ as shown in Figure P3.9.

3.29. Sketch the magnitude functions of the Fourier transforms of these amplitude-modulated signals, (a)–(e). Identify important values on the axes.
    (a) $x(t) = (e^{-10t}u(t))\cos(100t)$

    (b) $x(t) = (1 - |t|)p_2(t)\cos(10t)$

    (c) $x(t) = p_2(t)\cos(10t)$

    (d) $x(t) = 2\,\text{sinc}^2\left(\dfrac{t}{2\pi}\right)\cos(10t)$

    (e) $x(t) = 4\cos(10t)\cos(100t)$

3.30. Generate a MATLAB plot of the output of the envelope detector in the case when $x(t) = \cos\left(\dfrac{\pi}{3}t\right)$, $\omega_c = 30\pi$, $k = 0.5$, $R_s = 100$ ohms, $C = 5$ microfarads, and $R_L = 50{,}000$ ohms.

Fourier Analysis of Discrete-Time Signals

CHAPTER

4

The discrete-time counterpart to the Fourier theory developed in Chapter 3 is presented in this chapter. The development begins in Section 4.1 with the study of the discrete-time Fourier transform (DTFT), which is the discrete-time counterpart to the Fourier transform. As is the case for the Fourier transform of a continuous-time signal, the DTFT of a discrete-time signal is a function of a continuum of frequencies, but unlike the continuous-time case, the DTFT is always a periodic function with period 2π.

In Section 4.2 a transform of a discrete-time signal is defined that is a function of a finite number of frequencies. This transform is called the discrete Fourier transform (DFT). For time-limited discrete-time signals, it is shown that the DFT is equal to the DTFT with the frequency variable evaluated at a finite number of points. Hence, the DFT can be viewed as a "discretization in frequency" of the DTFT. Since the DFT is a function of a finite number of frequencies, it is the transform that is often used in practice. In particular, the DFT is used extensively in digital signal processing and digital communications.

In Section 4.3 the DFTs of truncated signals (corresponding to a given set of data values) are studied, and then in Section 4.4 a fast method for computing the DFT, called the fast Fourier transform (FFT) algorithm, is given. It is shown how the FFT algorithm can be utilized to compute the Fourier transform of a continuous-time signal and to compute the convolution of two discrete-time signals. In Section 4.5, the discrete Fourier transform is applied to data analysis, with the focus on the extraction of a signal embedded in noise, the analysis of sunspot data that can be downloaded from the Web, and the analysis of stock price data. Section 4.6 contains a summary of the chapter.

4.1   DISCRETE-TIME FOURIER TRANSFORM

In Section 3.4 the Fourier transform $X(\omega)$ of a continuous-time signal $x(t)$ was defined by

$$X(\omega) = \int_{-\infty}^{\infty} x(t)e^{-j\omega t}\, dt \qquad (4.1)$$

Given a discrete-time signal $x[n]$, the discrete-time Fourier transform (DTFT) of $x[n]$ is defined by

$$X(\Omega) = \sum_{n=-\infty}^{\infty} x[n]e^{-j\Omega n} \qquad (4.2)$$

The DTFT $X(\Omega)$ defined by (4.2) is, in general, a complex-valued function of the real variable Ω (the frequency variable). Note that (4.2) is a natural discrete-time counterpart of (4.1) in that the integral is replaced by a summation. The uppercase omega (Ω) is utilized for the frequency variable to distinguish between the continuous- and discrete-time cases.

A discrete-time signal $x[n]$ is said to have a DTFT in the *ordinary sense* if the bi-infinite sum in (4.2) converges (i.e., is finite) for all real values of Ω. A sufficient condition for $x[n]$ to have a DTFT in the ordinary sense is that $x[n]$ be absolutely summable; that is,

$$\sum_{n=-\infty}^{\infty} |x[n]| < \infty \tag{4.3}$$

If $x[n]$ is a time-limited discrete-time signal (i.e., there is a positive integer N such that $x[n] = 0$ for all $n \leq -N$ and $n \geq N$), then, obviously, the sum in (4.3) is finite, and thus any such signal has a DTFT in the ordinary sense.

Example 4.1   Computation of DTFT

Consider the discrete-time signal $x[n]$ defined by

$$x[n] = \begin{cases} 0, & n < 0 \\ a^n, & 0 \leq n \leq q \\ 0, & n > q \end{cases}$$

where a is a nonzero real constant and q is a positive integer. This signal is clearly time limited, and thus it has a DTFT in the ordinary sense. To compute the DTFT, insert $x[n]$ into (4.2), which yields

$$X(\Omega) = \sum_{n=0}^{q} a^n e^{-j\Omega n}$$

$$= \sum_{n=0}^{q} (ae^{-j\Omega})^n \tag{4.4}$$

This summation can be written in "closed form" by using the relationship

$$\sum_{n=q_1}^{q_2} r^n = \frac{r^{q_1} - r^{q_2+1}}{1 - r} \tag{4.5}$$

where q_1 and q_2 are integers with $q_2 > q_1$ and r is a real or complex number. (The reader is asked to prove (4.5) in Problem 4.2.) Then, using (4.5) with $q_1 = 0$, $q_2 = q$ and $r = ae^{-j\Omega}$, we can write (4.4) in the form

$$X(\Omega) = \frac{1 - (ae^{-j\Omega})^{q+1}}{1 - ae^{-j\Omega}} \tag{4.6}$$

For any discrete-time signal $x[n]$, the DTFT $X(\Omega)$ is a periodic function of Ω with period 2π, that is,

$$X(\Omega + 2\pi) = X(\Omega) \text{ for all } \Omega, -\infty < \Omega < \infty$$

To prove the periodicity property, note that

$$X(\Omega + 2\pi) = \sum_{n=-\infty}^{\infty} x[n]e^{-jn(\Omega+2\pi)}$$

$$= \sum_{n=-\infty}^{\infty} x[n]e^{-jn\Omega}e^{-jn2\pi}$$

But

$$e^{-jn2\pi} = 1 \text{ for all integers } n$$

and thus

$$X(\Omega + 2\pi) = X(\Omega) \text{ for all } \Omega$$

An important consequence of periodicity of $X(\Omega)$ is that $X(\Omega)$ is completely determined by the computation of $X(\Omega)$ over any 2π interval such as $0 \leq \Omega \leq 2\pi$ or $-\pi \leq \Omega \leq \pi$.

Given the discrete-time signal $x[n]$ with DTFT $X(\Omega)$, since $X(\Omega)$ is complex valued in general, $X(\Omega)$ can be expressed in either rectangular or polar form. Using Euler's formula yields the following rectangular form of $X(\Omega)$:

$$X(\Omega) = R(\Omega) + jI(\Omega) \tag{4.7}$$

Here, $R(\Omega)$ and $I(\Omega)$ are real-valued functions of Ω given by

$$R(\Omega) = \sum_{n=-\infty}^{\infty} x[n] \cos n\Omega$$

$$I(\Omega) = -\sum_{n=-\infty}^{\infty} x[n] \sin n\Omega$$

The polar form of $X(\Omega)$ is

$$X(\Omega) = |X(\Omega)| \exp[j\angle X(\Omega)] \tag{4.8}$$

where $|X(\Omega)|$ is the magnitude of $X(\Omega)$ and $\angle X(\Omega)$ is the angle of $X(\Omega)$. Note that, since $X(\Omega)$ is periodic with period 2π, both $|X(\Omega)|$ and $\angle X(\Omega)$ are periodic with period 2π. Thus, both $|X(\Omega)|$ and $\angle X(\Omega)$ need to be specified only over some interval of length 2π such as $0 \leq \Omega \leq 2\pi$ or $-\pi \leq \Omega \leq \pi$.

Assuming that $x[n]$ is real valued, the magnitude function $|X(\Omega)|$ is an even function of Ω and the angle function $X(\Omega)$ is an odd function of Ω; that is,

$$|X(-\Omega)| = |X(\Omega)| \quad \text{for all } \Omega \tag{4.9}$$

$$\angle X(-\Omega) = -\angle X(\Omega) \quad \text{for all } \Omega \tag{4.10}$$

To verify (4.9) and (4.10), first replace Ω by $-\Omega$ in (4.2), which gives

$$X(-\Omega) = \sum_{n=-\infty}^{\infty} x[n]e^{j\Omega n}$$
$$= \overline{X(\Omega)} \tag{4.11}$$

where $\overline{X(\Omega)}$ is the complex conjugate of $X(\Omega)$. Now, replacing Ω by $-\Omega$ in the polar form (4.8) gives

$$X(-\Omega) = |X(-\Omega)| \exp[j\angle X(-\Omega)] \tag{4.12}$$

and taking the complex conjugate of both sides of (4.8) gives

$$\overline{X(\Omega)} = |X(\Omega)| \exp[-j\angle X(\Omega)] \tag{4.13}$$

Finally, combining (4.11)–(4.13) yields

$$|X(-\Omega)| \exp[j\angle X(-\Omega)] = |X(\Omega)| \exp[-j\angle X(\Omega)]$$

Hence, it must be true that

$$|X(-\Omega)| = |X(\Omega)|$$

and

$$\angle X(-\Omega) = -\angle X(\Omega)$$

which verifies (4.9) and (4.10).

Note that as a result of the even symmetry of $|X(\Omega)|$ and the odd symmetry of $\angle X(\Omega)$, the magnitude $|X(\Omega)|$ and phase $\angle X(\Omega)$ need to be plotted only over the interval $0 \leq \Omega \leq \pi$.

If the DTFT is given in the rectangular form (4.7), it is possible to generate the polar form (4.8) by the use of the relationships

$$|X(\Omega)| = \sqrt{R^2(\Omega) + I^2(\Omega)} \tag{4.14}$$

$$\angle X(\Omega) = \begin{cases} \tan^{-1}\dfrac{I(\Omega)}{R(\Omega)} \text{ when } R(\Omega) \geq 0 \\ \pi + \tan^{-1}\dfrac{I(\Omega)}{R(\Omega)} \text{ when } R(\Omega) < 0 \end{cases} \tag{4.15}$$

Example 4.2   Rectangular and Polar Forms

Consider the discrete-time signal $x[n] = a^n u[n]$, where a is a nonzero real constant and $u[n]$ is the discrete-time unit-step function. For the case $a = 0.5$, the signal is displayed in Figure 4.1. The signal $x[n] = a^n u[n]$ is equal to the signal in Example 4.1 in the limit as $q \to \infty$. Hence, the DTFT $X(\Omega)$ of $x[n]$ is equal to the limit as $q \to \infty$ of the DTFT of the signal in Example 4.1. That is, using (4.6), we find that the DTFT is

$$X(\Omega) = \lim_{q \to \infty} \frac{1 - (ae^{-j\Omega})^{q+1}}{1 - ae^{-j\Omega}}$$

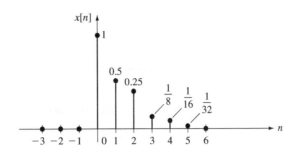

FIGURE 4.1
The signal $x[n] = (0.5)^n u[n]$.

Now, the preceding limit exists if and only if $|a| < 1$, in which case

$$\lim_{q \to \infty} (ae^{-j\Omega})^{q+1} = 0$$

Thus, when $|a| < 1$, the signal $x[n] = a^n u[n]$ has a DTFT in the ordinary sense given by

$$X(\Omega) = \frac{1}{1 - ae^{-j\Omega}} \tag{4.16}$$

When $|a| \geq 1$, $x[n] = a^n u[n]$ does not have a DTFT in the ordinary sense.

To express the DTFT given by (4.16) in rectangular form, first multiply the right-hand side of (4.16) by $[1 - a\exp(j\Omega)]/[1 - a\exp(j\Omega)]$, which gives

$$X(\Omega) = \frac{1 - ae^{j\Omega}}{(1 - ae^{-j\Omega})(1 - ae^{j\Omega})}$$

$$= \frac{1 - ae^{j\Omega}}{1 - a(e^{-j\Omega} + e^{j\Omega}) + a^2}$$

Using Euler's formula, we see that $X(\Omega)$ becomes

$$X(\Omega) = \frac{1 - a\cos\Omega - ja\sin\Omega}{1 - 2a\cos\Omega + a^2}$$

and thus the rectangular form of $X(\Omega)$ is

$$X(\Omega) = \frac{1 - a\cos\Omega}{1 - 2a\cos\Omega + a^2} + j\frac{-a\sin\Omega}{1 - 2a\cos\Omega + a^2} \tag{4.17}$$

To compute the polar form of $X(\Omega)$, first take the magnitude of both sides of (4.16), which gives

$$|X(\Omega)| = \frac{1}{|1 - ae^{-j\Omega}|}$$

$$= \frac{1}{|1 - a\cos\Omega + ja\sin\Omega|}$$

$$= \frac{1}{\sqrt{(1 - a \cos \Omega)^2 + a^2 \sin^2 \Omega}}$$

$$|X(\Omega)| = \frac{1}{\sqrt{1 - 2a \cos \Omega + a^2}}$$

Finally, taking the angle of the right-hand side of (4.16) yields

$$\angle X(\Omega) = -\angle(1 - ae^{-j\Omega})$$

$$= -\angle(1 - a \cos \Omega + ja \sin \Omega)$$

$$= \begin{cases} -\tan^{-1}\dfrac{a \sin \Omega}{1 - a \cos \Omega} & \text{when } 1 - a \cos \Omega \geq 0 \\ \pi - \tan^{-1}\dfrac{a \sin \Omega}{1 - a \cos \Omega} & \text{when } 1 - a \cos \Omega < 0 \end{cases}$$

Therefore, the polar form of $X(\Omega)$ is

$$X(\Omega) = \begin{cases} \dfrac{1}{\sqrt{1 - 2a \cos \Omega + a^2}}\exp\left(-j\tan^{-1}\dfrac{a \sin \Omega}{1 - a \cos \Omega}\right) & \text{when } 1 - a \cos \Omega \geq 0 \\ \dfrac{1}{\sqrt{1 - 2a \cos \Omega + a^2}}\exp\left(j\left[\pi - \tan^{-1}\dfrac{a \sin \Omega}{1 - a \cos \Omega}\right]\right) & \text{when } 1 - a \cos \Omega < 0 \end{cases} \tag{4.18}$$

Note that the polar form (4.18) could also have been determined directly from the rectangular form (4.17) by the use of the relationships (4.14) and (4.15). The reader is invited to check that this results in the same answer as (4.18).

For the case $a = 0.5$, the magnitude function $|X(\Omega)|$ and the angle function $\angle X(\Omega)$ of the DTFT are plotted in Figure 4.2. In these plots, the frequency is the normalized frequency (Ω/π radians per unit time), and so a normalized frequency of 1 corresponds to π radians per unit time.

4.1.1 Signals with Even or Odd Symmetry

A real-valued discrete-time signal $x[n]$ is an even function of n if $x[-n] = x[n]$ for all integers $n \geq 1$. If $x[n]$ is an even signal, it follows from Euler's formula that the DTFT $X(\Omega)$ given by (4.2) can be expressed in the form

$$X(\Omega) = x[0] + \sum_{n=1}^{\infty} 2x[n] \cos \Omega n \tag{4.19}$$

From (4.19) it is seen that $X(\Omega)$ is a real-valued function of Ω, and thus the DTFT of an even signal is always real valued.

If $x[n]$ is an odd signal; that is, $x[-n] = -x[n]$ for all integers $n \geq 1$, the DTFT $X(\Omega)$ can be written in the form

$$X(\Omega) = x[0] - \sum_{n=1}^{\infty} j2x[n] \sin \Omega n \tag{4.20}$$

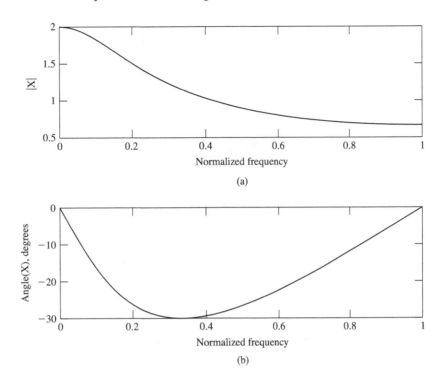

FIGURE 4.2
(a) Magnitude and (b) angle of the DTFT of the signal $x[n] = (0.5)^n u[n]$.

(The reader is asked to verify (4.20) [and (4.19)] in Problem 4.3.) From (4.20) it is seen that the DTFT of an odd signal is equal to the constant $x[0]$ plus a purely imaginary-valued function.

Example 4.3   DTFT of Rectangular Pulse

Given a positive integer q, let $p[n]$ denote the discrete-time rectangular pulse function defined by

$$p[n] = \begin{cases} 1, & n = -q, -q+1, \ldots, -1, 0, 1, \ldots, q \\ 0, & \text{all other } n \end{cases}$$

This signal is even, and thus the DTFT is a real-valued function of Ω. To compute the DTFT $P(\Omega)$ of the pulse $p[n]$, it turns out to be easier to use (4.2) instead of (4.19): Inserting $p[n]$ into (4.2) gives

$$P(\Omega) = \sum_{n=-q}^{q} e^{-j\Omega n} \tag{4.21}$$

Then, using (4.5) with $q_1 = -q$, $q_2 = q$, and $r = e^{-j\Omega}$, we find that (4.21) becomes

$$P(\Omega) = \frac{e^{j\Omega q} - e^{-j\Omega(q+1)}}{1 - e^{-j\Omega}} \tag{4.22}$$

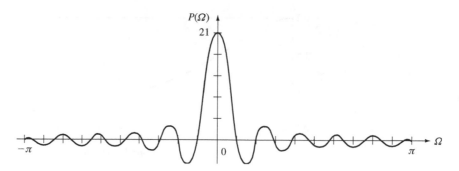

FIGURE 4.3
DTFT of the rectangular pulse $p[n]$ with $q = 10$.

Multiplying the top and bottom of the right-hand side of (4.22) by $e^{j\Omega/2}$ gives

$$P(\Omega) = \frac{e^{j\Omega(q+1/2)} - e^{-j\Omega(q+1/2)}}{e^{j(\Omega/2)} - e^{-j(\Omega/2)}} \tag{4.23}$$

Finally, using Euler's formula, we conclude that (4.23) reduces to

$$P(\Omega) = \frac{\sin[(q + 1/2)\Omega]}{\sin(\Omega/2)} \tag{4.24}$$

Hence, the DTFT of the rectangular pulse $p[n]$ is given by (4.24). It is interesting to note that, as the value of the integer q is increased, the plot of the DTFT $P(\Omega)$ looks more and more like a sinc function of the variable Ω. For example, in the case $q = 10$, $P(\Omega)$ is plotted in Figure 4.3 for $-\pi \leq \Omega \leq \pi$. The transform (4.24) is the discrete-time counterpart to the transform of the rectangular pulse in the continuous-time case (see Example 3.9 in Chapter 3).

4.1.2   Spectrum of a Discrete-Time Signal

Fourier analysis can be used to determine the frequency components of a discrete-time signal just as it was used for continuous-time signals. The decomposition of a periodic discrete-time signal $x[n]$ into sinusoidal components can be viewed as a generalization of the Fourier series representation of a periodic discrete-time signal. To keep the presentation as simple as possible, the discrete-time version of the Fourier series is not considered in this book.

For a discrete-time signal $x[n]$ that is not equal to a sum of sinusoids, the frequency spectrum consists of a continuum of frequency components that make up the signal. As in the continuous-time case, the DTFT $X(\Omega)$ displays the various sinusoidal components (with frequency Ω) that make up $x[n]$, and thus $X(\Omega)$ is called the *frequency spectrum* of $x[n]$. The magnitude function $|X(\Omega)|$ is called the *amplitude spectrum* of the signal, and the angle function $\angle X(\Omega)$ is called the *phase spectrum* of the signal. In this book the plots of $|X(\Omega)|$ and $\angle X(\Omega)$ will usually be specified over the interval $0 \leq \Omega \leq \pi$. The sinusoidal components included in $x[n]$ have positive frequencies ranging from 0 to π. Thus, the highest possible frequency that may be in the spectrum of $x[n]$ is $\Omega = \pi$ radians per unit time.

Example 4.4   Signal with Low-Frequency Components

Consider the discrete-time signal $x[n] = (0.5)^n u[n]$, which is plotted in Figure 4.1. The amplitude and phase spectrum of the signal were determined in Example 4.2, with the results plotted in Figure 4.2. Note that over the frequency range from 0 to π, most of the spectral content of the signal is concentrated near the zero frequency $\Omega = 0$. Thus, the signal has a preponderance of low-frequency components.

Example 4.5   Signal with High-Frequency Components

Now consider the signal $x[n] = (-0.5)^n u[n]$, which is plotted in Figure 4.4. Note that, due to the sign changes, the time variations of this signal are much greater than those of the signal in Example 4.4. Hence, it is expected that the spectrum of this signal should contain a much larger portion of high-frequency components in comparison with the spectrum of the signal in Example 4.4.

From the results in Example 4.2, the DTFT of $x[n] = (-0.5)^n u[n]$ is

$$X(\Omega) = \frac{1}{1 + 0.5e^{-j\Omega}}$$

and the amplitude and phase spectra are given by

$$|X(\Omega)| = \frac{1}{\sqrt{1.25 + \cos \Omega}} \tag{4.25}$$

$$\angle X(\Omega) = -\tan^{-1}\frac{-0.5 \sin \Omega}{1 + 0.5 \cos \Omega} \tag{4.26}$$

Plots of $|X(\Omega)|$ and $\angle X(\Omega)$ are shown in Figure 4.5. From the figure, note that over the frequency range from 0 to π, the spectral content of the signal is concentrated near the highest

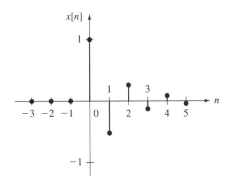

FIGURE 4.4
The signal $x[n] = (-0.5)^n u[n]$

175

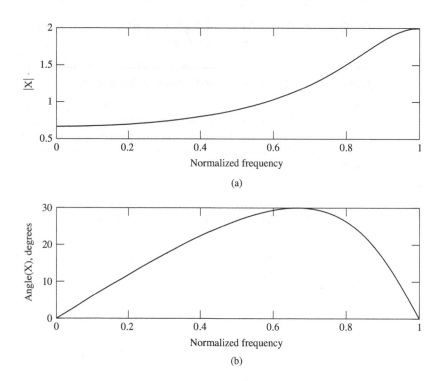

FIGURE 4.5

(a) Amplitude and (b) phase spectra of the signal $x[n] = (-0.5)^n u[n]$.

possible frequency $\Omega = \pi$, and therefore this signal has a preponderance of high-frequency components.

4.1.3   Inverse DTFT

Given a signal $x[n]$ with DTFT $X(\Omega)$, $x[n]$ can be recomputed from $X(\Omega)$ by application of the inverse DTFT to $X(\Omega)$. The inverse DTFT is defined by

$$x[n] = \frac{1}{2\pi} \int_0^{2\pi} X(\Omega) e^{jn\Omega} \, d\Omega \tag{4.27}$$

Since $X(\Omega)$ and $e^{jn\Omega}$ are both periodic functions of Ω with period 2π, the product $X(\Omega) \, e^{jn\Omega}$ is also a periodic function of Ω with period 2π. As a result, the integral in (4.27) can be evaluated over any interval of length 2π. For example,

$$x[n] = \frac{1}{2\pi} \int_{-\pi}^{\pi} X(\Omega) e^{jn\Omega} \, d\Omega \tag{4.28}$$

4.1.4   Generalized DTFT

As in the continuous-time Fourier transform, there are discrete-time signals that do not have a DTFT in the ordinary sense, but do have a generalized DTFT. One such signal is given in the following example:

Example 4.6   DTFT of Constant Signal

Consider the constant signal $x[n] = 1$ for all integers n. Since

$$\sum_{n=-\infty}^{\infty} x[n] = \infty$$

this signal does not have a DTFT in the ordinary sense. The constant signal does have a generalized DTFT that is defined to be the impulse train

$$X(\Omega) = \sum_{k=-\infty}^{\infty} 2\pi\delta(\Omega - 2\pi k)$$

This transform is displayed in Figure 4.6.

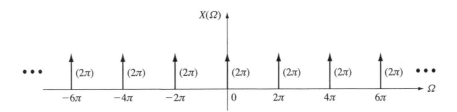

FIGURE 4.6
DTFT of the discrete-time constant signal.

The justification for taking the transform in Figure 4.6 to be the generalized DTFT of the constant signal follows from the property that the inverse DTFT of $X(\Omega)$ is equal to the constant signal. To see this, by (4.28),

$$x[n] = \frac{1}{2\pi} \int_{-\pi}^{\pi} X(\Omega)e^{jn\Omega}\, d\Omega$$

$$= \frac{1}{2\pi} \int_{-\pi}^{\pi} 2\pi\delta(\Omega)e^0\, d\Omega = \int_{-\pi}^{\pi} \delta(\Omega)\, d\Omega$$

$$= 1 \quad \text{for all } n$$

4.1.5   Transform Pairs

As in the Fourier transform theory of continuous-time signals, the transform pair notation

$$x[n] \leftrightarrow X(\Omega)$$

will be used to denote the fact that $X(\Omega)$ is the DTFT of $x[n]$, and conversely, that $x[n]$ is the inverse DTFT of $X(\Omega)$. For the convenience of the reader, a list of common DTFT pairs is given in Table 4.1.

TABLE 4.1   Common DTFT Pairs

$1, \text{all } n \leftrightarrow \displaystyle\sum_{k=-\infty}^{\infty} 2\pi\delta(\Omega - 2\pi k)$

$\text{sgn}[n] \leftrightarrow \dfrac{2}{1 - e^{-j\Omega}}, \text{ where sgn}[n] = \begin{cases} 1, & n = 0, 1, 2, \ldots \\ -1, & n = -1, -2, \ldots \end{cases}$

$u[n] \leftrightarrow \dfrac{1}{1 - e^{-j\Omega}} + \displaystyle\sum_{k=-\infty}^{\infty} \pi\delta(\Omega - 2\pi k)$

$\delta[n] \leftrightarrow 1$

$\delta[n - q] \leftrightarrow e^{-jq\Omega}, q = \pm 1, \pm 2, \ldots$

$a^n u[n] \leftrightarrow \dfrac{1}{1 - ae^{-j\Omega}}, \ |a| < 1$

$e^{j\Omega_0 n} \leftrightarrow \displaystyle\sum_{k=-\infty}^{\infty} 2\pi\delta(\Omega - \Omega_0 - 2\pi k)$

$p[n] \leftrightarrow \dfrac{\sin\left[\left(q + \frac{1}{2}\right)\Omega\right]}{\sin(\Omega/2)}$

$\dfrac{B}{\pi}\text{sinc}\left(\dfrac{B}{\pi}n\right) \leftrightarrow \displaystyle\sum_{k=-\infty}^{\infty} p_{2B}(\Omega + 2\pi k)$

$\cos \Omega_0 n \leftrightarrow \displaystyle\sum_{k=-\infty}^{\infty} \pi[\delta(\Omega + \Omega_0 - 2\pi k) + \delta(\Omega - \Omega_0 - 2\pi k)]$

$\sin \Omega_0 n \leftrightarrow \displaystyle\sum_{k=-\infty}^{\infty} j\pi[\delta(\Omega + \Omega_0 - 2\pi k) - \delta(\Omega - \Omega_0 - 2\pi k)]$

$\cos(\Omega_0 n + \theta) \leftrightarrow \displaystyle\sum_{k=-\infty}^{\infty} \pi[e^{-j\theta}\delta(\Omega + \Omega_0 - 2\pi k) + e^{j\theta}\delta(\Omega - \Omega_0 - 2\pi k)]$

4.1.6   Properties of the DTFT

The DTFT has several properties, most of which are discrete-time versions of the properties of the continuous-time Fourier transform (CTFT). The properties of the DTFT are listed in Table 4.2. Except for the last property in Table 4.2, the proofs of these properties closely resemble the proofs of the corresponding properties of the CTFT. The details are omitted.

It should be noted that, in contrast to the CTFT, there is no duality property for the DTFT. However, there is a relationship between the inverse DTFT and the inverse CTFT. This is the last property listed in Table 4.2. This property is stated and proved subsequently.

Given a discrete-time signal $x[n]$ with DTFT $X(\Omega)$, let $X(\omega)$ denote $X(\Omega)$ with Ω replaced by ω, and let $p_{2\pi}(\omega)$ denote the rectangular frequency function with width equal to 2π. Then, the product $X(\omega)p_{2\pi}(\omega)$ is equal to $X(\omega)$ for $-\pi \leq \omega < \pi$ and is equal to zero for all other values of ω. Let $\gamma(t)$ denote the inverse CTFT of $X(\omega)p_{2\pi}(\omega)$. Then, the last property in Table 4.2 states that $x[n] = \gamma(n)$. To prove this,

TABLE 4.2   Properties of the DTFT

Property	Transform Pair/Property
Linearity	$ax[n] + bv[n] \leftrightarrow aX(\Omega) + bV(\Omega)$
Right or left shift in time	$x[n - q] \leftrightarrow X(\Omega)e^{-jq\Omega}, q$ any integer
Time reversal	$x[-n] \leftrightarrow X(-\Omega) = \overline{X(\Omega)}$
Multiplication by n	$nx[n] \leftrightarrow j\dfrac{d}{d\Omega}X(\Omega)$
Multiplication by a complex exponential	$x[n]e^{jn\Omega_0} \leftrightarrow X(\Omega - \Omega_0), \Omega_0$ real
Multiplication by $\sin \Omega_0 n$	$x[n]\sin \Omega_0 n \leftrightarrow \dfrac{j}{2}[X(\Omega + \Omega_0) - X(\Omega - \Omega_0)]$
Multiplication by $\cos \Omega_0 n$	$x[n]\cos \Omega_0 n \leftrightarrow \dfrac{1}{2}[X(\Omega + \Omega_0) + X(\Omega - \Omega_0)]$
Convolution in the time domain	$x[n] * v[n] \leftrightarrow X(\Omega)V(\Omega)$
Summation	$\displaystyle\sum_{i=0}^{n} x[i] \leftrightarrow \dfrac{1}{1 - e^{-j\Omega}}X(\Omega) + \sum_{n=-\infty}^{\infty} \pi X(2\pi n)\delta(\Omega - 2\pi n)$
Multiplication in the time domain	$x[n]v[n] \leftrightarrow \dfrac{1}{2\pi}\displaystyle\int_{-\pi}^{\pi} X(\Omega - \lambda)V(\lambda)\,d\lambda$
Parseval's theorem	$\displaystyle\sum_{n=-\infty}^{\infty} x[n]v[n] = \dfrac{1}{2\pi}\int_{-\pi}^{\pi} \overline{X(\Omega)}V(\Omega)\,d\Omega$
Special case of Parseval's theorem	$\displaystyle\sum_{n=-\infty}^{\infty} x^2[n] = \dfrac{1}{2\pi}\int_{-\pi}^{\pi} |X(\Omega)|^2\,d\Omega$
Relationship to inverse CTFT	If $x[n] \leftrightarrow X(\Omega)$ and $\gamma(t) \leftrightarrow X(\omega)p_{2\pi}(\omega)$, then $x[n] = \gamma(t)|_{t=n} = \gamma(n)$

first observe that by definition of the inverse CTFT,

$$\gamma(t) = \frac{1}{2\pi}\int_{-\infty}^{\infty} X(\omega)p_{2\pi}(\omega)e^{j\omega t}\,d\omega \tag{4.29}$$

By the definition of $X(\omega)p_{2\pi}(\omega)$, (4.29) reduces to

$$\gamma(t) = \frac{1}{2\pi}\int_{-\pi}^{\pi} X(\omega)e^{j\omega t}\,d\omega \tag{4.30}$$

Setting $t = n$ in (4.30) gives

$$\gamma(t)|_{t=n} = \gamma(n) = \frac{1}{2\pi}\int_{-\pi}^{\pi} X(\omega)e^{j\omega n}\,d\omega \tag{4.31}$$

and replacing ω by Ω in (4.31) gives

$$\gamma(n) = \frac{1}{2\pi}\int_{-\pi}^{\pi} X(\Omega)e^{j\Omega n}\,d\Omega \tag{4.32}$$

The right-hand side of (4.32) is equal to the inverse DTFT of $X(\Omega)$, and thus $\gamma(n)$ is equal to $x[n]$.

The relationship between the inverse CTFT and inverse DTFT can be used to generate DTFT pairs from CTFT pairs, as illustrated in the following example:

Example 4.7   DTFT Pair from CTFT

Suppose that

$$X(\Omega) = \sum_{k=-\infty}^{\infty} p_{2B}(\Omega + 2\pi k)$$

where $B < \pi$. The transform $X(\Omega)$ is plotted in Figure 4.7. It is seen that

$$X(\omega)p_{2\pi}(\omega) = p_{2B}(\omega)$$

From Table 3.2, the inverse CTFT of $p_{2B}(\omega)$ is equal to

$$\frac{B}{\pi}\text{sinc}\left(\frac{B}{\pi}t\right), \qquad -\infty < t < \infty$$

Thus,

$$x[n] = \gamma[n] = \frac{B}{\pi}\text{sinc}\left(\frac{B}{\pi}n\right), \qquad n = 0, \pm1, \pm2, \ldots$$

which yields the DTFT pair

$$\frac{B}{\pi}\text{sinc}\left(\frac{B}{\pi}n\right) \leftrightarrow \sum_{k=-\infty}^{\infty} p_{2B}(\Omega + 2\pi k)$$

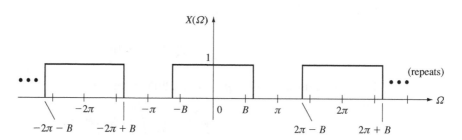

FIGURE 4.7
Transform in Example 4.7.

4.2   DISCRETE FOURIER TRANSFORM

Let $x[n]$ be a discrete-time signal with DTFT $X(\Omega)$. Since $X(\Omega)$ is a function of the continuous variable Ω, it cannot be stored in the memory of a digital computer unless $X(\Omega)$ can be expressed in a closed form. To implement DTFT techniques on a digital computer, it is necessary to discretize in frequency. This leads to the concept of the discrete Fourier transform, which is defined next.

Given a positive integer N, the N-point discrete Fourier transform (DFT) X_k of $x[n]$ over the time interval from $n = 0$ to $n = N - 1$ is defined by

$$X_k = \sum_{n=0}^{N-1} x[n] e^{-j2\pi kn/N}, \quad k = 0, 1, \ldots, N - 1 \tag{4.33}$$

From (4.33) it is seen that the DFT X_k is a function of the discrete (integer) variable k. Also note that, in contrast to the DTFT, the DFT X_k is completely specified by the N values $X_0, X_1, X_2, \ldots, X_{N-1}$. In general, these values are complex, and thus X_k can be expressed in either polar or rectangular form. The polar form is

$$X_k = |X_k| \exp[j \angle X_k], \quad k = 0, 1, \ldots, N - 1 \tag{4.34}$$

where $|X_k|$ is the magnitude of X_k and $\angle X_k$ is the angle of X_k. The rectangular form is

$$X_k = R_k + jI_k, \quad k = 0, 1, \ldots, N - 1 \tag{4.35}$$

where R_k is the real part of X_k given by

$$R_k = x[0] + \sum_{n=1}^{N-1} x[n] \cos \frac{2\pi kn}{N} \tag{4.36}$$

and I_k is the imaginary part of X_k given by

$$I_k = -\sum_{n=1}^{N-1} x[n] \sin \frac{2\pi kn}{N} \tag{4.37}$$

Since the summation in (4.33) is finite, the DFT X_k always exists. Further, X_k can be computed by simply evaluating the finite summation in (4.33). A MATLAB program for computing the DFT is given in Figure 4.8. In Section 4.4, the MATLAB command fft is considered, which implements a fast algorithm for computing the DFT.

```
%
% Discrete Fourier Transform
%
function Xk = dft(x)
[N,M] = size(x);
if M ~= 1          % makes sure that x is a column vector
        x = x´;
        N = M;
end
Xk = zeros(N,1);
n = 0:N-1
for k = 0:N-1,
        Xk(k + 1) = exp(-j*2*pi*k*n/N)*x;
end
```

FIGURE 4.8
MATLAB program for evaluating the DFT.

Example 4.8   Computation of DFT

Suppose that $x[0] = 1$, $x[1] = 2$, $x[2] = 2$, $x[3] = 1$, and $x[n] = 0$ for all other integers n. With $N = 4$, from (4.33) the DFT of $x[n]$ over the time interval $n = 0$ to $n = N - 1 = 3$ is

$$X_k = \sum_{n=0}^{3} x[n]e^{-j\pi kn/2}, \quad k = 0, 1, 2, 3$$

$$= x[0] + x[1]e^{-j\pi k/2} + x[2]e^{-j\pi k} + x[3]e^{-j\pi 3k/2}, \quad k = 0, 1, 2, 3$$

$$= 1 + 2e^{-j\pi k/2} + 2e^{-j\pi k} + e^{-j\pi 3k/2}, \quad k = 0, 1, 2, 3$$

The real part R_k of X_k is

$$R_k = 1 + 2\cos\frac{-\pi k}{2} + 2\cos(-\pi k) + \cos\frac{-\pi 3k}{2}, \quad k = 0, 1, 2, 3$$

Thus

$$R_k = \begin{cases} 6, & k = 0 \\ -1, & k = 1 \\ 0, & k = 2 \\ -1, & k = 3 \end{cases}$$

The imaginary part I_k of X_k is

$$I_k = -2\sin\frac{\pi k}{2} - 2\sin\pi k - \sin\frac{\pi 3k}{2}, \quad k = 0, 1, 2, 3$$

Hence,

$$I_k = \begin{cases} 0, & k = 0 \\ -1, & k = 1 \\ 0, & k = 2 \\ 1, & k = 3 \end{cases}$$

and the rectangular form of X_k is

$$X_k = \begin{cases} 6, & k = 0 \\ -1 - j, & k = 1 \\ 0, & k = 2 \\ -1 + j, & k = 3 \end{cases}$$

As a check, these values for X_k were also obtained by the MATLAB program in Figure 4.8 with the commands

```
x = [1 2 2 1];
Xk = dft(x)
```

The polar form of X_k is

$$X_k = \begin{cases} 6e^{j0}, & k = 0 \\ \sqrt{2}e^{j5\pi/4}, & k = 1 \\ 0e^{j0}, & k = 2 \\ \sqrt{2}e^{j3\pi/4}, & k = 3 \end{cases}$$

4.2.1   Symmetry

In Section 4.1 it was shown that the magnitude $|X(\Omega)|$ and phase $\angle X(\Omega)$ of the DTFT $X(\Omega)$ of a discrete-time signal $x[n]$ are even and odd functions of Ω, respectively. It turns out that the N-point DFT X_k of $x[n]$ also has symmetry properties, which are derived as follows. First, replacing k by $N - k$ in (4.33) yields

$$X_{N-k} = \sum_{n=0}^{N-1} x[n]e^{-j2\pi(N-k)n/N}, k = 0, 1, 2, \ldots, N - 1$$

$$X_{N-k} = \sum_{n=0}^{N-1} x[n]e^{j2\pi kn/N}e^{-j2\pi n}, k = 0, 1, 2, \ldots, N - 1 \qquad (4.38)$$

Since $e^{-j2\pi n} = 1$ for $n = 0, 1, 2, \ldots$, (4.38) reduces to

$$X_{N-k} = \overline{X_k}, k = 0, 1, 2, \ldots, N - 1 \qquad (4.39)$$

where $\overline{X_k}$ is the complex conjugate of X_k. It follows from (4.39) that $|X_k|$ is symmetrical about $k = (N/2)$ when N is even, and $\angle X_k$ has odd symmetry about $k = (N/2)$ when N is even. The relationship (4.39) is used later to express the given signal as a sum of sinusoids.

4.2.2   Inverse DFT

If X_k is the N-point DFT of $x[n]$, then $x[n]$ can be determined from X_k by applying the inverse DFT given by

$$x[n] = \frac{1}{N}\sum_{k=0}^{N-1} X_k e^{j2\pi kn/N}, n = 0, 1, \ldots, N - 1 \qquad (4.40)$$

Since the sum in (4.40) is finite, we can compute the inverse DFT by simply evaluating the summation in (4.40). A MATLAB® program for computing the inverse DFT is given in Figure 4.9.

```
%
% Inverse Discrete Fourier Transform
%
function x = idft(Xk)
[N,M] = size(Xk);
if M ~= 1        % makes sure that Xk is a column vector
  Xk = Xk.';    % .' takes the transpose without conjugation
  N = M;
end
x = zeros(N,1);
k = 0:N-1;
for n = 0:N-1,
  x(n + 1) = exp(j*2*pi*k*n/N)*Xk;
end
x = x/N;
```

FIGURE 4.9
MATLAB program for computing the inverse DFT.

Example 4.9   Computation of Inverse DFT

Again consider the signal in Example 4.8 with the rectangular form of the DFT given by

$$X_k = \begin{cases} 6, & k = 0 \\ -1 - j, & k = 1 \\ 0, & k = 2 \\ -1 + j, & k = 3 \end{cases}$$

Evaluating (4.40) with $N = 4$ gives

$$x[n] = \frac{1}{4}[X_0 + X_1 e^{j\pi n/2} + X_2 e^{j\pi n} + X_3 e^{j3\pi n/2}], \quad n = 0, 1, 2, 3$$

Thus,

$$x[0] = \frac{1}{4}[X_0 + X_1 + X_2 + X_3] = 1$$

$$x[1] = \frac{1}{4}[X_0 + jX_1 - X_2 - jX_3] = \frac{1}{4}[8] = 2$$

$$x[2] = \frac{1}{4}[X_0 - X_1 + X_2 - X_3] = 2$$

$$x[3] = \frac{1}{4}[X_0 - jX_1 - X_2 + jX_3] = \frac{1}{4}[4] = 1$$

These values are equal to the values of $x[n]$ specified in Example 4.8. Also, these values for $x[n]$ result when the program in Figure 4.9 is run with the commands

```
Xk = [6 -1-j 0 -1+j];
x = idft(Xk)
```

4.2.3   Sinusoidal Form

Equation (4.40) shows that, over the time interval from $n = 0$ to $n = N - 1$, $x[n]$ is equal to a sum of complex exponentials. Since the values of $x[n]$ are real numbers, it turns out that the right-hand side of (4.40) can be expressed as a sum of sinusoids. The derivation of this form is based on the relationship (4.39):

First, suppose that N is an odd integer with $N \geq 3$. Then $N - 1$ and $N + 1$ are even integers, and (4.40) can be written in the form

$$x[n] = \frac{1}{N}X_0 + \frac{1}{N}\sum_{k=1}^{(N-1)/2} X_k e^{j2\pi kn/N} + \frac{1}{N}\sum_{k=(N+1)/2}^{N-1} X_k e^{j2\pi kn/N},$$

$$n = 0, 1, \ldots, N - 1 \qquad (4.41)$$

Carrying out a change of index in the second summation on the right-hand side of (4.41) and using (4.39) yield the result

$$\frac{1}{N}\sum_{k=(N+1)/2}^{N-1} X_k e^{j2\pi kn/N} = \frac{1}{N}\sum_{k=1}^{(N-1)/2} \overline{X_k} e^{-j2\pi kn/N} \qquad (4.42)$$

(The reader is asked to verify (4.42) in Problem 4.10.)

Inserting (4.42) into (4.41) yields

$$x[n] = \frac{1}{N}X_0 + \frac{1}{N}\sum_{k=1}^{(N-1)/2} [X_k e^{j2\pi kn/N} + \overline{X_k} e^{-j2\pi kn/N}], \, n = 0, 1, \ldots, N - 1 \quad (4.43)$$

The term $\overline{X_k} e^{-j2\pi kn/N}$ is the complex conjugate of $X_k e^{j2\pi kn/N}$, and thus the sum $X_k e^{j2\pi kn/N} + \overline{X_k} e^{-j2\pi kn/N}$ is equal to 2 times the real part of $X_k e^{j2\pi kn/N}$. Using the rectangular form of X_k given by (4.35) and Euler's formula gives

$$X_k e^{j2\pi kn/N} + \overline{X_k} e^{-j2\pi kn/N} = 2\,\mathrm{Re}[X_k e^{j2\pi kn/N}]$$

$$= 2\,\mathrm{Re}\left[(R_k + jI_k)\left(\cos\frac{2\pi kn}{N} + j\sin\frac{2\pi kn}{N}\right)\right]$$

$$= 2\left[R_k \cos\frac{2\pi kn}{N} - I_k \sin\frac{2\pi kn}{N}\right]$$

Inserting this result into (4.43) yields the following sinusoidal form for $x[n]$:

$$x[n] = \frac{1}{N}X_0 + \frac{2}{N}\sum_{k=1}^{(N-1)/2}\left[R_k \cos\frac{2\pi kn}{N} - I_k \sin\frac{2\pi kn}{N}\right], \, n = 0, 1, \ldots, N - 1 \quad (4.44)$$

From (4.44), it can be seen that the constant (zero-frequency) component of $x[n]$ is equal to $(1/N)X_0$, and from (4.33),

$$X_0 = \sum_{n=0}^{N-1} x[n]$$

Hence, the constant component of $x[n]$ is equal to the average value of $x[n]$ over the interval from $n = 0$ to $n = N - 1$. From (4.44), it can also be seen that the first harmonic contained in $x[n]$ is the term $\dfrac{2}{N}\left[R_1 \cos\dfrac{2\pi n}{N} - I_1 \sin\dfrac{2\pi n}{N} \right]$, which has frequency $2\pi/N$. This is the lowest possible (nonzero) frequency component contained in $x[n]$. The highest possible frequency component contained in $x[n]$ has frequency $\dfrac{(N - 1)\pi}{N}$.

When N is even, the sinusoidal form of $x[n]$ is given by

$$x[n] = \frac{1}{N}X_0 + \frac{2}{N}\sum_{k=1}^{\frac{N}{2}-1}\left[R_k \cos\frac{2\pi kn}{N} - I_k \sin\frac{2\pi kn}{N} \right] + \frac{1}{N}R_{N/2}\cos\pi n \quad (4.45)$$

Note that in this case, the highest frequency component contained in $x[n]$ has frequency π if $R_{N/2} \neq 0$. The derivation of (4.45) is left to the reader. (See Problem 4.11.)

Example 4.10   Sinusoidal Form

Again consider the signal in Example 4.8 with the rectangular form of the DFT given by

$$X_k = \begin{cases} 6, & k = 0 \\ -1 - j, & k = 1 \\ 0, & k = 2 \\ -1 + j, & k = 3 \end{cases}$$

Since $N = 4$, N is even, and the sinusoidal form of $x[n]$ is given by (4.45). Inserting the values of R_k and I_k into (4.45) results in the following sinusoidal form of $x[n]$:

$$x[n] = \frac{6}{4} + \frac{2}{4}\left[R_1 \cos\frac{2\pi n}{4} - I_1 \sin\frac{2\pi n}{4} \right] + \frac{1}{4}R_2 \cos\pi n$$

$$= 1.5 - 0.5\cos\frac{\pi n}{2} + 0.5\sin\frac{\pi n}{2}$$

4.2.4   Relationship to DTFT

Given a discrete-time signal $x[n]$, let X_k denote the N-point DFT defined by (4.33), and let $X(\Omega)$ denote the DTFT of $x[n]$ defined by

$$X(\Omega) = \sum_{n=-\infty}^{\infty} x[n]e^{-j\Omega n} \quad (4.46)$$

If $x[n] = 0$ for $n < 0$ and $n \geq N$, (4.46) reduces to

$$X(\Omega) = \sum_{n=0}^{N-1} x[n]e^{-j\Omega n} \quad (4.47)$$

Comparing (4.33) and (4.47) reveals that

$$X_k = X(\Omega)|_{\Omega=2\pi k/N} = X\left(\frac{2\pi k}{N}\right), k = 0, 1, \ldots, N-1 \qquad (4.48)$$

Thus, if $x[n] = 0$ for $n < 0$ and $n \geq N$, the DFT X_k can be viewed as a frequency-sampled version of the DTFT $X(\Omega)$; more precisely, X_k is equal to $X(\Omega)$ with Ω evaluated at the frequency points $\Omega = 2\pi k/N$ for $k = 0, 1, \ldots, N-1$.

Example 4.11   DTFT and DFT of a Pulse

DTFT and DFT of a Pulse

With $p[n]$ equal to the rectangular pulse defined in Example 4.3, let $x[n] = p[n - q]$. Then by definition of $p[n]$,

$$x[n] = \begin{cases} 1, & n = 0, 1, 2, \ldots, 2q \\ 0, & \text{all other } n \end{cases}$$

From the result in Example 4.3,

$$p[n] \leftrightarrow \frac{\sin\left[\left(q + \frac{1}{2}\right)\Omega\right]}{\sin(\Omega/2)}$$

and by the time-shift property of the DTFT (see Table 4.2), the DTFT of $x[n] = p[n - q]$ is given by

$$X[\Omega] = \frac{\sin\left[\left(q + \frac{1}{2}\right)\Omega\right]}{\sin(\Omega/2)} e^{-jq\Omega}$$

Thus, the amplitude spectrum of $x[n]$ is

$$|X[\Omega]| = \frac{\left|\sin\left[\left(q + \frac{1}{2}\right)\Omega\right]\right|}{|\sin(\Omega/2)|}$$

In the case $q = 5$, $|X(\Omega)|$ is plotted in Figure 4.10. Note that $|X(\Omega)|$ is plotted for Ω ranging from 0 to 2π (as opposed to $-\pi$ to π). In this section the amplitude spectrum $|X(\Omega)|$ is displayed

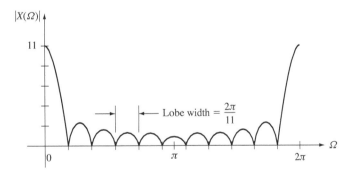

FIGURE 4.10
Amplitude spectrum in the case $q = 5$.

from 0 to 2π, since X_k for $k = 0, 1, \ldots, N - 1$ corresponds to the values of $X(\Omega)$ for values of Ω over the interval from 0 to 2π.

Now, letting X_k denote the N-point DFT of $x[n]$ with $N = 2q + 1$, from (4.48),

$$|X_k| = \left| X\left(\frac{2\pi k}{N}\right) \right| = \frac{\left|\sin\left[\left(q + \frac{1}{2}\right)(2\pi k/N)\right]\right|}{|\sin(\pi k/N)|}, \quad k = 0, 1, \ldots, 2q$$

Replacing N by $2q + 1$ gives

$$|X_k| = \frac{\left|\sin\left[\left(\dfrac{2q + 1}{2}\right)\dfrac{2\pi k}{2q + 1}\right]\right|}{\left|\sin\left(\dfrac{\pi k}{2q + 1}\right)\right|}, \quad k = 0, 1, \ldots, 2q$$

$$|X_k| = \frac{|\sin \pi k|}{\left|\sin\left(\dfrac{\pi k}{2q + 1}\right)\right|}, \quad k = 0, 1, \ldots, 2q$$

$$|X_k| = \begin{cases} 2q + 1, & k = 0 \\ 0, & k = 1, 2, \ldots, 2q \end{cases}$$

The value of $|X_k|$ for $k = 0$ was computed by the use of l'Hôpital's rule.

Note that, since $|X_k| = 0$ for $k = 1, 2, \ldots, 2q$, the sample values $X(2\pi k/N)$ of $X(\Omega)$ are all equal to zero for these values of k. This is a consequence of sampling $X(\Omega)$ at the zero points located between the sidelobes of $X(\Omega)$. (See Figure 4.10.) Since X_k is nonzero for only $k = 0$, X_k bears very little resemblance to the spectrum $X(\Omega)$ of the rectangular pulse $x[n] = p[n - q]$. However, by making N larger, so that the sampling frequencies $2\pi k/N$ are closer together, it is expected that the DFT X_k should be a better representation of the spectrum $X(\Omega)$. For example, when $N = 2(2q + 1)$, the resulting N-point DFT X_k is equal to the values of $X(\Omega)$ at the frequency points corresponding to the peaks and zero points of the sidelobes in $X(\Omega)$. A plot of the amplitude $|X_k|$ for $q = 5$ and $N = 2(2q + 1) = 22$ is given in Figure 4.11. Clearly, $|X_k|$ now bears some resemblance to $|X(\Omega)|$ displayed in Figure 4.10. To obtain an even closer correspondence, N can be increased again. For instance, $|X_k|$ is plotted in Figure 4.12 for

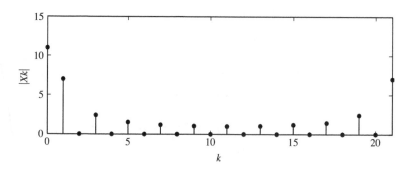

FIGURE 4.11
Amplitude of DFT in the case $q = 5$ and $N = 22$.

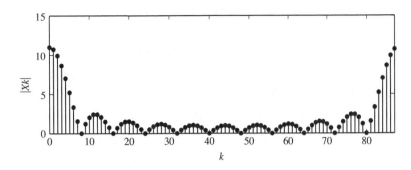

FIGURE 4.12
Amplitude of DFT in the case $q = 5$ and $N = 88$.

the case when $q = 5$ and $N = 88$. Here the DFT was computed by running the program in Figure 4.8 with

```
q = 5; N = 88;
x = [ones(1,2*q + 1) zeros(1,N-2*q-1)];
Xk = dft(x);
k = 0:N-1;
stem(k,abs(Xk),'filled')  % plots the magnitude
```

4.3   DFT OF TRUNCATED SIGNALS

In most applications, a discrete-time signal $x[n]$ is known over only a finite interval of time. To compute the Fourier transform in such cases, it is necessary to consider the truncated signal $\widetilde{x}[n]$ defined by

$$\widetilde{x}[n] = \begin{cases} x[n], & n = 0, 1, \dots, N - 1 \\ 0, & n < 0 \text{ and } n \geq N \end{cases} \tag{4.49}$$

where N is the number of values of $x[n]$ that are known. The integer N is often referred to as the *record length*. As a result of the truncation, the values of the DFT of the truncated signal $\widetilde{x}[n]$ are usually different from the values $X\left(\dfrac{2\pi k}{N}\right)$ of the DTFT of $x[n]$; in other words, the relationship (4.48) no longer holds, in general, when the DFT is applied to the truncated signal. This is shown to be the case by the following analysis:

Again consider the truncated signal given by (4.49), where now it is assumed that N is an odd integer; and consider the rectangular pulse $p[n]$ defined in Example 4.3. Setting $q = (N - 1)/2$ in the definition of the pulse and shifting the pulse to the right by $(N - 1)/2$ time units give

$$p\left[n - \frac{N - 1}{2}\right] = \begin{cases} 1, & n = 0, 1, \dots, N - 1 \\ 0, & \text{all other } n \end{cases}$$

Hence, the truncated signal given by (4.49) can be expressed in the form

$$\widetilde{x}[n] = x[n]p\left[n - \frac{N-1}{2}\right] \tag{4.50}$$

where $x[n]$ is the original discrete-time signal whose values are known only for $n = 0, 1, \ldots, N - 1$.

Now let $P(\Omega)$ denote the DTFT of the rectangular pulse $p\left[n - \dfrac{N-1}{2}\right]$. Setting $q = (N-1)/2$ in the result in Example 4.11 gives

$$P(\Omega) = \frac{\sin[N\Omega/2]}{\sin(\Omega/2)}e^{-j(N-1)\Omega/2}$$

Then, by the DTFT property involving multiplication of signals (see Table 4.2), taking the DTFT of both sides of (4.50) results in the DTFT $\widetilde{X}(\Omega)$ of the truncated signal $\widetilde{x}[n]$ given by

$$\widetilde{X}(\Omega) = X(\Omega) * P(\Omega) = \frac{1}{2\pi}\int_{-\pi}^{\pi} X(\Omega - \lambda)P(\lambda)\,d\lambda \tag{4.51}$$

where $X(\Omega)$ is the DTFT of $x[n]$. Thus, the N-point DFT $\widetilde{X}_k$ of the truncated signal $\widetilde{x}[n]$ [defined by (4.49)] is given by

$$\widetilde{X}_k = [X(\Omega) * P(\Omega)]_{\Omega=2\pi k/N}, \quad k = 0, 1, \ldots, N - 1 \tag{4.52}$$

By (4.52) it is seen that the distortion in $\widetilde{X}_k$ from the desired values $X(2\pi k/N)$ can be characterized in terms of the effect of convolving $P(\Omega)$ with the spectrum $X(\Omega)$ of the signal. If $x[n]$ is not suitably small for $n < 0$ and $n \geq N$, in general, the sidelobes that exist in the amplitude spectrum $|P(\Omega)|$ will result in sidelobes in the amplitude spectrum $|X(\Omega) * P(\Omega)|$. This effect is shown in the following example:

Example 4.12 N-Point DFT

Consider the discrete-time signal $x[n] = (0.9)^n u[n]$, $n \geq 0$, which is plotted in Figure 4.13. From the results in Example 4.2, the DTFT is

$$X(\Omega) = \frac{1}{1 - 0.9e^{-j\Omega}}$$

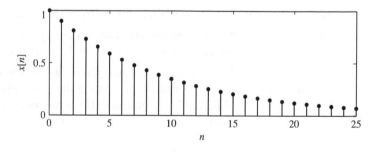

FIGURE 4.13
Signal in Example 4.12.

The amplitude spectrum $|X(\Omega)|$ is plotted in Figure 4.14. Note that, since the signal varies rather slowly, most of the spectral content over the frequency range from 0 to π is concentrated near the zero point $\Omega = 0$. For $N = 21$ the amplitude of the N-point DFT of the signal is shown in Figure 4.15. This plot was obtained by the commands

```
N = 21; n = 0:N-1;
x = 0.9.^n;
Xk = dft(x);
k = n;
stem(k,abs(Xk),'filled')
```

Comparing Figures 4.14 and 4.15, we see that the amplitude of the 21-point DFT is a close approximation to the amplitude spectrum $|X(\Omega)|$. This turns out to be the case since $x[n]$ is small for $n \geq 21$ and is zero for $n < 0$.

Now consider the truncated signal $x[n]$ shown in Figure 4.16. The amplitude of the 21-point DFT of the truncated signal is plotted in Figure 4.17. This plot was generated by the following commands:

```
N = 21; n = 0:N-1;
x = 0.9.^n;
x(12:21) = zeros(1,10);
Xk = dft(x);
k = n;
stem(k,abs(Xk),'filled')
```

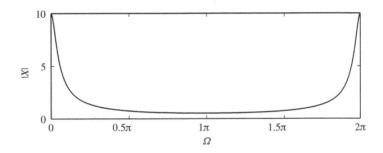

FIGURE 4.14
Amplitude spectrum of signal in Example 4.12.

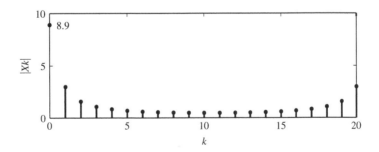

FIGURE 4.15
Amplitude of 21-point DFT.

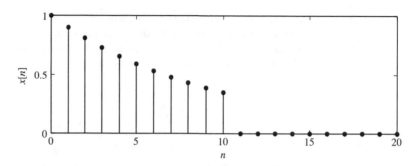

FIGURE 4.16
Truncated signal in Example 4.12.

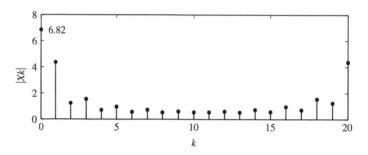

FIGURE 4.17
Amplitude of 21-point DFT of truncated signal.

Comparing Figures 4.17 and 4.15 reveals that the spectral content of the truncated signal has higher frequency components than those of the signal displayed in Figure 4.13. The reason for this is that the truncation at $n = 11$ causes an abrupt change in the signal magnitude, which introduces high-frequency components in the signal spectrum (as displayed by the DFT).

The next example shows that the sidelobes in the amplitude spectrum $|X(\Omega) * P(\Omega)|$ can produce a phenomenon whereby spectral components can "leak" into various frequency locations as a result of the truncation process.

Example 4.13   DFT of Truncated Sinusoid

Suppose that the signal $x[n]$ is the infinite-duration sinusoid $x[n] = (\cos \Omega_0 n)$, $-\infty < n < \infty$. From Table 4.1, we see that the DTFT of $x[n]$ is the impulse train

$$\sum_{i=-\infty}^{\infty} \pi[\delta(\Omega + \Omega_0 - 2\pi i) + \delta(\Omega - \Omega_0 - 2\pi i)]$$

A plot of the DTFT of $\cos \Omega_0 n$ for $-\pi \leq \Omega \leq \pi$ is shown in Figure 4.18. From the figure it is seen that, over the frequency range $-\pi \leq \Omega \leq \pi$, all the spectral content of the signal $\cos \Omega_0 n$ is concentrated at $\Omega = \Omega_0$ and $\Omega = -\Omega_0$. Now consider the truncated sinusoid $\widetilde{x}[n] = (\cos \Omega_0 n)$ $p\left[n - \dfrac{N-1}{2}\right]$, where $0 \leq \Omega_0 \leq \pi$ and $p\left[n - \dfrac{N-1}{2}\right]$ is the shifted rectangular pulse defined in Example 4.12, where N is an odd integer with $N \geq 3$. Then by definition of $p\left[n - \dfrac{N-1}{2}\right]$, the truncated signal is given by

$$\widetilde{x}[n] = \begin{cases} \cos \Omega_0 n, & n = 0, 1, \ldots, N-1 \\ 0, & \text{all other } n \end{cases}$$

As given in Example 4.12, the DTFT $P(\Omega)$ of the pulse $p\left[n - \dfrac{N-1}{2}\right]$ is

$$P(\Omega) = \frac{\sin[N\Omega/2]}{\sin(\Omega/2)} e^{-j(N-1)\Omega/2}$$

Then, by the DTFT property involving multiplication of signals, the DTFT $\widetilde{X}(\Omega)$ of the truncated signal $\widetilde{x}[n]$ is given by

$$\widetilde{X}(\Omega) = \frac{1}{2\pi} \int_{-\pi}^{\pi} P(\Omega - \lambda)\pi[\delta(\lambda + \Omega_0) + \delta(\lambda - \Omega_0)]\, d\lambda$$

Using the shifting property of the impulse (see Section 1.1) yields

$$\widetilde{X}(\Omega) = \frac{1}{2}[P(\Omega + \Omega_0) + P(\Omega - \Omega_0)]$$

Now the relationship (4.48) holds for the truncated signal $\widetilde{x}[n]$, and thus the N-point DFT $\widetilde{X}_k$ of $\widetilde{x}[n]$ is given by

$$\widetilde{X}_k = \widetilde{X}\left(\frac{2\pi k}{N}\right) = \frac{1}{2}\left[P\left(\frac{2\pi k}{N} + \Omega_0\right) + P\left(\frac{2\pi k}{N} - \Omega_0\right)\right], \quad k = 0, 1, \ldots, N-1$$

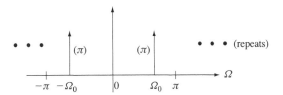

FIGURE 4.18
DTFT of $x[n] = \cos \Omega_0 n$ with $-\pi \leq \Omega \leq \pi$.

where

$$P\left(\frac{2\pi k}{N} \pm \Omega_0\right) = \frac{\sin\left[\left(\dfrac{N}{2}\right)\left(\dfrac{2\pi k}{N} \pm \Omega_0\right)\right]}{\sin\left[\left(\dfrac{2\pi k}{N} \pm \Omega_0\right)\Big/2\right]} \exp\left[-j\left(\frac{N-1}{2}\right)\left(\frac{2\pi k}{N} \pm \Omega_0\right)\right],$$

$$k = 0, 1, 2, \ldots, N-1$$

Suppose that $\Omega_0 = (2\pi r)/N$ for some integer r where $0 \le r \le N-1$. This is equivalent to assuming that $\cos \Omega_0 n$ goes through r complete periods as n is varied from $n = 0$ to $n = N-1$. Then

$$P\left(\frac{2\pi k}{N} \pm \Omega_0\right) = \frac{\sin\left[\left(\dfrac{N}{2}\right)\left(\dfrac{2\pi k \pm 2\pi r}{N}\right)\right]}{\sin\left[\left(\dfrac{2\pi k \pm 2\pi r}{N}\right)\Big/2\right]} \exp\left[-jq\,\frac{2\pi k \pm 2\pi r}{N}\right],$$

$$k = 0, 1, \ldots, N-1$$

$$= \frac{\sin(\pi k \pm \pi r)}{\sin\left(\dfrac{\pi k \pm \pi r}{N}\right)} \exp\left[-jq\,\frac{2\pi(k \pm r)}{N}\right], \quad k = 0, 1, \ldots, N-1$$

and thus

$$P\left(\frac{2\pi k}{N} - \Omega_0\right) = \begin{cases} N, & k = r \\ 0, & k = 0, 1, \ldots, r-1, r+1, \ldots, N-1 \end{cases}$$

$$P\left(\frac{2\pi k}{N} + \Omega_0\right) = \begin{cases} N, & k = N-r \\ 0, & k = 0, 1, \ldots, N-r-1, N-r+1, \ldots, N-1 \end{cases}$$

Finally, the DFT $\widetilde{X}_k$ is given by

$$\widetilde{X}_k = \begin{cases} \dfrac{N}{2}, & k = r \\[2mm] \dfrac{N}{2}, & k = N-r \\[2mm] 0, & \text{all other } k \text{ for } 0 \le k \le N-1 \end{cases}$$

Since $k = r$ corresponds to the frequency point $\Omega_0 = (2\pi r)/N$, this result shows that the portion of the spectrum corresponding to the frequency range from 0 to π is concentrated at the expected point. Hence, the DFT $\widetilde{X}_k$ of the truncated signal $\widetilde{x}[n]$ is a "faithful" representation of the DTFT of the infinite-duration sinusoid $\cos \Omega_0 n$. The DFT $\widetilde{X}_k$ is plotted in Figure 4.19 for the case when $N = 21$ and $\Omega_0 = 10\pi/21$ (which implies that $r = 5$). This plot was generated by applying the MATLAB command $\mathtt{dft}$ directly to $x[n]$. The MATLAB program is

```
q = 10; N = 2*q + 1;
Wo = 10*pi/21;
n = 0:N-1;
x = cos(Wo*n);
Xk = dft(x);
k = 0:N-1;
stem(k,abs(Xk),'filled')
```

Now, suppose that Ω_0 is not equal to $2\pi r/N$ for any integer r. Let β denote the integer for which

$$\left| \Omega_0 - \frac{2\pi\beta}{N} \right|$$

has the smallest possible value. Then the DFT $\widetilde{X}_k$ will have nonzero values distributed in a neighborhood of the point $k = \beta$. This characteristic is referred to as *leakage*, meaning that the spectral component concentrated at Ω_0 is spread (or "leaks") into the frequency components in a neighborhood of $2\pi\beta/N$. For the case $N = 21$ and $\Omega_0 = 9.5\pi/21$, the DFT $\widetilde{X}_k$ is plotted in Figure 4.20. Here, $\beta = 5$ and thus the values of $\widetilde{X}_k$ are distributed about the point $k = 5$ (and the corresponding point $k = N - r$). The plot given in Figure 4.20 was generated by modification of the preceding MATLAB commands with $N = 21$ and $\mathtt{Wo} = \mathtt{9.5*pi/21}$.

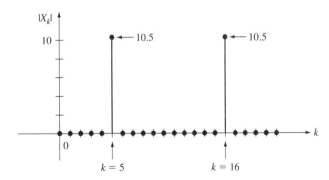

FIGURE 4.19

DFT $\widetilde{X}_k$ of the truncated signal $\widetilde{x}[n] = (\cos \Omega_0 n)p\left[n - \dfrac{N-1}{2}\right]$ when $N = 21$ and $\Omega_0 = 10\pi/21$.

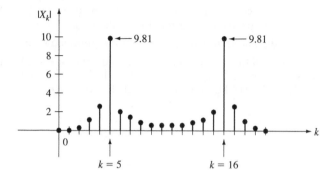

FIGURE 4.20

DFT $\widetilde{X}_k$ of the truncated signal $\widetilde{x}[n] = (\cos \Omega_0 n)p\left[n - \dfrac{N-1}{2}\right]$ when $N = 21$ and $\Omega_0 = 9.5\pi/21$.

4.4   FFT ALGORITHM

Given a discrete-time signal $x[n]$, in Section 4.2 the N-point DFT and inverse DFT were defined by

$$X_k = \sum_{n=0}^{N-1} x[n]e^{-j2\pi kn/N}, \quad k = 0, 1, \ldots, N-1 \tag{4.53}$$

$$x[n] = \frac{1}{N}\sum_{k=0}^{N-1} X_k e^{j2\pi kn/N}, \quad n = 0, 1, \ldots, N-1 \tag{4.54}$$

From (4.53) it is seen that, for each value of k, the computation of X_k from $x[n]$ requires N multiplications. Thus, the computation of X_k for $k = 0, 1, \ldots, N-1$ requires N^2 multiplications. Similarly, from (4.54) it follows that the computation of $x[n]$ from X_k also requires N^2 multiplications.

It should be mentioned that the multiplications in (4.53) and (4.54) are complex multiplications, in general; that is, the numbers being multiplied are complex numbers. The multiplication of two complex numbers requires four real multiplications. In the following analysis, the number of complex multiplications is counted. The number of additions required to compute the DFT or inverse DFT will not be considered.

Since the direct evaluation of (4.53) or (4.54) requires N^2 multiplications, this can result in a great deal of computation if N is large. It turns out that (4.53) or (4.54) can be computed with a fast Fourier transform (FFT) algorithm, which requires on the order of $(N \log_2 N)/2$ multiplications. This is a significant decrease in the N^2 multiplications required in the direct evaluation of (4.53) or (4.54). For instance, if $N = 1024$, the direct evaluation requires $N^2 = 1{,}048{,}576$ multiplications. In contrast, the FFT algorithm requires

$$\frac{1024(\log_2 1024)}{2} = 5120 \text{ multiplications}$$

There are different versions of the FFT algorithm. Here, the development is limited to one particular approach based on decimation in time. For an in-depth treatment of the FFT algorithm, the reader is referred to Brigham [1988] or Rabiner and Gold [1975].

The basic idea of the decimation-in-time approach is to subdivide the time interval into intervals having a smaller number of points. We illustrate this by first showing that the computation of X_k can be broken up into two parts. First, to simplify the notation, let W_N equal $\exp(-j2\pi/N)$. The complex number W_N is an Nth root of unity; that is,

$$W_N^N = e^{-j2\pi} = 1$$

It is assumed that $N > 1$, and thus $W_N \neq 1$. In terms of W_N, the N-point DFT and inverse DFT are given by

$$X_k = \sum_{n=0}^{N-1} x[n] W_N^{kn}, \ k = 0, 1, \ldots, N-1 \tag{4.55}$$

$$x[n] = \frac{1}{N} \sum_{k=0}^{N-1} X_k W_N^{-kn}, n = 0, 1, \ldots, N-1 \tag{4.56}$$

Now, let N be an even integer, so that $N/2$ is an integer. Given the signal $x[n]$, define the signals

$$a[n] = x[2n], \qquad n = 0, 1, 2, \ldots, \frac{N}{2} - 1$$

$$b[n] = x[2n + 1], \quad n = 0, 1, 2, \ldots, \frac{N}{2} - 1$$

Note that the signal $a[n]$ consists of the values of $x[n]$ at the even values of the time index n, while $b[n]$ consists of the values at the odd time points.

Let A_k and B_k denote the $(N/2)$-point DFTs of $a[n]$ and $b[n]$; that is,

$$A_k = \sum_{n=0}^{(N/2)-1} a[n] W_{N/2}^{kn}, \quad k = 0, 1, \ldots, \frac{N}{2} - 1 \tag{4.57}$$

$$B_k = \sum_{n=0}^{(N/2)-1} b[n] W_{N/2}^{kn}, \quad k = 0, 1, \ldots, \frac{N}{2} - 1 \tag{4.58}$$

Let X_k denote the N-point DFT of $x[n]$. Then it is claimed that

$$X_k = A_k + W_N^k B_k, \quad k = 0, 1, \ldots, \frac{N}{2} - 1 \tag{4.59}$$

$$X_{(N/2)+k} = A_k - W_N^k B_k, \quad k = 0, 1, \ldots, \frac{N}{2} - 1 \tag{4.60}$$

To verify (4.59), insert the expressions (4.57) and (4.58) for A_k and B_k into the right-hand side of (4.59). This gives

$$A_k + W_N^k B_k = \sum_{n=0}^{(N/2)-1} a[n]W_{N/2}^{kn} + \sum_{n=0}^{(N/2)-1} b[n]W_N^k W_{N/2}^{kn}$$

Now $a[n] = x[2n]$ and $b[n] = x[2n+1]$, and thus

$$A_k + W_N^k B_k = \sum_{n=0}^{(N/2)-1} x[2n]W_{N/2}^{kn} + \sum_{n=0}^{(N/2)-1} x[2n+1]W_N^k W_{N/2}^{kn}$$

Using the properties

$$W_{N/2}^{kn} = W_N^{2kn}, \quad W_N^k W_{N/2}^{kn} = W_N^{(1+2n)k}$$

yields the result

$$A_k + W_N^k B_k = \sum_{n=0}^{(N/2)-1} x[2n]W_N^{2kn} + \sum_{n=0}^{(N/2)-1} x[2n+1]W_N^{(1+2n)k} \tag{4.61}$$

Defining the change of index $\bar{n} = 2n$ in the first sum of the right-hand side of (4.61) and the change of index $\bar{n} = 2n + 1$ in the second sum yields

$$A_k + W_N^k B_k = \sum_{\substack{\bar{n}=0 \\ \bar{n} \text{ even}}}^{N-2} x[\bar{n}]W_N^{\bar{n}k} + \sum_{\substack{\bar{n}=0 \\ \bar{n} \text{ odd}}}^{N-1} x[\bar{n}]W_N^{\bar{n}k}$$

$$A_k + W_N^k B_k = \sum_{\bar{n}=0}^{N-1} x[\bar{n}]W_N^{\bar{n}k}$$

$$A_k + W_N^k B_k = X_k$$

Hence, (4.59) is verified. The proof of (4.60) is similar and is therefore omitted.

The computation of X_k by (4.59) and (4.60) requires $N^2/2 + N/2$ multiplications. To see this, first note that the computation of A_k requires $(N/2)^2 = N^2/4$ multiplications, as does the computation of B_k. The computation of the products $W_N^k B_n$ in (4.59) and (4.60) requires $N/2$ multiplications. So, the total number of multiplications is equal to $N^2/2 + N/2$. This is $N^2/2 - N/2$ multiplications less than N^2 multiplications. Therefore, when N is large, the computation of X_k by (4.59) and (4.60) requires significantly fewer multiplications than the computation of X_k by (4.55).

If $N/2$ is even, each of the signals $a[n]$ and $b[n]$ can be expressed in two parts, and then the process previously described can be repeated. If $N = 2^q$ for some positive integer q, the subdivision process can be continued until signals with only one nonzero value (with each value equal to one of the values of the given signal $x[n]$) are obtained.

In the case $N = 8$, a block diagram of the FFT algorithm is given in Figure 4.21. On the far left-hand side of the diagram, the values of the given signal $x[n]$ are inputted.

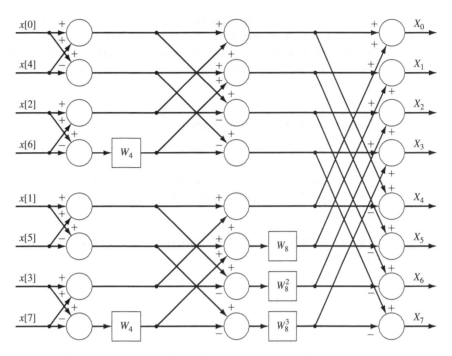

FIGURE 4.21
Block diagram of FFT algorithm when N = 8.

Note the order (in terms of row position) in which the signal values $x[n]$ are applied to the process. The order can be determined by a process called *bit reversing*. Suppose that $N = 2^q$. Given an integer n ranging from 0 to $N − 1$, the time index n can be represented by the q-bit binary word for the integer n. Reversing the bits that make up this word results in the integer corresponding to the reversed-bit word, which is the row at which the signal value $x[n]$ is applied to the FFT algorithm. For example, when $N = 8$, the binary words and bit-reversed words corresponding to the time index n are shown in Table 4.3. The last column in Table 4.3 gives the order for which the signal values are applied to the FFT algorithm shown in Figure 4.21.

TABLE 4.3   Bit Reversing in the Case $N = 8$

Time Point (n)	Binary Word	Reversed-Bit Word	Order
0	000	000	$x[0]$
1	001	100	$x[4]$
2	010	010	$x[2]$
3	011	110	$x[6]$
4	100	001	$x[1]$
5	101	101	$x[5]$
6	110	011	$x[3]$
7	111	111	$x[7]$

The MATLAB software package contains commands for computing the FFT and the inverse FFT, denoted by `fft` and `ifft`. The commands `fft` and `ifft` are interchangeable with the commands `dft` and `idft` used in the examples given in Section 4.2. Examples are given as follows and in the next section, demonstrating the use of these commands.

4.4.1   Applications of the FFT Algorithm

The FFT algorithm is very useful in a wide range of applications involving digital signal processing and digital communications. In this section it is first shown that the FFT algorithm can be used to compute the Fourier transform of a continuous-time signal. Then the FFT algorithm is applied to the problem of computing the convolution of two discrete-time signals.

Computation of the Fourier Transform via the FFT.   Let $x(t)$ be a continuous-time signal with Fourier transform $X(\Omega)$. It is assumed that $x(t) = 0$ for all $t < 0$ so that the Fourier transform $X(\Omega)$ of $x(t)$ is given by

$$X(\omega) = \int_0^\infty x(t)e^{-j\omega t}\, dt \tag{4.62}$$

Let Γ be a fixed positive real number, and let N be a fixed positive integer. It will be shown that by using the FFT algorithm, $X(\omega)$ can be computed for $\omega = k\Gamma$, $k = 0, 1, 2, \ldots, N - 1$.

Given a fixed positive number T, the integral in (4.62) can be written in the form

$$X(\omega) = \sum_{n=0}^\infty \int_{nT}^{nT+T} x(t)e^{-j\omega t}\, dt \tag{4.63}$$

Suppose that T is chosen small enough so that the variation in $x(t)$ is small over each T-second interval $nT \le t < nT + T$. Then the sum in (4.63) can be approximated by

$$\begin{aligned}
X(\omega) &= \sum_{n=0}^\infty \left(\int_{nT}^{nT+T} e^{-j\omega t}\, dt \right) x(nT) \\
&= \sum_{n=0}^\infty \left[\left(\frac{-1}{j\omega} e^{-j\omega t} \right)_{t=nT}^{t=nT+T} \right] x(nT) \\
&= \frac{1 - e^{-j\omega T}}{j\omega} \sum_{n=0}^\infty e^{-j\omega nT} x(nT)
\end{aligned} \tag{4.64}$$

Now suppose that for some sufficiently large positive integer N, the magnitude $|x(nT)|$ is small for all integers $n \ge N$. Then (4.64) becomes

$$X(\omega) = \frac{1 - e^{-j\omega T}}{j\omega} \sum_{n=0}^{N-1} e^{-j\omega nT} x(nT) \tag{4.65}$$

Evaluating both sides of (4.65) at $\omega = 2\pi k/NT$ gives

$$X\left(\frac{2\pi k}{NT}\right) = \frac{1 - e^{-j2\pi k/N}}{j2\pi k/NT} \sum_{n=0}^{N-1} e^{-j2\pi nk/N} x(nT) \qquad (4.66)$$

Now let X_k denote the N-point DFT of the sampled signal $x[n] = x(nT)$. By definition of the DFT,

$$X_k = \sum_{n=0}^{N-1} x[n] e^{-j2\pi kn/N}, \quad k = 0, 1, \ldots, N - 1 \qquad (4.67)$$

Comparing (4.66) and (4.67) reveals that

$$X\left(\frac{2\pi k}{NT}\right) = \frac{1 - e^{-j2\pi k/N}}{j2\pi k/NT} X_k \qquad (4.68)$$

Finally, letting $\Gamma = 2\pi/NT$, (4.68) can be rewritten in the form

$$X(k\Gamma) = \frac{1 - e^{-jk\Gamma T}}{jk\Gamma} X_k, \quad k = 0, 1, 2, \ldots, N - 1 \qquad (4.69)$$

By first calculating X_k via the FFT algorithm and then using (4.69), we can compute $X(k\Gamma)$ for $k = 0, 1, \ldots, N - 1$.

It should be stressed that the relationship (4.69) is an approximation, and so the values of $X(\omega)$ computed by (4.69) are only approximate values. We can obtain better accuracy by taking a smaller value for the sampling interval T and/or by taking a larger value for N. If the amplitude spectrum $|X(\omega)|$ is small for $\omega > B$, a good choice for T is the sampling interval π/B corresponding to the Nyquist sampling frequency $\omega_s = 2B$. (See Section 5.4.) If the given signal $x(t)$ is known only for the time interval $0 \le t \le t_1$, we can still select N to be as large as desired by taking the values of the sampled signal $x[n] = x(nT)$ to be zero for those values of n for which $nT > t_1$ (or $n > t_1/T$).

Example 4.14   Computation of Fourier Transform via the FFT

Consider the continuous-time signal $x(t)$ shown in Figure 4.22. The FFT program in MATLAB can be used to compute $X(\omega)$ via the following procedure. First, a sampled version of $x(t)$ is obtained and is denoted by $x(nT)$, where T is a small sampling interval and $n = 0, 1, \ldots, N - 1$. Then, the FFT X_k of $x[n] = x(nT)$ is determined. Finally, X_k is rescaled using (4.69) to obtain the approximation $X(k\Gamma)$ of the actual Fourier transform $X(\omega)$. The MATLAB commands for obtaining the approximation are given subsequently. For comparison's sake, the program also plots the actual Fourier transform $X(\omega)$, which can be computed as follows: Let $x_1(t) = t p_2(t)$, where $p_2(t)$ is the two-second rectangular pulse centered at the origin. Then $x(t) = x_1(t - 1)$, and from Example 3.16, the Fourier transform of $x_1(t)$ is

$$X_1(\omega) = j2 \frac{\omega \cos \omega - \sin \omega}{\omega^2}$$

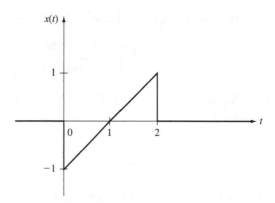

FIGURE 4.22
Continuous-time signal in Example 4.14.

Using the shift in time property, we find that

$$X(\omega) = j2\frac{\omega \cos \omega - \sin \omega}{\omega^2}e^{-j\omega}$$

Now the evaluation of the approximate and exact Fourier transforms of $x(t)$ can be carried out by the following MATLAB commands:

```
N = input('Input N: ');
T = input('Input T: ');
%
% compute the approximation of X(w)
t = 0:T:2;
x = [t-1 zeros(1,N-length(t))];
Xk = fft(x);
gam = 2*pi/N/T;
k = 0:10/gam;   % for plotting purposes
Xapp = (1-exp(-j*k*gam*T))/j/k/gam*Xk;
%
% compute the actual X(w)
w = 0.05:.05:10;
Xact = j*2*exp(-j*w).*(w.*cos(w)-sin(w))./(w.*w);
plot(k*gam,abs(Xapp(1:length(k))),'o',w,abs(Xact))
```

To run this program, the user first inputs the desired values for N and T, and then the program plots the approximate Fourier transform, denoted by Xapp, and the actual Fourier transform, denoted by Xact. The program was run with $N = 2^7 = 128$ and $T = 0.1$, in which case $\Gamma = 2\pi/NT = 0.4909$. The resulting amplitude spectra of the actual and the approximate Fourier transforms are plotted in Figure 4.23. Note that the approximation is reasonably accurate. More detail in the plot is achieved by an increase in NT, and more accuracy is achieved by a decrease in T. The program was rerun with $N = 2^9 = 512$ and $T = 0.05$ so that $\Gamma = 0.2454$. The resulting amplitude spectrum is displayed in Figure 4.24.

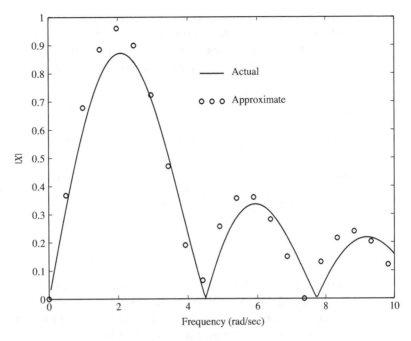

FIGURE 4.23
Amplitude spectrum in the case $N = 128$ and $T = 0.1$.

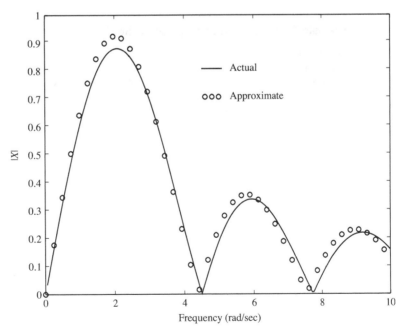

FIGURE 4.24
Amplitude spectrum in the case $N = 512$ and $T = 0.05$.

The computations required to perform the approximation of the Fourier transform for a generic signal are contained in an M-file named `contfft.m` that is given on the website that accompanies this text. We use this M-file by first defining the signal x and the time interval T. The Fourier transform is computed via the command

$$[\text{X,w}] = \text{contfft}(\text{x,T})$$

where $\text{x} = X(\omega)$ and $\text{w} = 2\pi k/NT$.

Performing Convolution by Use of the FFT.   The FFT algorithm can be used to compute the convolution of two discrete-time signals as follows. Given signals $x[n]$ and $v[n]$, with $x[n] = 0$ for $n < 0$ and $n \geq N$, and $v[n] = 0$ for $n < 0$ and $n > Q$, let r equal the smallest positive integer such that $N + Q \leq 2^r$, and let $L = 2^r$. The signals $x[n]$ and $v[n]$ can be "padded with zeros" so that

$$x[n] = 0, n = N, N + 1, \ldots, L - 1$$

$$v[n] = 0, n = Q + 1, Q + 2, \ldots, L - 1$$

Then the L-point DFTs of $x[n]$ and $v[n]$ can be computed by the FFT algorithm. With X_k and V_k equal to the DFTs, by the convolution property of the DTFT and (4.48), the convolution $v[n] * x[n]$ is equal to the inverse L-point DFT of the product $V_k X_k$, which also can be computed with the FFT algorithm. This approach requires on the order of $(1.5L) \log_2 L + L$ multiplications. In contrast, the computation of $v[n] * x[n]$ with the convolution sum requires on the order of $0.5L^2 + 1.5L$ multiplications.

Example 4.15   Convolution by Use of the FFT

Suppose that $v[n]$ is given by $v[n] = (0.8)^n u[n]$, which is plotted in Figure 4.25, and the signal $x[n]$ is the rectangular pulse shown in Figure 4.26. The convolution $v[n] * x[n]$ could be calculated by evaluating the convolution sum (see Chapter 2); however, here $v[n] * x[n]$ will be computed by using the FFT approach.

In this example there is no finite integer Q for which $v[n] = 0$ for all $n > Q$. However, from Figure 4.25 it is seen that $v[n]$ is very small for $n > 16$, and thus Q can be taken to be equal to 16. Since the signal $x[n]$ is zero for all $n \geq 10$, the integer N previously defined is equal to 10.

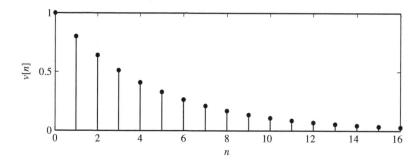

FIGURE 4.25
The signal $v[n]$ in Example 4.15.

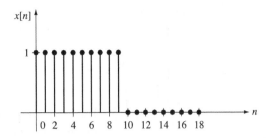

FIGURE 4.26
The signal $x[n]$ in Example 4.15.

Thus, with $Q = 16$, $N + Q = 26$, and the smallest integer value of r for which $N + Q \le 2^r$ is $r = 5$. With $L = 2^5 = 32$, the L-point DFT of the padded versions of $v[n]$ and $x[n]$ can be computed by the use of the MATLAB fft file. The MATLAB commands for generating the L-point DFTs are

```
n = 0:16; L = 32;
v = (.8).^n;
Vk = fft(v,L);
x = [ones(1,10)];
Xk = fft(x,L);
```

MATLAB plots of the magnitude and phase spectra of $v[n]$ and $x[n]$ are displayed in Figures 4.27 and 4.28, respectively. Letting $y[n] = v[n] * x[n]$, the L-point DFT Y_k of $y[n]$ and the inverse

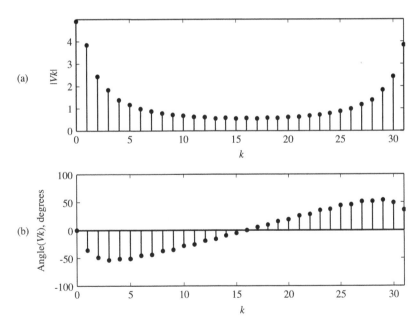

FIGURE 4.27
(a) Magnitude and (b) phase spectra of $v[n]$ in Example 4.15.

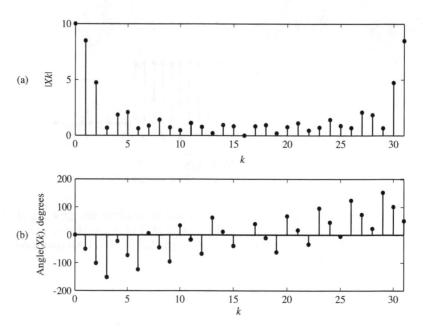

FIGURE 4.28
(a) Magnitude and (b) phase spectra of $x[n]$ in Example 4.15.

DFT of Y_k (by the inverse FFT algorithm) are computed by the use of the MATLAB commands

```
Yk = Vk.*Xk;
y = ifft(Yk, L);
```

A MATLAB plot of $y[n]$ is given in Figure 4.29.

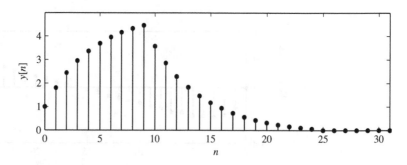

FIGURE 4.29
The convolution $y[n] = v[n] * x[n]$ in Example 4.15.

4.5   APPLICATION TO DATA ANALYSIS

An important part of data analysis is determining the dominant sinusoidal (or cyclic) components of a given signal $x[n]$ that is specified for $n = 0, 1, \ldots, N - 1$. Here, "dominant" refers to any sinusoidal components of $x[n]$ whose amplitudes are much larger than the amplitudes of most of the other sinusoidal components included in $x[n]$. The N-point DFT (or N-point FFT) can be applied to $x[n]$ to determine if there are any dominant cyclic components contained in the signal. This includes signals that contain noise. If the signal is given in continuous-time form $x(t)$, to apply the DFT it is first necessary to sample $x(t)$ to construct the discrete-time signal $x[n] = x(t)|_{t=nT} = x(nt)$. The sampling interval T must be selected so that $T < \pi/\omega_{max}$, where ω_{max} is the highest-frequency component contained in $x(t)$. A value for ω_{max} may not be known initially, in which case ω_{max} can be selected by "trial and error."

A key requirement in determining if there is a dominant sinusoidal component contained in a signal $x[n]$ is that the data must contain at least one full cycle of the component; that is, the number N of data values must be sufficiently large. To be precise, since the lowest-frequency component in the sum-of-sinusoids representation of $x[n]$ [see (4.44) or (4.45)] is the first harmonic with discrete-time frequency $2\pi/N$, it is necessary that $2\pi/N < \Omega_c$, where Ω_c is the discrete-time frequency in radians per unit of time of the cyclic component contained in $x[n]$. Thus, it is required that $N > 2\pi/\Omega_c$.

Examples on determining dominant cyclic components by use of the DFT/FFT are considered next.

4.5.1   Extraction of a Sinusoidal Component Embedded in Noise

Suppose that a signal $x(t)$ is the output of a sensor, but due to the nature of the sensing process, any dominant sinusoidal components contained in $x(t)$ are embedded in the noise. Applications where the dominant sinusoidal components of a signal (if they exist) may be extremely "weak" and are embedded in noise include signals generated by a radio telescope that is pointed at a specific location in deep space. This type of situation is easy to simulate with MATLAB. For example, consider the plot of 200 samples values of the signal $x(t)$ given in Figure 4.30. From the plot, it appears that the sampled signal $x[n]$ is varying randomly about zero, and there are no dominant cyclic components contained in $x[n]$.

The amplitude spectrum $|X_k|$ of $x[n]$ is given in Figure 4.31. The plot was generated by the MATLAB command abs(fft(x)), which does not require that N be a power of two. From Figure 4.31, a spike in the amplitude spectrum of $x[n]$ can be clearly seen, and thus we conclude that $x[n]$ does contain a dominant cyclic component. To determine the frequency corresponding to the spike in Figure 4.31, the values of $|X_k|$ are plotted for $k = 0$ to 25 by use of the stem plot. The result is displayed in Figure 4.32, which shows that the peak value of $|X_k|$ occurs at $k = 11$. Hence, the frequency of the dominant cyclic component is $2\pi k/N = 2\pi(11)/200 = (0.11)\pi$ radians per unit time. From Figure 4.32, it can also be seen that the values of $|X_k|$ for values of k in a neighborhood of $k = 11$ are comparable to the other values of $|X_k|$ for $k \neq 11$, and thus there does not appear to be leakage (see Example 4.13). This indicates that the frequency of $(0.11)\pi$ radians per unit time must be close to the actual frequency of

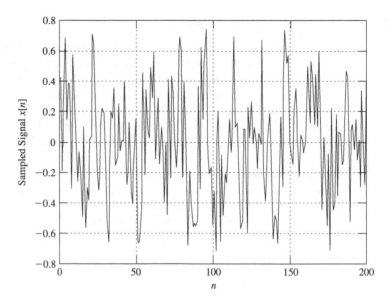

FIGURE 4.30
Sampled signal.

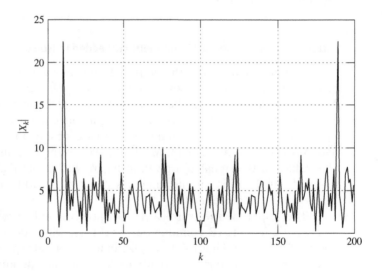

FIGURE 4.31
Amplitude spectrum of signal in Figure 4.30.

the dominant sinusoidal component; and in fact this is the case, as the sinusoidal term in $x[n]$ is equal to $(0.3) \sin[(0.112)\pi n]$. If the amplitude spectrum had indicated that there was leakage, a different value for N could have been used. This is illustrated in the next application.

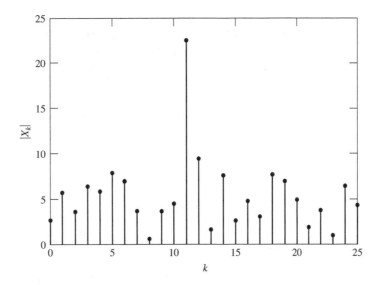

FIGURE 4.32
Stem plot of $|X_k|$ for $k = 0, 1, 2, \ldots, 25$.

4.5.2   Analysis of Sunspot Data

It is well known that the degree of sunspot activity on the sun follows an 11-year cycle. It turns out that sunspot data is available from the Royal Observatory of Belgium, which contains the Sunspot Index Data Center (SIDC). Sunspot data can be downloaded from this center by going to the following website:

http://sidc.oma.be/index.php3

To acquire the data, go to this website and click on "Sunspots download of data" located on the left-hand side of the Web page. Then when this page comes up, click on "monthly and monthly smoothed sunspot number," which will produce a table with the first column containing the year and month and the third column containing the average number of sunspots for that month. The fourth column is the "smoothed value" of the average number of sunspots for the month. Note that the data are available from January 1749 up to the present year and month. To analyze the data for any desired period of time, use the Microsoft copy command (if you are using MSDOS) and simply copy the data in the table for the time period and then paste the data into a Microsoft Excel spreadsheet. In the analysis considered here, January 1977 was selected as the start date, and December 2001 was selected as the end date for the period of interest. That covers 25 years of data, with 12 data values for each year, and so there are a total of 300 data values.

After copying the data for the desired period, open Microsoft Excel and paste the data into the spreadsheet. As a check, there should be 300 filled rows in the spreadsheet with all of the data located in the first column. Now it is necessary to put the data into a four-column format. To accomplish this, highlight the first column of the table in blue, then click on "data" in the Excel window, and click on "Text to Columns." In the box that appears, select "fixed width," and click on "Finish." After this step, the data in the spreadsheet should appear in a four-column format. Finally, click on "save as" and save the file under the name "sunspotdata.csv" in a subdirectory that contains your student version of MATLAB. Click "OK" and "yes" on the two boxes that appear.

To verify that the download process works, after the file sunspotdata.csv has been created, run the following MATLAB commands:

```
spd=csvread('sunspotdata.csv',0,2,[0 2 299 2]);
plot(spd)
grid
xlabel('Month')
ylabel('Average Number of Sunspots')
```

Note that in the csvread command just given, the first row that is read by MATLAB is Row 0, and the last row that is read is Row 299. Recall that MATLAB numbers the first row of a table of numerical data as Row 0. (See Section 1.2.)

As noted from the preceding MATLAB program, the sunspot data is denoted by *spd*, thus the average number of sunspots as a function of the Month *n* is denoted by *spd*[*n*], and the number of months varies from $n = 1$ to $n = 300$ (thus $N = 300$). Running the MATLAB program results in the MATLAB plot shown in Figure 4.33. From the plot it is clear that sunspot activity does cycle, but a precise value for the period of the cycle is not obvious from the plot due to the variability in the data. An estimate of the period can be computed from the plot by counting the number of months between the

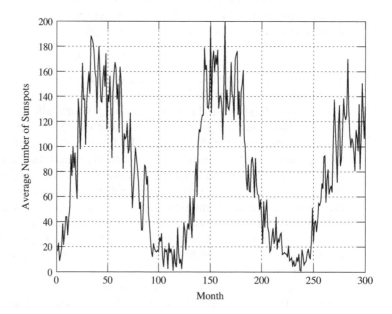

FIGURE 4.33

Average number of sunspots from January 1977 through December 2001.

occurrence of the first peak and the occurrence of the second peak in the data displayed in Figure 4.33. Checking the values of $spd[n]$ generated by the foregoing MATLAB commands reveals that the largest value of $spd[n]$ in forming the first peak is 188.4, which occurs on Month 33 (i.e., $n = 33$), and the largest value of $spd[n]$ in forming the second peak is 200.3, which occurs on Month 164. Hence, an estimate of the period is $164 - 33 = 131$ months, or $131/12 = 10.9$ years, which is close to the actual value of 11 years.

A DFT analysis of $spd[n]$ will now be carried out. First, from the plot of $spd[n]$ given in Figure 4.33, it is obvious that $spd[n]$ has a large constant component. As discussed in Section 4.2, the constant component is equal to the average value of $spd[n]$ over the interval from $n = 1$ to $n = 300$. To facilitate the analysis of $spd[n]$, the constant component will be subtracted out, which results in the signal

$$x[n] = spd[n] - \frac{1}{300}\sum_{i=1}^{300} spd[i] \qquad (4.70)$$

The amplitude spectrum $|X_k|$ of $x[n]$ is given in Figure 4.34. From Figure 4.34, a spike in the amplitude spectrum of $x[n]$ can be clearly seen, and thus $x[n]$ does contain a dominant cyclic component. To determine the frequency corresponding to the spike in Figure 4.34, the values of $|X_k|$ are plotted for $k = 0$ to 10 by use of the stem plot. The result is displayed in Figure 4.35, which shows that the sunspot data has two dominant sinusoidal components with frequencies $2\pi k/N = 2\pi(2)/300 = 4\pi/300$ radians per month and $2\pi k/N = 2\pi(3)/300 = 6\pi/300$ radians per month, with the frequencies

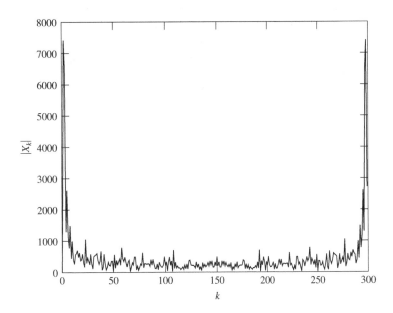

FIGURE 4.34
Amplitude spectrum of $x[n]$.

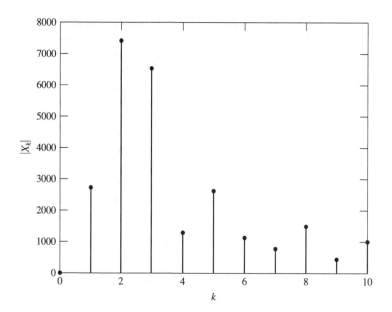

FIGURE 4.35
Stem plot of $|X_k|$ for $k = 0$ to $k = 10$.

corresponding to the spectral components at $k = 2$ and $k = 3$ shown in Figure 4.35. Also note that $|X_0| = 0$, since the constant component was subtracted out.

The large value of the spectral component at $k = 3$ is a result of leakage, since as can be seen from Figure 4.33, the range of the data ($n = 1$ to $n = 300$) does not cover an integer multiple of the period of the cyclic component in $spd[n]$ (or in $x[n]$). To eliminate the leakage, it is necessary to consider a range of data that covers an integer multiple of $(11)(12) = 132$ months, which is the period of the sunspot activity. Taking the integer multiple to be 2, the range of data that is selected runs from $n = 37$ to $n = 300$, which results in $N = 2(132) = 264$ data points. This range of the data can be generated by the MATLAB command `spd(37:300)`. Hence, the DFT will be applied to $spd[n]$ for $n = 37$ to $n = 300$. Denoting this signal by $v[n]$, it follows that

$$v[n] = spd[n + 36], n = 1, 2, \ldots, 264 \qquad (4.71)$$

Therefore, in going from $spd[n]$ to $v[n]$, the number of data points (the value of N) has been changed from 300 to 264.

Subtracting out the average value of $v[n]$ over the interval from $n = 1$ to $n = 264$ results in the signal $w[n]$ given by

$$w[n] = v[n] - \frac{1}{264}\sum_{i=1}^{264} v[i]$$

A stem plot of the amplitude spectrum $|W_k|$ of $w[n]$ is given in Figure 4.36 for $k = 0$ to $k = 10$. Note that the peak value of $|W_k|$ still occurs at $k = 2$, but now the value of $|W_2|$ is much larger than the values of $|W_k|$ for k in a neighborhood of the value 2.

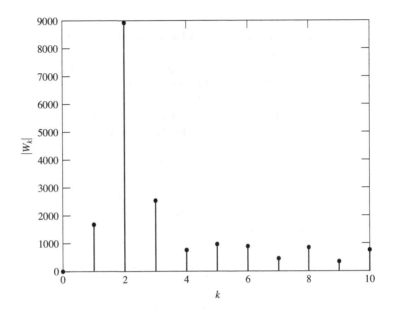

FIGURE 4.36
Stem plot of $|W_k|$ for $k = 0$ to $k = 10$.

This result shows that the sinusoidal component with frequency $\Omega = 2\pi k/N = 2\pi(2)/264$ radians per month is the only dominant cyclic component of $w[n]$ (or $v[n]$). This frequency corresponds to a period of $2\pi/\Omega = 132$ months, which is equal to 11 years. The values of $|W_k|$ for $k \neq 2$ are primarily due to the noise in the sunspot data. However, the larger values for $|W_1|$ and $|W_3|$ in comparison with $|W_k|$ for $k \neq 2$ (see Figure 4.36) indicate that the sunspot data may contain other sinusoidal components in addition to the dominant component with frequency $4\pi/264$, in which case sunspot activity is not a "pure sinusoid" consisting of the single frequency $4\pi/264$. A thorough evaluation of this is left to a more advanced treatment of data analysis.

4.5.3   Stock Price Analysis

Over long periods (at least 50 business days in length), stock prices often move up and down, forming cycles; or they can follow a ramp characteristic with short-term erratic movements. To determine if there are any dominant cyclic components in stock price data, it is first necessary to subtract out any ramp characteristic that may exist in the data. This is carried out next, and then a DFT analysis is applied to the transformed data. It will be shown how this process can be used to determine the trend of a stock price.

For $n = 1$ to $n = N$, let $c[n]$ denote the closing price of a stock (such as QQQQ). The first step in the analysis is to subtract a ramp from $c[n]$, which results in the signal $x[n]$ defined by

$$x[n] = c[n] - c[1] + \left(\frac{c[1] - c[N]}{N - 1}\right)(n - 1), n = 1, 2, \ldots, N \qquad (4.72)$$

Setting $n = 1$ in (4.72) shows that $x[1] = 0$, and setting $n = N$ in (4.72) shows that $x[N] = 0$. Thus, $x[1] = x[N] = 0$, which is a highly desirable property in applying a DFT analysis to $x[n]$.

As an example, consider the closing price $c[n]$ of QQQQ for the 50-business-day period from March 1, 2004, up to May 10, 2004 (see Example 1.4 in Chapter 1), and thus in this case, $N = 50$. Recall that a column vector containing the values of $c[n]$ can be generated from the MATLAB command `csvread('QQQQdata2.csv',1,4,[1 4 50 4])`. Then computing $x[n]$ given by (4.72) results in the MATLAB plot of $x[n]$ shown in Figure 4.37.

The amplitude spectrum of $x[n]$ is given in Figure 4.38. Note that $|X_1|$ is much larger than $|X_k|$ for k in a neighborhood of the value 1, and thus $x[n]$ has a dominant cyclic component with frequency $2\pi/N = 2\pi/50 = (0.04)\pi$. From the sinusoidal form of $x[n]$ given by (4.45), it can be seen that $x[n]$ can be approximated by the single sinusoidal component

$$\hat{x}[n] = \frac{2}{N}\left[R_1 \cos\frac{2\pi}{N}n - I_1 \sin\frac{2\pi}{N}n\right], n = 1, 2, \ldots, 50 \tag{4.73}$$

where $R_1 + jI_1 = X_1$. Taking the FFT of $x[n]$ by the use of MATLAB gives $X_1 = -9.3047 + j34.5869$, and thus, $R_1 = -9.3047$ and $I_1 = 34.5869$. Inserting these values for R_1 and I_1 and $N = 50$ into (4.73) gives

$$\hat{x}[n] = -0.372 \cos(0.04\,\pi n) - 1.383 \sin(0.04\,\pi n), n = 1, 2, \ldots, 50 \tag{4.74}$$

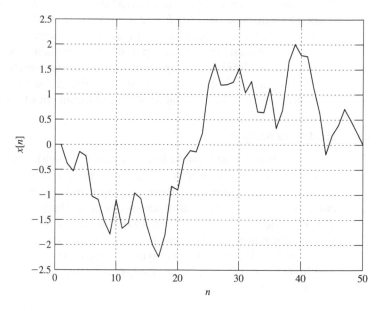

FIGURE 4.37
Plot of $x[n]$.

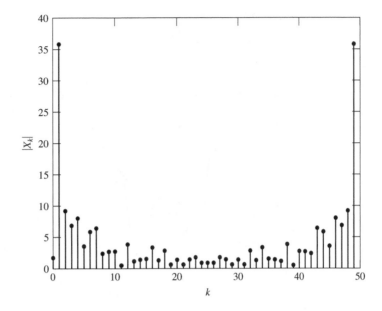

FIGURE 4.38
Amplitude spectrum of $x[n]$.

The approximation $\hat{x}[n]$ given by (4.74) and $x[n]$ are both plotted in Figure 4.39. The result in Figure 4.39 shows that $\hat{x}[n]$ is a good approximation to $x[n]$.

We can then generate a "smoothed version" of $c[n]$ by solving (4.72) for $c[n]$ and replacing $x[n]$ by $\hat{x}[n]$. Denoting the smoothed version of $c[n]$ by $\hat{c}[n]$ and using (4.72) gives

$$\hat{c}[n] = \hat{x}[n] + c[1] - \left(\frac{c[1] - c[50]}{49}\right)(n - 1), n = 1, 2, \ldots, 50 \qquad (4.75)$$

Both $c[n]$ and $\hat{c}[n]$ are plotted in Figure 4.40. The slope of the smoothed version $\hat{c}[n]$ of $c[n]$ gives an indication of the trend of $c[n]$.

In Figure 4.40, note that there is no time delay between $c[n]$ and $\hat{c}[n]$. This is in contrast to the time delay that occurs when $c[n]$ is filtered by MA or EWMA filters, as considered in Section 1.4 and Section 2.1. The DFT approach given here differs signif-icantly from filtering, since $\hat{c}[n]$ is not computed from $c[i]$ for $i = n, n - 1, \ldots$ as is the case in the MA and EWMA filters, or in other types of causal filters. The DFT ap-proach considered in this application is an example of *data smoothing*, not data filter-ing. In data smoothing, we compute the value $\hat{c}[i]$ of the smoothed signal at time i, where $0 \le i \le N - 1$, using all values of the data $c[n]$ for $n = 0, 1, \ldots, N - 1$. In gen-eral, the smoothed signal $\hat{c}[n]$ is closest to the true smooth part of $c[n]$ at the middle of the data range and is off somewhat at the end points where $n = 0$ and $n = N - 1$. Hence, the slope of $\hat{c}[n]$ may not be an accurate indication of the trend of $c[n]$ at $n = 0$ and $n = N - 1$.

Figure 4.40 reveals that the slope of $\hat{c}[n]$ is very negative at $n = 50$, which shows that there is a strong downward trend in the closing price of QQQQ at $n = 50$, and thus

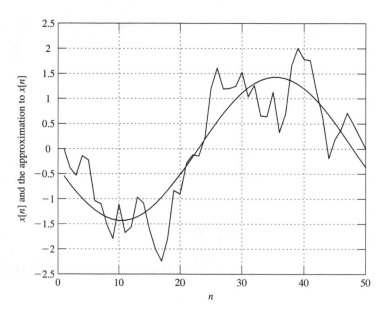

FIGURE 4.39
$x[n]$ and the approximation to $x[n]$.

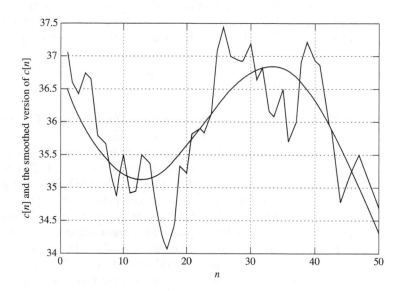

FIGURE 4.40
$c[n]$ and the smoothed version of $c[n]$.

the stock should not be purchased at the close on Day 50. The smoothed version of $c[n]$ can be recomputed as new values for $c[n]$ become available. A buy signal would be generated when the slope of $\hat{c}[n]$ becomes positive. A sell signal would be generated if the slope of $\hat{c}[n]$ has been positive and then becomes negative. This approach to trading does not always work well, since as noted previously, the slope of $\hat{c}[n]$ may not be an accurate indication of the trend of $c[n]$ at $n = N - 1$. The interested reader should practice this strategy extensively, using past data before attempting to use it for actual trading.

4.6   CHAPTER SUMMARY

The discrete-time Fourier transform (DTFT) is used to determine the frequency content of discrete-time signals. The DTFT of a signal $x[n]$ is a function $X(\Omega)$ of the frequency variable Ω defined as

$$X(\Omega) = \sum_{n=-\infty}^{\infty} x[n]e^{-j\Omega n}$$

The resulting DTFT is periodic with period 2π. For a real-valued signal $x[n]$, the magnitude function $|X(\Omega)|$ is an even function of Ω, and the angle function $\angle X(\Omega)$ is an odd function of Ω. Thus, the frequency spectrum can be plotted only over the interval $0 \le \Omega \le \pi$. The frequency shown in some of the plots is normalized by π, resulting in a frequency range of 0 to 1. Useful properties of the DTFT include linearity, left and right shifts in time, multiplication by a sinusoid, and convolution in the time domain.

When the discrete-time signal $x[n]$ is time-limited, that is, $x[n] = 0$ for $n < 0$ and $n \ge N$ for some positive integer N, the discrete Fourier transform (DFT) is equal to the DTFT at the discrete frequencies $\Omega = 2\pi k/N$, with the values of the transform given by

$$X_k = \sum_{n=0}^{N-1} x[n]e^{-j2\pi kn/N}, \quad k = 0, 1, \ldots, N - 1$$

The DFT is often used in place of the DTFT when the frequency content of a signal $x[n]$ needs to be computed numerically. This computation is actually an approximation of the DTFT when the signal is not time-limited and needs to be truncated in order to perform the DFT. The approximation errors can be made small if the values of the truncated part of the signal are negligible, or if the signal is periodic and the length of the truncated signal is an integer multiple of the period.

The DFT is very useful for numerical analysis of continuous-time data that has been recorded digitally by signal sampling. The FFT, a fast algorithm for computing the DFT, is used commonly in engineering applications where signal processing of measured data is performed. In these cases, the engineer or scientist is actually interested in the frequency content of a continuous-time signal. The FFT of the sampled signal can be used to approximate the continuous-time Fourier transform of the signal. The approximation improves if the sampling period T is decreased, and the resolution in the plot improves if the product NT is increased where N corresponds to the number of points in the data record. If the signal is periodic, then leakage in the spike about the fundamental frequency can be lessened by making N equal to an integer multiple of the fundamental period of the signal. This is illustrated in the example involving sunspot data in Section 4.5. In this section, it is also shown that the DFT (or FFT) can be used to extract the dominant cyclic components of a signal for the purpose of carrying out signal analysis.

PROBLEMS

4.1. Compute the DTFTs of the discrete-time signals shown in Figure P4.1. Express the DTFTs in the simplest possible form. Plot the amplitude and phase spectrum for each signal.

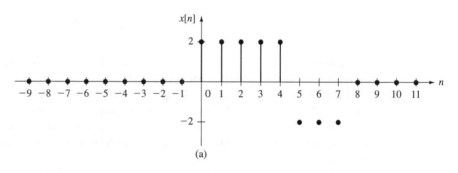

(a)

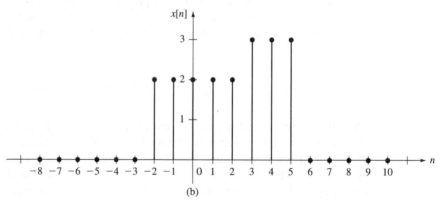

(b)

FIGURE P4.1

4.2. Prove the following relationship:

$$\sum_{n=q_1}^{q_2} r^n = \frac{r^{q_1} - r^{q_2+1}}{1 - r}$$

Hint: Multiply both sides of the equation by $(1 - r)$.

4.3. By breaking the DTFT $X(\Omega)$ into three summations, (from $n = -\infty$ to $n = -1$; $n = 0$; and $n = 1$ to $n = \infty$) and applying Euler's formula, prove the following:

(a) If $x[n]$ is an even function of n, then $X(\Omega) = x[0] + \sum_{n=1}^{\infty} 2x[n] \cos \Omega n$.

(b) If $x[n]$ is an odd function of n, then $X(\Omega) = x[0] - \sum_{n=1}^{\infty} j2x[n] \sin \Omega n$.

4.4. Compute the DTFT of the following discrete-time signals; plot the amplitude and the phase spectrum for each signal:

(a) $x[n] = (0.8)^n u[n]$

(b) $x[n] = (0.5)^n \cos 4n \, u[n]$

(c) $x[n] = n(0.5)^n u[n]$

(d) $x[n] = n(0.5)^n \cos 4n \, u[n]$

(e) $x[n] = 5(0.8)^n \cos 2n \, u[n]$

(f) $x[n] = (0.5)^{|n|}, \ -\infty < n < \infty$

(g) $x[n] = (0.5)^{|n|} \cos 4n, \ -\infty < n < \infty$

4.5. A discrete-time signal $x[n]$ has DTFT

$$X(\Omega) = \frac{1}{e^{j\Omega} + b}$$

where b is an arbitrary constant. Determine the DTFT $V(\Omega)$ of the following:

(a) $v[n] = x[n - 5]$

(b) $v[n] = x[-n]$

(c) $v[n] = nx[n]$

(d) $v[n] = x[n] - x[n - 1]$

(e) $v[n] = x[n] * x[n]$

(f) $v[n] = x[n] \cos 3n$

(g) $v[n] = x^2[n]$

(h) $v[n] = x[n]e^{j2n}$

4.6. Use (4.28) or the properties of the DTFT to compute the inverse DTFT of the frequency functions $X(\Omega)$ shown in Figure P4.6.

4.7. Determine the inverse DTFT of the following frequency functions:

(a) $X(\Omega) = \sin \Omega$

(b) $X(\Omega) = \cos \Omega$

(c) $X(\Omega) = \cos^2 \Omega$

(d) $X(\Omega) = \sin \Omega \cos \Omega$

4.8. The autocorrelation of a discrete-time signal $x[n]$ is defined by

$$R_x[n] = \sum_{i=-\infty}^{\infty} x[i]x[n + i]$$

Let $P_x(\Omega)$ denote the DTFT of $R_x[n]$.

(a) Derive an expression for $P_x(\Omega)$ in terms of the DTFT $X(\Omega)$ of $x[n]$.

(b) Derive an expression for $R_x[-n]$ in terms of $R_x[n]$.

(c) Express $P_x(0)$ in terms of $x[n]$.

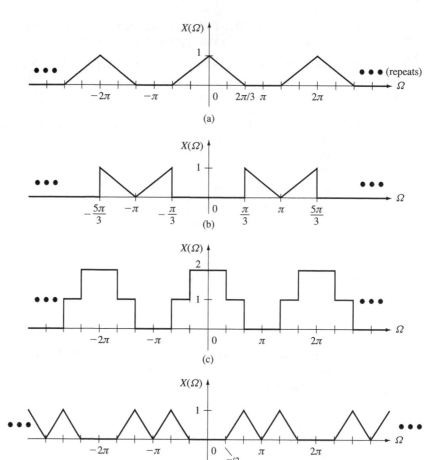

FIGURE P4.6

4.9. Compute the rectangular form of the four-point DFT of the following signals, all of which are zero for $n < 0$ and $n \geq 4$:

(a) $x[0] = 1, x[1] = 0, x[2] = 1, x[3] = 0$

(b) $x[0] = 1, x[1] = 0, x[2] = -1, x[3] = 0$

(c) $x[0] = 1, x[1] = 1, x[2] = -1, x[3] = -1$

(d) $x[0] = -1, x[1] = 1, x[2] = 1, x[3] = 1$

(e) $x[0] = -1, x[1] = 0, x[2] = 1, x[3] = 2$

(f) $x[0] = 1, x[1] = -1, x[2] = 1, x[3] = -1$

(g) Compute the DFT for each of the foregoing signals using the MATLAB M-file `dft`. Compare these results with the results obtained analytically in parts (a) to (f).

4.10. Verify the relationship given in Equation (4.42).

4.11. Derive the sinusoidal form (4.45) in the case when N is even.

4.12. Compute the sinusoidal form of the signals in Problem 4.9.

4.13. Using the MATLAB M-file dft, compute the 32-point DFT of the signals (a)–(f). Express your answer by plotting the amplitude $|X_k|$ and phase $\angle X_k$ of the DFTs.

    (a) $x[n] = 1, 0 \le n \le 10, x[n] = 0$ for all other n

    (b) $x[n] = 1, 0 \le n \le 10, x[n] = -1, 11 \le n \le 20, x[n] = 0$ for all other n

    (c) $x[n] = n, 0 \le n \le 20, x[n] = 0$ for all other n

    (d) $x[n] = n, 0 \le n \le 10, x[n] = 20 - n, 11 \le n \le 20, x[n] = 0$ for all other n

    (e) $x[n] = \cos(10\pi n/11), 0 \le n \le 10, x[n] = 0$ for all other n

    (f) $x[n] = \cos(9\pi n/11), 0 \le n \le 10, x[n] = 0$ for all other n

4.14. Using the MATLAB M-file dft, compute the magnitude of the 32-point DFT X_k of the following signals:

    (a) $x[n] = \begin{cases} 1, & n = 0 \\ \dfrac{1}{n}, & n = 1, 2, 3, \ldots, 31, \\ 0, & n = 32, 33, \ldots \end{cases}$

    (b) $x[n] = \begin{cases} 1, & n = 0 \\ \dfrac{1}{n^2}, & n = 1, 2, 3, \ldots, 31 \\ 0, & n = 32, 33, \ldots \end{cases}$

    (c) $x[n] = \begin{cases} 1, & n = 0 \\ \dfrac{1}{n!}, & n = 1, 2, 3, \ldots, 31 \\ 0, & n = 32, 33, \ldots \end{cases}$

    (d) Compare the results obtained for parts (a) to (c). Explain the differences in the results.

4.15. Consider the discrete-time signal

$$x[n] = \begin{cases} r[n] - 0.5, & n = 0, 1, 2, \ldots, 31 = N - 1 \\ 0, & \text{all other } n \end{cases}$$

where r is a sequence of random numbers uniformly distributed between 0 and 1. This sequence can be generated by the MATLAB command rand(N,1). The signal $x[n]$ can be interpreted as random noise. Using the dft M-file, compute the magnitude of the 32-point DFT of $x[n]$. What frequencies would you expect to see in the amplitude spectrum of $x[n]$? Explain.

4.16. Use the MATLAB M-file dft with $N = 10$ to approximate the DTFT of the signal plotted in Figure P4.1a. Plot the amplitude and phase spectrum for X_k versus $\Omega = 2\pi k/N$. Compare this result to the DTFT obtained in Problem 4.1 over the frequency range $\Omega = 0$ to $\Omega = 2\pi$. Repeat for $N = 20$.

4.17. Repeat Problem 4.16 for the signal plotted in Figure P4.1b.

4.18. To determine the effect of truncation in computing the approximation of a DTFT by a DFT, consider the signal defined by $x[n] = n(0.5)^n u[n]$.

    (a) Determine the minimum value of N so that the signal has magnitude $|x[n]| \le 20\%$ of its maximum value for all $n \ge N$.

(b) Use MATLAB to compute the 50-point DFT of the truncated signal $\tilde{x}[n]$ defined by

$$\tilde{x}[n] = \begin{cases} x[n], & 0 \le n \le N - 1 \\ 0, & \text{all other } n \end{cases}$$

where N was determined in part (a). Plot the amplitude and phase spectrum of X_k versus $\Omega = 2\pi k/50$.

(c) Compare the result obtained in part (b) with the DTFT of $x[n]$ found in Problem 4.4(c).

(d) Repeat parts (a) to (c) for the value of N such that the signal has magnitude $|x[n]| \le 5\%$ of its maximum value for all $n \ge N$.

4.19. This problem explores the use of the FFT in approximating the Fourier Transform of continuous-time signals.

(a) Compute the Fourier transform of $x(t) = 4e^{-4t}u(t)$.

(b) Create a sampled version of the signal $x(t)$ in MATLAB for the cases (i)–(iv), where T is the sampling time and N is the total number of points. Use the M-file contfft.m from the textbook to compute the approximation to $X(\omega)$. Plot $|X(\omega)|$ versus ω for the exact Fourier transform obtained in part (a) and for the approximated Fourier transform obtained from contfft, both on the same graph. (Use the range $0 \le \omega \le 50$ rad/sec.)

(i) $T = 0.5, N = 10$
(ii) $T = 0.1, N = 50$
(iii) $T = 0.05, N = 100$
(iv) $T = 0.05, N = 400$

(c) Identify the trends in accuracy and resolution in the plots, as T is decreased and as NT is increased.

4.20. Let $spd[n]$ denote the monthly sunspot data for the 33-year period from January 1875 through December 1907. Note that there are a total of $(12)(33) = 396$ data points.

(a) Download the data from the Web, using the procedure described in Section 4.5, and generate a MATLAB plot of $spd[n]$ for the 33-year period.

(b) Subtract out the constant component of $spd[n]$, and then generate a MATLAB plot of the amplitude spectrum of $spd[n]$.

(c) From your result in Part (b), verify that the frequency of the dominant cyclic component has the expected value.

(d) Determine a smoothed version of $spd[n]$, using only the constant component of $spd[n]$ and the dominant cyclic component of $spd[n]$. Generate a MATLAB plot of the smoothed version of $spd[n]$.

(e) Generate a MATLAB plot consisting of both $spd[n]$ and the smoothed version of $spd[n]$. How well does the smoothed version fit the data? Discuss your conclusions regarding the fit.

4.21. Using the method given in Section 4.5, determine a smoothed version of the closing price $c[n]$ of QQQQ for the 60-business-day period April 27, 2004, up to July 22, 2004.

CHAPTER 5

Fourier Analysis of Systems

In this chapter, Fourier analysis is applied to the study of linear time-invariant continuous-time and discrete-time systems. The development begins in the next section with the continuous-time case. In Section 5.1 the result that the Fourier transform of the response to an input is the product of the Fourier transform of the input and the Fourier transform of the impulse response is used to show that the output response resulting from a sinusoidal input is also a sinusoid having the same frequency as the input, but which is amplitude scaled and phase shifted. This leads to the notion of the frequency response function of a system. This frequency-domain description of system behavior gives a great deal of insight into how a system processes a given input to produce the resulting output.

In Section 5.2, the responses to periodic and nonperiodic input signals are studied by the use of the Fourier-domain representation. In Sections 5.3 and 5.4, the Fourier theory is applied to ideal filtering and sampling. Section 5.4 includes a proof of the famous sampling theorem. This very important result states that a bandlimited continuous-time signal can be completely reconstructed from a sampled version of the signal if the sampling frequency is suitably fast.

In Section 5.5 the discrete-time Fourier transform (DTFT) domain representation is generated for a linear time-invariant discrete-time system. The DTFT domain representation is the discrete-time counterpart of the Fourier transform representation of a linear time-invariant continuous-time system. In Section 5.6 the DTFT domain representation is illustrated by application of the theory to specific examples of digital filters. A summary of the chapter is given in Section 5.7.

5.1 FOURIER ANALYSIS OF CONTINUOUS-TIME SYSTEMS

Consider a linear time-invariant continuous-time system with impulse response $h(t)$. As discussed in Section 2.6, the output response $y(t)$ resulting from input $x(t)$ is given by the convolution relationship

$$y(t) = h(t) * x(t) = \int_{-\infty}^{\infty} h(\lambda)x(t - \lambda)\, d\lambda \tag{5.1}$$

In this chapter it is not assumed that the system is necessarily causal, and thus the impulse response $h(t)$ may be nonzero for $t < 0$. Throughout this chapter it is assumed that the impulse response $h(t)$ is absolutely integrable; that is,

$$\int_{-\infty}^{\infty} |h(t)|\, dt < \infty \tag{5.2}$$

From Chapter 5 of *Fundamentals of Signals and Systems Using the Web and MATLAB*, Third Edition. Edward W. Kamen, Bonnie S. Heck. Copyright © 2007 by Pearson Education, Inc. Publishing as Prentice Hall. All rights reserved.

The condition (5.2) is a type of *stability condition* on the given system. A key point to be made here is that the results in this chapter are *not valid in general* unless condition (5.2) is satisfied. As a result of the integrability condition (5.2), the ordinary Fourier transform $H(\omega)$ of the response $h(t)$ exists and is given by

$$H(\omega) = \int_{-\infty}^{\infty} h(t)e^{-j\omega t}\,dt \qquad (5.3)$$

As shown in Section 3.6, the Fourier transform of the convolution of two continuous-time signals is equal to the product of the Fourier transforms of the signals. Therefore, taking the Fourier transform of both sides of (5.1) gives

$$Y(\omega) = H(\omega)X(\omega) \qquad (5.4)$$

where $X(\omega)$ is the Fourier transform of the input $x(t)$. Equation (5.4) is the Fourier-domain (or ω-domain) representation of the given system. This is a *frequency-domain representation* of the given system, since the quantities in (5.4) are functions of the frequency variable ω.

From the ω-domain representation (5.4), it is seen that the frequency spectrum $Y(\omega)$ of the output is equal to the product of $H(\omega)$ with the frequency spectrum $X(\omega)$ of the input. Taking the magnitude and angle of both sides of (5.4) shows that the amplitude spectrum $|Y(\omega)|$ of the output response $y(t)$ is given by

$$|Y(\omega)| = |H(\omega)||X(\omega)| \qquad (5.5)$$

and the phase spectrum $\angle Y(\omega)$ is given by

$$\angle Y(\omega) = \angle H(\omega) + \angle X(\omega) \qquad (5.6)$$

Equation (5.5) shows that the amplitude spectrum of the output is equal to the product of $|H(\omega)|$ with the amplitude spectrum of the input, and (5.6) shows that the phase spectrum of the output is equal to the sum of $\angle H(\omega)$ and the phase spectrum of the input. As will be seen, these relationships provide a good deal of insight into how a system processes inputs.

5.1.1   Response to a Sinusoidal Input

Suppose that the input $x(t)$ to the system defined by (5.1) is the sinusoid

$$x(t) = A\cos(\omega_0 t + \theta), \; -\infty < t < \infty \qquad (5.7)$$

where A is a positive or negative real number, the frequency ω_0 is assumed to be non-negative, and the phase θ is arbitrary. To find the output response $y(t)$ resulting from $x(t)$, first note that from Table 3.2, the Fourier transform $X(\omega)$ of $x(t)$ is given by

$$X(\omega) = A\pi[e^{-j\theta}\delta(\omega + \omega_0) + e^{j\theta}\delta(\omega - \omega_0)]$$

Using (5.4) gives

$$Y(\omega) = A\pi H(\omega)[e^{-j\theta}\delta(\omega + \omega_0) + e^{j\theta}\delta(\omega - \omega_0)]$$

Now, $H(\omega)\delta(\omega + c) = H(-c)\delta(\omega + c)$ for any constant c, and thus

$$Y(\omega) = A\pi[H(-\omega_0)e^{-j\theta}\delta(\omega + \omega_0) + H(\omega_0)e^{j\theta}\delta(\omega - \omega_0)] \tag{5.8}$$

Since the impulse response $h(t)$ is real valued, by the results in Section 3.3, $|H(-\omega_0)| = |H(\omega_0)|$ and $\angle H(-\omega_0) = -\angle H(\omega_0)$, and thus $H(-\omega_0)$ and $H(\omega_0)$ have the polar forms

$$H(-\omega_0) = |H(\omega_0)|e^{-j\angle H(\omega_0)} \text{ and } H(\omega_0) = |H(\omega_0)|e^{j\angle H(\omega_0)} \tag{5.9}$$

Inserting (5.9) into (5.8) yields

$$Y(\omega) = A\pi|H(\omega_0)|[e^{-j(\angle H(\omega_0)+\theta)}\delta(\omega + \omega_0) + e^{j(\angle H(\omega_0)+\theta)}\delta(\omega - \omega_0)] \tag{5.10}$$

From Table 3.2, the inverse Fourier transform of (5.10) is

$$y(t) = A|H(\omega_0)| \cos(\omega_0 t + \theta + \angle H(\omega_0)), \quad -\infty < t < \infty \tag{5.11}$$

Hence, the response resulting from the sinusoidal input $x(t) = A\cos(\omega_0 t + \theta)$ is also a sinusoid with the same frequency ω_0, but with the amplitude scaled by the factor $|H(\omega_0)|$ and with the phase shifted by amount $\angle H(\omega_0)$. This is quite a remarkable result, and in fact it forms the basis for the frequency-domain approach to the study of linear time-invariant systems.

Since the magnitude and phase of the sinusoidal response are given directly in terms of $|H(\omega)|$ and $\angle H(\omega)$, $|H(\omega)|$ is called the *magnitude function* of the system and $\angle H(\omega)$ is called the *phase function* of the system. In addition, since the response of the system to a sinusoid with frequency ω_0 can be determined directly from $H(\omega)$, $H(\omega)$ is often referred to as the *frequency response function* (or *system function*) of the system.

Example 5.1   Response to Sinusoidal Inputs

Response to Sinu- soidal Inputs

Suppose that the frequency response function $H(\omega)$ is given by $|H(\omega)| = 1.5$ for $0 \le \omega \le 20$, $|H(\omega)| = 0$ for $\omega > 20$, and $\angle H(\omega) = -60°$ for all ω. Then if the input $x(t)$ is equal to $2\cos(10t + 90°) + 5\cos(25t + 120°)$ for $-\infty < t < \infty$, the response is given by $y(t) = 3\cos(10t + 30°)$ for $-\infty < t < \infty$.

Example 5.2   Frequency Analysis of an *RC* Circuit

Consider the *RC* circuit shown in Figure 5.1. As indicated in the figure, the input $x(t)$ is the voltage $v(t)$ applied to the circuit, and the output $y(t)$ is the voltage $v_C(t)$ across the capacitor. As shown in Section 2.4, the input/output differential equation of the circuit is

$$\frac{dy(t)}{dt} + \frac{1}{RC}y(t) = \frac{1}{RC}x(t) \tag{5.12}$$

Using the derivative property of the Fourier transform (see Table 3.1) and then taking the Fourier transform of both sides of (5.12) yield the following result:

$$j\omega Y(\omega) + \frac{1}{RC}Y(\omega) = \frac{1}{RC}X(\omega) \tag{5.13}$$

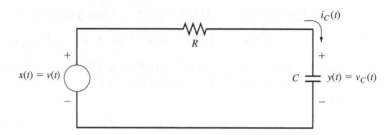

FIGURE 5.1
RC circuit in Example 5.2

Solving (5.13) for $Y(\omega)$ yields the ω-domain representation for the circuit:

$$Y(\omega) = \frac{1/RC}{j\omega + 1/RC} X(\omega) \tag{5.14}$$

Comparing (5.4) and (5.14) shows that the frequency response function $H(\omega)$ for the RC circuit is given by

$$H(\omega) = \frac{1/RC}{j\omega + 1/RC} \tag{5.15}$$

From (5.15), it is seen that the magnitude function $|H(\omega)|$ of the circuit is given by

$$|H(\omega)| = \frac{1/RC}{\sqrt{\omega^2 + (1/RC)^2}}$$

and the phase function $\angle H(\omega)$ is given by

$$\angle H(\omega) = -\tan^{-1} \omega RC$$

For any desired value of $1/RC$, we can compute the magnitude and phase functions by using MATLAB. For instance, in the case when $1/RC = 1000$, the MATLAB commands for generating the magnitude and phase (angle) functions are

```
RC = 0.001;
w = 0:50:5000;
H = (1/RC)./(j*w+1/RC);
magH = abs(H);
angH = 180*angle(H)/pi;
```

Using these commands and the plotting commands results in the plots of $|H(\omega)|$ and $\angle H(\omega)$ shown in Figure 5.2. From the figure, note that

$$|H(0)| = 1, \quad \angle H(0) = 0 \tag{5.16}$$

$$|H(1000)| = \frac{1}{\sqrt{2}} = 0.707, \quad \angle H(1000) = -45° \tag{5.17}$$

$$|H(3000)| = 0.316, \quad \angle H(3000) = -71.6° \tag{5.18}$$

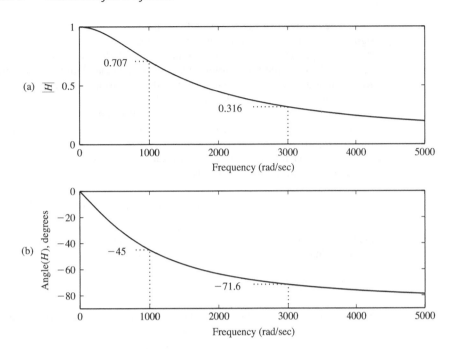

FIGURE 5.2
(a) Magnitude and (b) phase functions of the *RC* circuit in Example 5.2 for the case $1/RC = 1000$.

and

$$|H(\omega)| \to 0 \text{ as } \omega \to \infty, \quad \angle H(\omega) \to -90° \text{ as } \omega \to \infty \qquad (5.19)$$

Now, we can compute the output response $y(t)$ of the *RC* circuit resulting from a specific sinusoidal input $x(t) = A \cos(\omega_0 t + \theta)$ by inserting the appropriate values of $|H(\omega_0)|$ and $\angle H(\omega_0)$ into the expression (5.11). For example, suppose that $1/RC = 1000$. Then if $\omega_0 = 0$, by using (5.16) we find that the resulting output response is

$$y(t) = A(1)\cos(0t + \theta + 0) = A \cos \theta, \ -\infty < t < \infty \qquad (5.20)$$

If $\omega_0 = 1000$ rad/sec, using (5.17) yields the response

$$y(t) = A(0.707)\cos(1000t + \theta - 45°), \ -\infty < t < \infty \qquad (5.21)$$

and if $\omega_0 = 3000$ rad/sec, using (5.18) gives

$$y(t) = A(0.316)\cos(3000t + \theta - 71.6°), \ -\infty < t < \infty \qquad (5.22)$$

Finally, it follows from (5.19) that the output response $y(t)$ goes to zero as $\omega_0 \to \infty$.

From (5.20), it is seen that when $\omega_0 = 0$, so that the input is the constant signal $x(t) = A \cos \theta$, the response is equal to the input. Hence, the *RC* circuit passes a dc input without attenuation and without producing any phase shift. From (5.21), it is seen that when $\omega_0 = 1000$ rad/sec, the *RC* circuit attenuates the input sinusoid by a factor of 0.707, and phase shifts the input sinusoid by $-45°$; and by (5.22) it is seen that when $\omega_0 = 3000$ rad/sec, the attenuation factor is now 0.316 and the phase shift is $-71.6°$. Finally, as $\omega_0 \to \infty$, the magnitude of the output goes to zero while the phase shift goes to $-90°$.

The behavior of the *RC* circuit is summarized by observing that it passes low-frequency signals without any significant attenuation and without producing any significant phase shift. As the frequency increases, the attenuation and the phase shift become larger. Finally, as $\omega_0 \to \infty$, the *RC* circuit completely "blocks" the sinusoidal input. As a result of this behavior, the *RC* circuit is an example of a *lowpass filter*; that is, the circuit "passes," without much attenuation, input sinusoids whose frequency ω_0 is less than 1000 rad/sec, and it significantly attenuates input sinusoids whose frequency ω_0 is much above 1000 rad/sec. As discussed in Chapter 8, the frequency range from 0 to 1000 rad/sec (in the case $1/RC = 1000$) is called the 3dB *bandwidth* of the *RC* circuit.

To further illustrate the lowpass filter characteristic of the *RC* circuit, now suppose that the input is the sum of two sinusoids:

$$x(t) = A_1 \cos(\omega_1 t + \theta_1) + A_2 \cos(\omega_2 t + \theta_2)$$

Due to linearity of the *RC* circuit, the corresponding response $y(t)$ is the sum of the responses to the individual sinusoids:

$$y(t) = A_1 |H(\omega_1)| \cos(\omega_1 t + \theta_1 + \angle H(\omega_1)) + A_2 |H(\omega_2)| \cos(\omega_2 t + \theta_2 + \angle H(\omega_2))$$

To demonstrate the effect of the lowpass filtering, the response to the input

$$x(t) = \cos 100t + \cos 3000t$$

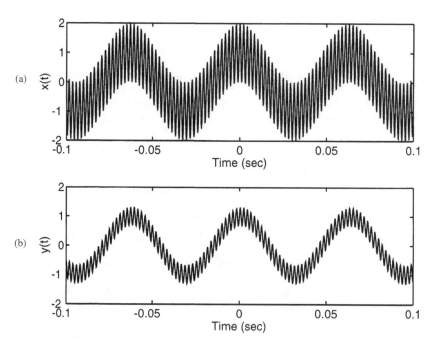

FIGURE 5.3
(a) Input and (b) output of *RC* circuit when $1/RC = 1000$.

will be calculated in the case when $1/RC = 1000$. The MATLAB commands used to generate $y(t)$ are

```
RC = 0.001;
t = -.1:.2/1000:.1;
w1 = 100; w2 = 3000;
Hw1 = (1/RC)/(j*w1+1/RC);
Hw2 = (1/RC)/(j*w2+1/RC);
x = cos(w1*t)+cos(w2*t);
 y = abs(Hw1)*cos(w1*t+angle(Hw1)) +
abs(Hw2)*cos(w2*t+angle(Hw2));
```

The plots for $x(t)$ and $y(t)$ are shown in Figure 5.3. Note that the amplitude of the low-frequency signal is approximately the same in both plots. However, the high-frequency component in $x(t)$ is very evident, but is much less significant in $y(t)$ due to the attenuation of high-frequency signals by the circuit.

Example 5.3   Mass–Spring–Damper System

Mass–
Spring–
Damper
System

Consider the mass–spring–damper system that was first defined in Section 1.4 of Chapter 1. (See Figure 1.25.) The input/output differential equation of the system is given by

$$M\frac{d^2y(t)}{dt^2} + D\frac{dy(t)}{dt} + Ky(t) = x(t) \qquad (5.23)$$

where M is the mass, D is the damping constant, K is the stiffness constant, $x(t)$ is the force applied to the mass, and $y(t)$ is the displacement of the mass relative to the equilibrium position. Taking the Fourier transform of both sides of (5.23) and solving for $Y(\omega)$ result in the following ω-domain representation for the mass-spring-damper system:

$$Y(\omega) = \frac{1}{M(j\omega)^2 + D(j\omega) + K}X(\omega) = \frac{1}{K - M\omega^2 + jD\omega}X(\omega)$$

Hence, the frequency response function of the system is

$$H(\omega) = \frac{1}{K - M\omega^2 + jD\omega}$$

The magnitude $|H(\omega)|$ of $H(\omega)$ can be determined over a frequency range from some start frequency to some stop frequency by inputting a sine sweep (a sinusoid whose frequency is varied from the start frequency to the stop frequency). This can be carried out with the mass–spring–damper system online demo on the Web. Trying this for various values of M, D, and K will result in different shapes for the magnitude function $|H(\omega)|$. In fact, for various values of M, D, and K, the magnitude function $|H(\omega)|$ will have a peak at some positive value of ω. As discussed in Chapter 8, the peak is due to a resonance in the mass–spring–damper system.

5.2   RESPONSE TO PERIODIC AND NONPERIODIC INPUTS

Suppose that the input $x(t)$ to the system defined by (5.1) is periodic so that $x(t)$ has the trigonometric Fourier series given in Equation (3.8):

$$x(t) = a_0 + \sum_{k=1}^{\infty} A_k \cos(k\omega_0 t + \theta_k), \quad -\infty < t < \infty$$

Here, ω_0 is the fundamental frequency of the signal, and a_0, A_k, and θ_k are constants with respect to t. It follows directly from (5.11) that the output response resulting from the sinusoidal input $A_k \cos(k\omega_0 t + \theta_k)$ is $A_k|H(k\omega_0)| \cos(k\omega_0 t + \theta_k + \angle H(k\omega_0))$. Similarly, the response due to a constant input a_0 is $a_0 H(0)$. Then by linearity, the response to the periodic input $x(t)$ is

$$y(t) = a_0 H(0) + \sum_{k=1}^{\infty} A_k|H(k\omega_0)| \cos(k\omega_0 t + \theta_k + \angle H(k\omega_0)),$$
$$-\infty < t < \infty \tag{5.24}$$

Since the right-hand side of this expression is a trigonometric form of a Fourier series, it follows that the response $y(t)$ is periodic. In addition, since the fundamental frequency of $y(t)$ is ω_0, which is the fundamental frequency of the input $x(t)$, the period of $y(t)$ is equal to the period of $x(t)$. Hence, the response to a periodic input with fundamental period T is periodic with fundamental period T. Now, let A_k^x and θ_k^x denote the coefficients of the trigonometric Fourier series for $x(t)$, and let A_k^y and θ_k^y denote the coefficients of the trigonometric Fourier series for the resulting output $y(t)$. From (5.24),

$$A_k^y = A_k^x|H(k\omega_0)| \quad \text{and} \quad \theta_k^y = \theta_k^x + \angle H(k\omega_0) \tag{5.25}$$

Thus, the Fourier series of the output can be computed directly from the coefficients of the input Fourier series.

As shown in Section 3.3, the complex Fourier series given in (3.19) is related to the trigonometric Fourier series through the relationship between the coefficients given in (3.27). Combining (3.27) and (5.25) gives the following expressions that can be plotted to obtain the line spectra of $y(t)$:

$$|c_k^y| = \tfrac{1}{2}|H(k\omega_0)| A_k^x \quad \text{and} \quad \angle c_k^y = \theta_k^x + \angle H(k\omega_o) \tag{5.26}$$

The process is illustrated by the following example:

Example 5.4   Response to a Rectangular Pulse Train

Response
to
Periodic
Inputs

Again, consider the RC circuit examined in Example 5.2 and shown in Figure 5.1. The objective is to determine the voltage $y(t)$ on the capacitor resulting from the rectangular pulse train $x(t)$ shown in Figure 5.4. From the results in Example 3.2, $x(t)$ has the trigonometric Fourier series

$$x(t) = a_0 + \sum_{k=1}^{\infty} a_k \cos(k\pi t), \quad -\infty < t < \infty$$

where $a_0 = 0.5$ and

$$a_k = \frac{2}{k\pi} \sin\left(\frac{k\pi}{2}\right) = \begin{cases} \dfrac{2}{k\pi}(-1)^{(k-1)/2}, & k = 1, 3, 5, \ldots \\ \\ 0, & k = 2, 4, 6, \ldots \end{cases}$$

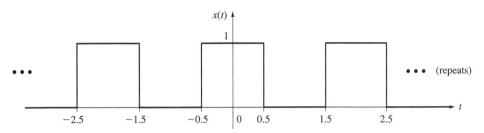

FIGURE 5.4
Periodic input signal in Example 5.4.

Using (3.23), we see that the magnitude $|c_k^x|$ of the coefficients of the complex Fourier series for the rectangular pulse train are given by

$$|c_k^x| = \begin{cases} 0.5, & k = 0 \\ \\ 0, & k = \pm 2, \pm 4, \pm 6, \ldots \\ \dfrac{1}{k\pi}, & k = \pm 1, \pm 3, \pm 5, \ldots \end{cases}$$

The amplitude spectrum (the plot of $|c_k^x|$ versus $\omega = k\omega_0 = k\pi$) of the rectangular pulse train is displayed in Figure 5.5. The plot shows that $x(t)$ has frequency components from dc all the way to infinite frequency, with the higher-frequency components having less significance. In particular, from the plot it is clear that most of the spectral content of $x(t)$ is contained in the frequency range 0 to 40 rad/sec. The "corners" encompassing the rectangular pulse train are a result of the spectral lines in the limit as $k \to \infty$.

As computed in Example 5.2, the frequency response function of the RC circuit is

$$H(\omega) = \frac{1/RC}{j\omega + 1/RC}$$

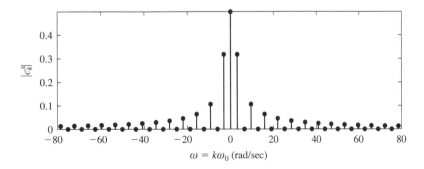

FIGURE 5.5
Amplitude spectrum of periodic input in Example 5.4.

Hence,

$$H(k\omega_0) = H(k\pi) = \frac{1/RC}{jk\pi + 1/RC}$$

and

$$|H(k\pi)| = \frac{1/RC}{\sqrt{k^2\pi^2 + (1/RC)^2}}$$

$$\angle H(k\pi) = -\tan^{-1}k\pi RC$$

Note that the magnitude function $|H(k\pi)|$ rolls off as the integer k is increased. This is, of course, due to the lowpass characteristic of the RC circuit. As discussed in Example 5.2, in the case when $1/RC = 1000$, the 3-dB bandwidth of the RC circuit is 1000 rad/sec. For arbitrary positive values of R and C, the 3-dB bandwidth is equal to $1/RC$ rad/sec. Frequency components of the input below $1/RC$ are passed without significant attenuation, while frequency components above $1/RC$ are attenuated. Thus, the larger the value of $1/RC$ is, the larger the bandwidth of the RC circuit will be, resulting in higher-frequency components of the input being passed through the circuit.

To find the coefficients of the cosine-with-phase form of the Fourier series (3.8), use (3.9) and (3.10), which gives

$$A_k^x = \begin{cases} \dfrac{2}{k\pi}, & k = 1, 3, 5, \ldots \\[2em] 0, & k = 2, 4, 6, \ldots \end{cases}$$

$$\theta_k^x = \begin{cases} \pi, & k = 3, 7, 11, \ldots \\ 0, & \text{all other } k \end{cases}$$

Then inserting the expressions for A_k^x, θ_k^x, $|H(k\pi)|$, and $\angle H(k\pi)$ into (5.25) yields the following expressions for the coefficients of the trigonometric Fourier series of the output:

$$a_0^y = H(0)a_0^x = 0.5 \tag{5.27}$$

$$A_k^y = \begin{cases} \dfrac{2}{k\pi}\dfrac{1/RC}{\sqrt{k^2\pi^2 + (1/RC)^2}}, & k \text{ odd} \\[2em] 0, & k \text{ even} \end{cases} \tag{5.28}$$

$$\theta_k^y = \begin{cases} \pi - \tan^{-1}k\pi RC, & k = 3, 7, 11, \ldots \\ -\tan^{-1}k\pi RC, & \text{all other } k \end{cases} \tag{5.29}$$

We can see the effect of the bandwidth of the circuit on the output by plotting the amplitude spectrum ($|c_k^y| = \frac{1}{2}A_k^y$ versus $\omega = k\pi$) of the output for various values of $1/RC$. The output amplitude spectrum is displayed in Figure 5.6a to 5.6c for the values $1/RC = 1$, $1/RC = 10$, and $1/RC = 100$. Comparing Figures 5.5 and 5.6a reveals that when the 3-dB bandwidth of the RC circuit is 1 rad/sec (i.e., $1/RC = 1$), the circuit attenuates much of the spectral content of the rectangular pulse train. On the other hand, comparing Figures 5.5 and 5.6c shows that there is

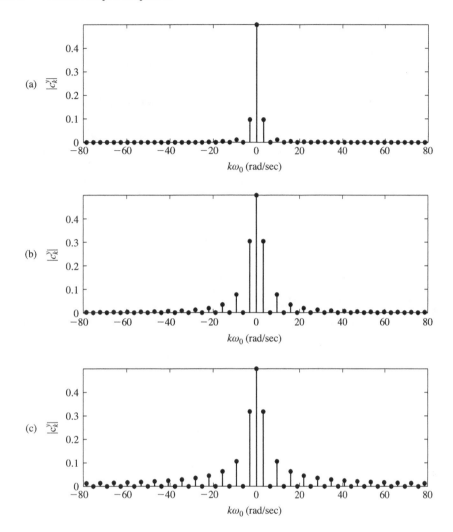

FIGURE 5.6
Amplitude spectrum of output when (a) $1/RC = 1$; (b) $1/RC = 10$; (c) $1/RC = 100$.

very little attenuation of the input spectral components when the 3-dB bandwidth is 100 rad/sec (i.e., $1/RC = 100$). It would therefore be expected that, when $1/RC = 1$ the circuit will significantly distort the pulse train, whereas when $1/RC = 100$ there should not be much distortion. To verify this, we will compute the output response by first computing the Fourier series representation of the output.

From (5.24), (5.27), and (5.28), the trigonometric Fourier series of the output is given by

$$y(t) = 0.5 + \sum_{\substack{k=1 \\ k \text{ odd}}}^{\infty} \frac{2}{k\pi} \frac{1/RC}{\sqrt{k^2\pi^2 + (1/RC)^2}} \cos(k\pi t + \theta_k^y) \tag{5.30}$$

By definition of the θ_k^y [see (5.29)], it follows that

$$\cos(k\pi t + \theta_k^y) = (-1)^{(k-1)/2}\cos(k\pi t - \tan^{-1}k\pi RC),\ k = 1, 3, 5, \ldots \tag{5.31}$$

Hence, inserting (5.31) into (5.30) yields the following form for the Fourier series of the output:

$$y(t) = 0.5 + \sum_{\substack{k=1 \\ k\ \text{odd}}}^{\infty} \frac{2}{k\pi}(-1)^{(k-1)/2}\frac{1/RC}{\sqrt{k^2\pi^2 + (1/RC)^2}}\cos(k\pi t - \tan^{-1}k\pi RC) \tag{5.32}$$

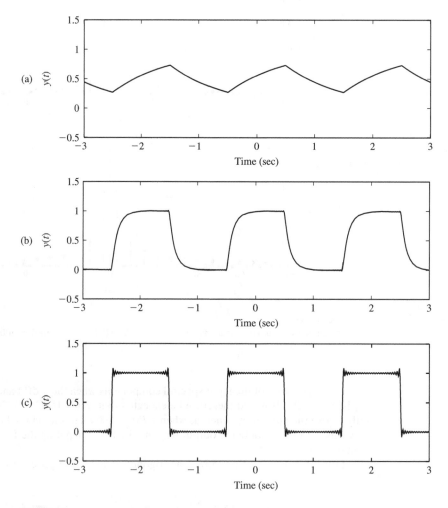

FIGURE 5.7
Plot of output when (a) 1/*RC* = 1; (b) 1/*RC* = 10; (c) 1/*RC* = 100.

Since the coefficients of the Fourier series (5.32) for $y(t)$ are getting very small as k increases, it is possible to determine the values of $y(t)$ by evaluating a suitable number of the terms comprising (5.32). The MATLAB commands used to obtain $y(t)$ for the case $1/RC = 1$ are

```
RC = 1; a0 = .5; H0 = 1; N = 50; w0=pi;
y = a0*H0*ones(1,length(t));
for k=1:2:N,
    Bk = 2/pi/k*(-1)^((k-1)/2);
    H = (1/RC)/(j*k*w0 + 1/RC);
    y = y + Bk*abs(H)*cos(k*w0*t+angle(H));
end
```

The value of $N = 50$ was chosen to be sufficiently large to achieve good accuracy in recovering the waveform of $y(t)$.

The response $y(t)$ is displayed in Figure 5.7 for the values $1/RC = 1$, $1/RC = 10$, and $1/RC = 100$. From the figure it is seen that the response more closely resembles the input pulse train as the bandwidth of the RC circuit is increased from 1 rad/sec (Figure 5.7a) to 100 rad/sec (Figure 5.7c). Again, this result is expected, since the circuit is passing more of the spectral content of the input pulse train as the bandwidth of the circuit is increased.

5.2.1   Response to Nonperiodic Inputs

We can compute the response $y(t)$ resulting from any input $x(t)$ by first finding the Fourier transform $Y(\omega)$ of $y(t)$ by using (5.4). Then we can compute the output by taking the inverse Fourier transform of $H(\omega)X(\omega)$; that is, $y(t)$ is given by

$$y(t) = \frac{1}{2\pi} \int_{-\infty}^{\infty} H(\omega)X(\omega)e^{j\omega t}\, d\omega \tag{5.33}$$

The computation of $y(t)$ by (5.33) is usually rather complicated to carry out, due to the integral form. In some cases it is possible to determine $y(t)$ by working with Fourier transform pairs from a table, rather than by evaluating the integral in (5.33).

Example 5.5   Response of *RC* Circuit to a Pulse

Again, consider the RC circuit shown in Figure 5.1. In this example the objective is to examine the response due to the rectangular pulse shown in Figure 5.8. The Fourier transform of $x(t)$ was found in Example 3.9 to be (setting $\tau = 1$)

$$X(\omega) = \mathrm{sinc}\frac{\omega}{2\pi} = \frac{\sin(\omega/2)}{\omega/2} = 2\frac{\sin(\omega/2)}{\omega}$$

The amplitude spectrum of the pulse is displayed in Figure 5.9.

From (5.4), the Fourier transform of the output is

$$Y(\omega) = X(\omega)H(\omega)$$

where, again,

$$H(\omega) = \frac{1/RC}{j\omega + 1/RC}$$

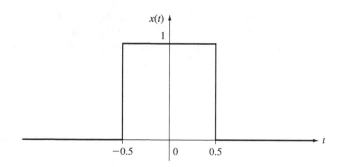

FIGURE 5.8
Input pulse in Example 5.5.

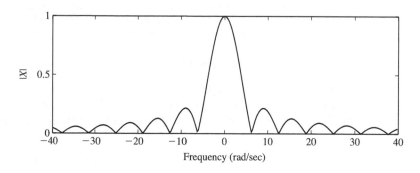

FIGURE 5.9
Amplitude spectrum of the input pulse.

The amplitude spectrum of the output response $y(t)$ is plotted in Figure 5.10a for the case $1/RC = 1$, and is plotted in Figure 5.10b for the case $1/RC = 10$. The amplitude spectrum $|Y(\omega)|$ was obtained by multiplying $|H(\omega)|$ by $|X(\omega)|$ at each frequency. The MATLAB commands to obtain $|Y(\omega)|$ when $1/RC = 1$ are

```
RC = 1;
w = -40:.3:40;
X = 2*sin(w/2)./w;
H = (1/RC)./(j*w+1/RC);
Y = X.*H;
magY = abs(Y);
```

As noted in Examples 5.2 and 5.4, the larger the value of $1/RC$ is, the larger the 3-dB bandwidth of the RC circuit will be, which means that the sidelobes of $X(\omega)$ are attenuated less in passing through the filter. In particular, from Figures 5.9 and 5.10a it can be seen that when $1/RC = 1$, there is a significant amount of attenuation of the main lobe and sidelobes of $|Y(\omega)|$ in comparison with those of $|X(\omega)|$, which is a result of the 3-dB bandwidth of the circuit being set too low

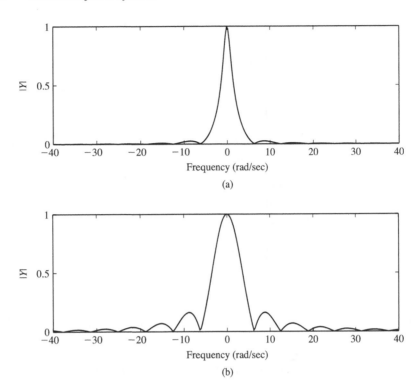

FIGURE 5.10
Amplitude spectrum of y(t) when (a) 1/RC = 1 and (b) 1/RC = 10.

(in the case when $1/RC = 1$). As seen from Figure 5.10b, increasing the bandwidth by setting $1/RC = 10$ results in much less attenuation of the sidelobes of $|Y(\omega)|$; that is, there is much less attenuation of the higher-frequency components of $x(t)$ in passing through the filter when the 3-dB bandwidth is 10 rad/sec. The effect of the filtering can be examined in the time domain by computing $y(t)$ for various values of $1/RC$.

We can compute the output $y(t)$ by taking the inverse Fourier transform of $H(\omega)X(\omega)$ as given by (5.33). But this computation is tedious to carry out for the present example. Instead, we can compute the response with the Symbolic Math Toolbox, using the following commands:

```
syms X H Y y w
X = 2*sin(w/2)./w;
H = (1/RC)./(j*w+1/RC);
Y = X.*H;
y = ifourier(Y);
ezplot(y,[-1 5])
axis([-1 5 0 1.5])
```

We can also compute the output response numerically by solving the differential equation for the circuit given in (5.12), using the methods described in Section 2.5, such as the command ode45. The time-domain responses for $1/RC = 1$ and for $1/RC = 10$ are shown in Figure 5.11. Note that in the higher-bandwidth case ($1/RC = 10$), when more of the main lobe and sidelobes of $X(\omega)$

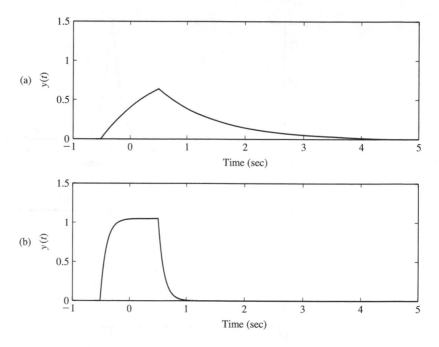

FIGURE 5.11
Output response when (a) $1/RC = 1$ and (b) $1/RC = 10$.

are passed through the circuit, the output response looks more like the input pulse. However, even in this case the cutoff of the high-frequency components of $X(\omega)$ causes the corners of the rectangular pulse to be smoothed in the output response.

5.3   ANALYSIS OF IDEAL FILTERS

Given a linear time-invariant continuous-time system with frequency function $H(\omega)$, in Section 5.1 it was shown that the output response $y(t)$ resulting from the sinusoidal input $x(t) = A\cos(\omega_0 t + \theta)$, $-\infty < t < \infty$, is given by

$$y(t) = A|H(\omega_0)|\cos(\omega_0 t + \theta + \angle H(\omega_0)), \quad -\infty < t < \infty \qquad (5.34)$$

From (5.34) it is clear that a sinusoid with a particular frequency ω_0 can be prevented from "going through" the system by selecting $H(\omega)$ so that $|H(\omega_0)|$ is zero or very small. The process of "rejecting" sinusoids having particular frequencies or range of frequencies is called *filtering*, and a system that has this characteristic is called a *filter*.

The concept of filtering was first considered in Chapter 1 in the discrete-time case, where the MA filter was studied. In the continuous-time case, the RC circuit considered in Sections 5.1 and 5.2 was shown to be an example of a filter. In this section ideal continuous-time filters are considered.

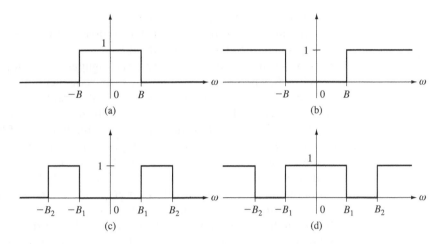

FIGURE 5.12
Magnitude functions of ideal filters: (a) lowpass; (b) highpass; (c) bandpass; (d) bandstop.

An *ideal filter* is a system that completely rejects sinusoidal inputs of the form $x(t) = A\cos \omega_0 t$, $-\infty < t < \infty$, for ω_0 in certain frequency ranges and does not attenuate sinusoidal inputs whose frequencies are outside these ranges. There are four basic types of ideal filters: lowpass, highpass, bandpass, and bandstop. The magnitude functions of these four types of filters are displayed in Figure 5.12. Mathematical expressions for these magnitude functions are as follows:

$$\text{Ideal lowpass:}\quad |H(\omega)| = \begin{cases} 1, & -B \leq \omega \leq B \\ 0, & |\omega| > B \end{cases} \tag{5.35}$$

$$\text{Ideal highpass:}\quad |H(\omega)| = \begin{cases} 0, & -B < \omega < B \\ 1, & |\omega| \geq B \end{cases} \tag{5.36}$$

$$\text{Ideal bandpass:}\quad |H(\omega)| = \begin{cases} 1, & B_1 \leq |\omega| \leq B_2 \\ 0, & \text{all other } \omega \end{cases} \tag{5.37}$$

$$\text{Ideal bandstop:}\quad |H(\omega)| = \begin{cases} 0, & B_1 \leq |\omega| \leq B_2 \\ 1, & \text{all other } \omega \end{cases} \tag{5.38}$$

The *stopband* of an ideal filter is defined to be the set of all frequencies ω_0 for which the filter completely stops the sinusoidal input $x(t) = A\cos \omega_0 t$, $-\infty < t < \infty$. The *passband* of the filter is the set of all frequencies ω_0 for which the input $x(t)$ is passed without attenuation.

From (5.34) and (5.35), it is seen that the ideal lowpass filter passes, with no attenuation, sinusoidal inputs with frequencies ranging from $\omega = 0$ (rad/sec) to $\omega = B$ (rad/sec), while it completely stops sinusoidal inputs with frequencies above $\omega = B$. The filter is said to be *lowpass*, since it passes low-frequency sinusoids and stops high-frequency sinusoids. The frequency range $\omega = 0$ to $\omega = B$ is the passband of the filter, and the range $\omega = B$ to $\omega = \infty$ is the stopband of the filter. The width B of the passband is defined to be the filter bandwidth.

From (5.34) and (5.36), it is seen that the highpass filter stops sinusoids with frequencies below B, while it passes sinusoids with frequencies above B; hence the term *highpass*. The stopband of the highpass filter is the frequency range from $\omega = 0$ to $\omega = B$, and the passband is the frequency range from $\omega = B$ to $\omega = \infty$.

Equation (5.37) shows that the passband of the bandpass filter is the frequency range from $\omega = B_1$ to $\omega = B_2$, while the stopband is the range from $\omega = 0$ to $\omega = B_1$ and the range from $\omega = B_2$ to $\omega = \infty$. The bandwidth of the bandpass filter is the width of the passband (i.e., $B_2 - B_1$). By (5.38) it is seen that the stopband of the band-stop filter is the range from $\omega = B_1$ to $\omega = B_2$, while the passband is the range from $\omega = 0$ to $\omega = B_1$ and the range from $\omega = B_2$ to $\omega = \infty$.

More complicated examples of ideal filters can be constructed by cascading ideal lowpass, highpass, bandpass, and bandstop filters. For instance, by cascading bandstop filters with various values of B_1 and B_2 it is possible to construct an ideal *comb* filter, whose magnitude function is illustrated in Figure 5.13.

5.3.1 Phase Function

In the prior discussion on ideal filters, nothing was said regarding the phase of the filters. It turns out that, to avoid phase distortion in the filtering process, a filter should have a linear phase characteristic over the passband of the filter. In other words, the phase function (in radians) should be of the form

$$\angle H(\omega) = -\omega t_d \text{ for all } \omega \text{ in the filter passband} \qquad (5.39)$$

where t_d is a fixed positive number. If ω_0 is in the passband of a linear phase filter, by (5.34) and (5.39) the response resulting from the input $x(t) = A \cos \omega_0 t$, $-\infty < t < \infty$, is given by

$$y(t) = A|H(\omega_0)| \cos(\omega_0 t - \omega_0 t_d), \quad -\infty < t < \infty$$

$$y(t) = A|H(\omega_0)| \cos[\omega_0(t - t_d)], \quad -\infty < t < \infty$$

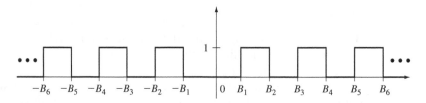

FIGURE 5.13
Magnitude function of an ideal comb filter.

Thus, the linear phase characteristic results in a time delay of t_d seconds through the filter.

Note that if the input is

$$x(t) = A_0 \cos \omega_0 t + A_1 \cos \omega_1 t, \quad -\infty < t < \infty \tag{5.40}$$

where ω_0 and ω_1 are in the passband of the filter, by linearity the response is

$$y(t) = A_0 |H(\omega_0)| \cos[\omega_0(t - t_d)] + A_1 |H(\omega_1)| \cos[\omega_1(t - t_d)], \quad -\infty < t < \infty$$

So again, the output is a t_d-second time delay of the input; in particular, there is no distortion of the input. In contrast, if the phase function is not linear, there will be phase distortion in the filter output. To see this, suppose that the phase function $\angle H(\omega)$ is equal to some nonzero constant C. In this case the response to the input (5.40) is

$$y(t) = A_0 |H(\omega_0)| \cos(\omega_0 t + C) + A_1 |H(\omega_1)| \cos(\omega_1 t + C), \quad -\infty < t < \infty$$

This output is not a time-delayed version of the input, so there is distortion in the filtering process. Therefore, for distortionless filtering, the phase function of the filter should be as close to linear as possible over the passband of the filter.

5.3.2   Ideal Linear-Phase Lowpass Filter

Consider the ideal lowpass filter with the frequency function

$$H(\omega) = \begin{cases} e^{-j\omega t_d}, & -B \le \omega \le B \\ 0, & \omega < -B, \omega > B \end{cases} \tag{5.41}$$

where t_d is a positive real number. Equation (5.41) is the polar-form representation of $H(\omega)$. From (5.41), we see that

$$|H(\omega)| = \begin{cases} 1, & -B \le \omega \le B \\ 0, & \omega < -B, \omega > B \end{cases}$$

and the phase in radians is

$$\angle H(\omega) = \begin{cases} -\omega t_d, & -B \le \omega \le B \\ 0, & \omega < -B, \omega > B \end{cases}$$

The phase function $\angle H(\omega)$ of the filter is plotted in Figure 5.14. Note that over the frequency range 0 to B, the phase function of the system is linear with slope equal to $-t_d$.

We can compute the impulse response of the lowpass filter defined by (5.41) by taking the inverse Fourier transform of the frequency function $H(\omega)$. First, using the definition of the rectangular pulse, we can express $H(\omega)$ in the form

$$H(\omega) = p_{2B}(\omega)e^{-j\omega t_d}, \quad -\infty < \omega < \infty \tag{5.42}$$

From Table 3.2 the following transform pair can be found:

$$\frac{\tau}{2\pi} \operatorname{sinc} \frac{\tau t}{2\pi} \leftrightarrow p_\tau(\omega) \tag{5.43}$$

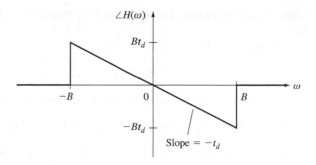

FIGURE 5.14
Phase function of ideal linear-phase lowpass filter defined by (5.41).

Setting $\tau = 2B$ in (5.43) gives

$$\frac{B}{\pi}\text{sinc}\left(\frac{B}{\pi}t\right) \leftrightarrow p_{2B}(\omega) \tag{5.44}$$

Applying the time-shift property to the transform pair (5.44) gives

$$\frac{B}{\pi}\text{sinc}\left[\frac{B}{\pi}(t - t_d)\right] \leftrightarrow p_{2B}(\omega)e^{-j\omega t_d} \tag{5.45}$$

Since the right-hand side of the transform pair (5.45) is equal to $H(\omega)$, the impulse response of the ideal lowpass filter is

$$h(t) = \frac{B}{\pi}\text{sinc}\left[\frac{B}{\pi}(t - t_d)\right], \quad -\infty < t < \infty \tag{5.46}$$

The impulse response $h(t)$ is plotted in Figure 5.15.

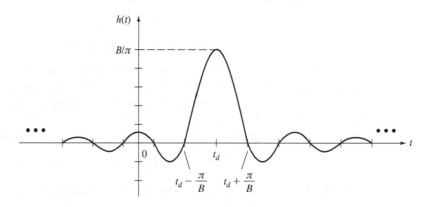

FIGURE 5.15
Impulse response of ideal linear-phase lowpass filter.

From Figure 5.15 it is clear that the impulse response $h(t)$ is not zero for $t < 0$, and thus the filter is noncausal. As a result, it is not possible to implement (in real time) an ideal lowpass filter. In fact, any ideal filter is noncausal and thus cannot be operated in real time. For real-time filtering, it is necessary to consider causal filters, such as the RC circuit given in the previous section.

5.3.3   Ideal Linear-Phase Bandpass Filters

The analysis previously given can be extended to the other types of ideal filters mentioned in the first part of this section. For example, the frequency function of an ideal linear-phase bandpass filter is given by

$$H(\omega) = \begin{cases} e^{-j\omega t_d}, & B_1 \leq |\omega| \leq B_2 \\ 0, & \text{all other } \omega \end{cases}$$

where t_d, B_1, and B_2 are positive real numbers. The magnitude function $|H(\omega)|$ is plotted in Figure 5.12c, and the phase function $\angle H(\omega)$ (in radians) is plotted in Figure 5.16. Since the passband of the filter is from B_1 to B_2, for any input signal $x(t)$ whose frequency components are contained in the region from B_1 to B_2, the filter will pass the signal with no distortion, although there will be a time delay of t_d seconds.

5.3.4   Causal Filters

The study of causal continuous-time filters is based on the transfer function formulation that is defined in terms of the Laplace transform, which is introduced in Chapter 6. Hence, the discussion of causal continuous-time filtering is delayed until a later chapter.

5.4   SAMPLING

An important application of the Fourier transform is in the study of sampling a continuous-time signal $x(t)$, which arises in various areas of technology such as communications, controls, and digital signal processing. In this section the sampling process is

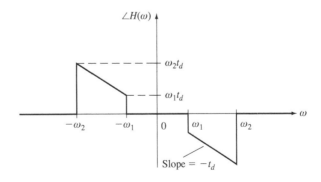

FIGURE 5.16
Phase function of ideal linear-phase bandpass filter.

characterized in terms of the Fourier transform, which reveals how a given continuous-time signal $x(t)$ can be reconstructed from a sampled version of the signal.

Sampling is part of a discretization process where a continuous-time signal $x(t)$ is converted into a discrete-time signal $x[n]$. In uniform sampling (the case of interest here), the sample values of $x(t)$ are equal to the values $x(nT)$, where n is the integer index, $n = 0, \pm 1, \pm 2, \ldots$, and T is the sampling interval. The corresponding discrete-time signal is defined as $x[n] = x(nT)$. To simplify the analysis of the sampling operation, the sampled version of $x(t)$ is often expressed in the form $x(t)p(t)$, where $p(t)$ is the impulse train given by

$$p(t) = \sum_{n=-\infty}^{\infty} \delta(t - nT) \tag{5.47}$$

Hence, the sampled waveform $x(t)p(t)$ is given by

$$x(t)p(t) = \sum_{n=-\infty}^{\infty} x(t)\delta(t - nT) = \sum_{n=-\infty}^{\infty} x(nT)\delta(t - nT) \tag{5.48}$$

Thus, the sampled waveform $x(t)p(t)$ is an impulse train whose weights (areas) are the sample values $x(nT)$ of the signal $x(t)$. The sampling process given by (5.48) is referred to as "idealized sampling."

To determine the Fourier transform of $x(t)p(t)$, first observe that since the impulse train $p(t)$ is a periodic signal with fundamental period T, $p(t)$ has the complex exponential Fourier series

$$p(t) = \sum_{k=-\infty}^{\infty} c_k e^{jk\omega_s t} \tag{5.49}$$

where $\omega_s = 2\pi/T$ is the *sampling frequency* in rad/sec. The coefficients c_k of the Fourier series are computed as follows:

$$c_k = \frac{1}{T} \int_{-T/2}^{T/2} p(t) e^{-jk\omega_s t} \, dt, \qquad k = 0, \pm 1, \pm 2, \ldots$$

$$= \frac{1}{T} \int_{-T/2}^{T/2} \delta(t) e^{-jk\omega_s t} \, dt$$

$$= \frac{1}{T} [e^{-jk\omega_s t}]_{t=0}$$

$$= \frac{1}{T}$$

Inserting $c_k = 1/T$ into (5.49) yields

$$p(t) = \sum_{k=-\infty}^{\infty} \frac{1}{T} e^{jk\omega_s t}$$

and thus

$$x(t)p(t) = \sum_{k=-\infty}^{\infty} \frac{1}{T} x(t) e^{jk\omega_s t} \tag{5.50}$$

We can then compute the Fourier transform of $x(t)p(t)$ by transforming the right-hand side of (5.50), using the property of the Fourier transform involving multiplication by a complex exponential. [See (3.50).] With $X(\omega)$ equal to the Fourier transform of $x(t)$, the result is

$$X_s(\omega) = \sum_{k=-\infty}^{\infty} \frac{1}{T} X(\omega - k\omega_s) \tag{5.51}$$

where $X_s(\omega)$ is the Fourier transform of the sampled waveform $x_s(t) = x(t)p(t)$.

From (5.51) it is seen that the Fourier transform $X_s(\omega)$ consists of a sum of magnitude-scaled replicas of $X(\omega)$ sitting at integer multiples $k\omega_s$ of ω_s for $k = 0, \pm 1, \pm 2, \ldots$. For example, suppose that $x(t)$ has the bandlimited Fourier transform $X(\omega)$ shown in Figure 5.17a. If $\omega_s - B > B$ or $\omega_s > 2B$, the Fourier transform $X_s(\omega)$ of the sampled signal $x_s(t) = x(t)p(t)$ is as shown in Figure 5.17b. Note that in this case, the replicas of $X(\omega)$ in $X_s(\omega)$ do not overlap in frequency. As a result, it turns out that we can reconstruct $x(t)$ from the sampled signal by using lowpass filtering. The reconstruction process is studied next.

5.4.1   Signal Reconstruction

Given a signal $x(t)$, the reconstruction of $x(t)$ from the sampled waveform $x(t)p(t)$ can be carried out as follows. First, suppose that $x(t)$ has bandwidth B; that is,

$$|X(\omega)| = 0 \text{ for } \omega > B$$

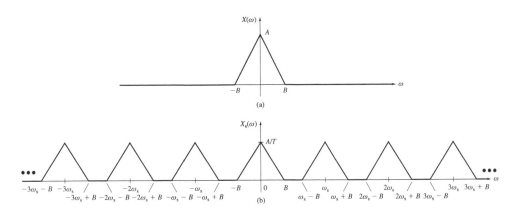

FIGURE 5.17
Fourier transform of (a) $x(t)$ and (b) $x_s(t) = x(t)p(t)$.

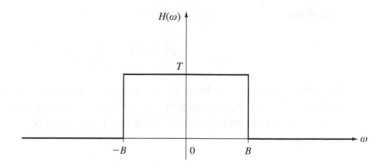

FIGURE 5.18
Frequency response function of ideal lowpass filter with bandwidth B.

Then if $\omega_s \geq 2B$, in the expression (5.51) for $X_s(\omega)$ the replicas of $X(\omega)$ do not overlap in frequency. For example, if $X(\omega)$ has the form shown in Figure 5.17a, then $X_s(\omega)$ has the form shown in Figure 5.17b in the case when $\omega_s \geq 2B$. Thus if the sampled signal $x_s(t)$ is applied to an ideal lowpass filter with the frequency function shown in Figure 5.18, the only component of $X_s(\omega)$ that is passed is $X(\omega)$. Hence, the output of the filter is equal to $x(t)$, which shows that the original signal $x(t)$ can be completely and exactly reconstructed from the sampled waveform $x_s(t)$. So, the reconstruction of $x(t)$ from the sampled signal $x_s(t) = x(t)p(t)$ can be accomplished by a simple lowpass filtering of the sampled signal. The process is illustrated in Figure 5.19. The filter in this figure is sometimes called an *interpolation filter*, since it reproduces $x(t)$ from the values of $x(t)$ at the time points $t = nT$.

By this result, which is called the *sampling theorem*, a signal with bandwidth B can be reconstructed completely and exactly from the sampled signal $x_s(t) = x(t)p(t)$ by lowpass filtering with cutoff frequency B if the sampling frequency ω_s is chosen to be greater than or equal to $2B$. The minimum sampling frequency $\omega_s = 2B$ is called the *Nyquist sampling frequency*.

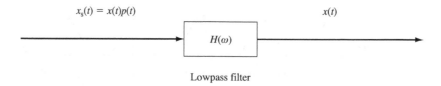

Lowpass filter

FIGURE 5.19
Reconstruction of $x(t)$ from $x_s(t) = x(t)p(t)$.

Example 5.6   Nyquist Sampling Frequency for Speech

The spectrum of a speech signal is essentially zero for all frequencies above 10 kHz, and so the bandwidth of a speech signal can be taken to be $2\pi \times 10^4$ rad/sec. Then the Nyquist sampling frequency for speech is

$$\omega_s = 2B = 4\pi \times 10^4 \text{ rad/sec}$$

Since $\omega_s = 2\pi/T$, the sampling interval T is equal to $2\pi/\omega_s = 50 \, \mu s$. So, the sampling interval corresponding to the Nyquist sampling rate is very small.

5.4.2   Interpolation Formula

From Figure 5.18 it is clear that the frequency response function $H(\omega)$ of the interpolating filter is given by

$$H(\omega) = \begin{cases} T, & -B \leq \omega \leq B \\ 0, & \text{all other } \omega \end{cases}$$

From the results in Section 5.3, the impulse $h(t)$ of this filter is given by

$$h(t) = \frac{BT}{\pi} \text{sinc}\left(\frac{B}{\pi}t\right), \qquad -\infty < t < \infty \tag{5.52}$$

and the output $y(t)$ of the interpolating filter is given by

$$y(t) = h(t) * x_s(t) = \int_{-\infty}^{\infty} x_s(\tau) h(t - \tau) \, d\tau \tag{5.53}$$

But we also have that

$$x_s(\tau) = x(\tau) p(\tau) = \sum_{n=-\infty}^{\infty} x(nT) \delta(\tau - nT)$$

and inserting this into (5.53) gives

$$y(t) = \int_{-\infty}^{\infty} \sum_{n=-\infty}^{\infty} x(nT) \delta(\tau - nT) h(t - \tau) \, d\tau$$

$$= \sum_{n=-\infty}^{\infty} \int_{-\infty}^{\infty} x(nT) \delta(\tau - nT) h(t - \tau) \, d\tau \tag{5.54}$$

From the sifting property of the impulse, (5.54) reduces to

$$y(t) = \sum_{n=-\infty}^{\infty} x(nT) h(t - nT) \tag{5.55}$$

Finally, inserting (5.52) into (5.55) gives

$$y(t) = \frac{BT}{\pi} \sum_{n=-\infty}^{\infty} x(nT) \, \text{sinc}\left[\frac{B}{\pi}(t - nT)\right] \tag{5.56}$$

But $y(t) = x(t)$, and thus (5.56) yields

$$x(t) = \frac{BT}{\pi} \sum_{n=-\infty}^{\infty} x(nT) \, \text{sinc}\left[\frac{B}{\pi}(t - nT)\right] \tag{5.57}$$

The expression (5.57) is called the *interpolation formula* for the signal $x(t)$. In particular, it shows how the original signal $x(t)$ can be reconstructed from the sample values $x(nT), n = 0, \pm1, \pm2, \ldots$.

5.4.3   Aliasing

Sampling and Aliasing

In Chapter 3 it was noted that a time-limited signal cannot be bandlimited. Since all actual signals are time limited, they cannot be bandlimited. Therefore, if a time-limited signal is sampled with sampling interval T, no matter how small T is, the replicas of $X(\omega)$ in (5.51) will overlap. As a result of the overlap of frequency components, it is not possible to reconstruct $x(t)$ exactly by lowpass filtering the sampled signal $x_s(t) = x(t)p(t)$.

Although time-limited signals are not bandlimited, the amplitude spectrum $|X(\omega)|$ of a time-limited signal $x(t)$ will be small for suitably large values of ω. Thus, for some finite B, all the significant components of $X(\omega)$ will be in the range $-B \leq \omega \leq B$. For instance, the signal $x(t)$ may have the amplitude spectrum shown in Figure 5.20. If B is chosen to have the value indicated, and if $x(t)$ is sampled with sampling frequency $\omega_s = 2B$, the amplitude spectrum of the resulting sampled signal $x_s(t)$ is as shown in Figure 5.21.

Now if the sampled signal $x_s(t)$ is lowpass filtered with cutoff frequency B, the output spectrum of the filter will contain high-frequency components of $x(t)$ transposed to low-frequency components. This phenomenon is called *aliasing*.

Aliasing will result in a distorted version of the original signal $x(t)$. It can be eliminated (theoretically) by first lowpass filtering $x(t)$ before $x(t)$ is sampled: If $x(t)$ is

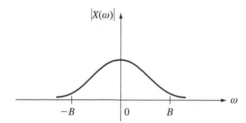

FIGURE 5.20
Amplitude spectrum of a time-limited signal.

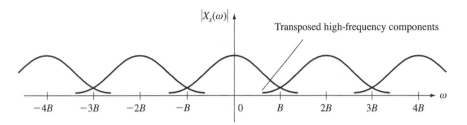

FIGURE 5.21
Amplitude spectrum of a sampled signal.

lowpass filtered so that all frequency components with values greater than B are removed, there will be no overlap of frequency components in the spectrum $X_s(\omega)$ of the sampled signal $x_s(t)$, assuming that $x(t)$ is sampled at the Nyquist rate $\omega_s = 2B$.

In practice, aliasing cannot be eliminated completely, since a lowpass filter that cuts off all frequency components above a certain frequency cannot be synthesized (i.e., built). However, the magnitude of the aliased components can be reduced if the signal $x(t)$ is lowpass filtered before sampling. This approach is feasible as long as lowpass filtering $x(t)$ does not remove the "information content" of the signal $x(t)$.

Example 5.7   Filtered Speech

Again, suppose that $x(t)$ is a speech waveform. Although a speech waveform may contain sizable frequency components above 4 kHz, voice recognition is possible for speech signals that have been filtered to a 4-kHz bandwidth. If B is chosen to be 4 kHz for filtered speech, the resulting Nyquist sampling frequency is

$$\omega_s = 2(2\pi)(4 \times 10^3) = 16\pi \times 10^3 \, \text{rad/sec}$$

For this sampling frequency, the sampling interval T is

$$T = \frac{2\pi}{\omega_s} = 0.125 \, \text{ms}$$

This is a much longer sampling interval than the 50-μs sampling interval required to transmit a 10-kHz bandwidth of speech. In general, the wider the bandwidth of a signal is, the more expensive it will be to transmit the signal. So it is much "cheaper" to send filtered speech over communication links.

In many applications, the signal $x(t)$ cannot be lowpass filtered without removing information contained in $x(t)$. In such cases, the bandwidth B of the signal must be taken to be sufficiently large so that the aliased components do not seriously distort the reconstructed signal. Equivalently, for a given value of B, sampling can be performed at a rate higher than the Nyquist rate. For example, in applications to sampled-data control, the sampling frequency may be as large as 10 or 20 times B, where B is the bandwidth of the system being controlled.

5.5   FOURIER ANALYSIS OF DISCRETE-TIME SYSTEMS

Consider a linear time-invariant discrete-time system with unit-pulse response $h[n]$. By the results in Chapter 2, the output response $y[n]$ resulting from the application of input $x[n]$ is given by

$$y[n] = h[n] * x[n] = \sum_{i=-\infty}^{\infty} h[i]x[n-i] \tag{5.58}$$

In this section it is not assumed that the system is necessarily causal, and thus $h[n]$ may be nonzero for values of $n < 0$. It is assumed that the unit-pulse response $h[n]$ satisfies the absolute summability condition

$$\sum_{n=-\infty}^{\infty} |h[n]| < \infty \tag{5.59}$$

As a result of the summability condition (5.59), the ordinary DTFT $H(\Omega)$ of the unit-pulse response $h[n]$ exists and is given by

$$H(\Omega) = \sum_{n=-\infty}^{\infty} h[n]e^{-j\Omega n}$$

Now, as given in Table 4.2, the DTFT of a convolution of two signals is equal to the product of the DTFTs of the two signals. Hence, taking the DTFT of both sides of the input/output relationship (5.58) gives

$$Y(\Omega) = H(\Omega)X(\Omega) \tag{5.60}$$

where $Y(\Omega)$ is the DTFT of the output $y[n]$ and $X(\Omega)$ is the DTFT of the input $x[n]$. Equation (5.50) is the DTFT domain (or Ω-domain) representation of the given discrete-time system.

The function $H(\Omega)$ in (5.60) is called the *frequency response function* of the system defined by (5.58). Thus, the DTFT of the unit-pulse response $h[n]$ is equal to the frequency response function of the system. The frequency function $H(\Omega)$ is the discrete-time counterpart of the frequency response function $H(\omega)$ of a linear time-invariant continuous-time system (as defined in Section 5.1).

Given a discrete-time system with frequency function $H(\Omega)$, the magnitude $|H(\Omega)|$ is the magnitude function of the system, and $\angle H(\Omega)$ is the phase function of the system. Taking the magnitude and angle of both sides of (5.60) yields

$$|Y(\Omega)| = |H(\Omega)||X(\Omega)| \tag{5.61}$$

$$\angle Y(\Omega) = \angle H(\Omega) + \angle X(\Omega) \tag{5.62}$$

By (5.61), the amplitude spectrum $|Y(\Omega)|$ of the output is the product of the amplitude spectrum $|X(\Omega)|$ of the input and the system's magnitude function $|H(\Omega)|$. By (5.62), the phase spectrum $\angle Y(\Omega)$ of the output is the sum of the phase spectrum $\angle X(\Omega)$ of the input and the system's phase function $\angle H(\Omega)$.

5.5.1 Response to a Sinusoidal Input

Suppose that the input $x[n]$ to the system defined by (5.58) is the sinusoid

$$x[n] = A \cos(\Omega_0 n + \theta), \quad n = 0, \pm1, \pm2, \ldots$$

where $\Omega_0 \geq 0$. To find the output response $y[n]$ resulting from $x[n]$, first note that from Table 4.1, the DTFT $X(\Omega)$ of $x[n]$ is given by

$$X(\Omega) = \sum_{k=-\infty}^{\infty} A\pi[e^{-j\theta}\delta(\Omega + \Omega_0 - 2\pi k) + e^{j\theta}\delta(\Omega - \Omega_0 - 2\pi k)]$$

From equation (5.60), we see that the DTFT $Y(\Omega)$ of $y[n]$ is equal to the product of $H(\Omega)$ and $X(\Omega)$, and thus

$$Y(\Omega) = \sum_{k=-\infty}^{\infty} A\pi H(\Omega)[e^{-j\theta}\delta(\Omega + \Omega_0 - 2\pi k) + e^{j\theta}\delta(\Omega - \Omega_0 - 2\pi k)]$$

Now, $H(\Omega)\delta(\Omega + c) = H(-c)\delta(\Omega + c)$ for any constant c, and thus

$$
\begin{aligned}
Y(\Omega) = \sum_{k=-\infty}^{\infty} A\pi[& H(-\Omega_0 + 2\pi k)e^{-j\theta}\delta(\Omega + \Omega_0 - 2\pi k) \\
& + H(\Omega_0 + 2\pi k)e^{j\theta}\delta(\Omega - \Omega_0 - 2\pi k)]
\end{aligned}
\tag{5.63}
$$

Since $H(\Omega)$ is periodic with period 2π, $H(-\Omega_0 + 2\pi k) = H(-\Omega_0)$ and $H(\Omega_0 + 2\pi k) = H(\Omega_0)$. In addition, since $h[n]$ is real valued, $|H(-\Omega)| = |H(\Omega)|$ and $\angle H(-\Omega) = -\angle H(\Omega)$, and thus the polar forms of $H(-\Omega_0)$ and $H(\Omega_0)$ are given by

$$H(-\Omega_0) = |H(\Omega_0)|e^{-j\angle H(\Omega_0)} \text{ and } H(\Omega_0) = |H(\Omega_0)|e^{j\angle H(\Omega_0)}$$

Hence, (5.63) becomes

$$
\begin{aligned}
Y(\Omega) = \sum_{k=-\infty}^{\infty} A\pi|H(\Omega_0)|[& e^{-j(\angle H(\Omega_0)+\theta)}\delta(\Omega + \Omega_0 - 2\pi k) \\
& + e^{j(\angle H(\Omega_0)+\theta)}\delta(\Omega - \Omega_0 - 2\pi k)]
\end{aligned}
\tag{5.64}
$$

Taking the inverse DTFT of (5.64) gives

$$y[n] = A|H(\Omega_0)| \cos(\Omega_0 n + \theta + \angle H(\Omega_0)), \quad n = 0, \pm1, \pm2, \ldots \tag{5.65}$$

Equation (5.65) is the discrete-time counterpart of the output response of a continuous-time system to a sinusoidal input as derived in Section 5.1. [See (5.11).]

Example 5.8   Response to a Sinusoidal Input

Suppose that $H(\Omega) = 1 + e^{-j\Omega}$ and that the objective is to find the output $y[n]$ resulting from the sinusoidal input $x[n] = 2 + 2 \sin\left(\dfrac{\pi}{2}n\right)$. By linearity, $y[n]$ is equal to the sum of the responses to $x_1[n] = 2$ and $x_2[n] = 2\sin\left(\dfrac{\pi}{2}n\right)$. The response to $x_1[n] = 2 = 2\cos(0n)$ is

$$y_1[n] = 2|H(0)|\cos(0n + \angle H(0)) = 4$$

The response to $x_2[n] = 2\sin\left(\dfrac{\pi}{2}n\right)$ is

$$y_2[n] = 2\left|H\left(\frac{\pi}{2}\right)\right|\cos\left(\frac{\pi}{2}n + \angle H\left(\frac{\pi}{2}\right)\right)$$

where $H\left(\dfrac{\pi}{2}\right) = 1 + e^{-j\pi/2} = \sqrt{2}e^{-j\pi/4}$. Hence, we see that

$$y_2[n] = 2\sqrt{2}\cos\left(\frac{\pi}{2}n - \frac{\pi}{4}\right)$$

Combining $y_1[n]$ and $y_2[n]$ yields

$$y[n] = 4 + 2\sqrt{2}\cos\left(\frac{\pi}{2}n - \frac{\pi}{4}\right)$$

Example 5.9   MA filter

Consider the N-point MA filter given by the input/output relationship

$$y[n] = \frac{1}{N}\left[x[n] + x[n-1] + x[n-2] + \cdots + x[n-N+1]\right] \tag{5.66}$$

Using the time shift property of the DTFT (see Table 4.2) and taking the DTFT of both sides of (5.66) give the following result:

$$Y(\Omega) = \frac{1}{N}[X(\Omega) + X(\Omega)e^{-j\Omega} + X(\Omega)e^{-j2\Omega}\cdots + X(\Omega)e^{-j(N-1)\Omega}]$$

$$Y(\Omega) = \frac{1}{N}[1 + e^{-j\Omega} + e^{-j2\Omega}\ldots + e^{-j(N-1)\Omega}]X(\Omega) \tag{5.67}$$

Comparing (5.60) and (5.67) shows that the frequency response function $H(\Omega)$ of the MA filter is given by

$$H(\Omega) = \frac{1}{N}[1 + e^{-j\Omega} + e^{-j2\Omega}\cdots + e^{-j(N-1)\Omega}] = \frac{1}{N}\sum_{q=0}^{N-1} e^{-jq\Omega} \qquad (5.68)$$

The summation in the right-hand side of (5.68) can be written in closed form by use of the relationship (4.5). This yields

$$H(\Omega) = \frac{1}{N}\left[\frac{1 - e^{-jN\Omega}}{1 - e^{-j\Omega}}\right]$$

$$H(\Omega) = \frac{1}{N}\left[\frac{e^{-jN\Omega/2}(e^{jN\Omega/2} - e^{-jN\Omega/2})}{e^{-j\Omega/2}(e^{j\Omega/2} - e^{-j\Omega/2})}\right]$$

$$H(\Omega) = \left[\frac{\sin(N\Omega/2)}{N\sin(\Omega/2)}\right]e^{-j(N-1)\Omega/2} \qquad (5.69)$$

From (5.69), it can be seen that the magnitude function $|H(\Omega)|$ of the MA filter is given by

$$|H(\Omega)| = \left|\frac{\sin(N\Omega/2)}{N\sin(\Omega/2)}\right| \qquad (5.70)$$

and the phase function $\angle H(\Omega)$ is given by

$$\angle H(\Omega) = -\frac{N-1}{2}\Omega, 0 \le \Omega < \frac{2\pi}{N} \qquad (5.71)$$

$$\angle H(\Omega) = -\frac{N-1}{2}\Omega + \pi, \frac{2\pi}{N} < \Omega < \frac{4\pi}{N} \qquad (5.72)$$

$$\angle H(\Omega) = -\frac{N-1}{2}\Omega, \frac{4\pi}{N} < \Omega < \frac{6\pi}{N} \qquad (5.73)$$

and so on. Note that the addition of π to the phase function in going from (5.71) to (5.72) is a result of the change in sign of $\sin(N\Omega/2)$. Note also that the phase is a linear function of Ω for $0 \le \Omega < \frac{2\pi}{N}$. It follows that when $0 \le \Omega_0 < \frac{2\pi}{N}$, the N-point MA filter delays the sinusoidal input $x[n] = A\cos(\Omega_0 n + \theta), n = 0, \pm1, \pm2, \ldots$ by $(N-1)/2$ units of time.

For any positive integer value of N, the magnitude and phase functions of the MA filter can be computed with MATLAB. For example, when $N = 2$, the MATLAB commands are

```
W=0:.01:1;
H=(1/2).*(1-exp(-j*2*pi*W))./(1-exp(-j*pi*W));
magH=abs(H);
angH=180*angle(H)/pi;
```

Using these commands results in the magnitude and phase plots shown in Figure 5.22. In these plots the frequency is the normalized frequency (Ω/π radians per unit time). From Figure 5.22a, it is seen that $H(\pi) = 0$, and thus if the input is $x[n] = A\cos(\pi n + \theta), n = 0, \pm1, \pm2, \ldots,$ the

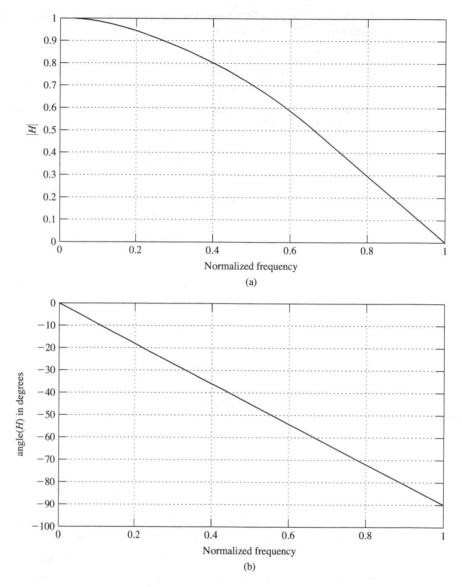

FIGURE 5.22

(a) Magnitude and (b) phase functions of the 2-point MA filter.

resulting output $y[n]$ is equal to zero for $n = 0, \pm1, \pm2, \ldots$. Figure 5.22a shows that the 2-point MA filter is a lowpass discrete-time filter, and Figure 5.22b shows that the filter has a linear phase. The linear phase characteristic also follows by setting $N = 2$ in (5.71). At $\Omega = \pi$, $\angle H(\Omega) = -90° = -\pi/2$ radians. Hence, the slope of the phase function is equal to $-\frac{1}{2}$, which shows that the 2-point MA filter delays any input $x[n]$ by $\frac{1}{2}$ time unit.

5.6   APPLICATION TO LOWPASS DIGITAL FILTERING

In Chapter 1 it was noted that a signal $x[n]$ with noise can often be expressed in the additive form $x[n] = s[n] + e[n]$, where $s[n]$ is the smooth part of $x[n]$ and $e[n]$ is the erratic or noisy part of $x[n]$. The spectrum $E(\Omega)$ of $e[n]$ consists primarily of high-frequency components whose frequencies are in a neighborhood of $\Omega = \pi$. The spectrum $S(\Omega)$ of the smooth part $s[n]$ consists primarily of low-frequency components whose values are in a neighborhood of $\Omega = 0$. Hence, if the objective is to filter $x[n]$ so that the noisy part $e[n]$ is removed, or at least greatly reduced in magnitude, a lowpass filter is required. In this section Fourier analysis is applied to noncausal and causal discrete-time lowpass filters.

This section begins with the analysis of a noncausal lowpass discrete-time filter and then specific examples of causal lowpass discrete-time filters are considered. A special type of causal filter is designed and is then applied to the problem of determining the trend of stock price data. In the text from here on, a discrete-time filter will be referred to as a *digital filter*.

5.6.1   Analysis of an Ideal Lowpass Digital Filter

As an illustration of its use, the DTFT representation will be applied to the study of an ideal lowpass digital filter. Consider the discrete-time system with the frequency function

$$H(\Omega) = \sum_{k=-\infty}^{\infty} p_{2B}(\Omega + 2\pi k) \tag{5.74}$$

where $B < \pi$. The function $H(\Omega)$ is plotted in Figure 5.23. Note that in this example, the magnitude function $|H(\Omega)|$ is equal to $H(\Omega)$, and the phase function $\angle H(\Omega)$ is identically zero.

The relationship (5.65) can be used to determine the output response $y[n]$ resulting from the sinusoidal input $x[n] = A \cos(\Omega_0 n)$, $n = 0, \pm 1, \pm 2, \ldots$. As seen from Figure 5.23, $H(\Omega) = 1$ for $0 \le \Omega_0 < B$, and thus from (5.65) the output response is $y[n] = A \cos(\Omega_0 n)$, $n = 0, \pm 1, \pm 2, \ldots$. Hence, $y[n] = x[n]$ when $0 \le \Omega_0 < B$, and in this case the filter passes the input with no attenuation and with no phase shift. From Figure 5.23, $H(\Omega) = 0$ for $B < \Omega_0 < \pi$, and therefore, using (5.65) gives $y[n] = 0$ for

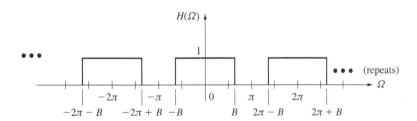

FIGURE 5.23
Frequency function $H(\Omega)$ given by (5.74).

$n = 0, \pm 1, \pm 2, \ldots$. Hence, in this case the system completely blocks the input. Summarizing these results yields the following expression for the output response:

$$y[n] = \begin{cases} A\cos(\Omega_0 n), & 0 \le \Omega < B \\ 0, & B < \Omega_0 < \pi \end{cases} \tag{5.75}$$

Equation (5.75) shows that the discrete-time system with the frequency response function $H(\Omega)$ given by (5.74) is an ideal lowpass digital filter. Here the term "ideal" refers to the sharp cutoff in $H(\Omega)$ at $\Omega = B$. The passband of the filter is the frequency range from $\Omega = 0$ to $\Omega = B$, and the stopband of the filter is the frequency range from $\Omega = B$ to $\Omega = \pi$.

As a result of the periodicity of $H(\Omega)$, the output response $y[n]$ is equal to $A\cos\Omega_0 n$ when

$$2\pi k - B \le \Omega_0 \le 2\pi k + B, k = 0, 1, 2, \ldots \tag{5.76}$$

For all other positive values of Ω_0, the response $y[n]$ is zero. It is important to note that, as a result of periodicity of the frequency response function $H(\Omega)$, this digital filter is not a "true" lowpass filter, since it passes input sinusoids $A\cos\Omega_0 n$ with Ω_0 belonging to the intervals given by (5.76). If Ω_0 is restricted to lie in the range $0 \le \Omega_0 < \pi$, the filter can be viewed as an ideal lowpass digital filter with bandwidth B.

5.6.2　Digital-Filter Realization of an Ideal Analog Lowpass Filter

The filter with frequency function (5.74) can be used as a digital-filter realization of an ideal lowpass zero-phase analog filter with bandwidth B. To see this, suppose that the input

$$x(t) = A\cos\omega_0 t, \; -\infty < t < \infty$$

is applied to an analog filter with the frequency function $p_{2B}(\omega)$. From the results in Section 5.3, the output of the filter is equal to $x(t)$ when $\omega_0 \le B$ and is equal to zero when $\omega_0 > B$.

Now suppose that the sampled version of the input

$$x[n] = x(t)|_{t=nT} = A\cos\Omega_0 n, \text{ where } \Omega_0 = \omega_0 T$$

is applied to the digital filter with frequency function (5.74). Then by (5.75), the output is equal to $x[n]$ when $\Omega_0 < B$ and is equal to zero when $B < \Omega_0 < \pi$. Thus, as long as $\Omega_0 < \pi$ or $\omega_0 < \pi/T$, the output $y[n]$ of the discrete-time filter will be equal to a sampled version of the output of the analog filter. An analog signal can then be generated from the sampled output by the use of a hold circuit, as discussed in Chapter 10. Hence, this results in a digital-filter realization of the given analog filter.

The requirement that the frequency ω_0 of the input sinusoid $A\cos\omega_0 t$ be less than π/T is not an insurmountable constraint, since π/T can be increased by decreasing the sampling interval T, which is equivalent to increasing the sampling frequency $\omega_s = 2\pi/T$. (See Section 5.4.) Therefore, as long as a suitably fast sample rate can be achieved, the upper bound π/T on the input frequency ω_0 is not a problem.

5.6.3   Unit-Pulse Response of Ideal Lowpass Filter

From the transform pairs in Table 4.1, the unit-pulse response $h[n]$ of the filter with the frequency function (5.74) is given by

$$h[n] = \frac{B}{\pi}\,\text{sinc}\!\left(\frac{B}{\pi}n\right), \quad n = 0, \pm 1, \pm 2, \ldots$$

The unit-pulse response is displayed in Figure 5.24. Note that the sinc function form of the unit-pulse response is very similar to the form of the impulse response of an ideal analog lowpass filter. (See Section 5.3.)

From Figure 5.24 it is seen that $h[n]$ is not zero for $n < 0$, and thus the filter is noncausal. Therefore, the filter cannot be implemented online (in real time); but it can be implemented off-line. In an off-line implementation, the filtering process is applied to the values of signals that have been stored in the memory of a digital computer or stored by some other means. So, in the discrete-time case, ideal filters can be used in practice, as long as the filtering process is carried out off-line.

5.6.4   Causal Lowpass Digital Filters

As noted previously, an ideal lowpass digital filter cannot be implemented in real time, since the filter is noncausal. For "real-time filtering" it is necessary to consider a causal lowpass digital filter. One very simple example is the 2-point MA filter, which is defined by the input/output equation

$$y[n] = \frac{1}{2}(x[n] + x[n-1])$$

The magnitude and phase functions of this filter are plotted in Figure 5.22. (See Example 5.9.) A nice feature of this lowpass filter is the linear phase characteristic shown in Figure 5.22b. As noted in Example 5.9, as a consequence of the linear phase characteristic, the 2-point MA filter delays any input $x[n]$ by $\frac{1}{2}$ time unit, and thus

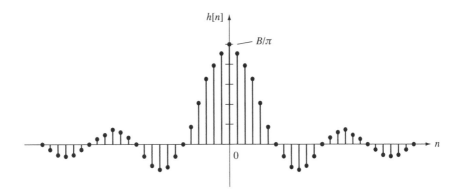

FIGURE 5.24
Unit-pulse response of ideal lowpass digital filter.

there is no distortion (just a time delay) of the input signal as it passes through the filter. However, as seen from Figure 5.22a, the 2-point MA filter does not provide much attenuation of sinusoids with high frequencies (above $\pi/2$ radians per unit time).

To achieve better attenuation of high-frequency components, one can attempt to use the N-point MA filter with $N > 2$. For example, the 3-point MA filter has the magnitude and phase functions displayed in Figure 5.25. From Figure 5.25a it is seen that the magnitude response function $|H(\Omega)|$ has a fairly sharp drop-off from $\Omega = 0$ to $\Omega = 2\pi/3$. However, as Ω is increased in value from $\Omega = 2\pi/3$ to $\Omega = \pi$, the magnitude function $|H(\Omega)|$ increases in value, which is not a desirable characteristic for a lowpass filter. Also, from (5.72) with N set equal to 3, and as can be seen from Figure 5.25b, the phase function $\angle H(\Omega)$ is linear when $0 \le \Omega < 2\pi/3$, but then there is a jump in the phase at $\Omega = 2\pi/3$ that corresponds to a sign change in $H(\Omega)$. Hence, the phase function is not linear over the entire interval from 0 to π.

Weighted Moving Average Filters.   A 3-point digital filter that does not have the "sidelobe" which appears in Figure 5.25a can be generated by a consideration of the digital filter with the input/output relationship

$$y[n] = cx[n] + dx[n - 1] + fx[n - 2] \tag{5.77}$$

where the weights c, d, and f are determined as follows: First, the system given by (5.77) is an example of a 3-point weighted moving average (WMA) filter if

$$c + d + f = 1 \tag{5.78}$$

The 3-point EWMA filter is a special case of a 3-point WMA filter; that is, (5.78) is satisfied for the 3-point EWMA filter.

Additional constraints will be placed on the weights c, d, and f in (5.77) in order to achieve desired filter characteristics. The analysis is based on the frequency response function $H(\Omega)$, which is given by

$$H(\Omega) = c + de^{-j\Omega} + fe^{-j2\Omega} \tag{5.79}$$

Setting $\Omega = \pi$ in (5.79) gives

$$H(\pi) = c - d + f \tag{5.80}$$

It is desirable to have $H(\pi) = 0$ so that the filter completely rejects any high-frequency component of the input at $\Omega = \pi$. Hence, using (5.80) yields

$$c - d + f = 0 \tag{5.81}$$

Subtracting (5.81) from (5.78) gives $2d = 1$, and thus $d = 0.5$. Adding (5.78) and (5.81) gives

$$2(c + f) = 1 \tag{5.82}$$

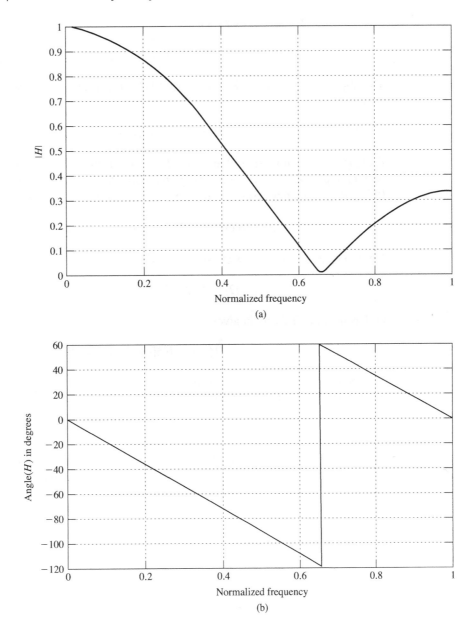

FIGURE 5.25
(a) Magnitude and (b) phase functions of the 3-point MA filter.

Setting $\Omega = \pi/2$ in (5.79) yields

$$H(\pi/2) = c - jd - f \tag{5.83}$$

The objective is to choose c and f so that $|H(\pi/2)|$ is as small as possible. From (5.83), we have that

$$|H(\pi/2)| = \sqrt{(c - f)^2 + d^2} \tag{5.84}$$

From (5.84), it can be seen that, since $d = 0.5$, the smallest possible value of $|H(\pi/2)|$ is obtained when $c = f$. Combining this with (5.82) gives $c = f = 0.25$. Then from (5.79), the frequency response function of the 3-point WMA filter is given by

$$H(\Omega) = 0.25 + 0.5e^{-j\Omega} + 0.25e^{-j2\Omega} \tag{5.85}$$

Note that (5.85) can be rewritten in the form

$$H(\Omega) = [0.25e^{j\Omega} + 0.5 + 0.25e^{-j\Omega}]e^{-j\Omega}$$

and using Euler's formula gives

$$H(\Omega) = 0.5[\cos \Omega + 1]e^{-j\Omega} \tag{5.86}$$

Since $0.5(\cos \Omega + 1) \geq 0$ for $0 \leq \Omega < \pi$, from (5.86) it can be seen that the magnitude function $|H(\Omega)|$ is given by

$$|H(\Omega)| = 0.5(\cos \Omega + 1), 0 \leq \Omega < \pi$$

Equation (5.86) also shows that the filter has a linear phase function and that the time delay through the filter is one unit of time. A plot of the magnitude function is given in Figure 5.26. Also displayed in Figure 5.26 is the magnitude function of the 2-point MA filter, which is given by the dotted curve. Comparing the two curves reveals that the 3-point WMA provides a sharper drop-off in magnitude as Ω is increased from $\Omega = 0$ to $\Omega = \pi$ than the 2-point MA filter does. However, there is a price to be paid for this; namely, the time delay through the 3-point WMA filter is one unit of time, which is twice that of the 2-point MA filter that has a time delay of $\frac{1}{2}$ time unit.

A Double WMA Filter.   To achieve a much sharper drop-off in the magnitude function, the 3-point WMA filter designed previously can be cascaded with itself. The resulting filter is sometimes called a *double WMA filter*. It is given by the cascade connection shown in Figure 5.27. As indicated in the figure, the output of the first filter is denoted by $v[n]$, and thus the DTFT representation of the first filter is $V(\Omega) = H(\Omega)X(\Omega)$, and the DTFT representation of the second filter is $Y(\Omega) = H(\Omega)V(\Omega)$. Combining the two representations gives $Y(\Omega) = H(\Omega)^2 X(\Omega)$. Thus, the frequency response function of the cascade connection is equal to $H(\Omega)^2$. From (5.86), we have that

$$H(\Omega)^2 = 0.25[\cos \Omega + 1]^2 e^{-j2\Omega} \tag{5.87}$$

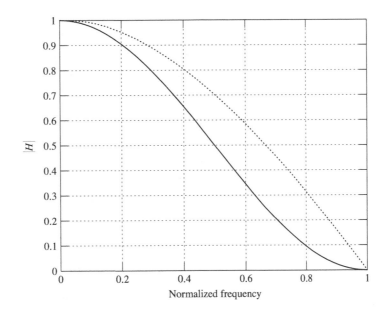

FIGURE 5.26
Magnitude functions of the 3-point WMA filter (solid line) and the 2-point MA filter (dotted line).

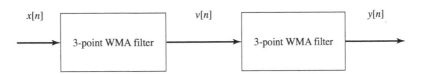

FIGURE 5.27
Cascade connection of two 3-point WMA filters.

Equation (5.87) shows that the double WMA filter also has a linear phase; however, the time delay through the filter is two units of time, as opposed to the one unit of time delay in the WMA filter. The magnitude response function $|H(\Omega)|^2$ of the double WMA filter is plotted in Figure 5.28. Also shown in Figure 5.28 is $|H(\Omega)|$, which is displayed with a dotted curve. Comparing the two curves shows that the magnitude function of the double WMA filter has a much sharper drop-off than the WMA filter. But again a price is paid for this in that the time delay through the double WMA filter is two units of time.

In order to implement the double WMA filter, it is first necessary to determine the unit-pulse response. Denoting the unit-pulse response function by $h2[n]$, since the frequency response function of the double WMA filter is equal to $H(\Omega)^2$, it follows from the convolution property of the DTFT (see Table 4.2) that $h2[n]$ is equal to the convolution $h[n] * h[n]$, where $h[n]$ is the unit-pulse response of the 3-point WMA filter. The convolution $h[n] * h[n]$ can be computed by the MATLAB commands

```
h = [0.25 0.5 0.25];
h2 = conv(h,h);
```

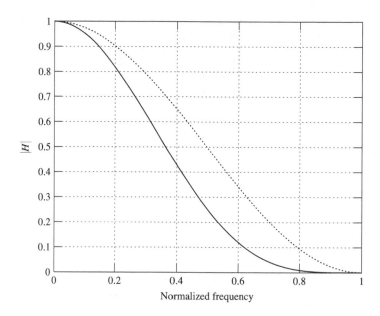

FIGURE 5.28
Magnitude response functions of the 3-point WMA filter (dotted line) and the double WMA filter (solid line).

Running these commands gives h2 = [0.0625 0.2500 0.3750 0.2500 0.0625]. Hence, the input/output relationship of the double WMA filter is

$$y[n] = 0.0625x[n] + 0.25x[n-1] + 0.375x[n-2]$$
$$+ 0.25x[n-3] + 0.0625x[n-4] \tag{5.88}$$

Note that the number of weights in (5.88) is equal to 5, the sum of the weights in the right-hand side of (5.88) is equal to 1, and thus the double WMA filter is a 5-point WMA filter.

Application to Stock Price Data.   Consider the closing price $c[n]$ of QQQQ for the 50-business-day period from March 1, 2004, up to May 10, 2004. (See Example 1.4 in Chapter 1.) The closing price is applied to the double WMA filter defined by (5.88). The output $y[n]$ of the filter and the closing price $c[n]$ are plotted for $n \geq 5$ in Figure 5.29. In the plot, the values of $c[n]$ are displayed by o's, and the values of $y[n]$ are displayed by ∗'s. The output $y[n]$ of the double WMA filter is an approximation of the smooth part of $c[n]$, delayed by two days. Comparing the filter response in Figure 5.29 with the response of the 11-day MA filter given in Section 1.4 reveals that the 11-day MA filter response is much smoother, although there is a 5-day time delay through the filter. In Section 7.5 it will be shown how a filtered response can be generated that is almost as smooth as that of the 11-day MA filter, but which has only a 2.5-day time delay through the filter.

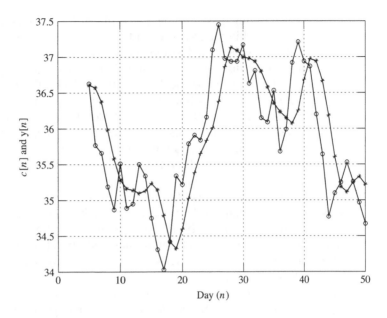

FIGURE 5.29
Plot of closing price and filtered closing price.

5.7   CHAPTER SUMMARY

Fourier analysis is used to examine how a system processes input signals of different frequencies. We can find a frequency domain representation of a system, known as the frequency response, for a continuous-time system by taking the magnitude of the Fourier transform of the impulse response, which results in the frequency response function $|H(\omega)|$. For a discrete-time system, we can find the frequency response by taking the magnitude of the discrete-time Fourier transform of the unit-pulse response, which yields the frequency response function $|H(\Omega)|$.

One of the important features of linear systems is that a sinusoidal input to the system results in a sinusoidal output at the same frequency, but with different amplitude and phase angle. The amplitude and phase angle of the output signal are found as functions of the frequency response of the system evaluated at the frequency of the input signal. In particular, the amplitude of the output sinusoid is the product of the amplitude of the input sinusoid and the magnitude of the frequency response at the input signal frequency. If the magnitude of the frequency response is less than one, the output sinusoidal signal is attenuated, compared with the input signal. The output signal is amplified if the magnitude of the frequency response is greater than one at that frequency. The response of a system to nonperiodic input signals is determined by $Y(\omega) = H(\omega)X(\omega)$ for continuous-time systems and $Y(\Omega) = H(\Omega)X(\Omega)$ for discrete-time systems.

Filters are systems that are designed to reject, or attenuate, input signals in certain frequency ranges and pass signals in other frequency ranges. The filtering characteristics of a system are determined by the shape of the magnitude function $|H(\omega)|$ for continuous-time systems or $|H(\Omega)|$ for discrete-time systems. Standard types of filters include lowpass, highpass, and bandpass filters. Ideal filters have magnitude functions that are equal to 1 in the passband and 0 in the stopband, and are allowed to transition between the passband and stopband regions discontinuously. Ideal filters are not causal due to the discontinuous change in the frequency response. Instead, causal filters are used typically in physical applications. Causal filters approximate the frequency response of ideal filters, but allow nonzero transition regions between the passband and stopband regions so that the magnitude of the frequency response is allowed to change continuously. Furthermore, the magnitude function of the filter is not required to be exactly 1 in the passband nor 0 in the stopband region. The moving average filter is an example of a causal discrete-time lowpass filter that is used commonly to filter high frequency noise in data.

Another important application of Fourier analysis in this chapter is the analysis of sampling. Sampling is the process by which a continuous-time signal is converted into a discrete-time signal. If the sampling rate is too low, then the continuous-time signal cannot be reconstructed from the sampled discrete-time signal due to aliasing. The effect of aliasing is easiest to view in the frequency domain. Sampling can be modeled as a multiplication of the signal by an impulse train, which is very similar to pulse amplitude modulation studied in Section 3.8. The frequency spectrum of the sampled signal is formed by placing copies of a scaled version of the original spectrum at integer multiples of the sampling frequency. If the original spectrum is not bandlimited to less than $\frac{1}{2}$ of the sampling frequency, then the copies of the spectrum will overlap, which will result in aliasing. In the case of an aliased signal, it is impossible to reconstruct the original signal from the aliased signal due to the overlap in the spectrum.

PROBLEMS

5.1. A linear time-invariant continuous-time system has the frequency response function

$$H(\omega) = \begin{cases} 1, & 2 \leq |\omega| \leq 7 \\ 0, & \text{all other } \omega \end{cases}$$

Compute the output response $y(t)$ resulting from the input $x(t)$ given by

(a) $x(t) = 2 + 3\cos(3t) - 5\sin(6t - 30°) + 4\cos(13t - 20°), \qquad -\infty < t < \infty$

(b) $x(t) = 1 + \sum_{k=1}^{\infty} \frac{1}{k}\cos(2kt), \qquad -\infty < t < \infty$

(c) $x(t)$ as shown in Figure P5.1

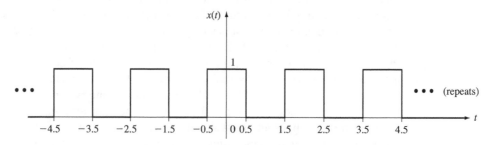

FIGURE P5.1

5.2. A linear time-invariant continuous-time system has the frequency response function

$$H(\omega) = \begin{cases} 2\exp(-|6-|\omega||)\exp(-j3\omega), & 4 \le |\omega| \le 12 \\ 0, & \text{all other } \omega \end{cases}$$

(a) Plot the magnitude and phase functions for $H(\omega)$.

(b) Compute and plot the output response $y(t)$ resulting from the input $x(t)$ defined in Figure P5.1.

(c) Plot the amplitude and phase spectra of $x(t)$ and $y(t)$ for $k = 0, \pm 1, \pm 2, \pm 3, \pm 4, \pm 5, \pm 6$.

5.3. A linear time-invariant continuous-time system has the frequency response function

$$H(\omega) = \frac{1}{j\omega + 1}$$

Compute the output response $y(t)$ for $-\infty < t < \infty$ when the input $x(t)$ is

(a) $x(t) = \cos t, -\infty < t < \infty$

(b) $x(t) = \cos(t + 45°), -\infty < t < \infty$

5.4. Consider the system with the frequency response given by

$$H(\omega) = \frac{10}{j\omega + 10}$$

(a) Give the output to $x(t) = 2 + 2\cos(50t + \pi/2)$.

(b) Sketch $|H(\omega)|$. What is the bandwidth of the filter?

(c) Sketch the response of the filter to an input of $x(t) = 2e^{-2t}\cos(4t)u(t) + e^{-2t}\cos(20t)u(t)$. (See Figure P5.4.)

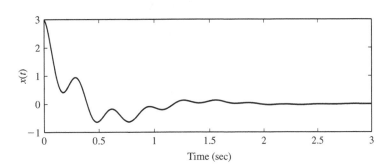

FIGURE P5.4

5.5. Repeat Problem 5.4 for the system given by

$$H(\omega) = \frac{40}{j\omega + 40}$$

5.6. A linear time-invariant continuous-time system receives the periodic signal $x(t)$ as shown in Figure P5.6. The frequency response function is given by

$$H(\omega) = \frac{j\omega}{j\omega + 2}$$

(a) Plot the amplitude and phase functions for $H(\omega)$.

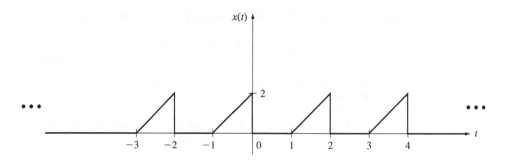

FIGURE P5.6

    (b) Compute the complex exponential Fourier series of the output response $y(t)$, and then sketch the amplitude and phase spectra for $k = 0, \pm1, \pm2, \pm3, \pm4, \pm5$ for both $x(t)$ and $y(t)$.

    (c) Plot an approximation for $y(t)$, using the truncated complex exponential Fourier series from $k = -5$ to $k = 5$.

5.7. A periodic signal $x(t)$ with period T has the constant component $c_o^x = 2$. The signal $x(t)$ is applied to a linear time-invariant continuous-time system with frequency response function

$$H(\omega) = \begin{cases} 10e^{-j5\omega}, & \omega > \dfrac{\pi}{T}, \omega < -\dfrac{\pi}{T} \\ \\ 0, & \text{all other } \omega \end{cases}$$

    (a) Show that the resulting output response $y(t)$ can be expressed in the form

$$y(t) = ax(t - b) + c$$

    Compute the constants a, b, and c.

    (b) Compute and plot the response of this system to the input $x(t)$ shown in Figure P5.1

5.8. The voltage $x(t)$ shown in Figure P5.8b is applied to the RL circuit shown in Figure P5.8a.

    (a) Find the value of L so that the peak of the largest ac component (harmonic) in the output response $y(t)$ is 1/30 of the dc component of the output.

    (b) Plot an approximation for $y(t)$, using the truncated complex exponential Fourier series from $k = -3$ to $k = 3$.

5.9. Consider the full-wave rectifier shown in Figure P5.9. The input voltage $v(t)$ is equal to $156\cos(120\,\pi t)$, $-\infty < t < \infty$. The voltage $x(t)$ is equal to $|v(t)|$.

    (a) Choose values for R and C such that the following two criteria are satisfied:

      (i) The dc component of $y(t)$ is equal to 90% of the dc component of the input $x(t)$.

      (ii) The peak value of the largest harmonic in $y(t)$ is $\frac{1}{30}$ of the dc component of $y(t)$.

    (b) Plot an approximation for $y(t)$, using the truncated complex exponential Fourier series from $k = -3$ to $k = 3$.

5.10. The input

$$x(t) = 1.5 + \sum_{k=1}^{\infty}\left(\frac{1}{k\pi}\sin k\pi t + \frac{2}{k\pi}\cos k\pi t\right), \quad -\infty < t < \infty$$

is applied to a linear time-invariant system with frequency function $H(\omega)$. This input produces the output response $y(t)$ shown in Figure P5.10. Compute $H(k\pi)$ for $k = 1, 2, 3, \ldots$.

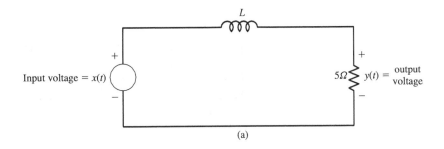

(a)

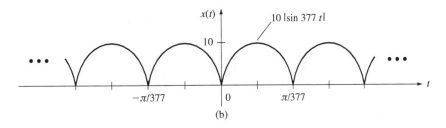

(b)

FIGURE P5.8

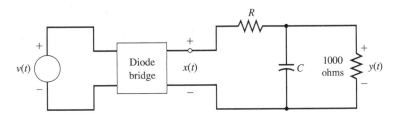

FIGURE P5.9

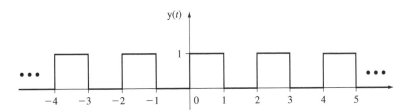

FIGURE P5.10

5.11. A linear time-invariant continuous-time system has frequency function $H(\omega)$ shown in Figure P5.11a. It is known that the system converts the sawtooth waveform in Figure P5.11b into the square waveform in Figure P5.11c, that is, the response to the sawtooth waveform is a square waveform. Compute the constants a and b in the plot of $H(\omega)$.

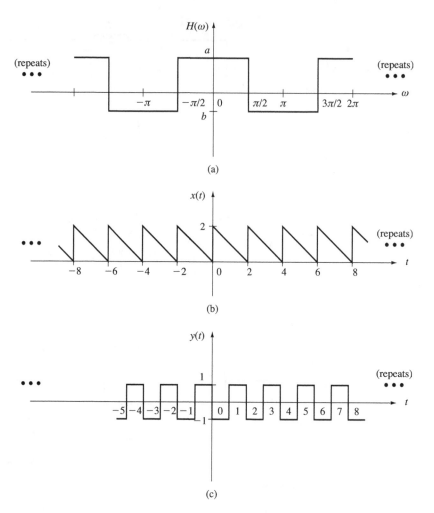

(a)

(b)

(c)

FIGURE P5.11

5.12. A linear time-invariant continuous-time system has the frequency function

$$H(\omega) = b - ae^{j\omega c}, \quad -\infty < \omega < \infty$$

where a, b, and c are constants (real numbers). The input $x(t)$ shown in Figure P5.12a is applied to the system. Determine the constants a, b, and c so that the output response $y(t)$ resulting from $x(t)$ is given by the plot in Figure P5.12b.

5.13. A linear time-invariant continuous-time system has the frequency function $H(\omega)$. It is known that the input

$$x(t) = 1 + 4\cos 2\pi t + 8\sin(3\pi t - 90°)$$

produces the response

$$y(t) = 2 - 2\sin 2\pi t$$

(a) For what values of ω is it possible to determine $H(\omega)$?

(b) Compute $H(\omega)$ for each of the values of ω determined in part (a).

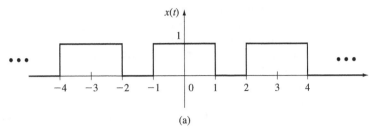

(a)

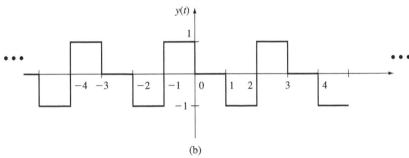

(b)

FIGURE P5.12

5.14. An ideal linear-phase lowpass filter has the frequency response function

$$H(\omega) = \begin{cases} e^{-j\omega}, & -2 < \omega < 2 \\ 0, & \text{all other } \omega \end{cases}$$

Compute the filter's output response $y(t)$ for the different inputs $x(t)$ as given next. Plot each input $x(t)$ and the corresponding output $y(t)$. Also plot the magnitude and phase functions for $X(\omega)$, $H(\omega)$, and $Y(\omega)$.

(a) $x(t) = 5\,\text{sinc}(3t/2\pi)$, $-\infty < t < \infty$

(b) $x(t) = 5\,\text{sinc}(t/2\pi)\cos(2t)$, $-\infty < t < \infty$

(c) $x(t) = \text{sinc}^2(t/2\pi)$, $-\infty < t < \infty$

(d) $x(t) = \displaystyle\sum_{k=1}^{\infty} \frac{1}{k}\cos\!\left(\frac{k\pi}{2}t + 30°\right)$, $\quad -\infty < t < \infty$

5.15. The triangular pulse shown in Figure P5.15 is applied to an ideal lowpass filter with frequency function $H(\omega) = p_{2B}(\omega)$. By using the Fourier transform approach and numerical integration, determine the filter output for the values of B given next. Express your results

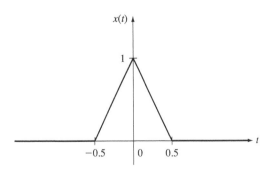

FIGURE P5.15

by plotting the output responses for $-1.5 \le t \le 1.5$. What do you conclude? You may wish to use the MATLAB M-file ode45, which can perform the integration for each value of t in the inverse Fourier transform or use the int command from the Symbolic Math Toolbox to compute the inverse Fourier transform as a function of t.

   (a) $B = 2\pi$

   (b) $B = 4\pi$

   (c) $B = 8\pi$

5.16. A lowpass filter has the frequency response function shown in Figure P5.16.

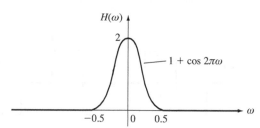

FIGURE P5.16

   (a) Compute the impulse response $h(t)$ of the filter.

   (b) Compute the response $y(t)$ when the input is $x(t) = \text{sinc}(t/2\pi)$, $-\infty < t < \infty$.

   (c) Compute the response $y(t)$ when $x(t) = \text{sinc}(t/4\pi)$, $-\infty < t < \infty$.

   (d) Compute the response $y(t)$ when $x(t) = \text{sinc}^2(t/2\pi)$, $-\infty < t < \infty$.

   (e) For parts (b)–(d), plot $x(t)$ and the corresponding $y(t)$.

5.17. A lowpass filter has the frequency response curves shown in Figure P5.17.

   (a) Compute the impulse response $h(t)$ of the filter.

   (b) Compute the response $y(t)$ when $x(t) = 3 \, \text{sinc}(t/\pi) \cos 4t$, $-\infty < t < \infty$.

   (c) Plot $x(t)$ and $y(t)$.

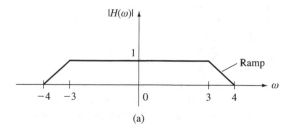

(a)

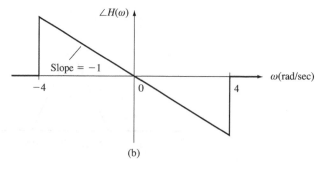

(b)

FIGURE P5.17

5.18. The input $x(t) = [\text{sinc}(t/\pi)](\cos 2t)$, $-\infty < t < \infty$, is applied to an ideal lowpass filter with frequency function $H(\omega) = 1$, $-a < \omega < a$, $H(\omega) = 0$ for all other ω. Determine the smallest possible value of a for which the resulting output response $y(t)$ is equal to the input $x(t) = [\text{sinc}(t/\pi)](\cos 2t)$.

5.19. An ideal linear-phase highpass filter has frequency response function

$$H(\omega) = \begin{cases} 6e^{-j2\omega}, & \omega > 3, \omega < -3 \\ 0, & \text{all other } \omega \end{cases}$$

(a) Compute the impulse response $h(t)$ of the filter.
(b) Compute the output response $y(t)$ when the input $x(t)$ is given by $x(t) = \text{sinc}(5t/\pi)$, $-\infty < t < \infty$. Plot $x(t)$ and $y(t)$.
(c) Compute the output response $y(t)$ when the input $x(t)$ is the periodic signal shown in Figure P5.19. Plot $y(t)$.

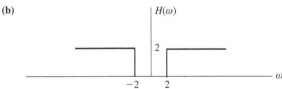

FIGURE P5.19

5.20. Given the input

$$x(t) = 4 + 2\cos(10t + \pi/4) + 3\cos(30t - \pi/2)$$

find the output $y(t)$ to each of the following filters:

(a)

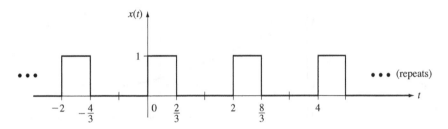

(b)

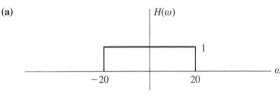

(c) $H(\omega) = \text{sinc}(\omega/20)$
(d) $H(\omega)$, as given in Figure P5.20.

5.21. Design a filter to give a response of $y(t) = 6\cos(30t)$ for the input given in Problem 5.20.

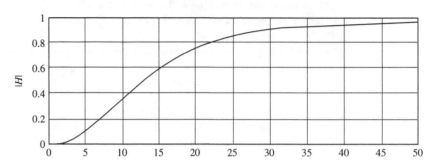

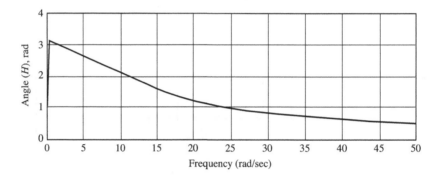

Frequency (rad/sec)

FIGURE P5.20

5.22. The input

$$x(t) = \text{sinc}\left(\frac{t}{2\pi}\right)(\cos 3t)^2 + \text{sinc}\left(\frac{t}{2\pi}\right)\cos t, \quad -\infty < t < \infty$$

is applied to a linear time-invariant continuous-time system with the frequency response function $H(\omega)$. Determine $H(\omega)$ so that the output response $y(t)$ resulting from this input is given by

$$y(t) = \text{sinc}\left(\frac{t}{2\pi}\right)$$

(a) Express your answer by giving $H(\omega)$ in analytical form.
(b) Plot $x(t)$ and $y(t)$ for $-30 < t < 30$ to see the filtering effect of $H(\omega)$ in the time domain. [To get sufficient resolution on $x(t)$, use a time increment of 0.1 second.]

5.23. An ideal linear-phase bandpass filter has frequency response

$$H(\omega) = \begin{cases} 10e^{-j4\omega}, & -4 < \omega < -2, 2 < \omega < 4 \\ 0, & \text{all other } \omega \end{cases}$$

Compute the output response $y(t)$ of the filter when the input $x(t)$ is
(a) $x(t) = \text{sinc}(2t/\pi), -\infty < t < \infty$
(b) $x(t) = \text{sinc}(3t/\pi), -\infty < t < \infty$
(c) $x(t) = \text{sinc}(4t/\pi), -\infty < t < \infty$

(d) $x(t) = \mathrm{sinc}(2t/\pi)\cos t, -\infty < t < \infty$

(e) $x(t) = \mathrm{sinc}(2t/\pi)\cos 3t, -\infty < t < \infty$

(f) $x(t) = \mathrm{sinc}(2t/\pi)\cos 6t, -\infty < t < \infty$

(g) $x(t) = \mathrm{sinc}^2(t/\pi)\cos 2t, -\infty < t < \infty$

Plot $x(t)$ and the corresponding output $y(t)$ for each of the cases computed (a)–(g). Use a small-enough time increment on the plot to capture the high-frequency content of the signal.

5.24. A linear time-invariant continuous-time system has the frequency response function $H(\omega) = p_2(\omega + 4) + p_2(\omega - 4)$. Compute the output response for the following inputs:

(a) $x(t) = \delta(t)$

(b) $x(t) = \cos t \sin \pi t, -\infty < t < \infty$

(c) $x(t) = \mathrm{sinc}(4t/\pi), -\infty < t < \infty$

(d) $x(t) = \mathrm{sinc}(4t/\pi)\cos 3t, -\infty < t < \infty$

Plot $x(t)$ and the corresponding output $y(t)$ for each of the cases computed (a)–(d). Use a small-enough time increment on the plot to capture the high-frequency content of the signal.

5.25. A periodic signal $x(t)$ with period $T = 2$ has the Fourier coefficients

$$c_k = \begin{cases} 0, & k = 0 \\ 0 & \text{if } k \text{ is even} \\ 1 & \text{if } k \text{ is odd} \end{cases}$$

The signal $x(t)$ is applied to a linear time-invariant continuous-time system with the magnitude and phase curves shown in Figure P5.25. Determine the system output.

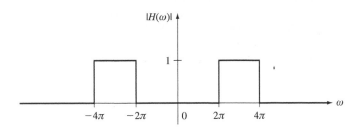

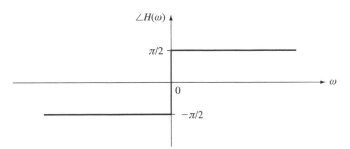

FIGURE P5.25

5.26. A linear time-invariant continuous-time system has frequency function $H(\omega) = 5\cos 2\omega$, $-\infty < \omega < \infty$.

(a) Sketch the system's magnitude function $|H(\omega)|$ and phase function $\angle H(\omega)$.

(b) Compute the system's impulse response $h(t)$.

(c) Derive an expression for the output response $y(t)$ resulting from an arbitrary input $x(t)$.

5.27. A *Hilbert transformer* is a linear time-invariant continuous-time system with impulse response $h(t) = 1/t$, $-\infty < t < \infty$. Using the Fourier transform approach, determine the output response resulting from the input $x(t) = A \cos \omega_0 t$, $-\infty < t < \infty$, where ω_0 is an arbitrary, strictly positive real number.

5.28. A linear time-invariant continuous-time system has frequency response function $H(\omega) = j\omega e^{-j\omega}$. The input $x(t) = \cos(\pi t/2) p_2(t)$ is applied to the system for $-\infty < t < \infty$.
(a) Determine the input spectrum $X(\omega)$ and the corresponding output spectrum $Y(\omega)$.
(b) Compute the output $y(t)$.

5.29. Consider the system in Figure P5.29, where $p(t)$ is an impulse train with period T and $H(\omega) = Tp_2(\omega)$. Compute $y(t)$ when

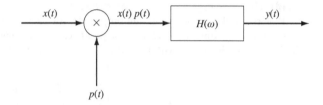

FIGURE P5.29

(a) $x(t) = \text{sinc}^2(t/2\pi)$ for $-\infty < t < \infty$, $T = \pi$
(b) $x(t) = \text{sinc}^2(t/2\pi)$ for $-\infty < t < \infty$, $T = 2\pi$
(c) For (a) and (b), compare the plots of $x(t)$ and the corresponding $y(t)$.
(d) Repeat part (a), using the interpolation formula to solve for $y(t)$, and plot your results with n ranging from $n = -5$ to $n = 5$.

5.30. Consider the signal whose Fourier transform is shown in Figure P5.30. Let $x_s(t) = x(t)p(t)$ represent the sampled signal. Draw $|X_s(\omega)|$ for the following cases:

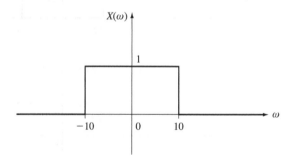

FIGURE P5.30

(a) $T = \pi/15$
(b) $T = 2\pi/15$

5.31. Repeat Problem 5.30 for the signal whose transform is shown in Figure P5.31.

5.32. Consider the signal with the amplitude spectrum shown in Figure P5.32. Let $x_s(t) = x(t)p(t)$ represent the sampled signal. Draw $|X_s(\omega)|$ for the following cases:
(a) $T = \pi/4$ sec
(b) $T = \pi/2$ sec
(c) $T = 2\pi/3$ sec

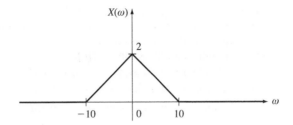

FIGURE P5.31

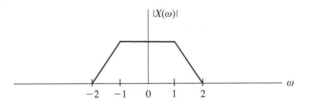

FIGURE P5.32

5.33. Repeat Problem 5.32, where now the signal is $x(t) = e^{-t/4}\cos(t)u(t)$. You can either sketch the plots by hand or use MATLAB for a more accurate plot. In order to examine the effects of aliasing in the time domain, plot $x(t)$ for each of the sampling times for $t = 0$ through 15 sec. In MATLAB, you can do this by defining your time vector with the time increment set to the desired sampling period. MATLAB then "reconstructs" the signal by connecting the sampled points with straight lines. (This procedure is known as a linear interpolation.) Compare your sampled/reconstructed signals with a signal that is more accurate, one that you create by using a very small sampling period (such as $T = 0.05$ sec), by plotting them on the same graph.

5.34. Consider the following sampling and reconstruction configuration:

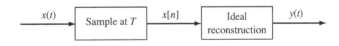

FIGURE P5.34

You can find the output $y(t)$ of the ideal reconstruction by sending the sampled signal $x_s(t) = x(t)p(t)$ through an ideal lowpass filter with the frequency response function

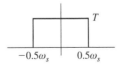

Let $x(t) = 2 + \cos(50\pi t)$ and $T = 0.01$ sec.

   (a) Draw $|X_s(\omega)|$, where $x_s(t) = x(t)p(t)$. Determine if aliasing occurs.

   (b) Determine the expression for $y(t)$.

   (c) Determine an expression for $x[n]$.

5.35. Repeat Problem 5.34 for $x(t) = 2 + \cos(50\pi t)$ and $T = 0.025$ sec.

5.36. Repeat Problem 5.34 for $x(t) = 1 + \cos(20\pi t) + \cos(60\pi t)$ and $T = 0.01$ sec.

5.37. Consider the following sampling and reconstruction configuration:

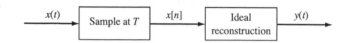

FIGURE P5.37

You can find the output $y(t)$ of the ideal reconstruction by sending the sampled signal $x_s(t) = x(t)p(t)$ through an ideal lowpass filter with the frequency response function

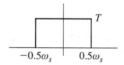

   (a) Let $x(t) = 1 + \cos(15\pi t)$ and $T = 0.1$ sec. Draw $|X_s(\omega)|$, where $x_s(t) = x(t)p(t)$. Determine the expression for $y(t)$.

   (b) Let $X(\omega) = 1/(j\omega + 1)$ and $T = 1$ sec. Draw $|X_s(\omega)|$, where $x_s(t) = x(t)p(t)$. Does aliasing occur? (Justify your answer.)

5.38. An ideal lowpass digital filter has the frequency function $H(\Omega)$ given by

$$H(\Omega) = \begin{cases} 1, & 0 \le |\Omega| \le \dfrac{\pi}{4} \\[2mm] 0, & \dfrac{\pi}{4} < |\Omega| \le \pi \end{cases}$$

   (a) Determine the unit-pulse response $h[n]$ of the filter.

   (b) Compute the output response $y[n]$ of the filter when the input $x[n]$ is given by

      (i) $x[n] = \cos(\pi n/8), n = 0, \pm 1, \pm 2, \ldots$

      (ii) $x[n] = \cos(3\pi n/4) + \cos(\pi n/16), n = 0, \pm 1, \pm 2, \ldots$

     (iii) $x[n] = \text{sinc}(n/2), n = 0, \pm 1, \pm 2, \ldots$

     (iv) $x[n] = \text{sinc}(n/4), n = 0, \pm 1, \pm 2, \ldots$

      (v) $x[n] = \text{sinc}(n/8) \cos(\pi n/8), n = 0, \pm 1, \pm 2, \ldots$

     (vi) $x[n] = \text{sinc}(n/8) \cos(\pi n/4), n = 0, \pm 1, \pm 2, \ldots$

   (c) For each signal defined in part (b), plot the input $x[n]$ and the corresponding output $y[n]$ to determine the effect of the filter.

5.39. An ideal linear-phase highpass digital filter has frequency function $H(\Omega)$, where for one period, $H(\Omega)$ is given by

$$H(\Omega) = \begin{cases} e^{-j3\Omega}, & \dfrac{\pi}{2} \leq |\Omega| \leq \pi \\ 0, & 0 \leq |\Omega| < \dfrac{\pi}{2} \end{cases}$$

(a) Determine the unit-pulse response $h[n]$ of the filter.

(b) Compute the output response $y[n]$ of the filter when the input $x[n]$ is given by

(i) $x[n] = \cos(\pi n/4), n = 0, \pm 1, \pm 2, \ldots$

(ii) $x[n] = \cos(3\pi n/4), n = 0, \pm 1, \pm 2, \ldots$

(iii) $x[n] = \text{sinc}(n/2), n = 0, \pm 1, \pm 2, \ldots$

(iv) $x[n] = \text{sinc}(n/4), n = 0, \pm 1, \pm 2, \ldots$

(v) $x[n] = \text{sinc}(n/4)\cos(\pi n/8), n = 0, \pm 1, \pm 2, \ldots$

(vi) $x[n] = \text{sinc}(n/2)\cos(\pi n/8), n = 0, \pm 1, \pm 2, \ldots$

(c) For each signal defined in part (b), plot the input $x[n]$ and the corresponding output $y[n]$ to determine the effect of the filter.

5.40. A linear time-invariant discrete-time system has the frequency response function $H(\Omega)$ shown in Figure P5.40.

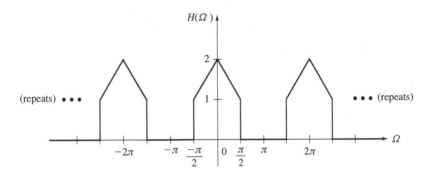

FIGURE P5.40

(a) Determine the unit-pulse response $h[n]$ of the system.

(b) Compute the output response $y[n]$ when the input $x[n]$ is equal to $\delta[n] - \delta[n-1]$.

(c) Compute the output response $y[n]$ when the input is $x[n] = 2 + \sin(\pi n/4) + 2\sin(\pi n/2)$.

(d) Compute the output response $y[n]$ when $x[n] = \text{sinc}(n/4), n = 0, \pm 1, \pm 2$.

(e) For the signals defined in parts (b) and (c), plot the input $x[n]$ and the corresponding output $y[n]$ to determine the effect of the filter.

5.41. As shown in Figure P5.41, a sampled version $x[n]$ of an analog signal $x(t)$ is applied to a linear time-invariant discrete-time system with frequency response function $H(\Omega)$.

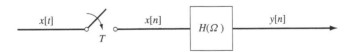

FIGURE P5.41

Choose the sampling interval T and determine the frequency response function $H(\Omega)$ so that

$$
y[n] = \begin{cases} x[n], & \text{when } x(t) = A \cos \omega_0 t, \quad 100 < \omega_0 < 1000 \\ 0, & \text{when } x(t) = A \cos \omega_0 t, \quad 0 \le \omega_0 \le 100 \end{cases}
$$

Express $H(\Omega)$ in analytical form.

5.42. Consider the ideal lowpass digital filter with frequency response function $H(\Omega)$ defined by

$$
H(\Omega) = \begin{cases} e^{-j10\Omega}, & 0 \le |\Omega| \le \dfrac{\pi}{4} \\ 0, & \dfrac{\pi}{4} < |\Omega| < \pi \end{cases}
$$

(a) The input $x[n] = u[n] - u[n - 10]$ is applied to the filter.

  (i) Using `fft` in MATLAB, compute the 32-point DFT of the resulting output response. *Note:* To calculate the DFT of the output, Y_k, write an M-file that carries out the multiplication of the DFT of the input, X_k, with $H(2\pi k/N)$, where $H(\Omega)$ is the frequency response function of the filter. Take $N = 32$.

  (ii) Using `ifft` in MATLAB, compute the output response $y[n]$ for $n = 0, 1, 2, \ldots, 31$.

(b) Repeat part (a) for the input $x[n] = u[n] - u[n - 5]$.

(c) Compare the output response obtained in parts (a) and (b). In what respects do the responses differ? Explain.

(d) Repeat part (a) for the input

$$
x[n] = \begin{cases} r[n] - 0.5, & n = 0, 1, 2, \ldots, 10 \\ 0, & \text{all other } n \end{cases}
$$

where $r[n]$ is a sequence of random numbers uniformly distributed between 0 and 1. How does the magnitude of the response compare with the magnitude of the input? Explain.

5.43. Repeat Problem 5.42 for the linear time-invariant discrete-time system with frequency response function

$$
H(\Omega) = \frac{0.04}{e^{j2\Omega} - 1.6e^{j\Omega} + 0.64}
$$

5.44. By definition of a WMA filter, the frequency response function $H(\Omega)$ is equal to 1 when $\Omega = 0$.

(a) Design a 4-point WMA digital filter so that the frequency response $H(\Omega)$ satisfies the following conditions: $H(\pi/2) = 0.2 - j0.2$, $H(\pi) = 0$. Express your answer by giving the input/output relationship of the filter.

(b) Plot the magnitude and phase functions of the filter designed in part (a).

5.45. Consider the discrete-time system given by the input/output difference equation

$$
y[n + 1] + 0.9y[n] = 1.9x[n + 1]
$$

(a) Show that the impulse response is given by $h[n] = 1.9(-0.9)^n u[n]$.

(b) Compute the frequency response function, and sketch the magnitude function $|H(\Omega)|$, for $-\pi \le \Omega \le \pi$.

(c) Compute the output response $y[n]$ to an input of $x[n] = 1 + \sin(\pi n/4) + \sin(\pi n/2)$.

(d) Compute the output response $y[n]$ resulting from the input $x[n] = u[n] - u[n - 3]$.

(e) Use the `fft` command to compute and plot the response of the system to the input given in part (d) for $n = 0$ through 30. Compare your numerical answer with the answer found in part (d). Does the response match what you might expect from the plot of the frequency response function determined in part (b)? Explain.

5.46. Consider the discrete-time system given by the input/output difference equation

$$y[n + 1] - 0.9y[n] = 0.1x[n + 1]$$

(a) Show that the impulse response is given by $h[n] = 0.1(0.9)^n u[n]$.

(b)–(e) Repeat Problem 5.45, parts (b)–(e) for this system.

CHAPTER 6

The Laplace Transform and the Transfer Function Representation

In this chapter the Laplace transform of a continuous-time signal is introduced, and then this operator is used to generate the transfer function representation of a causal linear time-invariant continuous-time system. It will be seen that the transfer function representation gives an algebraic relationship between the Laplace transforms of the input and output of a system; and in terms of this setup, the output response resulting from a large class of input signals can be computed by a purely algebraic procedure.

The Laplace transform is named after Pierre Simon Laplace (1749–1827), a French mathematician and astronomer. The chapter begins in Section 6.1 with the definition of the Laplace transform of a continuous-time signal. It is shown that the Laplace transform can be viewed as a generalization of the Fourier transform; more precisely, the addition of an exponential factor to the integrand in the definition of the Fourier transform results in the two-sided Laplace transform. The one-sided Laplace transform is then defined, which is the form of the transform that is studied in this book. In Section 6.2 the basic properties of the (one-sided) Laplace transform are given. By these properties it is shown that many new transforms can be generated from a small set of transforms. Then, in Section 6.3 the computation of the inverse Laplace transform is developed in terms of partial fraction expansions.

In Sections 6.4 through 6.6 the Laplace transform is applied to the study of causal linear time-invariant continuous-time systems. The development begins in Section 6.4 with systems defined by an input/output differential equation. For any such system we can generate the transfer function representation by taking the Laplace transform of the input/output differential equation. This results in an s-domain framework that can be used to solve the input/output differential equation via an algebraic procedure. In Section 6.5 we generate the transfer function representation by applying the Laplace transform to the input/output convolution relationship of the system. In Section 6.6, techniques for generating the transfer function model are given for *RLC* circuits, interconnections of integrators, and block diagrams. Section 6.7 contains a summary of the chapter.

6.1 LAPLACE TRANSFORM OF A SIGNAL

Given a continuous-time signal $x(t)$, in Chapter 3 the Fourier transform $X(\omega)$ of $x(t)$ was defined by

$$X(\omega) = \int_{-\infty}^{\infty} x(t)e^{-j\omega t}\,dt \qquad (6.1)$$

As discussed in Section 3.3, the Fourier transform $X(\omega)$ displays the frequency components comprising the signal $x(t)$.

It was also observed in Chapter 3 that for some common signals, the integral in (6.1) does not exist, and thus there is no Fourier transform (in the ordinary sense). An example is the unit-step function $u(t)$, for which (6.1) becomes

$$X(\omega) = \int_0^\infty e^{-j\omega t}\, dt \tag{6.2}$$

Although the integral in (6.2) does not exist, it is possible to circumvent this problem by addition of an exponential convergence factor $e^{-\sigma t}$ to the integrand, where σ is a real number. Then (6.2) becomes

$$X(\omega) = \int_0^\infty e^{-\sigma t} e^{-j\omega t}\, dt$$

which can be written as

$$X(\omega) = \int_0^\infty e^{-(\sigma + j\omega)t}\, dt \tag{6.3}$$

Now, $X(\omega)$ given by (6.3) is actually a function of the complex number $\sigma + j\omega$, so X should be expressed as a function of $\sigma + j\omega$ rather than ω. Then, rewriting (6.3) gives

$$X(\sigma + j\omega) = \int_0^\infty e^{-(\sigma + j\omega)t}\, dt \tag{6.4}$$

Evaluating the right-hand side of (6.4) gives

$$X(\sigma + j\omega) = -\frac{1}{\sigma + j\omega}[e^{-(\sigma + j\omega)t}]_{t=0}^{t=\infty} \tag{6.5}$$

Now, we see that

$$\lim_{t\to\infty} e^{-(\sigma + j\omega)t}$$

exists if and only if $\sigma > 0$, in which case

$$\lim_{t\to\infty} e^{-(\sigma + j\omega)t} = 0$$

and (6.5) reduces to

$$X(\sigma + j\omega) = -\frac{1}{\sigma + j\omega}[0 - e^{-(\sigma + j\omega)(0)}]$$

$$= \frac{1}{\sigma + j\omega} \tag{6.6}$$

Finally, by letting s denote the complex number $\sigma + j\omega$, we can rewrite (6.6) as

$$X(s) = \frac{1}{s} \tag{6.7}$$

The function $X(s)$ given by (6.7) is the Laplace transform of the unit-step function $u(t)$. Note that $X(s)$ is a complex-valued function of the complex number s. Here, complex valued means that if a particular complex number s is inserted into $X(s)$, the resulting value of $X(s)$ is, in general, a complex number. For example, inserting $s = 1 + j$ into (6.7) gives

$$X(1 + j) = \frac{1}{1 + j} = \frac{1 - j}{(1 + j)(1 - j)} = \frac{1}{2} - j\frac{1}{2} = \frac{1}{\sqrt{2}}e^{-j(\pi/4)}$$

so the value of $X(s)$ at $s = 1 + j$ is the complex number $\frac{1}{2} - j\frac{1}{2}$.

It is important to note that the function $X(s)$ given by (6.7) is defined for only those complex numbers s for which the real part of s (which is equal to σ) is strictly positive. In this particular example, the function $X(s)$ is not defined for $\sigma = 0$ or $\sigma < 0$, since the integral in (6.4) does not exist for such values of σ. The set of all complex numbers $s = \sigma + j\omega$ for which

$$\text{Re } s = \sigma > 0$$

is called the *region of convergence* of the Laplace transform of the unit-step function.

The preceding construction can be generalized to a very large class of signals $x(t)$ as follows: Given a signal $x(t)$, the exponential factor $e^{-\sigma t}$ is again added to the integrand in the definition (6.1) of the Fourier transform, which yields

$$X(\sigma + j\omega) = \int_{-\infty}^{\infty} x(t)e^{-(\sigma + j\omega)t} \, dt \tag{6.8}$$

With s equal to the complex number $\sigma + j\omega$, (6.8) becomes

$$X(s) = \int_{-\infty}^{\infty} x(t)e^{-st} \, dt \tag{6.9}$$

The function $X(s)$ given by (6.9) is the *two-sided* (or *bilateral*) *Laplace transform* of $x(t)$. As is the case for the unit-step function, the Laplace transform $X(s)$ is, in general, a complex-valued function of the complex number s.

Clearly, the two-sided Laplace transform $X(s)$ of a signal $x(t)$ can be viewed as a generalization of the Fourier transform of $x(t)$. More precisely, as was done previously, we can generate $X(s)$ directly from the definition of the Fourier transform by adding the exponential factor $e^{-\sigma t}$ to the integrand in the definition of the Fourier transform.

The *one-sided* (or *unilateral*) *Laplace transform* of $x(t)$, also denoted by $X(s)$, is defined by

$$X(s) = \int_{0}^{\infty} x(t)e^{-st} \, dt \tag{6.10}$$

From (6.10) it is clear that the one-sided transform depends only on the values of the signal $x(t)$ for $t \geq 0$. For this reason, the definition (6.10) of the transform is called the *one-sided Laplace transform*. The one-sided transform can be applied to signals $x(t)$ that are nonzero for $t < 0$; however, any nonzero values of $x(t)$ for $t < 0$ do not have any effect on the one-sided transform of $x(t)$. If $x(t)$ is zero for all $t < 0$, the expression (6.9) reduces to (6.10), and thus in this case the one- and two-sided Laplace transforms are the same.

As will be seen in the following section, the initial values of a signal and its derivatives can be explicitly incorporated into the s-domain framework by the one-sided Laplace transform (as opposed to the two-sided Laplace transform). This is particularly useful for some problems, such as solving a differential equation with initial conditions. Hence, in this book the development is limited to the one-sided transform, which will be referred to as the *Laplace transform*. In addition, as was the case for the Fourier transform, in this book the Laplace transform of a signal will always be denoted by an uppercase letter, with signals denoted by lowercase letters.

Given a signal $x(t)$, the set of all complex numbers s for which the integral in (6.10) exists is called the *region of convergence* of the Laplace transform $X(s)$ of $x(t)$. Hence, the Laplace transform $X(s)$ of $x(t)$ is well defined (i.e., exists) for all values of s belonging to the region of convergence. It should be stressed that the region of convergence depends on the given function $x(t)$. For example, when $x(t)$ is the unit-step function $u(t)$, as previously noted, the region of convergence is the set of all complex numbers s such that $\mathrm{Re}\, s > 0$. This is also verified in the example given next.

Example 6.1   Laplace Transform of Exponential Function

Let $x(t) = e^{-bt}u(t)$, where b is an arbitrary real number. The Laplace transform is

$$X(s) = \int_0^\infty e^{-bt}e^{-st}\, dt$$

$$= \int_0^\infty e^{-(s+b)t}\, dt$$

$$X(s) = -\frac{1}{s+b}[e^{-(s+b)t}]_{t=0}^{t=\infty} \tag{6.11}$$

To evaluate the right-hand side of (6.11), it is necessary to determine

$$\lim_{t\to\infty} e^{-(s+b)t} \tag{6.12}$$

Setting $s = \sigma + j\omega$ in (6.12) gives

$$\lim_{t\to\infty} e^{-(\sigma+j\omega+b)t} \tag{6.13}$$

The limit in (6.13) exists if and only if $\sigma + b > 0$, in which case the limit is zero, and from (6.11) the Laplace transform is

$$X(s) = \frac{1}{s+b} \tag{6.14}$$

The region of convergence of the transform $X(s)$ given by (6.14) is the set of all complex numbers s such that Re $s > -b$. Note that if $b = 0$, so that $x(t)$ is the unit-step function $u(t)$, then $X(s) = 1/s$ and the region of convergence is Re $s > 0$. This corresponds to the result that was obtained previously.

6.1.1   Relationship between the Fourier and Laplace Transforms

As shown, the two-sided Laplace transform of a signal $x(t)$ can be viewed as a generalization of the definition of the Fourier transform of $x(t)$; that is, the two-sided Laplace transform is the Fourier transform with the addition of an exponential factor. Given a signal $x(t)$ with $x(t) = 0$ for all $t < 0$, from the constructions previously given it may appear that the Fourier transform $X(\omega)$ can be computed directly from the (one-sided) Laplace transform $X(s)$ by setting $s = j\omega$. However, this is often not the case. To see this, let $x(t)$ be a signal that is zero for all $t < 0$, and suppose that $x(t)$ has the Laplace transform $X(s)$ given by (6.10). Since $x(t)$ is zero for $t < 0$, the Fourier transform $X(\omega)$ of $x(t)$ is given by

$$X(\omega) = \int_0^\infty x(t)e^{-j\omega t}\,dt \tag{6.15}$$

By comparing (6.10) and (6.15), it is tempting to conclude that the Fourier transform $X(\omega)$ is equal to the Laplace transform $X(s)$ with $s = j\omega$; or in mathematical terms,

$$X(\omega) = X(s)|_{s=j\omega} \tag{6.16}$$

However, (6.16) is valid if and only if the region of convergence for $X(s)$ includes $\sigma = 0$. For example, if $x(t)$ is the unit-step function $u(t)$, then (6.16) is *not* valid, since the region of convergence is Re $s > 0$, which does not include the point Re $s = \sigma = 0$. This is simply a restatement of the fact that the unit-step function has a Laplace transform, but does not have a Fourier transform (in the ordinary sense).

Example 6.2   Fourier Transform from Laplace Transform

Let $x(t) = e^{-bt}u(t)$, where b is an arbitrary real number. From the result in Example 6.1, the Laplace transform $X(s)$ of $x(t)$ is equal to $1/(s + b)$, and the region of convergence is Re $s > -b$. Thus, if $b > 0$, the region of convergence includes $\sigma = 0$, and the Fourier transform $X(\omega)$ of $x(t)$ is given by

$$X(\omega) = X(s)|_{s=j\omega} = \frac{1}{j\omega + b} \tag{6.17}$$

As was first observed in Example 3.8, when $b \leq 0$, $x(t)$ does not have a Fourier transform in the ordinary sense, but it does have the Laplace transform $X(s) = 1/(s + b)$.

As was done in the case of the Fourier transform, the transform pair notation

$$x(t) \leftrightarrow X(s)$$

will sometimes be used to denote the fact that $X(s)$ is the Laplace transform of $x(t)$, and conversely, that $x(t)$ is the inverse Laplace transform of $X(s)$. Some authors prefer to use the operator notation

$$X(s) = L[x(t)]$$

$$x(t) = L^{-1}[X(s)]$$

where L denotes the Laplace transform operator and L^{-1} denotes the inverse Laplace transform operator. In this book the transform pair notation will be used.

In giving a transform pair, the region of convergence will not be specified, since in most applications it is not necessary to consider the region of convergence (as long as the transform does have a region of convergence). An example of a transform pair arising from the result of Example 6.1 is

$$e^{-bt}u(t) \leftrightarrow \frac{1}{s + b} \qquad (6.18)$$

It should be noted that in the transform pair (6.18), the scalar b may be real or complex. The verification of this transform pair in the case when b is complex is an easy modification of the derivation given in Example 6.1. The details are omitted.

The Laplace transform of many signals of interest can be determined by table lookup. Hence, it is often not necessary to evaluate the integral in (6.10) in order to compute a transform. By use of the properties of the Laplace transform given in the next section, it will be shown that transform pairs for many common signals can be generated. These results will then be displayed in a table of transforms.

6.1.2   Laplace Transform Computation Using Symbolic Manipulation

The Symbolic Math Toolbox in MATLAB has a command `laplace` for computing the one-sided Laplace transform of signals. If x is a signal that is defined symbolically, the usage of the command is X = `laplace(x)`.

Example 6.3   Laplace Transforms Using Symbolic Manipulation

In Example 6.1, the Laplace transform of an exponential $x(t) = e^{-bt}u(t)$ was found analytically with the resulting expression given in (6.14). The Symbolic Math Toolbox in MATLAB computes this expression from the following commands:

```
syms x b t
x = exp(-b*t);
X = laplace(x)
```

Running these commands returns the expression

```
X =
1/(s+b)
```

Similarly, the Laplace transform of a unit-step function can be determined by typing

```
x = sym(1);
X = laplace(x)
```

which gives

```
X =
1/s
```

The command `sym(1)` is used here to create a symbolic object of the number 1.

6.2   PROPERTIES OF THE LAPLACE TRANSFORM

The Laplace transform satisfies a number of properties that are useful in a wide range of applications, such as the derivation of new transform pairs from a given transform pair. In this section various fundamental properties of the Laplace transform are presented. Most of these properties correspond directly to the properties of the Fourier transform that were studied in Section 3.6. The properties of the Laplace transform that correspond to properties of the Fourier transform can be proved simply by replacement of $j\omega$ by s in the proof of the Fourier transform property. Thus, the proofs of these properties follow easily from the constructions given in Section 3.6 and will not be considered here.

The Fourier transform does enjoy some properties for which there is no version in the Laplace transform theory. Two examples are the duality property and Parseval's theorem. Hence, the reader will notice that there are no versions of these properties stated in this section.

6.2.1   Linearity

The Laplace transform is a linear operation, as is the Fourier transform. Hence, if $x(t) \leftrightarrow X(s)$ and $v(t) \leftrightarrow V(s)$, then for any real or complex scalars a, b,

$$ax(t) + bv(t) \leftrightarrow aX(s) + bV(s) \tag{6.19}$$

Example 6.4   Linearity

Consider the signal $u(t) + e^{-t}u(t)$. Using the transform pair (6.18) and the property of linearity results in the transform pair

$$u(t) + e^{-t}u(t) \leftrightarrow \frac{1}{s} + \frac{1}{s+1} = \frac{2s+1}{s(s+1)} \tag{6.20}$$

6.2.2   Right Shift in Time

If $x(t) \leftrightarrow X(s)$, then for any positive real number c,

$$x(t-c)u(t-c) \leftrightarrow e^{-cs}X(s) \tag{6.21}$$

In (6.21), note that $x(t - c)u(t - c)$ is equal to the c-second right shift of $x(t)u(t)$. Here, multiplication of $x(t)$ by $u(t)$ is necessary to eliminate any nonzero values of $x(t)$ for $t < 0$.

From (6.21) it is seen that a c-second right shift in the time domain corresponds to multiplication by e^{-cs} in the Laplace transform domain (or *s-domain*). The proof of the right-shift property is analogous to the one given for the Fourier transform and is thus omitted.

Example 6.5   Laplace Transform of a Pulse

Let $x(t)$ denote the c-second rectangular pulse function defined by

$$x(t) = \begin{cases} 1, & 0 \le t < c \\ 0, & \text{all other } t \end{cases}$$

where c is an arbitrary positive real number. Expressing $x(t)$ in terms of the unit-step function $u(t)$ gives

$$x(t) = u(t) - u(t - c)$$

By linearity, the Laplace transform $X(s)$ of $x(t)$ is the difference of the transform of $u(t)$ and the transform of $u(t - c)$. Now $u(t - c)$ is the c-second right shift of $u(t)$, and thus by the right-shift property (6.21), the Laplace transform of $u(t - c)$ is equal to e^{-cs}/s. Hence,

$$u(t) - u(t - c) \leftrightarrow \frac{1}{s} - \frac{e^{-cs}}{s} = \frac{1 - e^{-cs}}{s} \tag{6.22}$$

It should be noted that there is no comparable result for a left shift in time. To see this, let c be an arbitrary positive real number, and consider the time-shifted signal $x(t + c)$. Since $c > 0$, $x(t + c)$ is a c-second left shift of the signal $x(t)$. The Laplace transform of $x(t + c)$ is equal to

$$\int_0^\infty x(t + c)e^{-st}\, dt \tag{6.23}$$

However, (6.23) cannot be expressed in terms of the Laplace transform $X(s)$ of $x(t)$. (Try it!) In particular, (6.23) is not equal to $e^{cs}X(s)$.

6.2.3   Time Scaling

If $x(t) \leftrightarrow X(s)$, for any positive real number a,

$$x(at) \leftrightarrow \frac{1}{a}X\left(\frac{s}{a}\right) \tag{6.24}$$

As discussed in Section 3.6, the signal $x(at)$ is a time-scaled version of $x(t)$. By (6.24) it is seen that time scaling corresponds to scaling the complex variable s by the factor of

$1/a$ in the Laplace transform domain (plus multiplication of the transform by $1/a$). The transform pair (6.24) can be proved in the same way that the corresponding transform pair in the Fourier theory was proved. (See Section 3.6.)

Example 6.6   Time Scaling

Consider the time-scaled unit-step function $u(at)$, where a is an arbitrary positive real number. By (6.24)

$$u(at) \leftrightarrow \frac{1}{a}\left(\frac{1}{s/a}\right) = \frac{1}{s}$$

This result is not unexpected, since $u(at) = u(t)$ for any real number $a > 0$.

6.2.4   Multiplication by a Power of t

If $x(t) \leftrightarrow X(s)$, then for any positive integer N,

$$t^N x(t) \leftrightarrow (-1)^N \frac{d^N}{ds^N} X(s) \tag{6.25}$$

In particular, for $N = 1$,

$$tx(t) \leftrightarrow -\frac{d}{ds} X(s) \tag{6.26}$$

and for $N = 2$,

$$t^2 x(t) \leftrightarrow \frac{d^2}{ds^2} X(s) \tag{6.27}$$

The proof of (6.26) is very similar to the proof of the multiplication-by-t property given in the Fourier theory, so the details are again omitted.

Example 6.7   Unit-Ramp Function

Consider the unit-ramp function $r(t) = tu(t)$. From (6.26), the Laplace transform $R(s)$ of $r(t)$ is given by

$$R(s) = -\frac{d}{ds}\frac{1}{s} = \frac{1}{s^2}$$

Generalizing Example 6.7 to the case $t^N u(t)$, $N = 1, 2, \ldots$, yields the transform pair

$$t^N u(t) \leftrightarrow \frac{N!}{s^{N+1}} \tag{6.28}$$

where $N!$ is N factorial.

Example 6.8   Multiplication of an Exponential by *t*

Let $v(t) = te^{-bt}u(t)$, where b is any real number. Using the transform pairs (6.18) and (6.26) yields

$$V(s) = -\frac{d}{ds}\frac{1}{s+b} = \frac{1}{(s+b)^2}$$

Generalizing Example 6.8 to the case $t^N e^{-bt}u(t)$ results in the transform pair

$$t^N e^{-bt}u(t) \leftrightarrow \frac{N!}{(s+b)^{N+1}} \tag{6.29}$$

6.2.5   Multiplication by an Exponential

If $x(t) \leftrightarrow X(s)$, then for any real or complex number a,

$$e^{at}x(t) \leftrightarrow X(s-a) \tag{6.30}$$

By the property (6.30), multiplication by an exponential function in the time domain corresponds to a shift of the s variable in the Laplace transform domain. The proof of (6.30) follows directly from the definition of the Laplace transform. The details are left to the reader.

Example 6.9   Multiplication by an Exponential

Let $v(t) = [u(t) - u(t-c)]e^{at}$, where c is a positive real number and a is any real number. The function $v(t)$ is the product of the c-second pulse $u(t) - u(t-c)$ and exponential function e^{at}. The function $v(t)$ is displayed in Figure 6.1 for the case $a < 0$. Now from the result in Example 6.5,

$$u(t) - u(t-c) \leftrightarrow \frac{1 - e^{-cs}}{s}$$

Then, using (6.30) yields

$$V(s) = \frac{1 - e^{-c(s-a)}}{s-a}$$

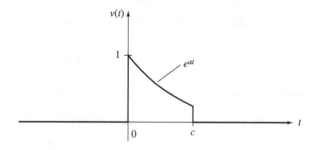

FIGURE 6.1
The function $v(t)$ in Example 6.9.

6.2.6   Multiplication by a Sinusoid

If $x(t) \leftrightarrow X(s)$, then for any real number ω,

$$x(t)\sin \omega t \leftrightarrow \frac{j}{2}[X(s + j\omega) - X(s - j\omega)] \tag{6.31}$$

$$x(t)\cos \omega t \leftrightarrow \frac{1}{2}[X(s + j\omega) + X(s - j\omega)] \tag{6.32}$$

The transform pairs (6.31) and (6.32) can be proved by first writing $x(t)\sin \omega t$ and $x(t)\cos \omega t$ in the form

$$x(t)\sin \omega t = \frac{j}{2}x(t)[e^{-j\omega t} - e^{j\omega t}] \tag{6.33}$$

$$x(t)\cos \omega t = \frac{1}{2}x(t)[e^{-j\omega t} + e^{j\omega t}] \tag{6.34}$$

By (6.30),

$$e^{\pm j\omega t}x(t) \leftrightarrow X(s \mp j\omega)$$

Combining this with (6.33) and (6.34) yields (6.31) and (6.32).

Example 6.10   Laplace Transform of a Cosine

Let $v(t) = (\cos \omega t)u(t)$. Now, $u(t) \leftrightarrow 1/s$, and using (6.32) with $x(t) = u(t)$ gives

$$V(s) = \frac{1}{2}\left(\frac{1}{s + j\omega} + \frac{1}{s - j\omega}\right)$$

$$= \frac{1}{2}\frac{s - j\omega + s + j\omega}{s^2 + \omega^2}$$

$$= \frac{s}{s^2 + \omega^2}$$

Example 6.10 yields the transform pair

$$(\cos \omega t)u(t) \leftrightarrow \frac{s}{s^2 + \omega^2} \tag{6.35}$$

Similarly, it is possible to verify the transform pair

$$(\sin \omega t)u(t) \leftrightarrow \frac{\omega}{s^2 + \omega^2} \tag{6.36}$$

Example 6.11   Multiplication of an Exponential by Cosine and Sine

Now let $v(t) = (e^{-bt} \cos \omega t)u(t)$. We can compute the Laplace transform of $v(t)$ by setting $x(t) = e^{-bt}u(t)$ and then using the multiplication by $\cos \omega t$ property. It is also possible to set $x(t) = (\cos \omega t)u(t)$ and use the multiplication by an exponential property. The latter is simpler to carry out, so it will be done that way. Replacing s by $s + b$ on the right-hand side of (6.35) results in the transform pair

$$(e^{-bt} \cos \omega t)u(t) \leftrightarrow \frac{s + b}{(s + b)^2 + \omega^2} \tag{6.37}$$

Similarly, it is possible to verify the transform pair

$$(e^{-bt} \sin \omega t)u(t) \leftrightarrow \frac{\omega}{(s + b)^2 + \omega^2} \tag{6.38}$$

Example 6.12   Multiplication by Sine

Let $v(t) = (\sin^2 \omega t)u(t)$. Setting $x(t) = (\sin \omega t)u(t)$ and using the multiplication by $\sin \omega t$ property yields

$$
\begin{aligned}
V(s) &= \frac{j}{2}\left[\frac{\omega}{(s + j\omega)^2 + \omega^2} - \frac{\omega}{(s - j\omega)^2 + \omega^2}\right] \\[2mm]
&= \frac{j}{2}\frac{\omega(s - j\omega)^2 + \omega^3 - \omega(s + j\omega)^2 - \omega^3}{(s + j\omega)^2(s - j\omega)^2 + \omega^2(s - j\omega)^2 + \omega^2(s + j\omega)^2 + \omega^4} \\[2mm]
&= \frac{j}{2}\frac{-j4\omega^2 s}{s^4 + 4\omega^2 s^2} \\[2mm]
&= \frac{2\omega^2}{s(s^2 + 4\omega^2)}
\end{aligned}
\tag{6.39}
$$

6.2.7   Differentiation in the Time Domain

If $x(t) \leftrightarrow X(s)$, then

$$\dot{x}(t) \leftrightarrow sX(s) - x(0) \tag{6.40}$$

where $\dot{x}(t) = dx(t)/dt$. Thus, differentiation in the time domain corresponds to multiplication by s in the Laplace transform domain [plus subtraction of the initial value $x(0)$]. We will prove the property (6.40) by computing the transform of the derivative of $x(t)$. The transform of $\dot{x}(t)$ is given by

$$\int_0^\infty \dot{x}(t)e^{-st}\, dt \tag{6.41}$$

The integral in (6.41) will be evaluated by parts: Let $v = e^{-st}$ so that $dv = -se^{-st}$, and let $w = x(t)$ so that $dw = \dot{x}(t)\, dt$. Then,

$$\int_0^\infty \dot{x}(t)e^{-st}\, dt = vw\big|_{t=0}^{t=\infty} - \int_0^\infty w\, dv$$

$$= e^{-st}x(t)\big|_{t=0}^{t=\infty} - \int_0^\infty x(t)(-s)e^{-st}\, dt$$

$$= \lim_{t\to\infty}[e^{-st}x(t)] - x(0) + sX(s) \qquad (6.42)$$

When $|x(t)| < ce^{at}, t > 0$, for some constants a and c, it follows that for any s such that Re $s > a$,

$$\lim_{t\to\infty} e^{-st}x(t) = 0$$

Thus, from (6.42),

$$\int_0^\infty \dot{x}(t)e^{-st}\, dt = -x(0) + sX(s)$$

which verifies (6.40).

It should be pointed out that, if $x(t)$ is discontinuous at $t = 0$ or if $x(t)$ contains an impulse or derivative of an impulse located at $t = 0$, it is necessary to take the initial time in (6.40) to be at 0^- (an infinitesimally small negative number). In other words, the transform pair (6.40) becomes

$$\dot{x}(t) \leftrightarrow sX(s) - x(0^-) \qquad (6.43)$$

Note that if $x(t) = 0$ for $t < 0$, then $x(0^-) = 0$ and

$$\dot{x}(t) \leftrightarrow sX(s) \qquad (6.44)$$

Example 6.13   Differentiation

Let $x(t) = u(t)$. Then $\dot{x}(t) = \delta(t)$. Since $\dot{x}(t)$ is the unit impulse located at $t = 0$, it is necessary to use (6.43) to compute the Laplace transform of $\dot{x}(t)$. This gives

$$\dot{x}(t) \leftrightarrow s\frac{1}{s} - u(0^-) = 1 - 0 = 1$$

Hence, the Laplace transform of the unit impulse $\delta(t)$ is equal to the constant function 1. We could have obtained this result by directly applying the definition (6.10) of the transform. Putting this result in the form of a transform pair yields

$$\delta(t) \leftrightarrow 1 \qquad (6.45)$$

The Laplace transform of the second- and higher-order derivatives of a signal $x(t)$ can also be expressed in terms of $X(s)$ and initial conditions. For example, the transform pair in the second-order case is

$$\frac{d^2x(t)}{dt^2} \leftrightarrow s^2X(s) - sx(0) - \dot{x}(0) \tag{6.46}$$

The transform pair (6.46) can be proved by integration by parts twice on the integral expression for the transform of the second derivative of $x(t)$. The details are omitted. It should be noted that if the second derivative of $x(t)$ is discontinuous or contains an impulse or a derivative of an impulse located at $t = 0$, it is necessary to take the initial conditions in (6.46) at time $t = 0^-$.

Now, let N be an arbitrary positive integer, and let $x^{(N)}(t)$ denote the Nth derivative of a given signal $x(t)$. Then, the transform of $x^{(N)}(t)$ is given by the transform pair

$$x^{(N)}(t) \leftrightarrow s^NX(s) - s^{N-1}x(0) - s^{N-2}\dot{x}(0) - \cdots - sx^{(N-2)}(0) - x^{(N-1)}(0) \tag{6.47}$$

6.2.8   Integration

If $x(t) \leftrightarrow X(s)$, then

$$\int_0^t x(\lambda)\, d\lambda \leftrightarrow \frac{1}{s}X(s) \tag{6.48}$$

By (6.48), the Laplace transform of the integral of $x(t)$ is equal to $X(s)$ divided by s. The transform pair (6.48) follows directly from the derivative property just given. To see this, let $v(t)$ denote the integral of $x(t)$ given by

$$v(t) = \begin{cases} \displaystyle\int_0^t x(\lambda)\, d\lambda, & t \geq 0 \\ 0, & t < 0 \end{cases}$$

Then $x(t) = \dot{v}(t)$ for $t > 0$, and since $v(t) = 0$ for $t < 0$, by (6.44), $X(s) = sV(s)$. Therefore, $V(s) = (1/s)X(s)$, which verifies (6.48).

Example 6.14   Integration

Let $x(t) = u(t)$. Then the integral of $x(t)$ is the unit-ramp function $r(t) = tu(t)$. By (6.48), the Laplace transform of $r(t)$ is equal to $1/s$ times the transform of $u(t)$. The result is the transform pair

$$r(t) \leftrightarrow \frac{1}{s^2}$$

Recall that this transform pair was derived previously by the multiplication-by-t property.

6.2.9   Convolution

Given two signals $x(t)$ and $v(t)$ with both $x(t)$ and $v(t)$ equal to zero for all $t < 0$, consider the convolution $x(t) * v(t)$ given by

$$x(t) * v(t) = \int_0^t x(\lambda)v(t - \lambda)\, d\lambda$$

Now with $X(s)$ equal to the Laplace transform of $x(t)$ and $V(s)$ equal to the Laplace transform of $v(t)$, it turns out that the transform of the convolution $x(t) * v(t)$ is equal to the product $X(s)V(s)$; that is, the following transform pair is valid:

$$x(t) * v(t) \leftrightarrow X(s)V(s) \tag{6.49}$$

By (6.49), convolution in the time domain corresponds to a product in the Laplace transform domain. It will be seen in Section 6.5 that this property results in an algebraic relationship between the transforms of the inputs and outputs of a linear time-invariant continuous-time system. The proof of (6.49) is very similar to the proof of the corresponding property in the Fourier theory, and thus will not be given.

The transform pair (6.49) yields a procedure for computing the convolution $x(t) * v(t)$ of two signals $x(t)$ and $v(t)$ [with $x(t) = v(t) = 0$ for all $t < 0$]: First, compute the Laplace transforms $X(s)$, $V(s)$ of $x(t)$, $v(t)$, and then compute the inverse Laplace transform of the product $X(s)V(s)$. The result is the convolution $x(t) * v(t)$. The process is illustrated by the following example:

Example 6.15   Convolution

Let $x(t)$ denote the one-second pulse given by $x(t) = u(t) - u(t - 1)$. The objective is to determine the convolution $x(t) * x(t)$ of this signal with itself. From Example 6.4 the transform $X(s)$ of $x(t)$ is $X(s) = (1 - e^{-s})/s$. Thus, from (6.49) the transform of the convolution $x(t) * x(t)$ is equal to $X^2(s)$, where

$$X^2(s) = \left(\frac{1 - e^{-s}}{s}\right)^2 = \frac{1 - 2e^{-s} + e^{-2s}}{s^2}$$

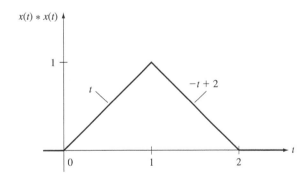

FIGURE 6.2
Plot of the convolution $x(t) * x(t)$.

Now, the convolution $x(t) * x(t)$ is equal to the inverse Laplace transform of $X^2(s)$, which can be computed in this case by using linearity, the right-shift property, and the transform pairs $u(t) \leftrightarrow 1/s$, $tu(t) \leftrightarrow 1/s^2$. The result is

$$x(t) * x(t) = tu(t) - 2(t-1)u(t-1) + (t-2)u(t-2)$$

The convolution $x(t) * x(t)$ is displayed in Figure 6.2. The plot shows that the convolution of a rectangular pulse with itself results in a triangular pulse. Note that this result was first observed in Chapter 2 by use of the MATLAB command conv.

6.2.10   Initial-Value Theorem

Given a signal $x(t)$ with transform $X(s)$, the initial values x(0) and $\dot{x}(0)$ can be computed by the expressions

$$x(0) = \lim_{s \to \infty} sX(s) \tag{6.50}$$

$$\dot{x}(0) = \lim_{s \to \infty} [s^2 X(s) - sx(0)] \tag{6.51}$$

In the general case, for an arbitrary positive integer N,

$$x^{(N)}(0) = \lim_{s \to \infty} [s^{N+1}X(s) - s^N x(0) - s^{N-1}\dot{x}(0) - \cdots - sx^{(N-1)}(0)] \tag{6.52}$$

It should be noted that the relationship (6.52) is not valid if the Nth derivative $x^{(N)}(t)$ contains an impulse or a derivative of an impulse at time $t = 0$.

The relationship (6.52) for $N = 0, 1, 2, \ldots$ is called the *initial-value theorem*. It will be proved for the case $N = 0$, assuming that $x(t)$ has the *Taylor series* expansion:

$$x(t) = \sum_{i=0}^{\infty} x^{(i)}(0) \frac{t^i}{i!} \tag{6.53}$$

where $i!$ is i factorial and $x^{(i)}(0)$ is the ith derivative of $x(t)$ evaluated at $t = 0$. Now,

$$sX(s) = \int_0^{\infty} sx(t)e^{-st} \, dt \tag{6.54}$$

and using (6.53) in (6.54) gives

$$sX(s) = \int_0^{\infty} \sum_{i=0}^{\infty} sx^{(i)}(0)\frac{t^i}{i!} e^{-st} \, dt \tag{6.55}$$

Interchanging the integral and summation in (6.55) and using the transform pair (6.28) yield

$$sX(s) = \sum_{i=0}^{\infty} x^{(i)}(0)\frac{s}{s^{i+1}} = \sum_{i=0}^{\infty} x^{(i)}(0)\frac{1}{s^i}$$

Then taking the limit as $s \to \infty$ gives

$$\lim_{s \to \infty} sX(s) = x(0)$$

which proves (6.50).

The initial-value theorem is useful, since it allows for computation of the initial values of a function $x(t)$ and its derivatives directly from the Laplace transform $X(s)$ of $x(t)$. If $X(s)$ is known, but $x(t)$ is not, it is possible to compute these initial values without having to compute the inverse Laplace transform of $x(t)$. It should also be noted that these initial values are at $t = 0$ or $t = 0^+$, not at $t = 0^-$. The initial values at $t = 0^-$ cannot be determined from the one-sided Laplace transform (unless the signal is continuous at $t = 0$), since the transform is based on the signal $x(t)$ for $t \geq 0$ only.

Example 6.16   Initial Value

Suppose that the signal $x(t)$ has the Laplace transform

$$X(s) = \frac{-3s^2 + 2}{s^3 + s^2 + 3s + 2}$$

Then,

$$\lim_{s \to \infty} sX(s) = \lim_{s \to \infty} \frac{-3s^3 + 2s}{s^3 + s^2 + 3s + 2} = \frac{-3}{1}$$

Thus, $x(0) = -3$.

6.2.11   Final-Value Theorem

Given the signal $x(t)$ with transform $X(s)$, suppose that $x(t)$ has a limit $x(\infty)$ as $t \to \infty$; that is,

$$x(\infty) = \lim_{t \to \infty} x(t)$$

The existence of the limit $x(\infty)$ turns out to be equivalent to requiring that the region of convergence of $sX(s)$ includes the value $s = 0$.

If $x(\infty)$ exists, the *final-value theorem* states that

$$\lim_{t \to \infty} x(t) = \lim_{s \to 0} sX(s) \tag{6.56}$$

To prove (6.56), first note that by the derivative property,

$$\int_0^\infty \dot{x}(t)e^{-st}\, dt = sX(s) - x(0) \tag{6.57}$$

Taking the limit as $s \to 0$ of both sides of (6.57) gives

$$\lim_{s \to 0} \int_0^\infty \dot{x}(t)e^{-st}\, dt = \int_0^\infty \dot{x}(t)\, dt = \lim_{s \to 0}[sX(s) - x(0)] \tag{6.58}$$

Now, if $x(t)$ has a limit $x(\infty)$ as $t \to \infty$ using integration by parts yields

$$\int_0^\infty \dot{x}(t)\, dt = x(\infty) - x(0) \tag{6.59}$$

Combining (6.58) and (6.59) gives (6.56).

The final-value theorem is a very useful property, since the limit as $t \to \infty$ of a time signal $x(t)$ can be computed directly from the Laplace transform $X(s)$. However, care must be used in applying the final-value theorem, since the limit of $sX(s)$ as $s \to 0$ may exist, even though $x(t)$ does not have a limit as $t \to \infty$. For example, suppose that

$$X(s) = \frac{1}{s^2 + 1}$$

Then,

$$\lim_{s \to 0} sX(s) = \lim_{s \to 0} \frac{s}{s^2 + 1} = 0$$

But $x(t) = \sin t$, and $\sin t$ does not have a limit as $t \to \infty$. In the next section it is shown that in many cases of interest, whether or not a signal $x(t)$ has a limit as $t \to \infty$ can be determined by a check of the transform $X(s)$.

For the convenience of the reader, the properties of the Laplace transform are summarized in Table 6.1. Table 6.2 contains a collection of common transform pairs, which includes the transform pairs that were derived in this section, by the properties of the Laplace transform.

6.3   COMPUTATION OF THE INVERSE LAPLACE TRANSFORM

Given a signal $x(t)$ with Laplace transform $X(s)$, we can compute $x(t)$ from $X(s)$ by taking the inverse Laplace transform of $X(s)$. The inverse transform operation is given by

$$x(t) = \frac{1}{2\pi j} \int_{c-j\infty}^{c+j\infty} X(s)e^{st}\, ds \tag{6.60}$$

The integral in (6.60) is evaluated along the path $s = c + j\omega$ in the complex plane from $c - j\infty$ to $c + j\infty$, where c is any real number for which the path $s = c + j\omega$ lies in the region of convergence of $X(s)$. For a detailed treatment of complex integration, see Churchill and Brown [2003].

The integral in (6.60) is usually difficult to evaluate, and thus it is desirable to avoid having to use (6.60) to compute the inverse transform. In this section an algebraic procedure is given for computing the inverse transform in the case when $X(s)$ is a rational function of s. The development begins subsequently with the definition of a rational Laplace transform.

TABLE 6.1   Properties of the Laplace Transform

Property	Transform Pair/Property
Linearity	$ax(t) + bv(t) \leftrightarrow aX(s) + bV(s)$
Right shift in time	$x(t - c)u(t - c) \leftrightarrow e^{-cs}X(s), c > 0$
Time scaling	$x(at) \leftrightarrow \dfrac{1}{a}X\left(\dfrac{s}{a}\right), a > 0$
Multiplication by a power of t	$t^N x(t) \leftrightarrow (-1)^N \dfrac{d^N}{ds^N}X(s), N = 1, 2, \ldots$
Multiplication by an exponential	$e^{at}x(t) \leftrightarrow X(s - a), a$ real or complex
Multiplication by $\sin \omega t$	$x(t) \sin \omega t \leftrightarrow \dfrac{j}{2}[X(s + j\omega) - X(s - j\omega)]$
Multiplication by $\cos \omega t$	$x(t) \cos \omega t \leftrightarrow \dfrac{1}{2}[X(s + j\omega) + X(s - j\omega)]$
Differentiation in the time domain	$\dot{x}(t) \leftrightarrow sX(s) - x(0)$
Second derivative	$\ddot{x}(t) \leftrightarrow s^2 X(s) - sx(0) - \dot{x}(0)$
Nth derivative	$x^{(N)}(t) \leftrightarrow s^N X(s) - s^{N-1}x(0) - s^{N-2}\dot{x}(0) - \cdots - sx^{(N-2)}(0) - x^{(N-1)}(0)$
Integration	$\displaystyle\int_0^t x(\lambda)\, d\lambda \leftrightarrow \dfrac{1}{s}X(s)$
Convolution	$x(t) * v(t) \leftrightarrow X(s)V(s)$
Initial-value theorem	$x(0) = \displaystyle\lim_{s \to \infty} sX(s)$
	$\dot{x}(0) = \displaystyle\lim_{s \to \infty}[s^2 X(s) - sx(0)]$
	$x^{(N)}(0) = \displaystyle\lim_{s \to \infty}[s^{N+1}X(s) - s^N x(0) - s^{N-1}\dot{x}(0) - \cdots - sx^{(N-1)}(0)]$
Final-value theorem	If $\displaystyle\lim_{t \to \infty} x(t)$ exists, then $\displaystyle\lim_{t \to \infty} x(t) = \lim_{s \to 0} sX(s)$

6.3.1   Rational Laplace Transforms

Suppose that $x(t)$ has Laplace transform $X(s)$ with

$$X(s) = \frac{B(s)}{A(s)} \tag{6.61}$$

where $B(s)$ and $A(s)$ are polynomials in the complex variable s given by

$$B(s) = b_M s^M + b_{M-1}s^{M-1} + \cdots + b_1 s + b_0 \tag{6.62}$$

$$A(s) = a_N s^N + a_{N-1}s^{N-1} + \cdots + a_1 s + a_0 \tag{6.63}$$

In (6.62) and (6.63), M and N are positive integers and the coefficients $b_M, b_{M-1}, \ldots, b_1, b_0$ and $a_N, a_{N-1}, \ldots, a_1, a_0$ are real numbers. Assuming that $b_M \neq 0$ and $a_N \neq 0$, the

TABLE 6.2   Common Laplace Transform Pairs

$$u(t) \leftrightarrow \frac{1}{s}$$

$$u(t) - u(t - c) \leftrightarrow \frac{1 - e^{-cs}}{s}, c > 0$$

$$t^N u(t) \leftrightarrow \frac{N!}{s^{N+1}}, N = 1, 2, 3, \ldots$$

$$\delta(t) \leftrightarrow 1$$

$$\delta(t - c) \leftrightarrow e^{-cs}, c > 0$$

$$e^{-bt} u(t) \leftrightarrow \frac{1}{s + b}, b \text{ real or complex}$$

$$t^N e^{-bt} u(t) \leftrightarrow \frac{N!}{(s + b)^{N+1}}, \quad N = 1, 2, 3, \ldots$$

$$(\cos \omega t) u(t) \leftrightarrow \frac{s}{s^2 + \omega^2}$$

$$(\sin \omega t) u(t) \leftrightarrow \frac{\omega}{s^2 + \omega^2}$$

$$(\cos^2 \omega t) u(t) \leftrightarrow \frac{s^2 + 2\omega^2}{s(s^2 + 4\omega^2)}$$

$$(\sin^2 \omega t) u(t) \leftrightarrow \frac{2\omega^2}{s(s^2 + 4\omega^2)}$$

$$(e^{-bt} \cos \omega t) u(t) \leftrightarrow \frac{s + b}{(s + b)^2 + \omega^2}$$

$$(e^{-bt} \sin \omega t) u(t) \leftrightarrow \frac{\omega}{(s + b)^2 + \omega^2}$$

$$(t \cos \omega t) u(t) \leftrightarrow \frac{s^2 - \omega^2}{(s^2 + \omega^2)^2}$$

$$(t \sin \omega t) u(t) \leftrightarrow \frac{2\omega s}{(s^2 + \omega^2)^2}$$

$$(te^{-bt} \cos \omega t) u(t) \leftrightarrow \frac{(s + b)^2 - \omega^2}{[(s + b)^2 + \omega^2]^2}$$

$$(te^{-bt} \sin \omega t) u(t) \leftrightarrow \frac{2\omega(s + b)}{[(s + b)^2 + \omega^2]^2}$$

degree of the polynomial $B(s)$ is equal to M, and the degree of the polynomial $A(s)$ is equal to N. The polynomial $B(s)$ is the "numerator polynomial" of $X(s)$, and $A(s)$ is the "denominator polynomial" of $X(s)$. It is always assumed that $B(s)$ and $A(s)$ do not have any common factors. If there are common factors, they should be divided out.

The transform $X(s) = B(s)/A(s)$ with $B(s)$ and $A(s)$ given by (6.62) and (6.63) is said to be a *rational function of s*, since it is a ratio of polynomials in s. The degree N of the denominator polynomial $A(s)$ is called the *order* of the rational function. For a large class of signals $x(t)$, the Laplace transform $X(s)$ is rational. For example, most of the signals in Table 6.2 have a rational Laplace transform. An exception is the c-second rectangular pulse $u(t) - u(t - c)$, whose Laplace transform is

$$\frac{1 - e^{-cs}}{s}$$

Due to the presence of the complex exponential e^{-cs}, this transform cannot be expressed as a ratio of polynomials in s, and thus the transform of the rectangular pulse is not a rational function of s.

Given a rational transform $X(s) = B(s)/A(s)$, let $p_1, p_2, \ldots, p_N$ denote the roots of the equation

$$A(s) = 0$$

Then $A(s)$ can be written in the factored form

$$A(s) = a_N(s - p_1)(s - p_2) \cdots (s - p_N) \tag{6.64}$$

The roots $p_1, p_2, \ldots, p_N$, which may be real or complex, are also said to be the *zeros* of the polynomial $A(s)$, since $A(s)$ is equal to zero when s is set equal to p_i for any value of i ranging from 1 to N. Note that if any one of the zeros (say, p_1) is complex, there must be another zero that is equal to the complex conjugate of p_1. In other words, complex zeros always appear in complex-conjugate pairs.

MATLAB can be used to find the zeros of a polynomial $A(s)$, by the command `roots`. For example, to find the zeros of

$$A(s) = s^3 + 4s^2 + 6s + 4 = 0$$

use the commands

```
A = [1 4 6 4]; % store the coefficients of A(s)
p = roots(A)
```

MATLAB returns the zeros:

```
p =
    -2.0000
    -1.0000 + 1.0000i
    -1.0000 - 1.0000i
```

Hence, $A(s)$ has the factored form

$$A(s) = (s + 2)(s + 1 - j)(s + 1 + j)$$

Now, given the rational transform $X(s) = B(s)/A(s)$, if $A(s)$ is specified by the factored form (6.64), the result is

$$X(s) = \frac{B(s)}{a_N(s - p_1)(s - p_2)\cdots(s - p_N)} \tag{6.65}$$

The p_i for $i = 1, 2, \ldots, N$ are called the *poles* of the rational function $X(s)$, since if the value $s = p_i$ is inserted into $X(s)$, the result is ∞. So, the poles of the rational function $X(s)$ are equal to the zeros (or roots) of the denominator polynomial $A(s)$.

We can compute the inverse Laplace transform of $X(s)$ by first carrying out a partial fraction expansion of (6.65). The procedure is described in the next subsection. In the development that follows it is assumed that $M < N$, that is, the degree of $B(s)$ is strictly less than the degree of $A(s)$. Such a rational function is said to be *strictly proper in s*. The case when $X(s)$ is not strictly proper is considered later.

6.3.2   Distinct Poles

The poles $p_1, p_2, \ldots, p_N$ of $X(s)$ are now assumed to be distinct (or nonrepeated); that is, $p_i \neq p_j$ when $i \neq j$. Then $X(s)$ has the partial fraction expansion

$$X(s) = \frac{c_1}{s - p_1} + \frac{c_2}{s - p_2} + \cdots + \frac{c_N}{s - p_N} \tag{6.66}$$

where

$$c_i = [(s - p_i)X(s)]_{s=p_i}, \quad i = 1, 2, \ldots, N \tag{6.67}$$

The expression (6.67) for the c_i can be verified by first multiplying both sides of (6.66) by $s - p_i$. This gives

$$(s - p_i)X(s) = c_i + \sum_{\substack{r=1 \\ r \neq i}}^{N} c_r \frac{s - p_i}{s - p_r} \tag{6.68}$$

Evaluating both sides of (6.68) at $s = p_i$ eliminates all terms inside the summation, which yields (6.67).

The constants c_i in (6.66) are called the *residues*, and the computation of the c_i by (6.67) is called the *residue method*. The constant c_i is real if the corresponding pole p_i is real. In addition, since the poles $p_1, p_2, \ldots, p_N$ appear in complex-conjugate pairs, the c_i must also appear in complex-conjugate pairs. Hence, if c_i is complex, one of the other constants must be equal to the complex conjugate of c_i.

It is worth noting that to compute the partial fraction expansion (6.66), it is not necessary to factor the numerator polynomial $B(s)$. However, it is necessary to compute the poles of $X(s)$, since the expansion is given directly in terms of the poles.

We can then determine the inverse Laplace transform $x(t)$ of $X(s)$ by taking the inverse transform of each term in (6.66) and using linearity of the inverse transform operation. The result is

$$x(t) = c_1 e^{p_1 t} + c_2 e^{p_2 t} + \cdots + c_N e^{p_N t}, \quad t \geq 0 \tag{6.69}$$

It is very important to note that the form of the time variation of the function $x(t)$ given by (6.69) is determined by the poles of the rational function $X(s)$; more precisely, $x(t)$ is a sum of exponentials in time whose exponents are completely specified in terms of the poles of $X(s)$. As a consequence, it is the poles of $X(s)$ that determine the characteristics of the time variation of $x(t)$. This fundamental result will be utilized extensively in Chapter 8 in the study of system behavior.

It should also be noted that if all the p_i are real, the terms that make up the function $x(t)$ defined by (6.69) are all real. However, if two or more of the p_i are complex, the corresponding terms in (6.69) will be complex, and thus in this case the complex terms must be combined to obtain a real form. This will be considered after the next example.

Given the rational function $X(s) = B(s)/A(s)$ with the polynomials $B(s)$ and $A(s)$ as previously defined, the MATLAB software can be used to compute the residues and the poles of $X(s)$. The commands are as follows:

```
num = [b_M b_{M-1} ... b_1 b_0];
den = [a_N a_{N-1} ... a_1 a_0];
[r,p] = residue(num,den);
```

The MATLAB program will then produce a vector r consisting of the residues and a vector p consisting of the corresponding poles. The process is illustrated in the example that follows.

Example 6.17   Distinct Pole Case

Suppose that

$$X(s) = \frac{s + 2}{s^3 + 4s^2 + 3s}$$

Here,

$$A(s) = s^3 + 4s^2 + 3s = s(s + 1)(s + 3)$$

The roots of $A(s) = 0$ are $0, -1, -3$, and thus the poles of $X(s)$ are $p_1 = 0$, $p_2 = -1$, $p_3 = -3$. Therefore,

$$X(s) = \frac{c_1}{s - 0} + \frac{c_2}{s - (-1)} + \frac{c_3}{s - (-3)}$$

$$X(s) = \frac{c_1}{s} + \frac{c_2}{s + 1} + \frac{c_3}{s + 3}$$

where

$$c_1 = [sX(s)]_{s=0} = \frac{s + 2}{(s + 1)(s + 3)}\bigg|_{s=0} = \frac{2}{3}$$

$$c_2 = [(s + 1)X(s)]_{s=-1} = \frac{s + 2}{s(s + 3)}\bigg|_{s=-1} = \frac{1}{-2}$$

$$c_3 = [(s + 3)X(s)]_{s=-3} = \frac{s + 2}{s(s + 1)}\bigg|_{s=-3} = \frac{-1}{6}$$

Hence, the inverse Laplace transform $x(t)$ of $X(s)$ is given by

$$x(t) = \frac{2}{3} - \frac{1}{2}e^{-t} - \frac{1}{6}e^{-3t}, \quad t \geq 0$$

The computation of the residues and poles can be checked by the MATLAB commands

```
num = [1 2];
den = [1 4 3 0];
[r,p] = residue(num,den);
```

The MATLAB program produces the vectors

```
r =                          p =
   -0.1667                      -3
   -0.5000                      -1
    0.6667                       0
```

which checks with the results obtained previously.

6.3.3   Distinct Poles with Two or More Poles Complex

It is still assumed that the poles of $X(s)$ are distinct, but now two or more of the poles of $X(s)$ are complex so that the corresponding exponentials in (6.69) are complex. As will be shown, it is possible to combine the complex terms in order to express $x(t)$ in real form.

Suppose that $p_1 = \sigma + j\omega$ is complex, so that $\omega \neq 0$. Then the complex conjugate $\overline{p}_1 = \sigma - j\omega$ is another pole of $X(s)$. Let p_2 denote this pole. Then the residue c_2 corresponding to the pole p_2 is equal to the conjugate $\overline{c}_1$ of the residue corresponding to the pole p_1, and $X(s)$ has the partial fraction expansion

$$X(s) = \frac{c_1}{s - p_1} + \frac{\overline{c}_1}{s - \overline{p}_1} + \frac{c_3}{s - p_3} + \cdots + \frac{c_N}{s - p_N}$$

where $c_1, c_3, \ldots, c_N$ are again given by (6.67). Hence, the inverse transform is

$$x(t) = c_1 e^{p_1 t} + \overline{c}_1 e^{\overline{p}_1 t} + c_3 e^{p_3 t} + \cdots + c_N e^{p_N t} \tag{6.70}$$

Now, the first two terms on the right-hand side of (6.70) can be expressed in real form as follows:

$$c_1 e^{p_1 t} + \overline{c}_1 e^{\overline{p}_1 t} = 2|c_1| e^{\sigma t} \cos(\omega t + \angle c_1) \tag{6.71}$$

Here, $|c_1|$ is the magnitude of the complex number c_1, and $\angle c_1$ is the angle of c_1. The verification of the relationship (6.71) is considered in the homework problems. (See Problem 6.9.)

Using (6.71), we find that the inverse transform of $X(s)$ is given by

$$x(t) = 2|c_1| e^{\sigma t} \cos(\omega t + \angle c_1) + c_3 e^{p_3 t} + \cdots + c_N e^{p_N t} \tag{6.72}$$

The expression (6.72) for $x(t)$ shows that if $X(s)$ has a pair of complex poles $p_1, p_2 = \sigma \pm j\omega$, the signal $x(t)$ contains a term of the form

$$ce^{\sigma t}\cos(\omega t + \theta)$$

Note that the coefficient σ of t in the exponential function is the real part of the pole $p_1 = \sigma + j\omega$, and the frequency ω of the cosine is equal to the imaginary part of pole p_1.

The computation of the inverse Laplace transform by (6.71) is illustrated by the following example:

Example 6.18   Complex Pole Case

Suppose that

$$X(s) = \frac{s^2 - 2s + 1}{s^3 + 3s^2 + 4s + 2}$$

Here,

$$A(s) = s^3 + 3s^2 + 4s + 2 = (s + 1 - j)(s + 1 + j)(s + 1)$$

The roots of $A(s) = 0$ are

$$p_1 = -1 + j, \qquad p_2 = -1 - j, \qquad p_3 = -1$$

Thus, $\sigma = -1$ and $\omega = 1$, and

$$X(s) = \frac{c_1}{s - (-1 + j)} + \frac{\bar{c}_1}{s - (-1 - j)} + \frac{c_3}{s - (-1)}$$

$$X(s) = \frac{c_1}{s + 1 - j} + \frac{\bar{c}_1}{s + 1 + j} + \frac{c_3}{s + 1}$$

where

$$c_1 = [(s + 1 - j)X(s)]_{s=-1+j} = \left.\frac{s^2 - 2s + 1}{(s + 1 + j)(s + 1)}\right|_{s=-1+j}$$

$$= \frac{-3}{2} + j2$$

$$c_3 = [(s + 1)X(s)]_{s=-1} = \left.\frac{s^2 - 2s + 1}{s^2 + 2s + 2}\right|_{s=-1} = 4$$

Now,

$$|c_1| = \sqrt{\frac{9}{4} + 4} = \frac{5}{2}$$

and since c_1 lies in the second quadrant, then

$$\angle c_1 = 180° + \tan^{-1}\frac{-4}{3} = 126.87°$$

Then, using (6.71) and (6.72) gives

$$x(t) = 5e^{-t}\cos(t + 126.87°) + 4e^{-t}, \quad t \geq 0$$

MATLAB will generate the residues and poles in the case when $X(s)$ has complex (and real) poles. In this example, the commands are

```
num = [1 -2 1];
den = [1 3 4 2];
[r,p] = residue(num,den);
```

which yield

```
r =                              p =
    -1.5000 + 2.0000i               -1.0000 + 1.0000i
    -1.5000 - 2.0000i               -1.0000 - 1.0000i
     4.0000                         -1.0000
```

This matches the poles and residues calculated previously.

When $X(s)$ has complex poles, we can avoid having to work with complex numbers by not factoring quadratic terms whose zeros are complex. For example, suppose that $X(s)$ is the second-order rational function given by

$$X(s) = \frac{b_1 s + b_0}{s^2 + a_1 s + a_0}$$

"Completing the square" in the denominator of $X(s)$ gives

$$X(s) = \frac{b_1 s + b_0}{(s + a_1/2)^2 + a_0 - a_1^2/4}$$

It follows from the quadratic formula that the poles of $X(s)$ are complex if and only if

$$a_0 - \frac{a_1^2}{4} > 0$$

in which case the poles of $X(s)$ are

$$p_1, p_2 = -\frac{a_1}{2} \pm j\omega$$

where

$$\omega = \sqrt{a_0 - \frac{a_1^2}{4}}$$

With $X(s)$ expressed in the form

$$X(s) = \frac{b_1 s + b_0}{(s + a_1/2)^2 + \omega^2} = \frac{b_1(s + a_1/2) + (b_0 - b_1 a_1/2)}{(s + a_1/2)^2 + \omega^2}$$

the inverse Laplace transform can be computed by table lookup. This is illustrated by the following example:

Example 6.19   Completing the Square

Suppose that

$$X(s) = \frac{3s + 2}{s^2 + 2s + 10}$$

Completing the square in the denominator of $X(s)$ gives

$$X(s) = \frac{3s + 2}{(s + 1)^2 + 9}$$

Then, since $9 > 0$, the poles of $X(s)$ are complex and are equal to $-1 \pm j3$. (Here, $\omega = 3$.) Now, $X(s)$ can be expressed in the form

$$X(s) = \frac{3(s + 1) - 1}{(s + 1)^2 + 9} = \frac{3(s + 1)}{(s + 1)^2 + 9} - \frac{1}{3}\frac{3}{(s + 1)^2 + 9}$$

and when the transform pairs in Table 6.2 are used, the inverse transform is

$$x(t) = 3e^{-t}\cos 3t - \frac{1}{3}e^{-t}\sin 3t, \quad t \geq 0$$

Finally, using the trigonometric identity

$$C\cos \omega t - D\sin \omega t = \sqrt{C^2 + D^2}\cos(\omega t + \theta), \quad \text{where } \theta = \begin{cases} \tan^{-1}\dfrac{D}{C}, C \geq 0 \\ \pi + \tan^{-1}\dfrac{D}{C}, C < 0 \end{cases} \quad (6.73)$$

we see that $x(t)$ can be written in the form

$$x(t) = ce^{-t}\cos(3t + \theta), t \geq 0$$

where

$$c = \sqrt{(3)^2 + \left(\tfrac{1}{3}\right)^2} = 3.018$$

and

$$\theta = \tan^{-1}\frac{1/3}{3} = 83.7°$$

Now, suppose that $X(s)$ has a pair of complex poles $p_1, p_2 = \sigma \pm j\omega$ and real distinct poles $p_3, p_4, \ldots, p_N$. Then,

$$X(s) = \frac{B(s)}{[(s - \sigma)^2 + \omega^2](s - p_3)(s - p_4)\cdots(s - p_N)}$$

which can be expanded into the form

$$X(s) = \frac{b_1 s + b_0}{(s - \sigma)^2 + \omega^2} + \frac{c_3}{s - p_3} + \frac{c_4}{s - p_4} + \cdots + \frac{c_N}{s - p_N} \qquad (6.74)$$

where the coefficients b_0 and b_1 of the second-order term are real numbers. The residues $c_3, c_4, \ldots, c_N$ are real numbers and are computed from (6.67) as before; however, b_0 and b_1 cannot be calculated from this formula. We can compute the constants b_0 and b_1 by putting the right-hand side of (6.74) over a common denominator and then equating the coefficients of the resulting numerator with the numerator of $X(s)$. The inverse Laplace transform can then be computed from (6.74). The process is illustrated by the following example:

Example 6.20   Equating Coefficients

Again, consider the rational function $X(s)$ in Example 6.18 given by

$$X(s) = \frac{s^2 - 2s + 1}{s^3 + 3s^2 + 4s + 2}$$

In this example,

$$A(s) = (s^2 + 2s + 2)(s + 1) = [(s + 1)^2 + 1](s + 1)$$

and thus $X(s)$ has the expansion

$$X(s) = \frac{b_1 s + b_0}{(s + 1)^2 + 1} + \frac{c_3}{s + 1} \qquad (6.75)$$

where

$$c_3 = [(s + 1)X(s)]_{s=-1} = 4$$

The right-hand side of (6.75) can be put over a common denominator, and then the resulting numerator can be equated to the numerator of $X(s)$. This yields

$$X(s) = \frac{(b_1 s + b_0)(s + 1) + 4[(s + 1)^2 + 1]}{[(s + 1)^2 + 1](s + 1)} = \frac{s^2 - 2s + 1}{s^3 + 3s^2 + 4s + 2}$$

and equating numerators yields

$$s^2 - 2s + 1 = (b_1 s + b_0)(s + 1) + 4(s^2 + 2s + 2)$$

$$s^2 - 2s + 1 = (b_1 + 4)s^2 + (b_1 + b_0 + 8)s + b_0 + 8$$

Hence,

$$b_1 + 4 = 1$$

and

$$b_0 + 8 = 1$$

which implies that $b_1 = -3$ and $b_0 = -7$. Thus, from (6.75),

$$X(s) = \frac{-3s - 7}{(s + 1)^2 + 1} + \frac{4}{s + 1}$$

Writing $X(s)$ in the form

$$X(s) = \frac{-3(s + 1)}{(s + 1)^2 + 1} - 4\frac{1}{(s + 1)^2 + 1} + \frac{4}{s + 1}$$

and using Table 6.2 result in the following inverse transform:

$$x(t) = -3e^{-t}\cos t - 4e^{-t}\sin t + 4e^{-t}, \quad t \geq 0$$

Finally, using the trigonometric identity (6.73) yields

$$x(t) = 5e^{-t}\cos(t + 126.87°) + 4e^{-t}, \quad t \geq 0$$

which agrees with the result obtained in Example 6.18.

6.3.4   Repeated Poles

Again, consider the general case where

$$X(s) = \frac{B(s)}{A(s)}$$

It is still assumed that $X(s)$ is strictly proper; that is, the degree M of $B(s)$ is strictly less than the degree N of $A(s)$. Now, suppose that pole p_1 of $X(s)$ is repeated r times and the other $N - r$ poles (denoted by $p_{r+1}, p_{r+2}, \ldots, p_N$) are distinct. Then, $X(s)$ has the partial fraction expansion

$$X(s) = \frac{c_1}{s - p_1} + \frac{c_2}{(s - p_1)^2} + \cdots + \frac{c_r}{(s - p_1)^r}$$

$$+ \frac{c_{r+1}}{s - p_{r+1}} + \cdots + \frac{c_N}{s - p_N} \tag{6.76}$$

In (6.76), the residues $c_{r+1}, c_{r+2}, \ldots, c_N$ are calculated as in the distinct-pole case; that is,

$$c_i = [(s - p_i)X(s)]_{s=p_i}, \quad i = r + 1, r + 2, \ldots, N$$

The constant c_r is given by

$$c_r = [(s - p_1)^r X(s)]_{s=p_1}$$

and the constants $c_1, c_2, \ldots, c_{r-1}$ are given by

$$c_{r-i} = \frac{1}{i!}\left[\frac{d^i}{ds^i}[(s - p_1)^r X(s)]\right]_{s=p_1}, \quad i = 1, 2, \ldots, r - 1 \tag{6.77}$$

In particular, setting the index i equal to 1, 2 in (6.77) gives

$$c_{r-1} = \left[\frac{d}{ds}[(s - p_1)^r X(s)]\right]_{s=p_1}$$

$$c_{r-2} = \frac{1}{2}\left[\frac{d^2}{ds^2}[(s - p_1)^r X(s)]\right]_{s=p_1}$$

We can also compute the constants $c_1, c_2, \ldots, c_{r-1}$ in (6.76) by putting the right-hand side of (6.76) over a common denominator and then equating coefficients of the resulting numerator with the numerator of $X(s)$.

If the poles of $X(s)$ are all real numbers, the inverse Laplace transform can then be determined by the transform pairs

$$\frac{t^{N-1}}{(N-1)!}e^{-at} \leftrightarrow \frac{1}{(s+a)^N}, \quad N = 1, 2, 3, \ldots$$

The process is illustrated by the following example:

Example 6.21   Repeated Poles

Consider the rational function

$$X(s) = \frac{5s - 1}{s^3 - 3s - 2}$$

The roots of $A(s) = 0$ are $-1, -1, 2$, so $r = 2$, and therefore the partial fraction expansion has the form

$$X(s) = \frac{c_1}{s + 1} + \frac{c_2}{(s + 1)^2} + \frac{c_3}{s - 2}$$

where

$$c_1 = \left[\frac{d}{ds}[(s + 1)^2 X(s)]\right]_{s=-1} = \left[\frac{d}{ds}\frac{5s - 1}{s - 2}\right]_{s=-1}$$

$$= \frac{-9}{(s - 2)^2}\bigg|_{s=-1} = -1$$

$$c_2 = [(s + 1)^2 X(s)]_{s=-1} = \frac{5s - 1}{s - 2}\bigg|_{s=-1} = 2$$

$$c_3 = [(s - 2)X(s)]_{s=2} = \frac{5s - 1}{(s + 1)^2}\bigg|_{s=2} = 1$$

Thus,

$$x(t) = -e^{-t} + 2te^{-t} + e^{2t}, \quad t \geq 0$$

Instead of having to differentiate with respect to s, we can compute the constant c_1 by putting the partial fraction expansion

$$X(s) = \frac{c_1}{s+1} + \frac{2}{(s+1)^2} + \frac{1}{s-2}$$

over a common denominator and then equating numerators. This yields

$$5s - 1 = c_1(s+1)(s-2) + 2(s-2) + (s+1)^2$$
$$5s - 1 = (c_1 + 1)s^2 + (-c_1 + 4)s + (-2c_1 - 3)$$

Hence, $c_1 = -1$, which is consistent with the preceding value.

MATLAB can also handle the case of repeated roots. In this example, the MATLAB commands are

```
num = [5 -1];
den = [1 0 -3 -2];
[r,p] = residue(num,den);
```

which yields

```
r =                         p =
    1.0000                      2.0000
   -1.0000                     -1.0000
    2.0000                     -1.0000
```

where the second residue, -1, corresponds to the $1/(s+1)$ term and the third residue, 2, corresponds to the $1/(s+1)^2$ term.

If $X(s)$ has repeated complex poles, we can avoid having to use complex arithmetic by expressing the complex part of $A(s)$ in terms of powers of quadratic terms. This solution technique is illustrated via the following example:

Example 6.22   Powers of Quadratic Terms

Suppose that

$$X(s) = \frac{s^3 + 3s^2 - s + 1}{s^5 + s^4 + 2s^3 + 2s^2 + s + 1}$$

Using the MATLAB command roots reveals that the poles of $X(s)$ are equal to $-1, -j, -j, j, j$. Thus, there are a pair of repeated complex poles corresponding to the factor

$$[(s+j)(s-j)]^2 = (s^2+1)^2$$

Therefore,

$$X(s) = \frac{s^3 + 3s^2 - s + 1}{(s^2+1)^2(s+1)}$$

and the expansion of $X(s)$ has the form

$$X(s) = \frac{cs+d}{s^2+1} + \frac{w(s)}{(s^2+1)^2} + \frac{c_5}{s+1}$$

where $w(s)$ is a polynomial in s. To be able to compute the inverse Laplace transform of $w(s)/(s^2 + 1)^2$ from the transform pairs in Table 6.2, it is necessary to write this term in the form

$$\frac{w(s)}{(s^2 + 1)^2} = \frac{g(s^2 - 1) + hs}{(s^2 + 1)^2}$$

for some real constants g and h. Table 6.2 yields the transform pair

$$\left(gt \cos t + \frac{h}{2}t \sin t \right)u(t) \leftrightarrow \frac{g(s^2 - 1) + hs}{(s^2 + 1)^2}$$

Now,

$$c_5 = [(s + 1)X(s)]_{s=-1} = \left.\frac{s^3 + 3s^2 - s + 1}{(s^2 + 1)^2}\right|_{s=-1} = 1$$

So,

$$X(s) = \frac{cs + d}{s^2 + 1} + \frac{g(s^2 - 1) + hs}{(s^2 + 1)^2} + \frac{1}{s + 1}$$

Putting the right-hand side over a common denominator and equating numerators give

$$s^3 + 3s^2 - s + 1 = (cs + d)(s^2 + 1)(s + 1)$$
$$+ [g(s^2 - 1) + hs](s + 1) + (s^2 + 1)^2$$
$$s^3 + 3s^2 - s + 1 = (c + 1)s^4 + (c + d + g)s^3 + (c + d + h + g + 2)s^2$$
$$+ (c + d - g + h)s + d - g + 1$$

Equating coefficients of the polynomials gives

$$c = -1$$
$$c + d + g = 1$$
$$c + d + h + g + 2 = 3$$
$$c + d - g + h = -1$$
$$d - g + 1 = 1$$

Solving these equations gives $d = 1$, $g = 1$, and $h = 0$, and thus the inverse transform is

$$x(t) = -\cos t + \sin t + t \cos t + e^{-t}, t \geq 0$$

6.3.5   Case when $M \geq N$

Consider the rational function $X(s) = B(s)/A(s)$, with the degree of $B(s)$ equal to M and the degree of $A(s)$ equal to N. If $M \geq N$, by long division, $X(s)$ can be written in the form

$$X(s) = Q(s) + \frac{R(s)}{A(s)}$$

where the quotient $Q(s)$ is a polynomial in s with degree $M - N$, and the remainder $R(s)$ is a polynomial in s with degree strictly less than N. The computation of the quotient $Q(s)$ and remainder $R(s)$ can be accomplished with MATLAB. In particular, the command conv can be used to multiply polynomials, and deconv used to divide polynomials. For example, consider a denominator polynomial given in factored form

$$A(s) = (s^2 + 3s)(s + 4)$$

Then, the command

```
den = conv([1 3 0],[1 4]);
```

multiplies the factors yielding den = [1 7 12 0], and thus $A(s) = s^3 + 7s^2 + 12s$. The command deconv is used in the following manner in order to divide polynomials that have degree $M > N$:

```
num = [b_M  b_{M-1} ... b_1 b_0];
den = [a_N a_{N-1} ... a_1 a_0];
[Q,R] = deconv(num,den)
```

Once $Q(s)$ and $R(s)$ are determined, we can then compute the inverse Laplace transform of $X(s)$ by determining the inverse Laplace transform of $Q(s)$ and the inverse Laplace transform of $R(s)/A(s)$. Since $R(s)/A(s)$ is strictly proper [i.e., degree $R(s) < N$], we can compute the inverse transform of $R(s)/A(s)$ by first expanding into partial fractions as given before. The residue command can be used to perform partial fraction expansion on $R(s)/A(s)$, as was done in the prior examples. The inverse transform of the quotient $Q(s)$ can be computed by the transform pair

$$\frac{d^N}{dt^N}\delta(t) \leftrightarrow s^N, \quad N = 1, 2, 3, \ldots$$

The process is illustrated by the following example:

Example 6.23 $M = 3, N = 2$

Suppose that

$$X(s) = \frac{s^3 + 2s - 4}{s^2 + 4s - 2}$$

Using the MATLAB commands

```
num = [1 0 2 -4];
den = [1 4 -2];
[Q,R] = deconv(num,den)
```

gives Q = [1 -4] and R = [20 -12]. Thus, the quotient is $Q(s) = s - 4$, and the remainder is $R(s) = 20s - 12$. Then,

$$X(s) = s - 4 + \frac{20s - 12}{s^2 + 4s - 2}$$

and thus

$$x(t) = \frac{d}{dt}\delta(t) - 4\delta(t) + v(t)$$

where $v(t)$ is the inverse Laplace transform of

$$V(s) = \frac{20s - 12}{s^2 + 4s - 2}$$

Using the MATLAB commands

```
num = [20 -12];
den = [1 4 -2];
[r,p] = residue(num,den);
```

results in the following residues and poles for $V(s)$:

```
r =                          p =
    20.6145                      -4.4495
    -0.6145                       0.4495
```

Hence, the partial fraction expansion of $V(s)$ is

$$V(s) = \frac{20.6145}{s + 4.4495} - \frac{0.6145}{s - 0.4495}$$

and the inverse Laplace transform of $V(s)$ is

$$v(t) = 20.6145e^{-4.4495t} - 0.6145e^{0.4495t}, \; t \geq 0$$

6.3.6   Pole Locations and the Form of a Signal

Given a signal $x(t)$ with rational Laplace transform $X(s) = B(s)/A(s)$, again suppose that $M < N$, where N is the degree of $A(s)$ and M is the degree of $B(s)$. As seen from the previous development, there is a direct relationship between the poles of $X(s)$ and the form of the signal $x(t)$. In particular, if $X(s)$ has a nonrepeated pole p that is real, then $x(t)$ contains a term of the form ce^{pt} for some constant c; and if the pole p is repeated twice, then $x(t)$ contains the term $c_1e^{pt} + c_2te^{pt}$ for some constants c_1 and c_2. If $X(s)$ has a nonrepeated pair $\sigma \pm j\omega$ of complex poles, then $x(t)$ contains a term of the form $ce^{\sigma t}\cos(\omega t + \theta)$ for some constants c and θ. If the complex pair $\sigma \pm j\omega$ is repeated twice, $x(t)$ contains the term $c_1e^{\sigma t}\cos(\omega t + \theta_1) + c_2te^{\sigma t}\cos(\omega t + \theta_2)$ for some constants c_1, c_2, θ_1, and θ_2.

As a result of these relationships, the form of a signal $x(t)$ can be determined directly from the poles of the transform $X(s)$. This is illustrated by the following example:

Example 6.24   General Form of a Signal

Consider the signal $x(t)$ with transform

$$X(s) = \frac{1}{s(s + 1)(s - 4)^2((s + 2)^2 + 3^2)}$$

The poles of $X(s)$ are $0, -1, 4, 4, -2 \pm 3j$, and thus the general form of the time signal $x(t)$ is

$$x(t) = c_1 + c_2 e^{-t} + c_3 e^{4t} + c_4 t e^{4t} + c_5 e^{-2t} \cos(3t + \theta)$$

where the c_i's and θ are constants. Note that the pole at the origin corresponds to the constant term (given by $c_1 e^{0t}$), the pole at -1 corresponds to the term $c_2 e^{-t}$, the repeated poles at -4 correspond to the $c_3 e^{4t}$ and $c_4 t e^{4t}$ terms, and the complex poles $-2 \pm 3j$ correspond to the term $c_5 e^{-2t} \cos(3t + \theta)$.

It is important to note that modifying the numerator of $X(s)$ does not change the general form of $x(t)$, but it will change the values of the constants (the coefficients of terms). For instance, in Example 6.24, if the numerator were changed from $B(s) = 1$ to any polynomial $B(s)$ with degree less than the denominator $A(s)$, the form of $x(t)$ remains the same.

A very important consequence of the relationship between the poles of $X(s)$ and the form of the signal $x(t)$ is that the behavior of the signal in the limit as $t \to \infty$ can be determined directly from the poles. In particular, it follows from the results given previously that $x(t)$ converges to 0 as $t \to \infty$ if and only if the poles $p_1, p_2, \ldots, p_N$ all have real parts that are strictly less than zero; or in mathematical terms,

$$\operatorname{Re}(p_i) < 0 \text{ for } i = 1, 2, \ldots, N \tag{6.78}$$

where $\operatorname{Re}(p_i)$ denotes the real part of the pole p_i.

It also follows from the relationship between poles and the form of the signal that $x(t)$ has a limit as $t \to \infty$ if and only if (6.78) is satisfied, except that one of the poles of $X(s)$ may be at the origin ($p = 0$). If $X(s)$ has a single pole at $s = 0$ [and all other poles satisfy (6.78)], the limiting value of $x(t)$ is equal to the value of the residue corresponding to the pole at 0; that is,

$$\lim_{t \to \infty} x(t) = [sX(s)]_{s=0}$$

This result is consistent with the final-value theorem [see (6.56)], since for any rational function $X(s)$,

$$\lim_{s \to 0} sX(s) = [sX(s)]_{s=0}$$

Example 6.25   Limiting Value

Suppose that

$$X(s) = \frac{2s^2 - 3s + 4}{s^3 + 3s^2 + 2s}$$

The poles of $X(s)$ are $0, -1, -2$, and thus (6.78) is satisfied for all poles of $X(s)$ except for one pole, which is equal to zero. Thus $x(t)$ has a limit at $t \to \infty$ and

$$\lim_{t \to \infty} x(t) = [sX(s)]_{s=0} = \left[\frac{2s^2 - 3s + 4}{s^2 + 3s + 2} \right]_{s=0} = \frac{4}{2} = 2$$

6.3.7 Inverse Laplace Transform Computation by the Use of Symbolic Manipulation

Given a rational transform $X(s) = B(s)/A(s)$ with the degree of $B(s)$ less than the degree of $A(s)$, the Symbolic Math Toolbox in MATLAB can be used to compute (and plot) the inverse Laplace transform $x(t)$. If X is a symbolic function of s, then the command `ilaplace(X)` returns the inverse Laplace transform of X. The command `ezplot(x)` plots $x(t)$.

Example 6.26   Use of MATLAB

Consider the Laplace transform given in Example 6.17:

$$X(s) = \frac{s + 2}{s^3 + 4s^2 + 3s}$$

To compute $x(t)$, use the commands

```
syms X s x
X = (s+2)/(s^3+4*s^2+3*s);
x = ilaplace(X)
```

The resulting expression for $x(t)$ is given by

```
x =
-1/6*exp(-3*t)-1/2*exp(-t)+2/3
```

This is consistent with the result found analytically in Example 6.17:

$$x(t) = \tfrac{2}{3} - \tfrac{1}{2}e^{-t} - \tfrac{1}{6}e^{-3t}, \quad t \geq 0$$

To plot the function $x(t)$ from $t = 0$ to $t = 10$, use the command

```
ezplot(x,[0,10])
```

The resulting plot is shown in Figure 6.3. From both the plot in Figure 6.3 and the preceding expression for $x(t)$, it is clear that $x(t)$ converges to the value $\tfrac{2}{3}$ as $t \rightarrow \infty$. Convergence of $x(t)$ to the value $\tfrac{2}{3}$ can also be verified by use of the final-value theorem, which is applicable here since the poles of $X(s)$ have negative real parts (except for the pole at $p = 0$). Applying the final-value theorem gives

$$\lim_{t \to \infty} x(t) = \lim_{s \to 0} sX(s)$$

$$= [sX(s)]_{s=0}$$

$$= \left[\frac{s + 2}{s^2 + 4s + 3} \right]_{s=0}$$

$$\lim_{t \to \infty} x(t) = \frac{2}{3}$$

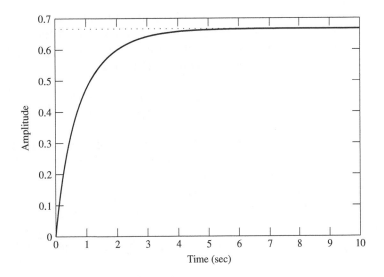

FIGURE 6.3
Plot of the inverse transform $x(t)$ in Example 6.26.

6.3.8   Transforms Containing Exponentials

In many cases of interest, a function $x(t)$ will have a transform $X(s)$ of the form

$$X(s) = \frac{B_0(s)}{A_0(s)} + \frac{B_1(s)}{A_1(s)}\exp(-h_1 s) + \cdots + \frac{B_q(s)}{A_q(s)}\exp(-h_q s) \qquad (6.79)$$

In (6.79) the h_i are distinct positive real numbers, the $A_i(s)$ are polynomials in s with real coefficients, and the $B_i(s)$ are polynomials in s with real coefficients. Here, it is assumed that $B_i(s) \neq 0$ for at least one value of $i \geq 1$.

The function $X(s)$ given by (6.79) is not rational in s. In other words, it is not possible to express $X(s)$ as a ratio of polynomials in s with real coefficients. This is a result of the presence of the exponential terms $\exp(-h_i s)$, which cannot be written as ratios of polynomials in s. Functions of the form (6.79) are examples of *irrational functions* of s. They are also called *transcendental functions* of s.

Functions $X(s)$ of the form (6.79) arise when the Laplace transform is applied to a piecewise-continuous function $x(t)$. For instance, as shown in Example 6.5 the transform of the c-second pulse $u(t) - u(t - c)$ is equal to

$$\frac{1}{s} - \frac{1}{s}e^{-cs}$$

Clearly, this transform is in the form (6.79). Take

$$B_0(s) = 1, \qquad B_1(s) = -1, \qquad A_0(s) = A_1(s) = s, \qquad h_1 = c$$

Since $X(s)$ given by (6.79) is not rational in s, it is not possible to apply the partial fraction expansion directly to (6.79). However, partial fraction expansions can still be used to compute the inverse transform of $X(s)$. The procedure is as follows:

First, $X(s)$ can be written in the form

$$X(s) = \frac{B_0(s)}{A_0(s)} + \sum_{i=1}^{q} \frac{B_i(s)}{A_i(s)} \exp(-h_i s) \tag{6.80}$$

Now, each $B_i(s)/A_i(s)$ in (6.80) is a rational function of s. If $\deg B_i(s) < \deg A_i(s)$ for $i = 0, 1, 2, \ldots, q$, each rational function $B_i(s)/A_i(s)$ can be expanded by partial fractions. In this way, the inverse Laplace transform of $B_i(s)/A_i(s)$ can be computed for $i = 0, 1, 2, \ldots, q$. Let $x_i(t)$ denote the inverse transform of $B_i(s)/A_i(s)$. Then, by linearity and the right-shift property, the inverse Laplace transform $x(t)$ is given by

$$x(t) = x_0(t) + \sum_{i=1}^{q} x_i(t - h_i)u(t - h_i), \quad t \geq 0$$

Example 6.27   Transform Containing an Exponential

Suppose that

$$X(s) = \frac{s + 1}{s^2 + 1} - \frac{1}{s + 1}e^{-s} + \frac{s + 2}{s^2 + 1}e^{-1.5s}$$

Using linearity and the transform pairs in Table 6.2 gives

$$(\cos t + \sin t)u(t) \leftrightarrow \frac{s + 1}{s^2 + 1}$$

$$(\cos t + 2\sin t)u(t) \leftrightarrow \frac{s + 2}{s^2 + 1}$$

Thus,

$$x(t) = \cos t + \sin t - \exp[-(t - 1)]u(t - 1) + [\cos(t - 1.5)$$
$$+ 2\sin(t - 1.5)]u(t - 1.5), \quad t \geq 0$$

6.4   TRANSFORM OF THE INPUT/OUTPUT DIFFERENTIAL EQUATION

The application of the Laplace transform to the study of causal linear time-invariant continuous-time systems is initiated in this section. The development begins with systems defined by an input/output differential equation. We can generate an "s-domain description" of any such system by taking the Laplace transform of the input/output differential equation. It will be shown that this yields an algebraic procedure to solve the input/output differential equation. Systems given by a first-order differential equation are considered first.

6.4.1   First-Order Case

Consider the linear time-invariant continuous-time system given by the first-order input/output differential equation

$$\frac{dy(t)}{dt} + ay(t) = bx(t) \tag{6.81}$$

where a and b are real numbers, $y(t)$ is the output, and $x(t)$ is the input. Taking the Laplace transform of both sides of (6.81) and using linearity and the differentiation-in-time property (6.43) give

$$sY(s) - y(0^-) + aY(s) = bX(s) \tag{6.82}$$

where $Y(s)$ is the Laplace transform of the output $y(t)$ and $X(s)$ is the Laplace transform of the input $x(t)$. Note that the initial condition $y(0^-)$ is at time $t = 0^-$.

Rearranging terms in (6.82) yields

$$(s + a)Y(s) = y(0^-) + bX(s)$$

and solving for $Y(s)$ gives

$$Y(s) = \frac{y(0^-)}{s + a} + \frac{b}{s + a}X(s) \tag{6.83}$$

Equation (6.83) is the *s-domain representation* of the system given by the input/output differential equation (6.81). The first term on the right-hand side of (6.83) is the Laplace transform of the part of the output response due to the initial condition $y(0^-)$, and the second term on the right-hand side of (6.83) is the Laplace transform of the part of the output response resulting from the input $x(t)$ applied for $t \geq 0$.

If the initial condition $y(0^-)$ is equal to zero, the transform of the output is given by

$$Y(s) = \frac{b}{s + a}X(s) \tag{6.84}$$

When we define

$$H(s) = \frac{b}{s + a}$$

(6.84) becomes

$$Y(s) = H(s)X(s) \tag{6.85}$$

We call the function $H(s)$ the *transfer function* of the system, since it specifies the transfer from the input to the output in the s-domain, assuming that the initial condition $y(0^-)$ is zero; and we refer to (6.85) as the *transfer function representation* of the system.

For any initial condition $y(0^-)$ and any input $x(t)$ with Laplace transform $X(s)$, we can compute the output $y(t)$ by taking the inverse Laplace transform of $Y(s)$ given by (6.83). The process is illustrated by the following example:

Example 6.28   RC Circuit

Consider the RC circuit in Figure 6.4, where the input $x(t)$ is the voltage applied to the circuit and the output $y(t)$ is the voltage across the capacitor. As shown in Section 2.4, the input/output differential equation of the circuit is

$$\frac{dy(t)}{dt} + \frac{1}{RC}y(t) = \frac{1}{RC}x(t) \tag{6.86}$$

Clearly, (6.86) has the form (6.81) with $a = 1/RC$ and $b = 1/RC$, and thus the s-domain representation (6.83) for the RC circuit is given by

$$Y(s) = \frac{y(0^-)}{s + 1/RC} + \frac{1/RC}{s + 1/RC}X(s) \tag{6.87}$$

Now for any input $x(t)$, we can determine the response $y(t)$ by first computing $Y(s)$ given by (6.87) and then taking the inverse Laplace transform. To illustrate this, suppose that the input $x(t)$ is the unit step $u(t)$ so that $X(s) = 1/s$. Then, (6.87) becomes

$$Y(s) = \frac{y(0^-)}{s + 1/RC} + \frac{1/RC}{(s + 1/RC)s} \tag{6.88}$$

Expanding the second term on the right-hand side of (6.88) gives

$$Y(s) = \frac{y(0^-)}{s + 1/RC} + \frac{1}{s} - \frac{1}{s + 1/RC}$$

Then, taking the inverse Laplace transform of $Y(s)$ yields the output response:

$$y(t) = y(0^-)e^{-(1/RC)t} + 1 - e^{-(1/RC)t}, t \geq 0 \tag{6.89}$$

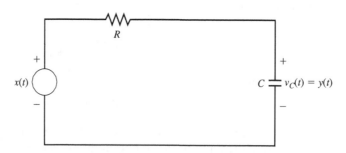

FIGURE 6.4
RC circuit in Example 6.28.

Note that if the initial condition $y(0^-)$ is zero, (6.89) reduces to

$$y(t) = 1 - e^{-(1/RC)t}, t \geq 0 \tag{6.90}$$

As first noted in Chapter 2, the response $y(t)$ given by (6.90) is called the *step response* of the circuit, since it is the output when $x(t)$ is the unit step $u(t)$ with zero initial condition.

6.4.2   Second-Order Case

Now consider the linear time-invariant continuous-time system given by the second-order input/output differential equation

$$\frac{d^2y(t)}{dt^2} + a_1\frac{dy(t)}{dt} + a_0y(t) = b_1\frac{dx(t)}{dt} + b_0x(t) \tag{6.91}$$

where a_1, a_0, b_1, and b_0 are real numbers. Assuming that $x(0^-) = 0$, we see that taking the Laplace transform of both sides of (6.91) gives

$$s^2Y(s) - y(0^-)s - \dot{y}(0^-) + a_1[sY(s) - y(0^-)] + a_0Y(s)$$
$$= b_1sX(s) + b_0X(s) \tag{6.92}$$

Solving (6.92) for $Y(s)$ gives

$$Y(s) = \frac{y(0^-)s + \dot{y}(0^-) + a_1y(0^-)}{s^2 + a_1s + a_0} + \frac{b_1s + b_0}{s^2 + a_1s + a_0}X(s) \tag{6.93}$$

Equation (6.93) is the s-domain representation of the system with the input/output differential equation (6.91). The first term on the right-hand side of (6.93) is the Laplace transform of the part of the output response resulting from initial conditions, and the second term is the transform of the part of the response resulting from the application of the input $x(t)$ for $t \geq 0$. If the initial conditions $y(0^-)$ and $\dot{y}(0^-)$ are equal to zero, (6.93) reduces to the following transfer function representation:

$$Y(s) = \frac{b_1s + b_0}{s^2 + a_1s + a_0}X(s) \tag{6.94}$$

In this case, the transfer function $H(s)$ of the system is the second-order rational function of s given by

$$H(s) = \frac{b_1s + b_0}{s^2 + a_1s + a_0} \tag{6.95}$$

Example 6.29   Second-Order Case

Consider the system given by the input/output differential equation

$$\frac{d^2y(t)}{dt^2} + 6\frac{dy(t)}{dt} + 8y(t) = 2x(t)$$

Hence, the input/output differential equation is in the form (6.91) with $a_1 = 6, a_0 = 8, b_1 = 0,$ and $b_0 = 2$; and from (6.95), the transfer function is

$$H(s) = \frac{2}{s^2 + 6s + 8}$$

Now, to compute the step response of the system, set $x(t) = u(t)$ so that $X(s) = 1/s$, and set the initial conditions to zero. Then the transform of the step response is

$$Y(s) = H(s)X(s) = \frac{2}{s^2 + 6s + 8} \frac{1}{s}$$

Expanding $Y(s)$ gives

$$Y(s) = \frac{0.25}{s} - \frac{0.5}{s + 2} + \frac{0.25}{s + 4}$$

and thus the step response is

$$y(t) = 0.25 - 0.5e^{-2t} + 0.25e^{-4t}, t \geq 0$$

Now suppose that $x(t) = u(t)$ with the initial conditions $y(0^-) = 1$ and $\dot{y}(0^-) = 2$. In this case, the initial conditions are not zero, and thus it is *not* true that $Y(s) = H(s)X(s)$. To compute $Y(s)$, it is necessary to use the s-domain representation (6.93). This gives

$$Y(s) = \frac{s + 8}{s^2 + 6s + 8} + \frac{2}{s^2 + 6s + 8} \frac{1}{s}$$

$$= \frac{s^2 + 8s + 2}{s(s^2 + 6s + 8)}$$

Expanding yields

$$Y(s) = \frac{0.25}{s} + \frac{2.5}{s + 2} - \frac{1.75}{s + 4}$$

Thus,

$$y(t) = 0.25 + 2.5e^{-2t} - 1.75e^{-4t}, t \geq 0$$

6.4.3   Nth-Order Case

Now, consider the general case where the system is given by the Nth-order input/output differential equation

$$\frac{d^N y(t)}{dt^N} + \sum_{i=0}^{N-1} a_i \frac{d^i y(t)}{dt^i} = \sum_{i=0}^{M} b_i \frac{d^i x(t)}{dt^i} \tag{6.96}$$

where $M \leq N$. It is assumed that $x^{(i)}(0^-) = 0$ for $i = 0, 1, 2, \ldots, M - 1$. By taking the Laplace transform of both sides of (6.96) with initial conditions at time $t = 0^-$, it is possible to express the Laplace transform $Y(s)$ of the output $y(t)$ in the form

$$Y(s) = \frac{C(s)}{A(s)} + \frac{B(s)}{A(s)} X(s) \tag{6.97}$$

where $B(s)$ and $A(s)$ are polynomials in s given by

$$B(s) = b_M s^M + b_{M-1} s^{M-1} + \cdots + b_1 s + b_0$$

$$A(s) = s^N + a_{N-1} s^{N-1} + \cdots + a_1 s + a_0$$

The numerator $C(s)$ of the first term on the right-hand side of (6.97) is also a polynomial in s whose coefficients are determined by the initial conditions $y(0^-)$, $y^{(1)}(0^-)$, $\ldots, y^{(N-1)}(0^-)$. For example, if $N = 2$, then by the foregoing results,

$$C(s) = y(0^-)s + \dot{y}(0^-) + a_1 y(0^-)$$

Equation (6.97) is the s-domain representation of the system with the Nth-order input/output differential equation (6.96).

Since $x^{(i)}(0^-) = 0$ for $i = 0, 1, 2, \ldots, M - 1$, if the initial conditions $y(0^-)$, $y^{(1)}(0^-), \ldots, y^{(N-1)}(0^-)$ are zero, which is equivalent to the condition that $C(s) = 0$, then the transform $Y(s)$ of the output response is given by

$$Y(s) = \frac{B(s)}{A(s)} X(s) = \frac{b_M s^M + \cdots + b_1 s + b_0}{s^N + a_{N-1} s^{N-1} + \cdots + a_1 s + a_0} X(s) \qquad (6.98)$$

The transfer function $H(s)$ of the system is the Nth-order rational function in s given by

$$H(s) = \frac{b_M s^M + \cdots + b_1 s + b_0}{s^N + a_{N-1} s^{N-1} + \cdots + a_1 s + a_0} \qquad (6.99)$$

Combining (6.98) and (6.99) results in the transfer function model $Y(s) = H(s)X(s)$. It is important to stress that the transfer function model expresses the relationship between $X(s)$ and $Y(s)$ when we assume that all initial conditions at time $t = 0$ (or $t = 0^-$) are equal to zero.

6.4.4   Computation of Output Response

Again, consider the transfer function representation (6.98). If the transform $X(s)$ of the input $x(t)$ is a rational function of s, the product $H(s)X(s)$ is a rational function of s. In this case the output $y(t)$ can be computed by first expanding $H(s)X(s)$ by partial fractions. The process is illustrated by the following example:

Example 6.30   Computation of Output Response

Consider the system with transfer function

$$H(s) = \frac{s^2 + 2s + 16}{s^3 + 4s^2 + 8s}$$

We will compute the output response $y(t)$ resulting from input $x(t) = e^{-2t}u(t)$, assuming the initial conditions are zero at time $t = 0$. The transform of $x(t)$ is

$$X(s) = \frac{1}{s + 2}$$

and thus

$$Y(s) = H(s)X(s) = \frac{s^2 + 2s + 16}{(s^3 + 4s^2 + 8s)(s + 2)} = \frac{s^2 + 2s + 16}{[(s + 2)^2 + 4]s(s + 2)}$$

Expanding by partial fractions yields

$$Y(s) = \frac{cs + d}{(s + 2)^2 + 4} + \frac{c_3}{s} + \frac{c_4}{s + 2} \tag{6.100}$$

where

$$c_3 = [sY(s)]_{s=0} = \frac{16}{2(8)} = 1$$

$$c_4 = [(s + 2)Y(s)]_{s=-2} = \frac{(-2)^2 - (2)(2) + 16}{(-2)(4)} = -2$$

Putting the right-hand side of (6.100) over a common denominator and equating numerators give

$$s^2 + 2s + 16 = (cs + d)s(s + 2) + c_3[(s + 2)^2 + 4](s + 2) + c_4[(s + 2)^2 + 4]s$$

Collecting terms with like powers of s yields

$$s^3 - 2s^3 + cs^3 = 0$$
$$6s^2 - 8s^2 + (d + 2c)s^2 = s^2$$

which implies that $c = 1$ and $d = 1$. Hence,

$$Y(s) = \frac{s + 1}{(s + 2)^2 + 4} + \frac{1}{s} + \frac{-2}{s + 2}$$

$$Y(s) = \frac{s + 2}{(s + 2)^2 + 4} + \frac{-1}{(s + 2)^2 + 4} + \frac{1}{s} + \frac{-2}{s + 2}$$

Using Table 6.1 gives

$$y(t) = e^{-2t}\cos 2t - \tfrac{1}{2}e^{-2t}\sin 2t + 1 - 2e^{-2t}, \quad t \geq 0$$

and when we use the trigonometric identity (6.73), $y(t)$ can be expressed in the form

$$y(t) = \frac{\sqrt{5}}{2}e^{-2t}\cos(2t + 26.565°) + 1 - 2e^{-2t}, \quad t \geq 0$$

6.5   TRANSFORM OF THE INPUT/OUTPUT CONVOLUTION INTEGRAL

If a linear time-invariant system is given by an input/output differential equation, then, as we saw in Section 6.4, we can generate an s-domain representation (or transfer function representation) of the system by taking the Laplace transform of the input/output differential equation. As will be shown subsequently, we can generate the transfer function representation for any causal linear time-invariant system by taking the Laplace transform of the input/output convolution expression given by

$$y(t) = h(t) * x(t) = \int_0^t h(\lambda) x(t - \lambda) \, d\lambda, t \geq 0 \qquad (6.101)$$

where $h(t)$ is the impulse response of the system, the input $x(t)$ is assumed to be zero for all $t < 0$, and all initial conditions at time $t = 0$ (or $t = 0^-$) are assumed to be equal to zero. From the results in Chapter 2, we see that $y(t)$ given by (6.101) is the output response resulting from the application of the input $x(t)$ for $t \geq 0$. Also recall that causality implies that $h(t)$ is zero for $t < 0$, and thus the lower limit of the integral in (6.101) is taken at $\lambda = 0$. Finally, it is important to note that, in contrast to the Fourier transform approach developed in Chapter 5, here there is no requirement that the impulse response $h(t)$ be absolutely integrable. [See (5.2).]

Since the input $x(t)$ is zero for all $t < 0$, the (one-sided) Laplace transform can be applied to both sides of (6.101), which results in the transfer function representation

$$Y(s) = H(s)X(s) \qquad (6.102)$$

where the transfer function $H(s)$ is the Laplace transform of the impulse response $h(t)$. The relationship between the impulse response $h(t)$ and the transfer function $H(s)$ can be expressed in terms of the transform pair notation

$$h(t) \leftrightarrow H(s) \qquad (6.103)$$

The transform pair (6.103) is of fundamental importance. In particular, it provides a bridge between the time domain representation given by the convolution relationship (6.101) and the s-domain representation (6.102) given in terms of the transfer function $H(s)$.

It is important to note that if the input $x(t)$ is not the zero function, so that $X(s)$ is not zero, both sides of (6.102) can be divided by $X(s)$, which yields

$$H(s) = \frac{Y(s)}{X(s)} \qquad (6.104)$$

From (6.104) it is seen that the transfer function $H(s)$ is equal to the ratio of the transform $Y(s)$ of the output and the transform $X(s)$ of the input.

Since $H(s)$ is the transform of the impulse response $h(t)$, and a system has only one $h(t)$, each system has a unique transfer function. Therefore, although $Y(s)$ will change as the input $x(t)$ ranges over some collection of signals, by (6.104) the ratio $Y(s)/X(s)$ cannot change. It also follows from (6.104) that the transfer function $H(s)$ can be determined from knowledge of the response $y(t)$ to any nonzero input signal

$x(t)$. It should be stressed that this result is valid only if it is known that the given system is both linear and time invariant. If the system is time varying or nonlinear, there is no transfer function, and thus (6.104) has no meaning in such cases.

Example 6.31   Determining the Transfer Function

Suppose that the input $x(t) = e^{-t}u(t)$ is applied to a causal linear time-invariant continuous-time system, and that the resulting output response is

$$y(t) = 2 - 3e^{-t} + e^{-2t} \cos 2t, t \geq 0$$

Then

$$Y(s) = \frac{2}{s} - \frac{3}{s+1} + \frac{s+2}{(s+2)^2 + 4}$$

and

$$X(s) = \frac{1}{s+1}$$

Inserting these expressions for $Y(s)$ and $X(s)$ into (6.104) gives

$$H(s) = \frac{\dfrac{2}{s} - \dfrac{3}{s+1} + \dfrac{s+2}{(s+2)^2 + 4}}{\dfrac{1}{s+1}}$$

$$= \frac{2(s+1)}{s} - 3 + \frac{(s+1)(s+2)}{(s+2)^2 + 4}$$

$$= \frac{[2(s+1) - 3s][(s+2)^2 + 4] + s(s+1)(s+2)}{s[(s+2)^2 + 4]}$$

$$= \frac{s^2 + 2s + 16}{s^3 + 4s^2 + 8s}$$

6.5.1   Finite-Dimensional Systems

A causal linear time-invariant continuous-time system is said to be *finite dimensional* if the transfer function $H(s)$ has the rational form

$$H(s) = \frac{b_M s^M + \cdots + b_1 s + b_0}{s^N + a_{N-1} s^{N-1} + \cdots + a_1 s + a_0} \tag{6.105}$$

In this case, the degree N of the denominator polynomial in (6.105) is called the *order* of the system. It turns out that any finite-dimensional system has an input/output differential equation. To see this, multiply both sides of (6.105) by the denominator polynomial, which gives

$$(s^N + a_{N-1}s^{N-1} + \cdots + a_1 s + a_0)Y(s) = (b_M s^M + \cdots + b_1 s + b_0)X(s) \qquad (6.106)$$

Inverse transforming both sides of (6.106) yields

$$\frac{d^N y(t)}{dt^N} + \sum_{i=0}^{N-1} a_i \frac{d^i y(t)}{dt^i} = \sum_{i=0}^{M} b_i \frac{d^i x(t)}{dt^i} \qquad (6.107)$$

Thus, the system can be described by an input/output differential equation.

Hence, if a system is finite dimensional, it is possible to go directly from the transfer function representation to the input/output differential equation, and from the input/output differential equation to the transfer function representation. This produces another fundamental link between the time domain and the s-domain. Note also that, by the foregoing result, a causal linear time-invariant system is finite-dimensional if and only if the system can be described by an input/output differential equation.

Poles and zeros of a system.   Given a finite-dimensional system with the transfer function

$$H(s) = \frac{b_M s^M + b_{M-1}s^{M-1} + \cdots + b_1 s + b_0}{s^N + a_{N-1}s^{N-1} + \cdots + a_1 s + a_0} \qquad (6.108)$$

$H(s)$ can be expressed in the factored form

$$H(s) = \frac{b_M(s - z_1)(s - z_2)\cdots(s - z_M)}{(s - p_1)(s - p_2)\cdots(s - p_N)} \qquad (6.109)$$

where $z_1, z_2, \ldots, z_M$ are the zeros of $H(s)$ and $p_1, p_2, \ldots, p_N$ are the poles of $H(s)$. The zeros of $H(s)$ are said to be the *zeros of the system*, and the poles of $H(s)$ are said to be the *poles of the system*. Note that the number of poles of the system is equal to the order N of the system. From (6.109) it is seen that, except for the constant b_M, the transfer function is determined completely by the values of the poles and zeros of the system. The poles and zeros of a finite-dimensional system are often displayed in a *pole–zero diagram,* which is a plot in the complex plane that shows the location of all the poles (marked by $\times$) and all zeros (marked by O). As will be seen in Chapter 8, the location of the poles and zeros is of fundamental importance in determining the behavior of the system.

Example 6.32   Mass–Spring–Damper System

Mass–
Spring–
Damper
System

For the mass–spring–damper system defined in Chapter 1 (see Section 1.4), the input/output differential equation of the system is given by

$$M\frac{d^2 y(t)}{dt^2} + D\frac{dy(t)}{dt} + Ky(t) = x(t)$$

where M is the mass, D is the damping constant, K is the stiffness constant, $x(t)$ is the force applied to the mass, and $y(t)$ is the displacement of the mass relative to the equilibrium position. Then, using the relationship between (6.105) and (6.107) reveals that the transfer function $H(s)$ of the mass–spring–damper system is given by

$$H(s) = \frac{1}{Ms^2 + Ds + K} = \frac{1/M}{s^2 + (D/M)s + (K/M)}$$

The system is a second-order system; that is, the system has two poles, p_1 and p_2. From the quadratic formula, p_1 and p_2 are given by

$$p_1 = -\frac{D}{2M} + \frac{1}{2}\sqrt{\frac{D^2}{M^2} - 4\frac{K}{M}}$$

$$p_2 = -\frac{D}{2M} - \frac{1}{2}\sqrt{\frac{D^2}{M^2} - 4\frac{K}{M}}$$

It turns out that the poles p_1 and p_2 may be real numbers, or they may be complex numbers with one pole equal to the complex conjugate of the other. The poles are real if and only if

$$\frac{D^2}{M^2} - 4\frac{K}{M} \geq 0$$

which is equivalent to the condition $D^2 \geq 4KM$. The poles are real and distinct (nonrepeated) if and only if $D^2 > 4KM$. Assuming that both the poles are real and distinct, the Laplace transform $Y(s)$ of the output response $y(t)$ resulting from the constant input $x(t) = A, t \geq 0$, is given by

$$Y(s) = \frac{1/M}{(s - p_1)(s - p_2)}\frac{A}{s} = \frac{A/(p_1 p_2 M)}{s} + \frac{A/[M(p_1 - p_2)]}{s - p_1} + \frac{A/[M(p_2 - p_1)]}{s - p_2}$$

Taking the inverse Laplace transform results in the output response

$$y(t) = \frac{A}{p_1 p_2 M} + \frac{A}{(p_1 - p_2)M}[e^{p_1 t} - e^{p_2 t}], t \geq 0$$

If M, D, and K are strictly positive (greater than 0) and $D^2 > 4KM$, it turns out that p_1 and p_2 are strictly negative real numbers (less than 0), and thus the response $y(t)$ converges to the constant $A/(p_1 p_2 M)$ as $t \to \infty$. The reader can see an animation of this behavior by running the online demo on the website.

Example 6.33   Third-Order System

Consider the system with the transfer function

$$H(s) = \frac{2s^2 + 12s + 20}{s^3 + 6s^2 + 10s + 8}$$

Factoring $H(s)$ gives

$$H(s) = \frac{2(s + 3 - j)(s + 3 + j)}{(s + 4)(s + 1 - j)(s + 1 + j)}$$

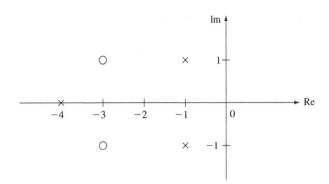

FIGURE 6.5
Pole-zero diagram in Example 6.33.

Thus, the zeros of the system are

$$z_1 = -3 + j \quad \text{and} \quad z_2 = -3 - j$$

and the poles of the system are

$$p_1 = -4, \qquad p_2 = -1 + j, \qquad p_3 = -1 - j$$

The pole–zero diagram is shown in Figure 6.5.

Computation of output response with MATLAB.   Given any finite-dimensional system, the system's response to an arbitrary input can be calculated by several methods in MATLAB, which differ depending on whether the Student Version or the professional version with appropriate toolboxes is used. First, consider the methods described in Section 2.5. These methods model the system in differential equation form and then solve the differential equation by use of either the symbolic manipulator or numerical integration methods, such as Euler integration and the ODE solvers in MATLAB. As mentioned before, a transfer function can be converted into a differential equation and thus solved by the differential equation solution methods.

Using the transfer function approach as described in this chapter to solve for the response can also be done either symbolically or numerically. To use the symbolic method, closed-form expressions must be known for both the input $X(s)$ and the transfer function $H(s)$. The output response is calculated from $Y(s) = H(s)X(s)$.

Example 6.34   Symbolic Manipulator Solution

Suppose that the transfer function and the Laplace transform of the input are given as

$$H(s) = \frac{2}{s^2 + 4s + 4}, \quad X(s) = \frac{1}{s}$$

The commands in the Symbolic Math Toolbox used to solve for $y(t)$ and plot it are

```
syms s H X Y
H = 2/(s^2+4*s+4);
X = 1/s;
Y = H*X;
y = ilaplace(Y) % use simplify(y) to simplify result
ezplot(y,[0,10]);
```

Running these commands returns

```
y =
    1/2+(-1/2-t)*exp(-2*t)
```

The solution for an arbitrary input can be computed numerically in the full version of MATLAB with the Control System Toolbox by the commands tf and lsim. This method is known as *numerical simulation*. Any transfer function $H(s)$ given by (6.108) is represented in MATLAB as a transfer function object, which can be created by the tf command. The numerator and denominator polynomials are represented in MATLAB as vectors containing the coefficients of the polynomials in descending powers of *s*. Hence, the transfer function (6.108) is stored in MATLAB via the commands

```
num = [b_M b_{M-1} ... b_0];
den = [1 a_{N-1} ... a_0];
H = tf(num,den);
```

To compute the system output resulting from an input $x(t)$, a time vector is defined that contains the values of *t* for which $y(t)$ will be computed. The use of the command lsim is illustrated in the following example:

Example 6.35   Output Response from lsim

Consider the system with the transfer function given in Example 6.30:

$$H(s) = \frac{s^2 + 2s + 16}{s^3 + 4s^2 + 8s}$$

The response of this system to an exponential input is obtained from the following commands:

```
num = [1 2 16];
den = [1 4 8 0];
H = tf(num,den);
t = 0:10/300:10;
x = exp(-2*t);
y = lsim(H,x,t);
```

Running the MATLAB program results in the output response vector y that is plotted in Figure 6.6 along with the exact response calculated from Example 6.30. The response obtained from the numerical simulation and the exact response are indistinguishable.

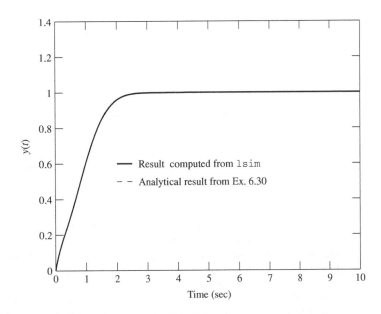

FIGURE 6.6
Output response of system in Example 6.35.

The step response and the impulse response of the system with the aforementioned transfer function $H(s)$ are obtained by replacing the lsim command used previously with

```
y = impulse(H,t);
y = step(H,t);
```

The full version of MATLAB has a toolbox available called Simulink® that is included in the Student Version of MATLAB. This toolbox allows the user to build a model graphically and then to simulate the system. Simulink is started by the typing of simulink at the MATLAB prompt. A window opens that contains a library of available blocks. To build a new model, the user clicks on "File" and then clicks on "New," which opens a new window for the model. The model is built by the dragging of menu items, called blocks, from the library window onto the new model window. Each block can be customized by a double click to set parameters for a particular model.

Example 6.36   Simulink Model

Consider again the system given in Example 6.30. Start Simulink as described previously. You will find the block for a transfer function of a continuous-time system by clicking on the menu item "Continuous." Drag the transfer function block onto the model window, and double click on it. Set the numerator and denominator coefficients the same way as the command tf was demonstrated in Example 6.35. Input signals are found under the menu item "Sources." These include some common signals, such as a step function, or the option to use a signal defined by a

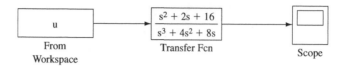

FIGURE 6.7
Simulink model of the system described in Example 6.36.

vector created in the normal MATLAB workspace. For example, to use the exponential signal used in Example 6.35, create the array

```
u = [t',x'];
```

from the standard MATLAB prompt. Drag the block described as "From Workspace" onto the model window, and set the parameters of the block by double-clicking on it. In particular, enter the name of the input vector, u in this case, in the data field. Drag a block from the "Sink" menu to the model window in order to display or save the output response. The Scope block is a Sink menu item that allows the user to see the output response plotted as the simulation is progressing. Double-click on the Scope block in the model window to view it. Connect the blocks together by clicking on the output port of one block and dragging the mouse to the input port of another block. Click on "Simulation," then "Start" to simulate the response. The resulting Simulink model for the system described in Example 6.30 with the exponential input is shown in Figure 6.7.

6.6   DIRECT CONSTRUCTION OF THE TRANSFER FUNCTION

The transfer function of a system is often determined directly from a wiring diagram of the system, so it is not always necessary to determine the impulse response or the input/output differential equation in order to generate the transfer function representation. This can be done for RLC circuits and systems consisting of interconnections of integrators. The following development begins with RLC circuits:

6.6.1   RLC Circuits

In Section 2.4 the resistor, capacitor, and inductor were defined in terms of the voltage–current relationships

$$v(t) = Ri(t) \tag{6.110}$$

$$\frac{dv(t)}{dt} = \frac{1}{C}i(t) \tag{6.111}$$

$$v(t) = L\frac{di(t)}{dt} \tag{6.112}$$

In (6.110)–(6.112), $i(t)$ is the current into the circuit element, and $v(t)$ is the voltage across the element. (See Figure 2.10.) We can express the voltage–current relationships in the s-domain by taking the Laplace transform of both sides of (6.110)–(6.112). Using the differentiation property of the Laplace transform yields

$$V(s) = RI(s) \tag{6.113}$$

$$sV(s) - v(0) = \frac{1}{C}I(s) \quad \text{or} \quad V(s) = \frac{1}{Cs}I(s) + \frac{1}{s}v(0) \tag{6.114}$$

$$V(s) = LsI(s) - Li(0) \tag{6.115}$$

where $V(s)$ is the Laplace transform of the voltage and $I(s)$ is the Laplace transform of the current. In (6.114), $v(0)$ is the initial voltage on the capacitor at time $t = 0$; and in (6.115), $i(0)$ is the initial current in the inductor at time $t = 0$.

Using (6.113)–(6.115) results in the s-domain representations of the resistor, capacitor, and inductor shown in Figure 6.8. Here, the circuit elements are represented in terms of their impedances; that is, the resistor has impedance R, the capacitor has (complex) impedance $1/Cs$, and the inductor has (complex) impedance Ls. Note that the initial voltage on the capacitor and the initial current in the inductor are treated as voltage sources in the s-domain representations.

Now, given an interconnection of RLCs, we can construct an s-domain representation of the circuit by taking the Laplace transform of the voltages and currents in the circuit and by expressing resistors, capacitors, and inductors in terms of their s-domain representations. The resulting s-domain representation satisfies the same circuit laws as a purely resistive circuit with voltage and current sources. In particular, the voltage

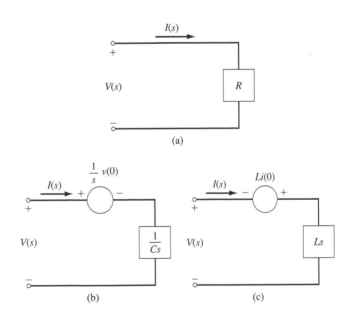

FIGURE 6.8
s-Domain representations: (a) resistor; (b) capacitor; (c) inductor.

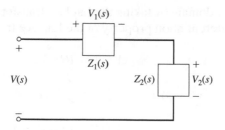

FIGURE 6.9
Series connection of two impedances.

and current division rules for resistive circuits can be applied. For example, consider two impedances $Z_1(s)$ and $Z_2(s)$ connected in series, as shown in Figure 6.9. With $V(s)$ equal to the Laplace transform of the applied voltage, and with $V_1(s)$ and $V_2(s)$ equal to the Laplace transforms of the voltages across the impedances $Z_1(s)$ and $Z_2(s)$, by the voltage division rule,

$$V_1(s) = \frac{Z_1(s)}{Z_1(s) + Z_2(s)} V(s)$$

$$V_2(s) = \frac{Z_2(s)}{Z_1(s) + Z_2(s)} V(s)$$

Now suppose that the two impedances $Z_1(s)$ and $Z_2(s)$ are connected in parallel, as shown in Figure 6.10. With $I(s)$ equal to the transform of the current into the connection and with $I_1(s)$ and $I_2(s)$ equal to the transforms of the currents in the impedances, by the current division rule,

$$I_1(s) = \frac{Z_2(s)}{Z_1(s) + Z_2(s)} I(s)$$

$$I_2(s) = \frac{Z_1(s)}{Z_1(s) + Z_2(s)} I(s)$$

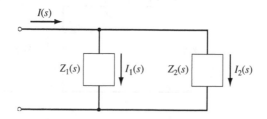

FIGURE 6.10
Two impedances in parallel.

By using the voltage and current division rules and other basic circuit laws, we can determine the transfer function of an *RLC* circuit directly from the *s*-domain representation of the circuit. In computing the transfer function, we assume that the initial conditions are zero at time $t = 0$; and thus we assume that the initial voltages on the capacitors and the initial currents in the inductors are all zero.

The construction of the transfer function is illustrated by the following two examples:

Example 6.37   Series *RLC* Circuit

Consider the series *RLC* circuit shown in Figure 6.11. As shown, the input is the voltage $x(t)$ applied to the series connection, and the output is the voltage $v_c(t)$ across the capacitor. Given that the initial voltage on the capacitor and the initial current in the inductor are both zero, the *s*-domain representation of the circuit is shown in Figure 6.12. Working with the *s*-domain representation and using voltage division give

$$V_c(s) = \frac{1/Cs}{Ls + R + (1/Cs)} X(s) = \frac{1/LC}{s^2 + (R/L)s + (1/LC)} X(s) \tag{6.116}$$

Comparing (6.116) with the general form (6.102) of the transfer function representation reveals that the transfer function $H(s)$ of the circuit is

$$H(s) = \frac{1/LC}{s^2 + (R/L)s + (1/LC)} \tag{6.117}$$

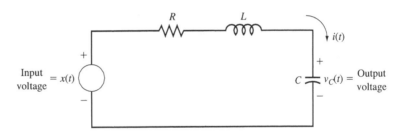

FIGURE 6.11
Series *RLC* circuit.

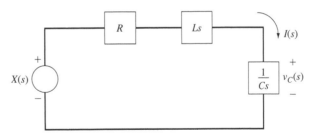

FIGURE 6.12
Representation of series *RLC* circuit in *s*-domain.

It is interesting to note that, if a different choice for the input and output of the circuit had been taken, the transfer function would not equal the result given in (6.117). For instance, if the definition of the input is kept the same, but the output is taken to be the voltage $v_R(t)$ across the resistor, by voltage division,

$$V_R(s) = \frac{R}{Ls + R + (1/Cs)}X(s) = \frac{(R/L)s}{s^2 + (R/L)s + (1/LC)}X(s)$$

The resulting transfer function is

$$H(s) = \frac{(R/L)s}{s^2 + (R/L)s + (1/LC)}$$

which differs from (6.117).

Example 6.38   Computation of Transfer Function

In the circuit shown in Figure 6.13, the input $x(t)$ is the applied voltage and the output $y(t)$ is the current in the capacitor with capacitance C_1. If it is assumed that the initial voltages on the capacitors and the initial currents in the inductors are all zero, then the s-domain representation of the circuit is as shown in Figure 6.14. The impedance of the parallel connection consisting of the capacitance C_1 and the inductance L_2 in series with the capacitance C_2 is equal to

$$\frac{(1/C_1s)[L_2s + (1/C_2s)]}{(1/C_1s) + L_2s + (1/C_2s)} = \frac{C_2L_2s^2 + 1}{C_1C_2L_2s^3 + (C_1 + C_2)s}$$

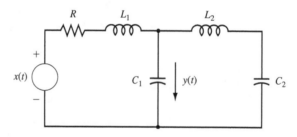

FIGURE 6.13
Circuit in Example 6.38.

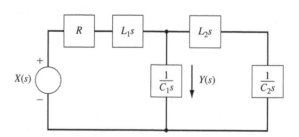

FIGURE 6.14
s-Domain representation of circuit in Example 6.38.

Letting $V_1(s)$ denote the transform of the voltage across the capacitance C_1 [with $V_1(s)$ defined so that $Y(s) = C_1 s V_1(s)$], we obtain, by voltage division,

$$V_1(s) = \frac{\dfrac{C_2 L_2 s^2 + 1}{C_1 C_2 L_2 s^3 + (C_1 + C_2)s}}{R + L_1 s + \dfrac{C_2 L_2 s^2 + 1}{C_1 C_2 L_2 s^3 + (C_1 + C_2)s}} X(s)$$

$$V_1(s) = \frac{C_2 L_2 s^2 + 1}{(R + L_1 s)(C_1 C_2 L_2 s^3 + (C_1 + C_2)s) + C_2 L_2 s^2 + 1} X(s)$$

$$V_1(s) = \frac{C_2 L_2 s^2 + 1}{C_1 C_2 L_1 L_2 s^4 + R C_1 C_2 L_2 s^3 + [L_1(C_1 + C_2) + L_2 C_2]s^2 + R(C_1 + C_2)s + 1} X(s)$$

Finally, since $Y(s) = C_1 s V_1(s)$, then

$$Y(s) = \frac{C_1 C_2 L_2 s^3 + C_1 s}{C_1 C_2 L_1 L_2 s^4 + R C_1 C_2 L_2 s^3 + [L_1(C_1 + C_2) + L_2 C_2]s^2 + R(C_1 + C_2)s + 1} X(s)$$

and thus the transfer function is

$$H(s) = \frac{C_1 C_2 L_2 s^3 + C_1 s}{C_1 C_2 L_1 L_2 s^4 + R C_1 C_2 L_2 s^3 + [L_1(C_1 + C_2) + L_2 C_2]s^2 + R(C_1 + C_2)s + 1}$$

Interconnections of Integrators.   Continuous-time systems are sometimes given in terms of an interconnection of integrators, adders, subtracters, and scalar multipliers. These basic system components are illustrated in Figure 6.15. As shown in Figure 6.15a, the output $y(t)$ of the integrator is equal to the initial value of $y(t)$ plus the integral of the input (hence the term integrator). In mathematical terms,

$$y(t) = y(0) + \int_0^t x(\lambda)\, d\lambda \tag{6.118}$$

Differentiating both sides of (6.118) yields the input/output differential equation

$$\frac{dy(t)}{dt} = x(t) \tag{6.119}$$

From (6.119) it is seen that, if the input to the integrator is the derivative of a signal $v(t)$, the resulting output is $v(t)$. This makes sense, since integration "undoes" differentiation.

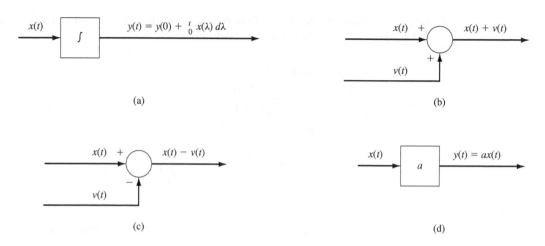

FIGURE 6.15
Basic system components: (a) integrator; (b) adder; (c) subtracter; (d) scalar multiplier.

As shown in Figures 6.15b and 6.15c, the adder simply adds the inputs and the subtracter subtracts the inputs. As the name implies, the scalar multiplier scales the input by the factor a, where a is any real number.

Now, taking the Laplace transform of both sides of (6.119) results in the s-domain representation of the integrator:

$$sY(s) - y(0) = X(s)$$

or

$$Y(s) = \frac{1}{s}y(0) + \frac{1}{s}X(s)$$

If $y(0) = 0$, the s-domain representation reduces to

$$Y(s) = \frac{1}{s}X(s) \qquad (6.120)$$

From (6.120) it is seen that the integrator has transfer function $1/s$.

Now consider an interconnection of integrators, adders, subtracters, and scalar multipliers. To compute the transfer function, we can first redraw the interconnection in the s-domain by taking transforms of all signals in the interconnection and by representing integrators by $1/s$. An equation for the Laplace transform of the output of each integrator in the interconnection can then be written. An equation for the transform of the output can also be written in terms of the transforms of the outputs of the integrators. These equations can then be combined algebraically to derive the transfer function relationship. The procedure is illustrated by the following example:

Example 6.39   Integrator Interconnection

Consider the system shown in Figure 6.16. The output of the first integrator is denoted by $q_1(t)$, and the output of the second integrator is denoted by $q_2(t)$. Assuming that $q_1(0) = q_2(0) = 0$, we note the s-domain representation of the system shown in Figure 6.17. Then,

$$sQ_1(s) = -4Q_1(s) + X(s) \tag{6.121}$$

$$sQ_2(s) = Q_1(s) - 3Q_2(s) + X(s) \tag{6.122}$$

$$Y(s) = Q_2(s) + X(s) \tag{6.123}$$

Solving (6.121) for $Q_1(s)$ gives

$$Q_1(s) = \frac{1}{s+4}X(s) \tag{6.124}$$

Solving (6.122) for $Q_2(s)$ and using (6.124) yield

$$Q_2(s) = \frac{1}{s+3}[Q_1(s) + X(s)] = \frac{1}{s+3}\left(\frac{1}{s+4} + 1\right)X(s)$$

$$= \frac{s+5}{(s+3)(s+4)}X(s) \tag{6.125}$$

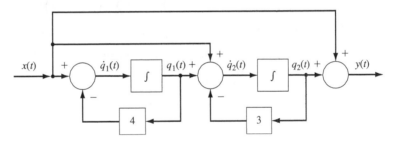

FIGURE 6.16
System with two integrators.

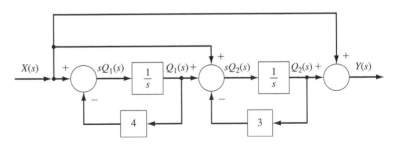

FIGURE 6.17
Representation of system in s-domain.

Inserting the expression (6.125) for $Q_2(s)$ into (6.123) gives

$$Y(s) = \left[\frac{s + 5}{(s + 3)(s + 4)} + 1 \right] X(s) = \frac{s^2 + 8s + 17}{(s + 3)(s + 4)} X(s)$$

Thus, the transfer function $H(s)$ is

$$H(s) = \frac{s^2 + 8s + 17}{(s + 3)(s + 4)} = \frac{s^2 + 8s + 17}{s^2 + 7s + 12}$$

TRANSFER FUNCTION OF BLOCK DIAGRAMS

A linear time-invariant continuous-time system is sometimes specified by a block diagram consisting of an interconnection of "blocks," with each block represented by a transfer function. The blocks can be thought of as subsystems that make up the given system. We can determine the transfer function of a system given by a block diagram by combining blocks in the diagram. In this section the transfer functions for three basic types of interconnections are considered.

Parallel Interconnection.   Consider a parallel interconnection of two linear time-invariant continuous-time systems with transfer functions $H_1(s)$ and $H_2(s)$. The interconnection is shown in Figure 6.18. The Laplace transform $Y(s)$ of the output of the parallel connection is given by

$$Y(s) = Y_1(s) + Y_2(s) \tag{6.126}$$

If the initial conditions in each system are equal to zero, then

$$Y_1(s) = H_1(s)X(s) \text{ and } Y_2(s) = H_2(s)X(s)$$

Inserting these expressions into (6.126) gives

$$Y(s) = H_1(s)X(s) + H_2(s)X(s)$$

$$= (H_1(s) + H_2(s))X(s) \tag{6.127}$$

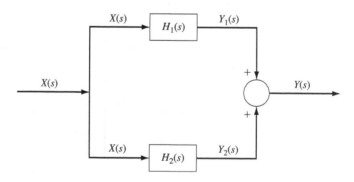

FIGURE 6.18
Parallel interconnection of two systems.

From (6.127) it is seen that the transfer function $H(s)$ of the parallel interconnection is equal to the sum of the transfer functions of the systems in the connection; that is,

$$H(s) = H_1(s) + H_2(s) \tag{6.128}$$

Series Connection. Now consider the series connection (also called the cascade connection) shown in Figure 6.19. It is assumed that the initial conditions are zero in both systems and that the second system does not load the first system. No loading means that

$$Y_1(s) = H_1(s)X(s) \tag{6.129}$$

If $y_1(t)$ is a voltage waveform, it may be assumed that there is no loading if the output impedance of the first system is much less than the input impedance of the second system.
Now, since

$$Y_2(s) = H_2(s)Y_1(s)$$

using (6.129) gives

$$Y(s) = Y_2(s) = H_2(s)H_1(s)X(s) \tag{6.130}$$

From (6.130) it follows that the transfer function $H(s)$ of the series connection is equal to the product of the transfer functions of the systems in the connection; that is,

$$H(s) = H_2(s)H_1(s)$$

Since $H_1(s)$ and $H_2(s)$ are scalar-valued functions of s,

$$H_2(s)H_1(s) = H_1(s)H_2(s)$$

and thus $H(s)$ can also be expressed in the form

$$H(s) = H_1(s)H_2(s) \tag{6.131}$$

Feedback Connection. Now consider the interconnection shown in Figure 6.20. In this connection the output of the first system is fed back to the input through the second system, and thus the connection is called a *feedback connection*. Note that if the feedback loop is disconnected, the transfer function from $X(s)$ to $Y(s)$ is $H_1(s)$. The system with transfer function $H_1(s)$ is called the *open-loop system*, since the transfer function from $X(s)$ to $Y(s)$ is equal to $H_1(s)$ if the feedback is disconnected. [Some authors refer to $H_1(s)H_2(s)$ as the *open-loop transfer function*.] The system with transfer function $H_2(s)$ is called the *feedback system*, and the feedback connection is

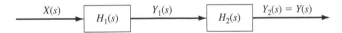

FIGURE 6.19
Series connection.

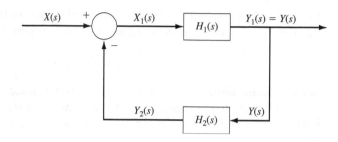

FIGURE 6.20
Feedback connection.

called the *closed-loop system*. The objective here is to compute the transfer function of the closed-loop system.

It is assumed that the initial conditions in either system are zero and that the feedback system does not load the open-loop system. Then from Figure 6.20,

$$Y(s) = H_1(s)X_1(s) \tag{6.132}$$

$$X_1(s) = X(s) - Y_2(s) = X(s) - H_2(s)Y(s) \tag{6.133}$$

Inserting the expression (6.133) for $X_1(s)$ into (6.132) yields

$$Y(s) = H_1(s)[X(s) - H_2(s)Y(s)] \tag{6.134}$$

Rearranging terms in (6.134) gives

$$[1 + H_1(s)H_2(s)]Y(s) = H_1(s)X(s) \tag{6.135}$$

Solving (6.135) for $Y(s)$ gives

$$Y(s) = \frac{H_1(s)}{1 + H_1(s)H_2(s)}X(s) \tag{6.136}$$

From (6.136) it is seen that the transfer function $H(s)$ of the feedback connection is given by

$$H(s) = \frac{H_1(s)}{1 + H_1(s)H_2(s)} \tag{6.137}$$

It follows from (6.137) that the closed-loop transfer function $H(s)$ is equal to the open-loop transfer function $H_1(s)$ divided by 1 plus the product $H_1(s)H_2(s)$ of the transfer functions of the open-loop system and feedback system. Note that, if the subtracter in Figure 6.20 were changed to an adder, the transfer function $H(s)$ of the closed-loop system would change to

$$H(s) = \frac{H_1(s)}{1 - H_1(s)H_2(s)} \tag{6.138}$$

It is worth noting that MATLAB can be used to compute the transfer function for feedback, series, or parallel connections. Some details on this can be found in the tutorial that is available from the text website.

6.7 CHAPTER SUMMARY

The Laplace transform is a powerful tool for solving differential equations, finding the response of a system, and performing general analysis of continuous-time systems. The one-sided Laplace transform is defined by the expression

$$X(s) = \int_0^\infty x(t)e^{-st}\, dt$$

There are numerous common functions that have well-defined Laplace transforms, such as the step function, impulse function, exponentials, sinusoids, and damped sinusoids. These signals and their Laplace transforms form a set of common Laplace transform pairs. We usually find the inverse Laplace transform of a complicated expression by using partial fraction expansion to rewrite the expression into a sum of simpler terms from the set of common pairs. We then take the inverse Laplace transform term by term, by utilizing a table of common pairs. This procedure utilizes the linearity property of the Laplace transform and its inverse. Some other useful properties of the Laplace transform include right shift in time, differentiation, integration, convolution, and the final-value theorem.

The differentiation property of the Laplace transform can be used when we solve a linear differential equation. The solution procedure is as follows: First take the Laplace transform of both sides of the differential equation, using the differentiation property as needed to find the Laplace transform of the derivatives of the output $y(t)$ or the input $x(t)$. The output $Y(s)$ is solved in terms of the initial conditions and the input $X(s)$. For a specific $X(s)$, find the output $y(t)$ by taking the inverse Laplace transform of $Y(s)$. Thus, a differential equation in t is transformed into an algebraic equation in s, which is easier to solve in part because of the existence of Laplace transform tables of common pairs.

The convolution property is used to find the response of a linear time-invariant system to an arbitrary input. The convolution integral studied in Chapter 2 is an often-tedious way of solving for a system response to an arbitrary input, given the impulse response. Taking the Laplace transform of the convolution integral results in the expression $Y(s) = H(s)X(s)$, where $H(s)$, the Laplace transform of the impulse response, is known as the transfer function of the system. Alternatively, we can find the transfer function by dividing the Laplace transform of an output of a system by the Laplace transform of its corresponding input when there are zero initial conditions applied to the system. Using this procedure, we may find the transfer function of a system by manipulating the Laplace transform of the system's differential equation. We can find the transfer function of a circuit directly from the circuit diagram by using an impedance method where the circuit elements are replaced by their s-domain counterparts.

A block diagram is a graphical representation of an interconnection of subsystems, where each subsystem is represented by a block that contains the transfer function of the subsystem. Common interconnections include the series connection, where the output of one system is the input to the other system; the parallel connection, where the outputs of two systems are added together; and the feedback connection, where the output of one system is fed back through another system to its own input. Each of these types of connections can be reduced to a single block that contains the transfer function of the combined system. A more complex block diagram can be similarly reduced to one block by a successive reduction of any series, parallel, or feedback connections in the block diagram to one block.

PROBLEMS

6.1. Determine the Laplace transform of the following signals:
   (a) $\cos(3t)u(t)$
   (b) $e^{-10t}u(t)$
   (c) $e^{-10t}\cos(3t)u(t)$
   (d) $e^{-10t}\cos(3t-1)u(t)$
   (e) $(2 - 2e^{-4t})u(t)$
   (f) $(t - 1 + e^{-10t}\cos(4t - \pi/3))u(t)$

6.2. A continuous-time signal $x(t)$ has the Laplace transform

$$X(s) = \frac{s+1}{s^2 + 5s + 7}$$

Determine the Laplace transform $V(s)$ of the following signals:
   (a) $v(t) = x(3t - 4)u(3t - 4)$
   (b) $v(t) = tx(t)$
   (c) $v(t) = \dfrac{d^2x(t)}{dt^2}$
   (d) $v(t) = \displaystyle\int_0^t x(\tau)\,d\tau$
   (e) $v(t) = x(t)\sin 2t$
   (f) $v(t) = e^{-3t}x(t)$
   (g) $v(t) = x(t) * x(t)$

6.3. Compute the Laplace transform of each of the signals displayed in Figure P6.3.

6.4. Using the transform pairs in Table 6.2 and the properties of the Laplace transform in Table 6.1, determine the Laplace transform of the following signals:
   (a) $x(t) = (e^{-bt}\cos^2\omega t)u(t)$
   (b) $x(t) = (e^{-bt}\sin^2\omega t)u(t)$
   (c) $x(t) = (t\cos^2\omega t)u(t)$
   (d) $x(t) = (t\sin^2\omega t)u(t)$
   (e) $x(t) = (\cos^3\omega t)u(t)$
   (f) $x(t) = (\sin^3\omega t)u(t)$

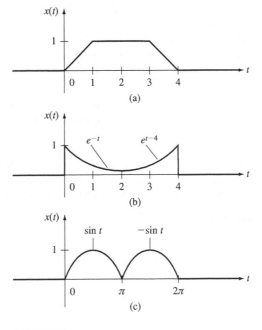

FIGURE P6.3

(g) $x(t) = (t^2 \cos \omega t)u(t)$

(h) $x(t) = (t^2 \sin \omega t)u(t)$

6.5. Determine the final values $[\lim_{t \to \infty} x(t)]$ of each of the signals whose Laplace transforms are given next. If there is no final value, state why not. Do not attempt to compute the inverse Laplace transforms.

(a) $X(s) = \dfrac{4}{s^2 + s}$

(b) $X(s) = \dfrac{3s + 4}{s^2 + s}$

(c) $X(s) = \dfrac{4}{s^2 - s}$

(d) $X(s) = \dfrac{3s^2 + 4s + 1}{s^3 + 2s^2 + s + 2}$

(e) $X(s) = \dfrac{3s^2 + 4s + 1}{s^3 + 3s^2 + 3s + 2}$

(f) $X(s) = \dfrac{3s^2 + 4s + 1}{s^4 + 3s^3 + 3s^2 + 2s}$

6.6. Determine the initial values $x(0)$ for each of the signals whose transforms are given in Problem 6.5.

6.7. A signal $x(t)$ that is zero for all $t < 0$ repeats itself every T seconds for $t \geq 0$; that is, $x(t + T) = x(t)$ for all $t \geq 0$. Let $x_0(t) = x(t)[u(t) - u(t - T)]$, and suppose that the Laplace transform of $x_0(t)$ is $X_0(s)$. Derive a closed-form expression for the Laplace transform $X(s)$ of $x(t)$ in terms of $X_0(s)$.

6.8. By using the Laplace transform, compute the convolution $x(t) * v(t)$, where the following conditions apply:

(a) $x(t) = e^{-t}u(t), v(t) = (\sin t)u(t)$

(b) $x(t) = (\cos t)u(t), v(t) = (\sin t)u(t)$

(c) $x(t) = (\sin t)u(t), v(t) = (t \sin t)u(t)$

(d) $x(t) = (\sin^2 t)u(t), v(t) = tu(t)$

6.9. Let p be a complex number given by $p = \sigma + j\omega$. Use Euler's formula and trigonometric identities to verify the following expression:

$$ce^{pt} + \bar{c}e^{\bar{p}t} = 2|c|e^{\sigma t}\cos(\omega t + \angle c)$$

6.10. Determine the inverse Laplace transform of each of the functions that follow. Compute the partial fraction expansion analytically for each case. You may use the MATLAB command residue to check your answers for parts (a) to (e).

(a) $X(s) = \dfrac{s + 2}{s^2 + 7s + 12}$

(b) $X(s) = \dfrac{s + 1}{s^3 + 5s^2 + 7s}$

(c) $X(s) = \dfrac{2s^2 - 9s - 35}{s^2 + 4s + 2}$

(d) $X(s) = \dfrac{3s^2 + 2s + 1}{s^3 + 5s^2 + 8s + 4}$

(e) $X(s) = \dfrac{s^2 + 1}{s^5 + 18s^3 + 81s}$

(f) $X(s) = \dfrac{s + e^{-s}}{s^2 + s + 1}$

(g) $X(s) = \dfrac{s}{s + 1} + \dfrac{se^{-s} + e^{-2s}}{s^2 + 2s + 1}$

6.11. Compute the inverse Laplace transform of the signals defined in Problem 6.10, using the Symbolic Math Toolbox. Plot the results, and compare them with those found analytically in Problem 6.10.

6.12. Determine the inverse Laplace transform of each of the functions that follow. Compute the partial fraction expansion analytically for each case. You may use residue to check your answers for parts (a) to (h).

(a) $X(s) = \dfrac{s^2 - 2s + 1}{s(s^2 + 4)}$

(b) $X(s) = \dfrac{s^2 - 2s + 1}{s(s^2 + 4)^2}$

(c) $X(s) = \dfrac{s^2 - 2s + 1}{s^2(s^2 + 4)}$

(d) $X(s) = \dfrac{s^2 - 2s + 1}{s^2(s^2 + 4)^2}$

(e) $X(s) = \dfrac{s^2 - 2s + 1}{(s + 2)^2 + 4}$

(f) $X(s) = \dfrac{s^2 - 2s + 1}{s[(s + 2)^2 + 4]}$

(g) $X(s) = \dfrac{s^2 - 2s + 1}{[(s + 2)^2 + 4]^2}$

(h) $X(s) = \dfrac{s^2 - 2s + 1}{s[(s + 2)^2 + 4]^2}$

(i) $X(s) = \dfrac{1}{s + se^{-s}}$

(j) $X(s) = \dfrac{1}{(s + 1)(1 + e^{-s})}$

6.13. Compute the inverse Laplace transform of the signals defined in Problem 6.12, using the Symbolic Math Toolbox. Plot the results, and compare them with those found analytically in Problem 6.12.

6.14. Use Laplace transforms to compute the solution to the following differential equations:

(a) $\dfrac{dy}{dt} + 2y = u(t), \quad y(0) = 0$

(b) $\dfrac{dy}{dt} - 2y = u(t), \quad y(0) = 1$

(c) $\dfrac{dy}{dt} + 10y = 4\sin(2t)u(t), \quad y(0) = 1$

(d) $\dfrac{dy}{dt} + 10y = 8e^{-10t}u(t), \quad y(0) = 0$

(e) $\dfrac{d^2y}{dt^2} + 6\dfrac{dy}{dt} + 8y = u(t), \quad y(0) = 0, \dot{y}(0) = 1$

(f) $\dfrac{d^2y}{dt^2} + 6\dfrac{dy}{dt} + 9y = \sin(2t)u(t), \quad y(0) = 0, \dot{y}(0) = 0$

(g) $\dfrac{d^2y}{dt^2} + 6\dfrac{dy}{dt} + 13y = u(t), \quad y(0) = 1, \dot{y}(0) = 1$

6.15. A continuous-time system is given by the input/output differential equation

$$\frac{d^2y(t)}{dt^2} + 2\frac{dy(t)}{dt} + 3y(t) = \frac{dx(t)}{dt} + x(t - 2)$$

(a) Compute the transfer function $H(s)$ of the system.
(b) Compute the impulse response $h(t)$.
(c) Compute the step response.
(d) Verify the results of (b) and (c) by simulation, using lsim (or step and impulse) or Simulink.

6.16. A continuous-time system is given by the input/output differential equation

$$\frac{d^2y(t)}{dt^2} + 4\frac{dy(t)}{dt} + 3y(t) = 2\frac{d^2x(t)}{dt^2} - 4\frac{dx(t)}{dt} - x(t)$$

In each of the following parts, compute the response $y(t)$ for all $t \geq 0$.
(a) $y(0^-) = -2, \dot{y}(0^-) = 1, x(t) = 0$ for all $t \geq 0^-$.
(b) $y(0^-) = 0, \dot{y}(0^-) = 0, x(t) = \delta(t), \delta(t) = $ unit impulse
(c) $y(0^-) = 0, \dot{y}(0^-) = 0, x(t) = u(t)$
(d) $y(0^-) = -2, \dot{y}(0^-) = 1, x(t) = u(t)$
(e) $y(0^-) = -2, \dot{y}(0^-) = 1, x(t) = u(t + 1)$

6.17. Consider the field-controlled dc motor shown in Figure P6.17. The input/output differential equation of the motor is

$$IL_f \frac{d^3 y(t)}{dt^3} + (k_d L_f + R_f I)\frac{d^2 y(t)}{dt^2} + R_f k_d \frac{dy(t)}{dt} = kx(t)$$

where the input $x(t)$ is the voltage applied to the field winding and the output $y(t)$ is the angle of the motor shaft and load, I is the moment of inertia of the motor and load, k_d is the viscous friction coefficient of the motor and load, and k is a constant.

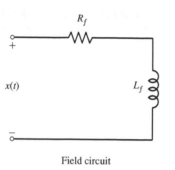

Field circuit   Load

FIGURE P6.17

(a) Determine the transfer function of the system.

(b) Find the impulse response $h(t)$ of the system.

6.18. Consider the single-eye system studied in Problem 2.35. The model for eye movement is given by the equations

$$T_e \frac{d\theta_E(t)}{dt} + \theta_E(t) = R(t)$$

$$R(t) = b\theta_T(t - d) - b\theta_T(t - d - c) + \theta_T(t - d)$$

where the input is the angular position $\theta_T(t)$ of the target and the output is the angular position $\theta_E(t)$ of the eye. (See Figure P2.35.)

(a) Determine the transfer function $H(s)$ of the system.

(b) Find the impulse response $h(t)$ of the system.

(c) Using the transfer function representation, compute $\theta_E(t)$ for all $t > 0$ when $\theta_T(t) = Au(t)$ and $\theta_E(0) = 0$.

(d) Repeat part (c) when $\theta_T(t) = Atu(t)$ and $\theta_E(0) = 0$.

(e) Sketch the output $\theta_E(t)$ found in parts (c) and (d), assuming that $T_e = c = 0.1$, $d = 0.2$, $A = 1$, and $b = 0.58$. Does the eye lock onto the target? Discuss your results.

6.19. As discussed in McClamroch [1980], the ingestion and metabolism of a drug in a human is described by the equations

$$\frac{dq(t)}{dt} = -k_1 q(t) + x(t)$$

$$\frac{dy(t)}{dt} = k_1 q(t) - k_2 y(t)$$

where the input $x(t)$ is the ingestion rate of the drug, the output $y(t)$ is the mass of the drug in the bloodstream, and $q(t)$ is the mass of the drug in the gastrointestinal tract. The constants k_1 and k_2 are metabolism rates that satisfy the inequality $k_1 > k_2 > 0$. The constant k_2 characterizes the excretory process of the individual.

(a) Determine the transfer function $H(s)$.

(b) Determine the impulse response $h(t)$. Assume that $k_1 \neq k_2$.

(c) By using the Laplace transform, compute $y(t)$ for $t > 0$ when $x(t) = 0$ for $t \geq 0$, $q(0) = M_1$ and $y(0) = M_2$. Assume that $k_1 \neq k_2$.

(d) Sketch your answer in part (c), assuming that $M_1 = 100$ mg, $M_2 = 10$ mg, $k_1 = 0.05$, and $k_2 = 0.02$. Does your result "make sense"? Explain.

(e) By using the Laplace transform, compute $y(t)$ for $t > 0$ when $x(t) = e^{-at}u(t)$, $q(0) = y(0) = 0$. Assume that $a \neq k_1 \neq k_2$.

(f) Sketch your answer in part (e), assuming that $a = 0.1$, $k_1 = 0.05$, and $k_2 = 0.02$. When is the mass of the drug in the bloodstream equal to its maximum value? When is the mass of the drug in the gastrointestinal tract equal to its maximum value?

6.20. Determine the transfer function of the mass/spring system in Problem 2.23.

6.21. For each of the continuous-time systems defined next, determine the system's transfer function $H(s)$ if the system has a transfer function. If there is no transfer function, state why not.

(a) $\dfrac{dy(t)}{dt} + e^{-t}y(t) = x(t)$

(b) $\dfrac{dy(t)}{dt} + v(t) * y(t) = x(t)$, where $v(t) = (\sin t)u(t)$

(c) $\dfrac{d^2y(t)}{dt^2} + \displaystyle\int_0^t y(\lambda)\, d\lambda = \dfrac{dx(t)}{dt} - x(t)$

(d) $\dfrac{dy(t)}{dt} = y(t) * x(t)$

(e) $\dfrac{dy(t)}{dt} - 2y(t) = tx(t)$

6.22. A linear time-invariant continuous-time system has transfer function $H(s) = (s + 7)/(s^2 + 4)$. Derive an expression for the output response $y(t)$ in terms of $y(0^-)$, $\dot{y}(0^-)$, and the input $x(t)$. Assume that $x(0^-) = 0$.

6.23. The input $x_1(t) = e^{-t}u(t)$ is applied to a linear time-invariant continuous-time system with nonzero initial conditions $y(0)$, $\dot{y}(0)$. The resulting response is $y_1(t) = 3t + 2 - e^{-t}$, $t \geq 0$. A second input $x_2(t) = e^{-2t}u(t)$ is applied to the system with the *same* initial conditions $y(0)$, $\dot{y}(0)$. The resulting response is $y_2(t) = 2t + 2 - e^{-2t}$, $t \geq 0$. Compute $y(0)$, $\dot{y}(0)$, and the impulse response $h(t)$ of the system.

6.24. The input

$$x(t) = \begin{cases} \sin t, & 0 \leq t \leq \pi \\ -\sin t, & \pi \leq t \leq 2\pi \\ 0, & \text{all other } t \end{cases}$$

is applied to a linear time-invariant continuous-time system with zero initial conditions in the system at time $t = 0$. The resulting response is displayed in Figure P6.24. Determine the transfer function $H(s)$ of the system.

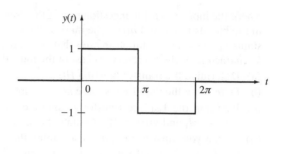

FIGURE P6.24

6.25. Using the *s*-domain representation, compute the transfer functions of the *RC* circuits shown in Figure P6.25.

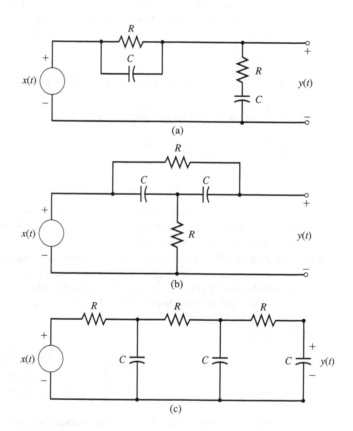

FIGURE P6.25

6.26. Using the *s*-domain representation, compute the transfer functions of the circuits shown in Figure P6.26.

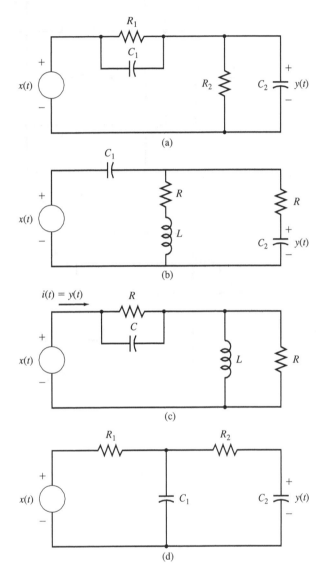

FIGURE P6.26

6.27. For the circuit in Figure P6.26c, determine all values of *R, L,* and *C* such that $H(s) = K$, where *K* is a constant.

6.28. Using the *s*-domain representation, compute the transfer function for the systems displayed in Figure P6.28.

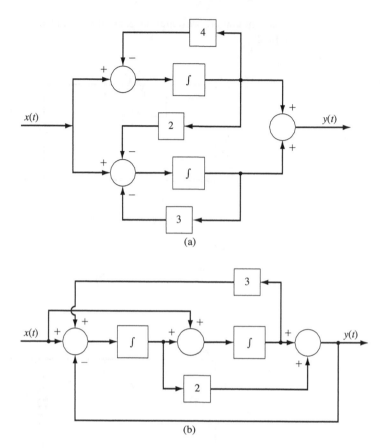

FIGURE P6.28

6.29. A linear time-invariant continuous-time system has the impulse response $h(t) = [\cos 2t + 4 \sin 2t]u(t)$.

    (a) Determine the transfer function $H(s)$ of the system.

    (b) By using the Laplace transform, compute the output response $y(t)$ when the input $x(t)$ is equal to $\frac{5}{7}e^{-t} - \frac{12}{7}e^{-8t}$ for $t \geq 0$ with zero initial conditions in the system at time $t = 0$.

    (c) With MATLAB, find $y(t)$ by using the Symbolic Math Toolbox, and compare this answer with the answer found in (b).

    (d) Use MATLAB to find $y(t)$ numerically, either with lsim or Simulink, and compare the simulated response with the response found analytically in either part (b) or (c), by plotting both answers.

6.30. A linear time-invariant continuous-time system has impulse response $h(t)$ given by

$$h(t) = \begin{cases} e^{-t}, & 0 \leq t \leq 2 \\ e^{t-4}, & 2 \leq t \leq 4 \\ 0, & \text{all other } t \end{cases}$$

(a) Determine the transfer function $H(s)$ of the system.

(b) By using the Laplace transform, compute the output response $y(t)$ resulting from the input $x(t) = (\sin t)u(t)$ with zero initial conditions.

(c) With MATLAB, find $y(t)$ by using the Symbolic Math Toolbox, and compare this answer with the answer found in (b).

(d) With MATLAB, find $y(t)$ numerically, either using `lsim` or Simulink, and compare the simulated response with the response found analytically in either part (b) or (c).

6.31. In this problem the objective is to design the oscillator illustrated in Figure P6.31. Using two integrators and subtracters, adders, and scalar multipliers, design the oscillator so that when the switch is closed at time $t = 0$, the output voltage $v(t)$ is $\sin 200t, t \geq 0$. We are assuming that the initial condition of the oscillator is zero at time $t = 0$.

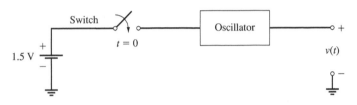

FIGURE P6.31

6.32. A linear time-invariant continuous-time system has impulse response $h(t) = [e^{-t}\cos(2t - 45°)]u(t) - tu(t)$. Determine the input/output differential equation of the system.

6.33. A system has the following transfer function:

$$H(s) = \frac{s + 2}{(s + 1)^2 + 4}$$

Use the Symbolic Math Toolbox to find the response to the following inputs:

(a) $x(t) = \delta(t)$

(b) $x(t) = u(t)$

(c) $x(t) = u(t) - u(t - 5)$

(d) $x(t) = \sin(2t)u(t)$

6.34. Repeat Problem 6.33, except use numerical simulation to solve for $y(t)$. You may use either Simulink or `lsim`.

The z-Transform and Discrete-Time Systems

This chapter deals with the z-transform, which is the discrete-time counterpart of the Laplace transform. The z-transform operates on a discrete-time signal $x[n]$ in contrast to the Laplace transform, which operates on a continuous-time or analog signal $x(t)$. In Sections 7.1 and 7.2, the z-transform of a discrete-time signal $x[n]$ is defined, and then the basic properties of the z-transform are studied. In Section 7.3 the computation of the inverse z-transform is considered, and then in Section 7.4 the z-transform is applied to the study of causal linear time-invariant discrete-time systems. The development begins in Section 7.4 with the generation of the z-domain representation from the input/output difference equation, and then the transfer function representation is generated by applying the z-transform to the input/output convolution sum representation of a system. The transfer function of interconnections containing unit-delay elements and interconnections of blocks is also studied in Section 7.4. In Section 7.5, the transfer function representation is utilized in the study of stability and in the study of the frequency response behavior of a discrete-time system. A summary of Chapter 7 is presented in Section 7.6.

The theory of the z-transform and its application to causal linear time-invariant discrete-time systems closely resembles the theory of the Laplace transform and its application to causal linear time-invariant continuous-time systems. In particular, results and techniques in this chapter closely parallel the results and techniques given in Chapters 6 and 8 on the Laplace transform. However, there are some differences between the transform theory in the continuous-time case and the transform theory in the discrete-time case, although for the most part, these differences are minor. In reading this chapter and Chapters 6 and 8, the reader should look for the similarities and differences in the two cases.

7.1   z-TRANSFORM OF A DISCRETE-TIME SIGNAL

Given the discrete-time signal $x[n]$, recall that in Chapter 4 the discrete-time Fourier transform (DTFT) was defined by

$$X(\Omega) = \sum_{n=-\infty}^{\infty} x[n]e^{-j\Omega n} \tag{7.1}$$

Recall that $X(\Omega)$ is, in general, a complex-valued function of the frequency variable Ω.

The z-transform of the signal $x[n]$ is generated by the addition of the factor ρ^{-n} to the summation in (7.1), where ρ is a real number. The factor ρ^{-n} plays the same role as the exponential factor $e^{-\sigma t}$ that was added to the Fourier transform to generate the Laplace transform in the continuous-time case. Inserting ρ^{-n} in (7.1) gives

$$X(\Omega) = \sum_{n=-\infty}^{\infty} x[n]\rho^{-n}e^{-j\Omega n} \tag{7.2}$$

which can be rewritten as

$$X(\Omega) = \sum_{n=-\infty}^{\infty} x[n](\rho e^{j\Omega})^{-n} \tag{7.3}$$

The function $X(\Omega)$ given by (7.3) is now a function of the complex number

$$z = \rho e^{j\Omega}$$

So, X should be written as a function of z, which gives

$$X(z) = \sum_{n=-\infty}^{\infty} x[n]z^{-n} \tag{7.4}$$

The function $X(z)$ given by (7.4) is the *two-sided* z-transform of the discrete-time signal $x[n]$. The *one-sided* z-transform of $x[n]$, also denoted by $X(z)$, is defined by

$$X(z) = \sum_{n=0}^{\infty} x[n]z^{-n} = x[0] + x[1]z^{-1} + x[2]z^{-2} + \cdots \tag{7.5}$$

As seen from (7.5), the one-sided z-transform is a power series in z^{-1} whose coefficients are the values of the signal $x[n]$.

Note that if $x[n] = 0$ for $n = -1, -2, \ldots$, the one- and two-sided z-transforms are identical. The one-sided z-transform can be applied to signals $x[n]$ that are nonzero for $n = -1, -2, \ldots$, but any nonzero values of $x[n]$ for $n < 0$ cannot be recovered from the one-sided z-transform. In this book, only the one-sided z-transform is pursued, which will be referred to as the z-transform.

Given a discrete-time signal $x[n]$ with z-transform $X(z)$, the set of all complex numbers z such that the summation on the right-hand side of (7.5) converges (i.e., exists) is called the *region of convergence* of the z-transform $X(z)$. The z-transform $X(z)$ exists (is well defined) for any value of z belonging to the region of convergence.

Example 7.1   z-Transform of Unit Pulse

Let $\delta[n]$ denote the unit pulse concentrated at $n = 0$ given by

$$\delta[n] = \begin{cases} 1, & n = 0 \\ 0, & n \neq 0 \end{cases}$$

Since $\delta[n]$ is zero for all n except $n = 0$, the z-transform is

$$\sum_{n=0}^{\infty} \delta[n]z^{-n} = \delta[0]z^{-0} = 1 \tag{7.6}$$

Thus, the z-transform of the unit pulse $\delta[n]$ is equal to 1. In addition, it is obvious that the summation in (7.6) exists for any value of z, and thus the region of convergence of the z-transform of the unit pulse is the set of all complex numbers.

Note that the unit pulse $\delta[n]$ is the discrete-time counterpart to the unit impulse $\delta(t)$ in the sense that the z-transform of $\delta[n]$ is equal to 1 and the Laplace transform of $\delta(t)$ is also equal to 1. However, as noted in Section 1.2, the pulse $\delta[n]$ is not a sampled version of $\delta(t)$.

Example 7.2   z-Transform of Shifted Pulse

Given a positive integer q, consider the unit pulse $\delta[n - q]$ located at $n = q$. For example, when $q = 2$, $\delta[n - 2]$ is the pulse shown in Figure 7.1. For any positive integer value of q, the z-transform of $\delta[n - q]$ is

$$\sum_{n=0}^{\infty}\delta[n - q]z^{-n} = \delta[0]z^{-q} = z^{-q} = \frac{1}{z^{q}}$$

The region of convergence is the set of all complex numbers z such that z is not zero.

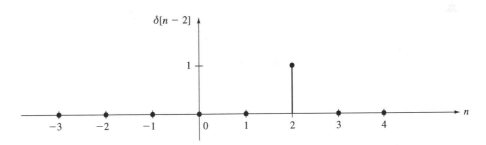

FIGURE 7.1
Unit pulse $\delta[n - 2]$ located at $n = 2$.

Example 7.3   Unit-Step Function

Consider the discrete-time unit-step function $u[n]$ given by

$$u[n] = \begin{cases} 1, & n = 0, 1, 2, \ldots \\ 0, & n = -1, -2, \ldots \end{cases}$$

The z-transform $U(z)$ is

$$U(z) = \sum_{n=0}^{\infty}u[n]z^{-n}$$

$$= \sum_{n=0}^{\infty}z^{-n}$$

$$= 1 + z^{-1} + z^{-2} + z^{-3} + \cdots \tag{7.7}$$

The transform $U(z)$ can be expressed as a rational function of z: Multiplying both sides of (7.7) by $z - 1$ gives

$$(z - 1)U(z) = (z + 1 + z^{-1} + z^{-2} + \cdots) - (1 + z^{-1} + z^{-2} + \cdots) \tag{7.8}$$
$$= z$$

Dividing both sides of (7.8) by $z - 1$ yields

$$U(z) = \frac{z}{z - 1} = \frac{1}{1 - z^{-1}} \tag{7.9}$$

Note that the form of the z-transform $U(z)$ of the discrete-time unit-step function $u[n]$ is different from the form of the Laplace transform $1/s$ of the continuous-time unit-step function $u(t)$.

The region of convergence for the z-transform $U(z)$ given by (7.9) includes the set of all complex numbers z such that $|z| > 1$. This follows from the result that

$$\left| \sum_{n=0}^{\infty} (1)z^{-n} \right| < \infty \qquad \text{if } |z| > 1 \tag{7.10}$$

To prove (7.10), first note that using (4.5) (see Problem 4.2) for any positive integer q results in

$$\sum_{n=0}^{q} z^{-n} = \frac{(1/z)^{q+1} - 1}{(1/z) - 1} \tag{7.11}$$

where it is assumed that $z \neq 1$. Then, using (7.11)

$$\left| \sum_{n=0}^{q} z^{-n} \right| = \left| \frac{(1/z)^{q+1} - 1}{(1/z) - 1} \right| \leq \frac{(1/|z|)^{q+1} + 1}{|(1/z) - 1|} \tag{7.12}$$

and using (7.12), if $|z| > 1$, yields

$$\left| \sum_{n=0}^{\infty} z^{-n} \right| = \lim_{q \to \infty} \left| \sum_{n=0}^{q} z^{-n} \right| \leq \frac{1}{|(1/z) - 1|} < \infty$$

Thus, (7.10) is verified.

Example 7.4   z-Transform of $a^n u[n]$

Given a real or complex number a, let $x[n] = a^n u[n]$. The z-transform $X(z)$ of $x[n]$ is given by

$$X(z) = \sum_{n=0}^{\infty} a^n z^{-n}$$

$$= 1 + az^{-1} + a^2 z^{-2} + \cdots \tag{7.13}$$

This transform can also be written as a rational function in z: Multiplying both sides of (7.13) by $z - a$ gives

$$(z - a)X(z) = (z + a + a^2 z^{-1} + a^3 z^{-2} + \cdots) - (a + a^2 z^{-1} + a^3 z^{-2} + \cdots)$$
$$= z$$

Hence,

$$X(z) = \frac{z}{z-a} = \frac{1}{1-az^{-1}} \tag{7.14}$$

Note that if $a = 1$, (7.14) is the same as (7.9). The region of convergence for the transform $X(z) = z/(z-a)$ includes the set of all complex numbers z such that $|z| > |a|$. This follows from an argument similar to that given in Example 7.3. The details are left to the reader.

7.1.1   Relationship between the DTFT and the z-Transform

As shown, the z-transform $X(z)$ of a discrete-time signal $x[n]$ can be viewed as a generalization of the discrete-time Fourier transform (DTFT) $X(\Omega)$. In fact, from (7.5) it appears that

$$X(\Omega) = X(z)|_{z=e^{j\Omega}} \tag{7.15}$$

However, (7.15) is *not valid*, in general, unless the region of convergence of $X(z)$ includes all complex numbers z such that $|z| = 1$. If this is the case, the DTFT $X(\Omega)$ of $x[n]$ is given by (7.15). For example, suppose that $x[n] = a^n u[n]$, where a is a real or complex number. In Example 7.4 it was shown that the z-transform of $x[n]$ is $X(z) = z/(z-a)$ and that the region of convergence of the z-transform includes the set of all complex numbers z such that $|z| > |a|$. Thus, if $|a| < 1$, the DTFT of $x[n]$ exists (in the ordinary sense) and is given by

$$X(\Omega) = X(z)|_{z=e^{j\Omega}} = \frac{e^{j\Omega}}{e^{j\Omega}-a} = \frac{1}{1-ae^{-j\Omega}} \tag{7.16}$$

Given a signal $x[n]$ with z-transform $X(z)$, the transform pair notation

$$x[n] \leftrightarrow X(z)$$

will sometimes be used to denote the fact that $X(z)$ is the z-transform of $x[n]$, and conversely, that $x[n]$ is the inverse z-transform of $X(z)$. The transform pairs derived in the preceding examples are shown in Table 7.1.

As seen in Table 7.1, the z-transform can sometimes be expressed as a ratio of polynomials in z or z^{-1}. In this book, preference will be given to expressing z-transforms in terms of positive powers of z, as opposed to negative powers of z.

TABLE 7.1   Basic z-Transform Pairs

$\delta[n] \leftrightarrow 1$

$\delta[n-q] \leftrightarrow \dfrac{1}{z^q} = z^{-q}$

$u[n] \leftrightarrow \dfrac{z}{z-1} = \dfrac{1}{1-z^{-1}}$

$a^n u[n] \leftrightarrow \dfrac{z}{z-a} = \dfrac{1}{1-az^{-1}},\ a$ real or complex

---

7.2   PROPERTIES OF THE z-TRANSFORM

The z-transform possesses a number of properties that are useful in deriving transform pairs and in the application of the transform to the study of causal linear time-invariant discrete-time systems. These properties are very similar to the properties of the Laplace transform that were given in Section 6.2. In this section the properties of the z-transform are stated and proved. As an illustration of the use of the properties, a collection of common transform pairs is generated from the basic set of pairs given in Table 7.1.

7.2.1   Linearity

The z-transform is a linear operation, as is the Laplace transform. Hence, if $x[n] \leftrightarrow X(z)$ and $v[n] \leftrightarrow V(z)$, then for any real or complex scalars a, b,

$$ax[n] + bv[n] \leftrightarrow aX(z) + bV(z) \tag{7.17}$$

The proof of (7.17) follows directly from the definition of the z-transform. The details are omitted.

Example 7.5   Linearity

Let $x[n] = u[n]$ and $v[n] = a^n u[n]$, where $a \neq 1$. From Table 7.1,

$$u[n] \leftrightarrow \frac{z}{z-1} \quad \text{and} \quad a^n u[n] \leftrightarrow \frac{z}{z-a}$$

Hence, by linearity,

$$u[n] + a^n u[n] \leftrightarrow \frac{z}{z-1} + \frac{z}{z-a} = \frac{2z^2 - (1+a)z}{(z-1)(z-a)}$$

7.2.2   Right Shift of $x[n]u[n]$

Suppose that $x[n] \leftrightarrow X(z)$. Given a positive integer q, consider the discrete-time signal $x[n-q]u[n-q]$, which is the q-step right shift of $x[n]u[n]$. Then

$$x[n-q]u[n-q] \leftrightarrow z^{-q}X(z) \tag{7.18}$$

To prove this property, first note that by definition of the z-transform,

$$x[n-q]u[n-q] \leftrightarrow \sum_{n=0}^{\infty} x[n-q]u[n-q]z^{-n}$$

Then, since $u[n-q] = 1$ for $n \geq q$ and $u[n-q] = 0$ for $n < q$,

$$x[n-q]u[n-q] \leftrightarrow \sum_{n=q}^{\infty} x[n-q]z^{-n} \tag{7.19}$$

Consider a change of index in the summation in (7.19): With $\bar{n} = n - q$ so that $n = \bar{n} + q$, then $\bar{n} = 0$ when $n = q$ and $\bar{n} = \infty$ when $n = \infty$. Hence, we see that

$$\sum_{n=q}^{\infty} x[n-q]z^{-n} = \sum_{\bar{n}=0}^{\infty} x[\bar{n}]z^{-(\bar{n}+q)}$$

$$= z^{-q} \sum_{\bar{n}=0}^{\infty} x[\bar{n}]z^{-\bar{n}}$$

$$= z^{-q}X(z)$$

Therefore, combining this with (7.19) yields the transform pair (7.18).

Example 7.6   z-Transform of a Pulse

Given a positive integer q, the objective is to determine the z-transform of the pulse $p[n]$ defined by

$$p[n] = \begin{cases} 1, & n = 0, 1, 2, \ldots, q-1 \\ 0, & \text{all other } n \end{cases}$$

Writing $p[n]$ in terms of the unit-step function $u[n]$ gives

$$p[n] = u[n] - u[n-q]$$

From Table 7.1, we find that the z-transform of $u[n]$ is $z/(z-1)$, and thus by the right-shift property (7.18), the z-transform of $u[n-q]$ is equal to

$$z^{-q}\frac{z}{z-1} = \frac{z^{-q+1}}{z-1}$$

Thus, by linearity the z-transform of the pulse $p[n]$ is

$$\frac{z}{z-1} - \frac{z^{-q+1}}{z-1} = \frac{z(1-z^{-q})}{z-1} = \frac{z^q - 1}{z^{q-1}(z-1)}$$

7.2.3   Right Shift of x[n]

Suppose that $x[n] \leftrightarrow X(z)$. Then

$$x[n-1] \leftrightarrow z^{-1}X(z) + x[-1] \tag{7.20}$$

$$x[n-2] \leftrightarrow z^{-2}X(z) + x[-2] + z^{-1}x[-1] \tag{7.21}$$

$$\vdots$$

$$x[n-q] \leftrightarrow z^{-q}X(z) + x[-q] + z^{-1}x[-q+1] + \cdots + z^{-q+1}x[-1] \tag{7.22}$$

Note that if $x[n] = 0$ for $n = -1, -2, \ldots, -q$, the transform pair (7.22) reduces to

$$x[n - q] \leftrightarrow z^{-q}X(z) \qquad (7.23)$$

which is identical to the transform pair (7.18).

To prove the transform pair (7.20), first note that by definition of the z-transform,

$$x[n - 1] \leftrightarrow \sum_{n=0}^{\infty} x[n - 1]z^{-n} \qquad (7.24)$$

Defining the change of index $\bar{n} = n - 1$ in the summation in (7.24) gives

$$x[n - 1] \leftrightarrow \sum_{\bar{n}=-1}^{\infty} x[\bar{n}]z^{-(\bar{n} + 1)} = \sum_{\bar{n}=0}^{\infty} x[\bar{n}]z^{-(\bar{n}+1)} + x[-1]$$

$$\leftrightarrow z^{-1} \sum_{\bar{n}=0}^{\infty} x[\bar{n}]z^{-\bar{n}} + x[-1]$$

$$\leftrightarrow z^{-1}X(z) + x[-1]$$

Thus, the transform pair (7.20) is verified. The verification of (7.21) and (7.22) for $q > 2$ can be demonstrated in a similar manner. The details are left to the interested reader.

7.2.4   Left Shift in Time

In contrast to the Laplace transform, the z-transform does have a left-shift property, as follows. Given the discrete-time signal $x[n]$ and a positive integer q, the q-step left shift of $x[n]$ is the signal $x[n + q]$. Now, suppose that $x[n] \leftrightarrow X(z)$. Then,

$$x[n + 1] \leftrightarrow zX(z) - x[0]z \qquad (7.25)$$

$$x[n + 2] \leftrightarrow z^2X(z) - x[0]z^2 - x[1]z \qquad (7.26)$$

$$\vdots$$

$$x[n + q] \leftrightarrow z^qX(z) - x[0]z^q - x[1]z^{q-1} - \cdots - x[q - 1]z \qquad (7.27)$$

To prove (7.25), first observe that

$$x[n + 1] \leftrightarrow \sum_{n=0}^{\infty} x[n + 1]z^{-n} \qquad (7.28)$$

Defining the change of index $\bar{n} = n + 1$ in the summation in (7.28) gives

$$x[n + 1] \leftrightarrow \sum_{\bar{n}=1}^{\infty} x[\bar{n}]z^{-(\bar{n}-1)}$$

$$\leftrightarrow z \sum_{\bar{n}=1}^{\infty} x[\bar{n}]z^{-\bar{n}} = z\left[\sum_{\bar{n}=0}^{\infty} x[\bar{n}]z^{-\bar{n}} - x[0]\right]$$

$$\leftrightarrow z[X(z) - x[0]]$$

Hence, (7.25) is verified.

Example 7.7   Left Shift of Unit-Step Function

Consider the one-step left shift $u[n + 1]$ of the discrete-time unit-step function $u[n]$. By the left-shift property (7.25), the z-transform of $u[n + 1]$ is equal to

$$zU(z) - u[0]z = \frac{z^2}{z - 1} - z = \frac{z^2 - z(z - 1)}{z - 1} = \frac{z}{z - 1}$$

Hence, the z-transform of $u[n + 1]$ is equal to the z-transform of $u[n]$. This result is not unexpected, since $u[n + 1] = u[n]$ for $n = 0, 1, 2, \ldots$

7.2.5   Multiplication by n and n^2

If $x[n] \leftrightarrow X(z)$, then

$$nx[n] \leftrightarrow -z\frac{d}{dz}X(z) \tag{7.29}$$

and

$$n^2x[n] \leftrightarrow z\frac{d}{dz}X(z) + z^2\frac{d^2}{dz^2}X(z) \tag{7.30}$$

To prove (7.29), first recall the definition of the z-transform:

$$X(z) = \sum_{n=0}^{\infty} x[n]z^{-n} \tag{7.31}$$

Taking the derivative with respect to z of both sides of (7.31) yields

$$\frac{d}{dz}X(z) = \sum_{n=0}^{\infty} (-n)x[n]z^{-n-1}$$

$$= -z^{-1}\sum_{n=0}^{\infty} nx[n]z^{-n} \tag{7.32}$$

Thus,

$$-z\frac{d}{dz}X(z) = \sum_{n=0}^{\infty} nx[n]z^{-n} \tag{7.33}$$

Now, the right-hand side of (7.33) is equal to the z-transform of the signal $nx[n]$, and thus (7.29) is verified. Taking the second derivative of $X(z)$ with respect to z proves (7.30). The details are left to the reader.

Example 7.8   z-Transform of $na^n u[n]$

Let $x[n] = a^n u[n]$, where a is any nonzero real or complex number. From Table 7.1, we see that

$$X(z) = \frac{z}{z - a}$$

Then,

$$z \frac{d}{dz} X(z) = z \left[\frac{-z}{(z - a)^2} + \frac{1}{z - a} \right] = \frac{-az}{(z - a)^2}$$

which gives the transform pair

$$na^n u[n] \leftrightarrow \frac{az}{(z - a)^2} \qquad (7.34)$$

Note that when $a = 1$, (7.34) becomes

$$nu[n] \leftrightarrow \frac{z}{(z - 1)^2} \qquad (7.35)$$

Example 7.9   z-Transform of $n^2 a^n u[n]$

To compute the z-transform of the signal $n^2 a^n u[n]$, first set $x[n] = a^n u[n]$, so that $X(z) = z/(z - a)$. Then,

$$\frac{d^2}{dz^2} X(z) = \frac{2a}{(z - a)^3}$$

and thus, using the results in Example 7.8 and the transform pair (7.30) gives

$$n^2 a^n u[n] \leftrightarrow \frac{-az}{(z - a)^2} + \frac{2az^2}{(z - a)^3}$$

$$\leftrightarrow \frac{az(z + a)}{(z - a)^3} \qquad (7.36)$$

Setting $a = 1$ in (7.36) results in the transform pair

$$n^2 u[n] \leftrightarrow \frac{z(z + 1)}{(z - 1)^3} \qquad (7.37)$$

7.2.6   Multiplication by a^n

If $x[n] \leftrightarrow X(z)$, then for any nonzero real or complex number a,

$$a^n x[n] \leftrightarrow X\left(\frac{z}{a}\right) \qquad (7.38)$$

By (7.38), multiplication by a^n in the time domain corresponds to scaling of the z variable in the transform domain. To prove (7.38), observe that

$$a^n x[n] \leftrightarrow \sum_{n=0}^{\infty} a^n x[n] z^{-n}$$

$$\leftrightarrow \sum_{n=0}^{\infty} x[n] \left(\frac{z}{a}\right)^{-n} = X\left(\frac{z}{a}\right)$$

Example 7.10   z-Transform of $a^n p[n]$

Let $p[n]$ denote the pulse defined by $p[n] = u[n] - u[n-q]$, where q is a positive integer. From Example 7.6, the z-transform of the pulse was found to be

$$\frac{z(1 - z^{-q})}{z - 1}$$

Then, using (7.38) results in the transform pair

$$a^n p[n] \leftrightarrow \frac{(z/a)[1 - (z/a)^{-q}]}{(z/a) - 1}$$

$$\leftrightarrow \frac{z(1 - a^q z^{-q})}{z - a}$$

7.2.7   Multiplication by cos Ωn and sin Ωn

If $x[n] \leftrightarrow X(z)$, then for any positive real number Ω,

$$(\cos \Omega n) x[n] \leftrightarrow \frac{1}{2}[X(e^{j\Omega} z) + X(e^{-j\Omega} z)] \tag{7.39}$$

and

$$(\sin \Omega n) x[n] \leftrightarrow \frac{j}{2}[X(e^{j\Omega} z) - X(e^{-j\Omega} z)] \tag{7.40}$$

To prove (7.39) and (7.40), first note that using Euler's formula yields

$$(\cos \Omega n) x[n] = \frac{1}{2}[e^{-j\Omega n} x[n] + e^{j\Omega n} x[n]] \tag{7.41}$$

$$(\sin \Omega n) x[n] = \frac{j}{2}[e^{-j\Omega n} x[n] - e^{j\Omega n} x[n]] \tag{7.42}$$

By (7.38),

$$e^{-j\Omega n} x[n] \leftrightarrow X(e^{j\Omega} z) \quad \text{and} \quad e^{j\Omega n} x[n] \leftrightarrow X(e^{-j\Omega} z) \tag{7.43}$$

Then, using (7.43) with (7.41) and (7.42) yields (7.39) and (7.40).

Example 7.11   z-Transform of Sinusoids

Let $v[n] = (\cos \Omega n)u[n]$. With $x[n]$ set equal to the unit step $u[n]$, $X(z) = z/(z-1)$, and using (7.39) gives

$$(\cos \Omega n)u[n] \leftrightarrow \frac{1}{2}\left(\frac{e^{j\Omega}z}{e^{j\Omega}z-1} + \frac{e^{-j\Omega}z}{e^{-j\Omega}z-1}\right)$$

$$\leftrightarrow \frac{1}{2}\left[\frac{e^{j\Omega}z(e^{-j\Omega}z-1) + e^{-j\Omega}z(e^{j\Omega}z-1)}{(e^{j\Omega}z-1)(e^{-j\Omega}z-1)}\right]$$

$$\leftrightarrow \frac{1}{2}\left[\frac{z^2 - e^{j\Omega}z + z^2 - e^{-j\Omega}z}{z^2 - (e^{j\Omega} + e^{-j\Omega})z + 1}\right]$$

$$\leftrightarrow \frac{z^2 - (\cos \Omega)z}{z^2 - (2\cos \Omega)z + 1} \tag{7.44}$$

Similarly, using (7.40) results in the transform pair

$$(\sin \Omega n)u[n] \leftrightarrow \frac{(\sin \Omega)z}{z^2 - (2\cos \Omega)z + 1} \tag{7.45}$$

Example 7.12 a^n Times a Sinusoid

Now let $v[n] = a^n(\cos \Omega n)u[n]$. We can compute the z-transform of $v[n]$ by setting $x[n] = a^n u[n]$ and using the multiplication by $\cos \Omega n$ property. However, it is easier to set $x[n]$ equal to $(\cos \Omega n)u[n]$ and then apply the multiplication by a^n property. Using (7.38) and the transform pair (7.44) gives

$$a^n(\cos \Omega n)u[n] \leftrightarrow \frac{(z/a)^2 - (\cos \Omega)(z/a)}{(z/a)^2 - (2\cos \Omega)(z/a) + 1}$$

$$\leftrightarrow \frac{z^2 - (a\cos \Omega)z}{z^2 - (2a\cos \Omega)z + a^2} \tag{7.46}$$

Using (7.38) and the transform pair (7.45) results in the transform pair

$$a^n(\sin \Omega n)u[n] \leftrightarrow \frac{(a\sin \Omega)z}{z^2 - (2a\cos \Omega)z + a^2} \tag{7.47}$$

7.2.8   Summation

Given the discrete-time signal $x[n]$ with $x[n] = 0$ for $n = -1, -2, \ldots$, let $v[n]$ denote the sum of $x[n]$, defined by

$$v[n] = \sum_{i=0}^{n} x[i] \tag{7.48}$$

To derive an expression for the z-transform of $v[n]$, first note that $v[n]$ can be expressed in the form

$$v[n] = \sum_{i=0}^{n-1} x[i] + x[n]$$

and using the definition (7.48) of $v[n]$ gives

$$v[n] = v[n-1] + x[n] \tag{7.49}$$

Then, taking the z-transform of both sides of (7.49) and using the right-shift property yield

$$V(z) = z^{-1}V(z) + X(z)$$

and solving for $V(z)$ gives

$$V(z) = \frac{1}{1 - z^{-1}} X(z)$$

$$= \frac{z}{z - 1} X(z) \tag{7.50}$$

Hence the z-transform of the sum of a signal $x[n]$ is equal to $z/(z - 1)$ times the z-transform of the signal.

Example 7.13 z-Transform of $(n + 1)u[n]$

Let $x[n] = u[n]$. Then the sum is

$$v[n] = \sum_{i=0}^{n} u[i] = (n + 1)u[n]$$

and thus the sum of the step is a ramp. By (7.50), the transform of the sum is

$$V(z) = \frac{z}{z - 1} X(z) = \frac{z^2}{(z - 1)^2}$$

This yields the transform pair

$$(n + 1)u[n] \leftrightarrow \frac{z^2}{(z - 1)^2} \tag{7.51}$$

7.2.9 Convolution

Given two discrete-time signals $x[n]$ and $v[n]$ with both signals equal to zero for $n = -1, -2, \ldots$, in Chapter 2 the convolution of $x[n]$ and $v[n]$ was defined by

$$x[n] * v[n] = \sum_{i=0}^{n} x[i]v[n - i]$$

Note that, since $v[n] = 0$ for $n = -1, -2, \ldots$, the convolution sum can be taken from $i = 0$ to $i = \infty$; that is, the convolution operation is given by

$$x[n] * v[n] = \sum_{i=0}^{\infty} x[i]v[n - i] \tag{7.52}$$

Taking the z-transform of both sides of (7.52) yields the transform pair

$$x[n] * v[n] \leftrightarrow \sum_{n=0}^{\infty} \left[\sum_{i=0}^{\infty} x[i]v[n-i] \right] z^{-n}$$

$$\leftrightarrow \sum_{i=0}^{\infty} x[i] \left[\sum_{n=0}^{\infty} v[n-i]z^{-n} \right] \tag{7.53}$$

Using the change of index $\bar{n} = n - i$ in the second summation of (7.53) gives

$$x[n] * v[n] \leftrightarrow \sum_{i=0}^{\infty} x[i] \left[\sum_{\bar{n}=-i}^{\infty} v[\bar{n}]z^{-\bar{n}-i} \right]$$

$$\leftrightarrow \sum_{i=0}^{\infty} x[i] \left[\sum_{\bar{n}=0}^{\infty} v[\bar{n}]z^{-\bar{n}-i} \right], \text{ since } v[\bar{n}] = 0 \text{ for } \bar{n} < 0$$

$$\leftrightarrow \left[\sum_{i=0}^{\infty} x[i]z^{-i} \right] \left[\sum_{\bar{n}=0}^{\infty} v[\bar{n}]z^{-\bar{n}} \right]$$

$$\leftrightarrow X(z)V(z) \tag{7.54}$$

From (7.54), it is seen that the z-transform of the convolution $x * v$ is equal to the product $X(z)V(z)$, where $X(z)$ and $V(z)$ are the z-transforms of $x[n]$ and $v[n]$, respectively. Therefore, convolution in the discrete-time domain corresponds to a product in the z-transform domain. This result is obviously analogous to the result in the continuous-time framework where convolution corresponds to multiplication in the s-domain. Examples of the use of the transform pair (7.54) will be given in Section 7.4 when the transfer function representation is developed for linear time-invariant discrete-time systems.

7.2.10   Initial-Value Theorem

If $x[n] \leftrightarrow X(z)$, the initial values of $x[n]$ can be computed directly from $X(z)$ by use of the relationships

$$x[0] = \lim_{z \to \infty} X(z) \tag{7.55}$$

$$x[1] = \lim_{z \to \infty} [zX(z) - zx[0]]$$

$$\vdots$$

$$x[q] = \lim_{z \to \infty} [z^q X(z) - z^q x[0] - z^{q-1}x[1] - \cdots - zx[q-1]] \tag{7.56}$$

To prove (7.55), first note that

$$z^{-n} \to 0 \text{ as } z \to \infty \text{ for all } n \geq 1$$

and thus,

$$x[n]z^{-n} \to 0 \text{ as } z \to \infty \text{ for all } n \geq 1$$

Therefore, taking the limit as $z \to \infty$ of both sides of

$$X(z) = \sum_{n=0}^{\infty} x[n]z^{-n}$$

yields (7.55).

In the next section it will be shown that, if the transform $X(z)$ is a rational function of z, the initial values of $x[n]$ can be calculated by a long-division operation.

7.2.11 Final-Value Theorem

Given a discrete-time signal $x[n]$ with z-transform $X(z)$, suppose that $x[n]$ has a limit as $n \to \infty$. Then the final-value theorem states that

$$\lim_{n \to \infty} x[n] = \lim_{z \to 1}(z - 1)X(z) \tag{7.57}$$

The proof of (7.57) is analogous to the proof that was given of the final-value theorem in the continuous-time case. The details are not pursued here.

As in the continuous-time case, we must exercise care in using the final-value theorem, since the limit on the right-hand side of (7.57) may exist even though $x[n]$ does not have a limit as $n \to \infty$. Existence of the limit of $x[n]$ as $n \to \infty$ can readily be checked if the transform $X(z)$ is rational in z, that is, $X(z)$ can be written in the form $X(z) = B(z)/A(z)$, where $B(z)$ and $A(z)$ are polynomials in z with real coefficients. Here, it is assumed that $B(z)$ and $A(z)$ do not have any common factors; if there are common factors, they should be canceled.

Now, letting $p_1, p_2, \ldots, p_N$ denote the poles of $X(z) = B(z)/A(z)$ [i.e., the roots of $A(z) = 0$], we find that $x[n]$ has a limit as $n \to \infty$ if and only if the magnitudes $|p_1|, |p_2|, \ldots, |p_N|$ are all strictly less than 1, except that one of the p_i's may be equal to 1. This is equivalent to the condition that all the poles of $(z - 1)X(z)$ have magnitudes strictly less than 1. The proof that the pole condition on $(z - 1)X(z)$ is necessary and sufficient for the existence of the limit follows from the results given in the next section. If this condition is satisfied, the limit of $x[n]$ as $n \to \infty$ is given by

$$\lim_{n \to \infty} x[n] = [(z - 1)X(z)]_{z=1} \tag{7.58}$$

As in the continuous-time case, the relationship (7.58) makes it possible to determine the limiting value of a time signal directly from the transform of the signal (without having to compute the inverse transform).

Example 7.14   Limiting Value

Suppose that

$$X(z) = \frac{3z^2 - 2z + 4}{z^3 - 2z^2 + 1.5z - 0.5}$$

In this case $X(z)$ has a pole at $z = 1$, and thus there is a pole–zero cancellation in $(z - 1)X(z)$. Performing the cancellation gives

$$(z - 1)X(z) = \frac{3z^2 - 2z + 4}{z^2 - z + 0.5}$$

Using the MATLAB command roots reveals that the poles of $(z - 1)X(z)$ are $z = 0.5 \pm j0.5$. The magnitude of both these poles is equal to 0.707, and therefore $x[n]$ has a limit as $n \to \infty$. From (7.58), the limit is

$$\lim_{n \to \infty} x[n] = [(z - 1)X(z)]_{z=1} = \left[\frac{3z^2 - 2z + 4}{z^2 - z + 0.5}\right]_{z=1} = \frac{5}{0.5} = 10$$

The aforementioned properties of the z-transform are summarized in Table 7.2. In Table 7.3, a collection of common z-transform pairs is given, which includes the transform pairs that were previously derived by the use of the properties of the z-transform.

7.3   COMPUTATION OF THE INVERSE z-TRANSFORM

If $X(z)$ is the z-transform of the discrete-time signal $x[n]$, we can compute the signal from $X(z)$ by taking the inverse z-transform of $X(z)$ given by

$$x[n] = \frac{1}{j2\pi} \int X(z)z^{n-1} \, dz \tag{7.59}$$

We can evaluate the integral in (7.59) by integrating along a counterclockwise closed circular contour that is contained in the region of convergence of $X(z)$.

When the transform $X(z)$ is a rational function of z, we can compute the inverse z-transform (thank goodness!) without having to evaluate the integral in (7.59). The computation of $x[n]$ from a rational $X(z)$ is considered in this section. When $X(z)$ is rational, $x[n]$ can be computed by expansion of $X(z)$ into a power series in z^{-1} or by expansion of $X(z)$ into partial fractions. The following development begins with the power-series expansion approach.

7.3.1   Expansion by Long Division

Let $X(z)$ be given in the rational form $X(z) = B(z)/A(z)$ with the polynomials $B(z)$ and $A(z)$ written in descending powers of z. To compute the inverse z-transform $x[n]$ for a finite range of values of n, we can expand $X(z)$ into a power series in z^{-1} by dividing

$A(z)$ into $B(z)$ using long division. The values of the signal $x[n]$ are then "read off" from the coefficients of the power-series expansion. The process is illustrated by the following example:

TABLE 7.2 Properties of the z-Transform

Property	Transform Pair/Property
Linearity	$ax[n] + bv[n] \leftrightarrow aX(z) + bV(z)$
Right shift of $x[n]u[n]$	$x[n - q]u[n - q] \leftrightarrow z^{-q}X(z)$
Right shift of $x[n]$	$x[n - 1] \leftrightarrow z^{-1}X(z) + x[-1]$
	$x[n - 2] \leftrightarrow z^{-2}X(z) + x[-2] + z^{-1}x[-1]$
	$\vdots$
	$x[n - q] \leftrightarrow z^{-q}X(z) + x[-q] + z^{-1}x[-q + 1] + \cdots + z^{-q+1}x[-1]$
Left shift in time	$x[n + 1] \leftrightarrow zX(z) - x[0]z$
	$x[n + 2] \leftrightarrow z^2X(z) - x[0]z^2 - x[1]z$
	$x[n + q] \leftrightarrow z^qX(z) - x[0]z^q - x[1]z^{q-1} - \cdots - x[q - 1]z$
Multiplication by n	$nx[n] \leftrightarrow -z\dfrac{d}{dz}X(z)$
Multiplication by n^2	$n^2x[n] \leftrightarrow z\dfrac{d}{dz}X(z) + z^2\dfrac{d^2}{dz^2}X(z)$
Multiplication by a^n	$a^nx[n] \leftrightarrow X\left(\dfrac{z}{a}\right)$
Multiplication by $\cos \Omega n$	$(\cos \Omega n)x[n] \leftrightarrow \dfrac{1}{2}[X(e^{j\Omega}z) + X(e^{-j\Omega}z)]$
Multiplication by $\sin \Omega n$	$(\sin \Omega n)x[n] \leftrightarrow \dfrac{j}{2}[X(e^{j\Omega}z) - X(e^{-j\Omega}z)]$
Summation	$\displaystyle\sum_{i=0}^{n}x[i] \leftrightarrow \dfrac{z}{z - 1}X(z)$
Convolution	$x[n] * v[n] \leftrightarrow X(z)V(z)$
Initial-value theorem	$x[0] = \lim_{z\to\infty} X(z)$
	$x[1] = \lim_{z\to\infty} [zX(z) - zX[0]]$
	$\vdots$
	$x[q] = \lim_{z\to\infty} [z^qX(z) - z^qx[0] - z^{q-1}x[1] - \cdots - zx[q - 1]]$
Final-value theorem	If $X(z)$ is rational and the poles of $(z - 1)X(z)$ have magnitudes <1, then
	$\lim_{n\to\infty} x[n] = [(z - 1)X(z)]_{z=1}$

TABLE 7.3   Common z-Transform Pairs

$\delta[n] \leftrightarrow 1$

$\delta[n - q] \leftrightarrow \dfrac{1}{z^q}, \quad q = 1, 2, \ldots$

$u[n] \leftrightarrow \dfrac{z}{z - 1}$

$u[n] - u[n - q] \leftrightarrow \dfrac{z^q - 1}{z^{q-1}(z - 1)}, \quad q = 1, 2, \ldots$

$a^n u[n] \leftrightarrow \dfrac{z}{z - a}, \, a \text{ real or complex}$

$nu[n] \leftrightarrow \dfrac{z}{(z - 1)^2}$

$(n + 1)u[n] \leftrightarrow \dfrac{z^2}{(z - 1)^2}$

$n^2 u[n] \leftrightarrow \dfrac{z(z + 1)}{(z - 1)^3}$

$na^n u[n] \leftrightarrow \dfrac{az}{(z - a)^2}$

$n^2 a^n u[n] \leftrightarrow \dfrac{az(z + a)}{(z - a)^3}$

$n(n + 1)a^n u[n] \leftrightarrow \dfrac{2az^2}{(z - a)^3}$

$(\cos \Omega n)u[n] \leftrightarrow \dfrac{z^2 - (\cos \Omega)z}{z^2 - (2 \cos \Omega)z + 1}$

$(\sin \Omega n)u[n] \leftrightarrow \dfrac{(\sin \Omega)z}{z^2 - (2 \cos \Omega)z + 1}$

$a^n(\cos \Omega n)u[n] \leftrightarrow \dfrac{z^2 - (a \cos \Omega)z}{z^2 - (2a \cos \Omega)z + a^2}$

$a^n(\sin \Omega n)u[n] \leftrightarrow \dfrac{(a \sin \Omega)z}{z^2 - (2a \cos \Omega)z + a^2}$

Example 7.15   Inverse z-Transform via Long Division

Suppose that

$$X(z) = \frac{z^2 - 1}{z^3 + 2z + 4}$$

Dividing $A(z)$ into $B(z)$ gives

$$
z^3 + 2z + 4 \overline{) z^2 - 1}
$$

$$
\begin{array}{r}
z^{-1} + 0z^{-2} - 3z^{-3} - 4z^{-4} + \cdots \\
\hline
z^2 + 2 \quad\; + 4z^{-1} \\
\hline
-3 \quad\;\; - 4z^{-1} \\
-3 \qquad\qquad\quad - 6z^{-2} - 12z^{-3} \\
\hline
-4z^{-1} + 6z^{-2} + 12z^{-3} \\
-4z^{-1} \qquad\quad - 8z^{-3} - 16z^{-4} \\
\hline
6z^{-2} + 20z^{-3} + 16z^{-4} \\
\vdots
\end{array}
$$

Thus,

$$
X(z) = z^{-1} - 3z^{-3} - 4z^{-4} \cdots \tag{7.60}
$$

By definition of the z-transform,

$$
X(z) = x[0] + x[1]z^{-1} + x[2]z^{-2} + \cdots \tag{7.61}
$$

Equating (7.60) and (7.61) yields the following values for $x[n]$:

$$
x[0] = 0, \quad x[1] = 1, \quad x[2] = 0, \quad x[3] = -3, \quad x[4] = -4, \ldots
$$

From the results in Example 7.15, we see that we can compute the initial values $x[0], x[1], x[2], \ldots$ of a signal $x[n]$ by carrying out the first few steps of the expansion of $X(z) = B(z)/A(z)$, using long division. In particular, note that the initial value $x[0]$ is nonzero if and only if the degree of $B(z)$ is equal to the degree of $A(z)$. If the degree of $B(z)$ is strictly less than the degree of $A(z)$ minus 1, both $x[0]$ and $x[1]$ are zero, and so on.

Instead of carrying out the long division by hand, we can use MATLAB to compute $x[n]$, applying either the symbolic manipulator or the command `filter` that arises in the transfer function formulation (considered in the next section). Consider a z-transform of the form

$$
X(z) = \frac{b_M z^M + b_{M-1} z^{M-1} + \cdots + b_0}{a_N z^N + a_{N-1} z^{N-1} + \cdots + a_0}
$$

where $M \leq N$. To solve for the signal values $x[n]$ symbolically, use the command `iztrans`. To solve for the inverse z transform numerically for $n = 0$ to $n = q$, use the commands

```
num = [b_M b_{M-1} ... b_0];
den = [a_N a_{N-1} ... a_0];
x = filter(num,den,[1 zeros(1,q)]);
```

The computation of $x[n]$ from $X(z)$ by MATLAB is illustrated in the following example:

Example 7.16   Inverse z-Transform Using MATLAB

Consider the z-transform:

$$X(z) = \frac{8z^3 + 2z^2 - 5z}{z^3 - 1.75z + .75}$$

To solve for $x[n]$ symbolically, use the commands

```
syms X x z
X = (8*z^3+2*z^2-5*z)/(z^3-1.75*z+.75);
x = iztrans(X)
```

which returns

```
x =
2*(1/2)^n+2*(-3/2)^n+4
```

To evaluate $x[n]$ numerically for $n = 0$ to $n = 10$, we apply the following MATLAB commands:

```
num = [8 2 -5 0];
den = [1 0 -1.75 .75];
x = filter(num,den,[1 zeros(1,9)])
```

Running the program results in the following vector (the first element of which is $x[0]$):

```
8    2    9    -2.5    14.25    -11.125    26.8125    -30.1563    55.2656
```

Note that the values of $|x[n]|$ appear to be growing without bound as n increases. As will be seen from the development that follows, the unbounded growth of the magnitude of $x[n]$ is a result of $X(z)$ having a pole with magnitude >1.

7.3.2   Inversion via Partial Fraction Expansion

Using long division as previously described, we can compute the inverse z-transform $x[n]$ of $X(z)$ for any finite range of integer values of n. However, if an analytical (closed-form) expression for $x[n]$ is desired that is valid for all $n \geq 0$, it is necessary to use partial fraction expansion, as was done in the Laplace transform theory. The steps are as follows.

Again, suppose that $X(z)$ is given in the rational form $X(z) = B(z)/A(z)$. If the degree of $B(z)$ is equal to $A(z)$, the partial fraction expansion in Section 6.3 cannot be applied directly to $X(z)$. However, dividing $A(z)$ into $B(z)$ yields the following form for $X(z)$:

$$X(z) = x[0] + \frac{R(z)}{A(z)}$$

Here, $x[0]$ is the initial value of the signal $x[n]$ at time $n = 0$, and $R(z)$ is a polynomial in z whose degree is strictly less than that of $A(z)$. The rational function $R(z)/A(z)$ can then be expanded by partial fractions.

There is another approach that avoids having to divide $A(z)$ into $B(z)$; namely, first expand

$$\frac{X(z)}{z} = \frac{B(z)}{zA(z)}$$

The rational function $X(z)/z$ can be expanded into partial fractions, since the degree of $B(z)$ is strictly less than the degree of $zA(z)$ in the case when $B(z)$ and $A(z)$ have the same degrees. After $X(z)/z$ has been expanded, the result can be multiplied by z to yield an expansion for $X(z)$. The inverse z-transform of $X(z)$ can then be computed term by term. There are two cases to consider.

Distinct Poles.   Suppose that the poles $p_1, p_2, \ldots, p_N$ of $X(z)$ are distinct and are all nonzero. Then, $X(z)/z$ has the partial fraction expansion

$$\frac{X(z)}{z} = \frac{c_0}{z} + \frac{c_1}{z - p_1} + \frac{c_2}{z - p_2} + \cdots + \frac{c_N}{z - p_N} \tag{7.62}$$

where c_0 is the real number given by

$$c_0 = \left[z \frac{X(z)}{z} \right]_{z=0} = X(0) \tag{7.63}$$

and the other residues in (7.62) are real or complex numbers given by

$$c_i = \left[(z - p_i) \frac{X(z)}{z} \right]_{z=p_i}, \quad i = 1, 2, \ldots, N \tag{7.64}$$

Multiplying both sides of (7.62) by z yields the following expansion for $X(z)$:

$$X(z) = c_0 + \frac{c_1 z}{z - p_1} + \frac{c_2 z}{z - p_2} + \cdots + \frac{c_N z}{z - p_N} \tag{7.65}$$

Then, taking the inverse z-transform of each term in (7.65) and using Table 7.3 give

$$x[n] = c_0 \delta[n] + c_1 p_1^n + c_2 p_2^n + \cdots + c_N p_N^n, \quad n = 0, 1, 2, \ldots \tag{7.66}$$

From (7.66), it is clear that the form of the time variation of the signal $x[n]$ is determined by the poles $p_1, p_2, \ldots, p_N$ of the rational function $X(z)$. Hence, the poles of $X(z)$ determine the characteristics of the time variation of $x[n]$. The reader will recall that this is also the case in the Laplace transform theory, except that here the terms included in $x[n]$ are of the form cp^n; whereas in the continuous-time case, the terms are of the form ce^{pt}.

If all the poles of $X(z)$ are real, the terms comprising the signal defined by (7.66) are all real. However, if two or more of the poles are complex, the corresponding terms in (7.66) will be complex. Such terms can be combined to yield a real form. To see this, suppose that the pole $p_1 = a + jb$ is complex, so that $b \neq 0$. Then, one of the other poles of $X(z)$ must be equal to the complex conjugate $\bar{p}_1$ of p_1. Suppose that $p_2 = \bar{p}_1$; then in (7.66) it must be true that $c_2 = \bar{c}_1$. Hence, the second and third terms of the right-hand side of (7.66) are equal to

$$c_1 p_1^n + \bar{c}_1 \bar{p}_1^n \tag{7.67}$$

This term can be expressed in the form

$$2|c_1| \sigma^n \cos(\Omega n + \angle c_1) \tag{7.68}$$

where

$$\sigma = |p_1| = \text{magnitude of the pole } p_1$$

and

$$\Omega = \angle p_1 = \text{angle of } p_1$$

The verification that (7.67) and (7.68) are equivalent is left to the homework problems. (See Problem 7.7.) Using (7.68) in (7.66) results in the following expression for $x[n]$:

$$x[n] = c_0\delta[n] + 2|c_1|\sigma^n \cos(\Omega n + \angle c_1) + c_3 p_3^n$$
$$+ \cdots + c_n p_N^n, \, n = 0, 1, 2, \ldots \tag{7.69}$$

The expression (7.69) shows that, if $X(z)$ has a pair of complex poles p_1, p_2 with magnitude σ and angle $\pm\Omega$, the signal $x[n]$ contains a term of the form

$$2|c|\sigma^n \cos(\Omega n + \angle c)$$

The computation of the inverse z-transform by the foregoing procedure is illustrated in the following example:

Example 7.17   Complex Pole Case

Suppose that

$$X(z) = \frac{z^3 + 1}{z^3 - z^2 - z - 2}$$

Here,

$$A(z) = z^3 - z^2 - z - 2$$

Using the MATLAB command `roots` reveals that the roots of $A(z)$ are

$$p_1 = -0.5 - j0.866$$
$$p_2 = -0.5 + j0.866$$
$$p_3 = 2$$

Then, expanding $X(z)/z$ gives

$$\frac{X(z)}{z} = \frac{c_0}{z} + \frac{c_1}{z + 0.5 + j0.866} + \frac{\bar{c}_1}{z + 0.5 - j0.866} + \frac{c_3}{z - 2}$$

where

$$c_0 = X(0) = \frac{1}{-2} = -0.5$$

$$c_1 = \left[(z + 0.5 + j0.866)\frac{X(z)}{z} \right]_{z=-0.5-j0.866} = 0.429 + j0.0825$$

$$c_3 = \left[(z - 2)\frac{X(z)}{z} \right]_{z=2} = 0.643$$

From (7.66), we find that the inverse z-transform is

$$x[n] = -0.5\delta[n] + c_1(-0.5 - j0.866)^n + \bar{c}_1(-0.5 + j0.866)^n + 0.643(2)^n, \quad n = 0, 1, 2, \ldots$$

We can write the second and third terms in $x[n]$ in real form by using the form (7.68). Here, the magnitude and angle of p_1 are given by

$$|p_1| = \sqrt{(0.5)^2 + (0.866)^2} = 1$$

$$\angle p_1 = \pi + \tan^{-1}\frac{0.866}{0.5} = \frac{4\pi}{3}\text{rad}$$

and the magnitude and angle of c_1 are given by

$$|c_1| = \sqrt{(0.429)^2 + (0.0825)^2} = 0.437$$

$$\angle c_1 = \tan^{-1}\frac{0.0825}{0.429} = 10.89°$$

Then, rewriting $x[n]$ in the form (7.69) yields

$$x[n] = -0.5\delta[n] + 0.874\cos\left(\frac{4\pi}{3}n + 10.89°\right) + 0.643(2)^n, \quad n = 0, 1, 2, \ldots$$

It should be noted that the poles and the associated residues (the c_i) for $X(z)/z$ can be computed by use of the MATLAB commands

```
num = [1 0 0 1];
den = [1 -1 -1 -2 0];
[r,p] = residue(num,den)
```

Running the program yields

```
r =                      p =
   0.6429                   2.0000
   0.4286 - 0.0825i        -0.5000 + 0.8660i
   0.4286 + 0.0825i        -0.5000 - 0.8660i
  -0.5000                   0
```

which matches with the poles and residues previously computed.

The inverse z-transform of $X(z)$ can also be computed by symbolic manipulation or the numerical method given in Example 7.16: The following commands compute $x[n]$ for $n = 0$ to $n = 20$.

```
num = [1 0 0 1];
den = [1 -1 -1 -2];
x = filter(num,den,[1 zeros(1,19)]);
```

The reader is invited to plot this response and compare it with the response calculated analytically.

Repeated Poles.   Again, let $p_1, p_2, \ldots, p_N$ denote the poles of $X(z) = B(z)/A(z)$, and assume that all the p_i are nonzero. Suppose that the pole p_1 is repeated r times and that the other $N - r$ poles are distinct. Then, $X(z)/z$ has the partial fraction expansion

$$\frac{X(z)}{z} = \frac{c_0}{z} + \frac{c_1}{z - p_1} + \frac{c_2}{(z - p_1)^2} + \cdots + \frac{c_r}{(z - p_1)^r} + \frac{c_{r+1}}{z - p_{r+1}}$$

$$+ \cdots + \frac{c_N}{z - p_N} \tag{7.70}$$

In (7.70), $c_0 = X(0)$ and the residues $c_{r+1}, c_{r+2}, \ldots, c_N$ are computed in the same way as in the distinct pole case. [See (7.64).] The constants $c_r, c_{r-1}, \ldots, c_1$ are given by

$$c_r = \left[(z - p_1)^r \frac{X(z)}{z} \right]_{z=p_1}$$

$$c_{r-1} = \left[\frac{d}{dz} (z - p_1)^r \frac{X(z)}{z} \right]_{z=p_1}$$

$$c_{r-2} = \frac{1}{2!} \left[\frac{d^2}{dz^2} (z - p_1)^r \frac{X(z)}{z} \right]_{z=p_1}$$

$$\vdots$$

$$c_{r-i} = \frac{1}{i!} \left[\frac{d^i}{dz^i} (z - p_1)^r \frac{X(z)}{z} \right]_{z=p_1}$$

Then, multiplying both sides of (7.70) by z gives

$$X(z) = c_0 + \frac{c_1 z}{z - p_1} + \frac{c_2 z}{(z - p_1)^2} + \cdots + \frac{c_r z}{(z - p_1)^r} + \frac{c_{r+1} z}{z - p_{r+1}}$$

$$+ \cdots + \frac{c_N z}{z - p_N} \tag{7.71}$$

We can compute the inverse z-transform of the terms

$$\frac{c_i z}{(z - p_1)^i} \tag{7.72}$$

for $i = 2$ and 3 by using the transform pairs in Table 7.2. This results in the transform pairs

$$c_2 n (p_1)^{n-1} u[n] \leftrightarrow \frac{c_2 z}{(z - p_1)^2} \tag{7.73}$$

$$\frac{1}{2} c_3 n(n - 1)(p_1)^{n-2} u[n] \leftrightarrow \frac{c_3 z}{(z - p_1)^3} \tag{7.74}$$

We can compute the inverse transform of (7.72) for $i = 4, 5, \ldots$ by repeatedly using the multiplication by n property of the z-transform. This results in the transform pair

$$\frac{c_i}{(i - 1)!} n(n - 1) \cdots (n - i + 2)(p_1)^{n-i-1} u[n - i + 2] \leftrightarrow \frac{c_i z}{(z - p_1)^i}, \quad i = 4, 5, \ldots$$

Example 7.18   Repeated Pole Case

Suppose that

$$X(z) = \frac{6z^3 + 2z^2 - z}{z^3 - z^2 - z + 1}$$

Then,

$$\frac{X(z)}{z} = \frac{6z^2 + 2z - 1}{z^3 - z^2 - z + 1} = \frac{6z^2 + 2z - 1}{(z-1)^2(z+1)}$$

Note that the common factor of z in the numerator and denominator of $X(z)/z$ has been canceled. To eliminate unnecessary computations, any common factors in $X(z)/z$ should be canceled before a partial fraction expansion is performed.

Now the poles of $X(z)/z$ are $p_1 = 1$, $p_2 = 1$, $p_3 = -1$, and thus the expansion has the form

$$\frac{X(z)}{z} = \frac{c_1}{z-1} + \frac{c_2}{(z-1)^2} + \frac{c_3}{z+1}$$

where

$$c_2 = \left[(z-1)^2 \frac{X(z)}{z}\right]_{z=1} = \frac{6+2-1}{2} = 3.5$$

$$c_1 = \left[\frac{d}{dz}(z-1)^2 \frac{X(z)}{z}\right]_{z=1} = \left[\frac{d}{dz}\frac{6z^2 + 2z - 1}{z+1}\right]_{z=1}$$

$$= \frac{(z+1)(12z+2) - (6z^2 + 2z - 1)(1)}{(z+1)^2}\Bigg|_{z=1}$$

$$= \frac{2(14) - 7}{4} = 5.25$$

and

$$c_3 = \left[(z+1)\frac{X(z)}{z}\right]_{z=-1} = \frac{6-2-1}{(-2)^2} = 0.75$$

Hence,

$$X(z) = \frac{5.25z}{z-1} + \frac{3.5z}{(z-1)^2} + \frac{0.75z}{z+1}$$

Using the transform pair (7.73) results in the following inverse transform:

$$x[n] = 5.25(1)^n + 3.5n(1)^{n-1} + 0.75(-1)^n, \quad n = 0, 1, 2, \ldots$$

$$= 5.25 + 3.5n + 0.75(-1)^n, \quad n = 0, 1, 2, \ldots$$

As a check on the foregoing results, we can compute the poles and residues of $X(z)/z$ by using the commands

```
num = [6 2 -1];
den = [1 -1 -1 1];
[r,p] = residue(num,den)
```

Running the program yields

```
r =              p =
   5.2500          1.0000
   3.5000          1.0000
   0.7500         -1.0000
```

This matches with the poles and residues previously computed. Note that the residue 5.25 for the first occurrence (in the preceding list) of the pole at $p = 1$ corresponds to the term $c_1/(z - 1)$ in the expansion of $X(z)/z$, and the constant 3.5 for the second occurrence of $p = 1$ corresponds to the term $c_2/(z - 1)^2$.

As in Example 7.17, we can compute the inverse z-transform symbolically or numerically by using the following MATLAB commands:

```
num = [6 2 -1 0];
den = [1 -1 -1 1];
x = filter(num,den,[1 zeros(1,19)]);
```

Pole Locations and the Form of a Signal.   Given a discrete-time signal $x[n]$ with rational z-transform $X(z) = B(z)/A(z)$, by the foregoing results we see that there is a direct relationship between the poles of $X(z)$ and the form of the time variation of the signal $x[n]$. In particular, if $X(z)$ has a nonrepeated real pole p, then $x[n]$ contains a term of the form $c(p)^n$ for some constant c; and if the pole p is repeated twice, $x[n]$ contains the terms $c_1(p)^n$ and $c_2 n(p)^n$ for some constants c_1 and c_2. If $X(z)$ has a nonrepeated pair $a \pm jb$ of complex poles with magnitude σ and angles $\pm\Omega$, then $x[n]$ contains a term of the form $c\sigma^n \cos(\Omega n + \theta)$ for some constants c and θ. If the complex pair $a \pm jb$ is repeated twice, $x[n]$ contains the terms $c_1\sigma^n \cos(\Omega n + \theta_1) + c_2 n\sigma^n \cos(\Omega n + \theta_2)$ for some constants $c_1, c_2, \theta_1, \theta_2$.

Note that these relationships between signal terms and poles are analogous to those in the Laplace transform theory of continuous-time signals. In fact, as was the case for continuous-time signals, the behavior of a discrete-time signal as $n \to \infty$ can be determined directly from the poles of $X(z)$. In particular, it follows from the preceding results that $x[n]$ converges to 0 as $n \to \infty$ if and only if all the poles $p_1, p_2, \ldots, p_N$ of $X(z)$ have magnitudes that are strictly less than 1; that is,

$$|p_i| < 1 \qquad \text{for } i = 1, 2, \ldots, N \tag{7.75}$$

In addition, it follows that $x[n]$ converges (as $n \to \infty$) to a finite constant if and only if (7.75) is satisfied, except that one of the p_i may be equal to 1. If this is the case, $x[n]$ converges to the value of the residue in the expansion of $X(z)/z$ corresponding to the pole at 1. In other words,

$$\lim_{n\to\infty} x[n] = \left[(z - 1)\frac{X(z)}{z} \right]_{z=1} = [(z - 1)X(z)]_{z=1}$$

Note that this result is consistent with the final-value theorem given in Section 7.2.

7.4   TRANSFER FUNCTION REPRESENTATION

In this section the transfer function representation is generated for the class of causal linear time-invariant discrete-time systems. The development begins with discrete-time systems defined by an input/output difference equation. Systems given by a first-order input/output difference equation are considered first.

7.4.1   First-Order Case

Consider the linear time-invariant discrete-time system given by the first-order input/output difference equation

$$y[n] + ay[n - 1] = bx[n] \tag{7.76}$$

where a and b are real numbers, $y[n]$ is the output, and $x[n]$ is the input. Taking the z-transform of both sides of (7.76) and using the right-shift property (7.20) give

$$Y(z) + a[z^{-1}Y(z) + y[-1]] = bX(z) \tag{7.77}$$

where $Y(z)$ is the z-transform of the output response $y[n]$ and $X(z)$ is the z-transform of the input $x[n]$. Solving (7.77) for $Y(z)$ yields

$$Y(z) = -\frac{ay[-1]}{1 + az^{-1}} + \frac{b}{1 + az^{-1}}X(z) \tag{7.78}$$

and multiplying the terms on the right-hand side of (7.78) by z/z gives

$$Y(z) = -\frac{ay[-1]z}{z + a} + \frac{bz}{z + a}X(z) \tag{7.79}$$

Equation (7.79) is the *z-domain representation* of the discrete-time system defined by the input/output difference equation (7.76). The first term on the right-hand side of (7.79) is the z-transform of the part of the output response resulting from the initial condition $y[-1]$, and the second term on the right-hand side of (7.79) is the z-transform of the part of the output response resulting from the input $x[n]$ applied for $n = 0, 1, 2, \ldots$.

If the initial condition $y[-1]$ is equal to zero, (7.79) reduces to

$$Y(z) = \frac{bz}{z + a}X(z) \tag{7.80}$$

If we define

$$H(z) = \frac{bz}{z + a}$$

then (7.80) becomes

$$Y(z) = H(z)X(z) \tag{7.81}$$

We call the function $H(z)$ the *transfer function* of the system, since it specifies the transfer from the input to the output in the z-domain, assuming zero initial condition ($y[-1] = 0$). Equation (7.81) is the *transfer function representation* of the system.

For any initial condition $y[-1]$ and any input $x[n]$ with rational z-transform $X(z)$, we can compute the output $y[n]$ by taking the inverse z-transform of $Y(z)$ given by (7.79). The procedure is illustrated by the following example:

Example 7.19   Step Response

For the system given by (7.76), suppose that $a \neq -1$ and $x[n]$ is equal to the unit-step function $u[n]$. Then, $X(z) = z/(z-1)$, and from (7.79) the z-transform of the output response is

$$Y(z) = -\frac{ay[-1]z}{z+a} + \frac{bz}{z+a}\left(\frac{z}{z-1}\right)$$

$$= -\frac{ay[-1]z}{z+a} + \frac{bz^2}{(z+a)(z-1)} \tag{7.82}$$

Expanding

$$\frac{bz^2}{(z+a)(z-1)}\frac{1}{z} = \frac{ab/(a+1)}{z+a} + \frac{b/(a+1)}{z-1}$$

and taking the inverse z-transform of both sides of (7.82) give

$$y[n] = -ay[-1](-a)^n + \frac{b}{a+1}[a(-a)^n + (1)^n]$$

$$= -ay[-1](-a)^n + \frac{b}{a+1}[-(-a)^{n+1} + 1], \quad n = 0, 1, 2, \ldots \tag{7.83}$$

If the initial condition $y[-1]$ is zero, (7.83) reduces to

$$y[n] = \frac{b}{a+1}[-(-a)n + 1], \quad n = 0, 1, 2, \ldots \tag{7.84}$$

The output $y[n]$ given by (7.84) is called the *step response*, since it is the output response when the input $x[n]$ is the unit step $u[n]$ with zero initial condition.

7.4.2   Second-Order Case

Now consider the discrete-time system given by the second-order input/output difference equation

Pole
Posi-
tions
and Im-
pulse
Re-
sponse

$$y[n] + a_1 y[n-1] + a_2 y[n-2] = b_0 x[n] + b_1 x[n-1] \tag{7.85}$$

Taking the z-transform of both sides of (7.85) and using the right-shift properties (7.20) and (7.21) give (assuming that $x[-1] = 0$)

$$Y(z) + a_1[z^{-1}Y(z) + y[-1]] + a_2[z^{-2}Y(z) + z^{-1}y[-1] + y[-2]]$$

$$= b_0 X(z) + b_1 z^{-1} X(z)$$

Solving for $Y(z)$ gives

$$Y(z) = \frac{-a_2 y[-2] - a_1 y[-1] - a_2 y[-1]z^{-1}}{1 + a_1 z^{-1} + a_2 z^{-2}} + \frac{b_0 + b_1 z^{-1}}{1 + a_1 z^{-1} + a_2 z^{-2}} X(z) \tag{7.86}$$

Multiplying both sides of (7.86) by z^2/z^2 yields

$$Y(z) = \frac{-(a_1 y[-1] + a_2 y[-2])z^2 - a_2 y[-1]z}{z^2 + a_1 z + a_2} + \frac{b_0 z^2 + b_1 z}{z^2 + a_1 z + a_2} X(z) \quad (7.87)$$

Equation (7.87) is the z-domain representation of the discrete-time system given by the second-order input/output difference equation (7.85). The first term on the right-hand side of (7.87) is the z-transform of the part of the output response resulting from the initial conditions $y[-1]$ and $y[-2]$, and the second term on the right-hand side of (7.87) is the z-transform of the part of the output response resulting from the input $x[n]$ applied for $n \geq 0$.

If $y[-1] = y[-2] = 0$, (7.87) reduces to the transfer function representation

$$Y(z) = \frac{b_0 z^2 + b_1 z}{z^2 + a_1 z + a_2} X(z) \quad (7.88)$$

where the transfer function $H(z)$ of the system is given by

$$H(z) = \frac{b_0 z^2 + b_1 z}{z^2 + a_1 z + a_2} \quad (7.89)$$

Note that $H(z)$ is a second-order rational function of z.

Example 7.20   Second-Order System

Consider the discrete-time system given by the input/output difference equation

$$y[n] + 1.5y[n - 1] + 0.5y[n - 2] = x[n] - x[n - 1]$$

By (7.89), the transfer function of the system is

$$H(z) = \frac{z^2 - z}{z^2 + 1.5z + 0.5}$$

Suppose that the goal is to compute the output response $y[n]$ when $y[-1] = 2$, $y[-2] = 1$, and the input $x[n]$ is the unit step $u[n]$. Then, by (7.87), the z-transform of the response is

$$Y(z) = \frac{-[(1.5)(2) + (0.5)(1)]z^2 - (0.5)(2)z}{z^2 + 1.5z + 0.5} + \frac{z^2 - z}{z^2 + 1.5z + 0.5}\left(\frac{z}{z - 1}\right)$$

$$= \frac{-3.5z^2 - z}{z^2 + 1.5z + 0.5} + \frac{z^2}{z^2 + 1.5z + 0.5}$$

$$= \frac{-2.5z^2 - z}{z^2 + 1.5z + 0.5}$$

$$= \frac{0.5z}{z + 0.5} - \frac{3z}{z + 1}$$

Then, taking the inverse z-transform gives

$$y[n] = 0.5(-0.5)^n - 3(-1)^n, \quad n = 0, 1, 2, \ldots$$

We can obtain the response to a step input when $y[-1] = 2$ and $y[-2] = 1$ by using the command `filter` in MATLAB. The initial conditions required by this command are related to, but not equal to $y[-1]$ and $y[-2]$. For the general second-order difference equation given in (7.85), define an initial condition vector to be $\mathtt{zi = [-a_1*y[-1] - a_2*y[-2]}$, $\mathtt{-a_2*y[-1]]}$ if $x = 0$ for $n < 0$. The use of the filter command is demonstrated as follows:

```
num = [1 -1 0];
den = [1 1.5 .5];
n = 0:20;
x = ones(1,length(n));
zi = [-1.5*2-0.5*1,-0.5*2];
y = filter(num,den,x,zi);
```

The response of this system is shown in Figure 7.2. Note that this response matches the result obtained analytically, where the term $0.5(0.5)^n$ decays to zero quickly and the term $-3(-1)^n$ simply oscillates between -3 and 3.

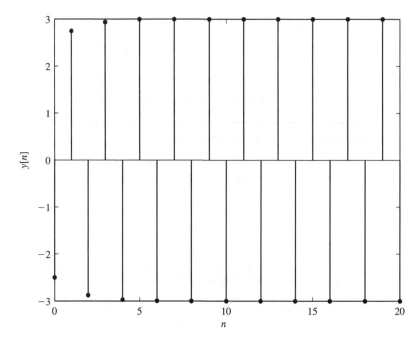

FIGURE 7.2
Output response in Example 7.20.

7.4.3   Nth-Order Case

Now suppose that the discrete-time system under study is specified by the Nth-order input/output difference equation

$$y[n] + \sum_{i=1}^{N} a_i y[n-i] = \sum_{i=1}^{M} b_i x[n-i] \tag{7.90}$$

Taking the z-transform of both sides of (7.90) and multiplying the resulting equation by z^N/z^N yield the z-domain representation of the system, given by

$$Y(z) = \frac{C(z)}{A(z)} + \frac{B(z)}{A(z)}X(z) \tag{7.91}$$

where

$$B(z) = b_0 z^N + b_1 z^{N-1} + \cdots + b_M z^{N-M}$$

and

$$A(z) = z^N + a_1 z^{N-1} + \cdots + a_{N-1}z + a_N$$

and where $C(z)$ is a polynomial in z whose coefficients are determined by the initial conditions $y[-1], y[-2], \ldots, y[-N]$ (assuming that $x[n] = 0$ for $n = -1, -2, \ldots, -M$). If all the initial conditions are zero, then $C(z) = 0$ and (7.91) reduces to the transfer function representation

$$Y(z) = \frac{B(z)}{A(z)}X(z) \tag{7.92}$$

where the transfer function $H(z)$ of the system is given by

$$H(z) = \frac{B(z)}{A(z)} = \frac{b_0 z^N + b_1 z^{N-1} + \cdots + b_M z^{N-M}}{z^N + a_1 z^{N-1} + \cdots + a_{N-1}z + a_N} \tag{7.93}$$

7.4.4   Transform of the Input/Output Convolution Sum

Suppose that a causal linear time-invariant discrete-time system is given by the input/output convolution relationship

$$y[n] = h[n] * x[n] = \sum_{i=0}^{n} h[i]x[n-i], \quad n = 0, 1, 2, \ldots \tag{7.94}$$

where $h[n]$ is the unit-pulse response of the system, the input is zero for $n < 0$, and all initial conditions are zero. (See Chapter 2.) The z-transform can be applied to both sides of (7.94), which results in the transfer function representation

$$Y(z) = H(z)X(z) \tag{7.95}$$

where the transfer function $H(z)$ is the z-transform of the unit-pulse response $h[n]$. The relationship between the unit-pulse response $h[n]$ and the transfer function $H(z)$ can be expressed in terms of the transform pair notation

$$h[n] \leftrightarrow H(z) \tag{7.96}$$

The transform pair (7.96) is analogous to the relationship generated in Section 6.5 between the impulse response and the transfer function in the continuous-time case. As in the continuous-time theory, (7.96) provides a major link between the time domain and the transform domain in the study of discrete-time systems.

As shown previously, if the discrete-time system is given by an input/output difference equation, the transfer function is a rational function of z. It turns out that the converse is also true; that is, if $H(z)$ is rational, the system can be described by an input/output difference equation. To see this, suppose that $H(z)$ can be expressed in the rational form (7.93). Then multiplying both sides of (7.92) by $A(z)$ gives

$$A(z)Y(z) = B(z)X(z)$$

and taking the inverse z-transform of this results in the difference equation (7.90).

A causal linear time-invariant discrete-time system is said to be *finite dimensional* if the transfer function $H(z)$ is a rational function of z; that is, $H(z)$ can be expressed in the form (7.93). Hence, as shown before, a discrete-time system is finite-dimensional if and only if it can be described by an input/output difference equation of the form (7.90).

If a given discrete-time system is finite dimensional, so that $H(z)$ is rational, the poles and zeros of the system are defined to be the poles and zeros of $H(z)$. It is also possible to define the pole–zero diagram of a discrete-time system as was done in the continuous-time case. (See Section 6.5.)

It is important to observe that, if the input $x[n]$ is nonzero for at least one positive value of n, so that $X(z) \neq 0$, then both sides of the transfer function representation (7.95) can be divided by $X(z)$, which gives

$$H(z) = \frac{Y(z)}{X(z)} \qquad (7.97)$$

Thus, the transfer function $H(z)$ is equal to the ratio of the z-transforms of the output and input. Note that, since $H(z)$ is unique, the ratio $Y(z)/X(z)$ cannot change as the input $x[n]$ ranges over some collection of input signals. From (7.97) it is also seen that $H(z)$ can be determined from the output response to any input that is not identically zero for $n \geq 0$.

Example 7.21   Computation of Transfer Function

A linear time-invariant discrete-time system has unit-pulse response

$$h[n] = 3(2^{-n}) \cos\left(\frac{\pi n}{6} + \frac{\pi}{12}\right), \quad n = 0, 1, 2, \ldots \qquad (7.98)$$

where the argument of the cosine in (7.98) is in radians. The transfer function $H(z)$ of the system is equal to the z-transform of $h[n]$ given by (7.98). To compute the transform of $h[n]$, first expand the cosine in (7.98) by using the trigonometric identity

$$\cos(a + b) = (\cos a)(\cos b) - (\sin a)(\sin b) \qquad (7.99)$$

Applying (7.99) to (7.98) gives

$$h[n] = 3(2^{-n})\left[\cos\left(\frac{\pi n}{6}\right)\cos\left(\frac{\pi}{12}\right) - \sin\left(\frac{\pi n}{6}\right)\sin\left(\frac{\pi}{12}\right)\right], \quad n \geq 0$$

$$= 2.898\left(\frac{1}{2}\right)^n \cos\left(\frac{\pi n}{6}\right) - 0.776\left(\frac{1}{2}\right)^n \sin\left(\frac{\pi n}{6}\right), \quad n \geq 0 \qquad (7.100)$$

Taking the z-transform of (7.100) yields

$$H(z) = 2.898 \frac{z^2 - [0.5 \cos(\pi/6)]z}{z^2 - [\cos(\pi/6)]z + 0.25} - 0.776 \frac{[0.5 \sin(\pi/6)]z}{z^2 - [\cos(\pi/6)]z + 0.25}$$

$$= \frac{2.898z^2 - 1.449z}{z^2 - 0.866z + 0.25}$$

If both $H(z)$ and $X(z)$ are rational functions of z, we can compute the output response by first expanding the product $H(z)X(z)$ [or $H(z)X(z)/z$] by partial fractions. The process is illustrated by the following example:

Example 7.22   Computation of Step Response

Suppose that the objective is to compute the step response of the system in Example 7.21, where the step response is the output resulting from the step input $x[n] = u[n]$ with zero initial conditions. Inserting $X(z) = z/(z - 1)$ and $H(z)$ into $Y(z) = H(z)X(z)$ gives

$$Y(z) = \frac{2.898z^3 - 1.449z^2}{(z - 1)(z^2 - 0.866z + 0.25)}$$

Since the zeros of $z^2 - 0.866z + 0.25$ are complex, we avoid complex arithmetic by expanding $Y(z)/z$ into the form

$$\frac{Y(z)}{z} = \frac{cz + d}{z^2 - 0.866z + 0.25} + \frac{c_3}{z - 1}$$

where

$$c_3 = \left[(z - 1) \frac{Y(z)}{z} \right]_{z=1} = \frac{2.898 - 1.449}{1 - 0.866 + 0.25} = 3.773$$

Hence,

$$Y(z) = \frac{cz^2 + dz}{z^2 - 0.866z + 0.25} + \frac{3.773z}{z - 1} \tag{7.101}$$

Putting the right-hand side of (7.101) over a common denominator and equating coefficients give

$$c + 3.773 = 2.898$$

$$d - c - (0.866)(3.773) = -1.449$$

Solving for c and d yields

$$c = -0.875 \quad \text{and} \quad d = 0.943$$

Now, to determine the inverse z-transform of the first term on the right-hand side of (7.101), set

$$z^2 - 0.866z + 0.25 = z^2 - (2a \cos \Omega)z + a^2$$

Then,

$$a = \sqrt{0.25} = 0.5$$

$$\Omega = \cos^{-1}\left(\frac{0.866}{2a} \right) = \frac{\pi}{6} \text{rad}$$

and thus

$$\frac{cz^2 + dz}{z^2 - 0.866z + 0.25} = \frac{-0.875z^2 + 0.943z}{z^2 - (\cos \pi/6)z + 0.25} \tag{7.102}$$

Expressing the right-hand side of (7.102) in the form

$$\frac{\alpha(z^2 - 0.5(\cos \pi/6)z)}{z^2 - (\cos \pi/6)z + 0.25} + \frac{\beta(\sin \pi/6)z}{z^2 - (\cos \pi/6)z + 0.25}$$

results in $\alpha = -0.875$ and

$$-0.5\alpha \cos\frac{\pi}{6} + \beta \sin\frac{\pi}{6} = 0.943 \tag{7.103}$$

Solving (7.103) for β yields $\beta = 1.128$. Thus,

$$Y(z) = \frac{-0.875(z^2 - 0.5(\cos \pi/6)z)}{z^2 - (\cos \pi/6)z + 0.25} + \frac{1.128(\sin \pi/6)z}{z^2 - (\cos \pi/6)z + 0.25} + \frac{3.773z}{z - 1}$$

Finally, using Table 7.3 gives

$$y[n] = -0.875\left(\frac{1}{2}\right)^n \cos\frac{\pi n}{6} + 2.26\left(\frac{1}{2}\right)^n \sin\frac{\pi n}{6} + 3.773, \quad n = 0, 1, 2, \ldots$$

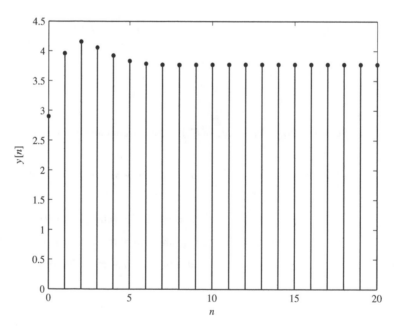

FIGURE 7.3
Step response in Example 7.22.

We can also compute the step response numerically by using the MATLAB command `filter`. For this example, the following commands compute and plot the step response:

```
num = [2.898 -1.449 0];
den = [1 -.866 .25];
n = 0:20;
u = ones(length(n));
y = filter(num,den,u);
stem(n,y,'filled')
```

The resulting plot of the step response is given in Figure 7.3. Note that the response settles to a steady-state value of 3.773, which matches the limiting value of $y[n]$ obtained by an examination of the preceding expression for $y[n]$.

Note that the poles of $(z - 1)Y(z)$ are $0.433 \pm j0.25$, which have a magnitude of 0.5. Then, since the poles of $(z - 1)Y(z)$ have magnitude less than 1, the final-value theorem can be applied to compute the limiting value of $y[n]$. The result is

$$\lim_{n \to \infty} y[n] = [(z - 1)Y(z)]_{z=1} = \left[\frac{2.898z^3 - 1.449z^2}{z^2 - 0.866z + 0.25} \right]_{z=1} = \frac{1.449}{0.384} = 3.773$$

which checks with the foregoing value.

7.4.5   Transfer Function of Interconnections

The transfer function of a linear time-invariant discrete-time system can be computed directly from a signal-flow diagram of the system. This approach is pursued here for systems given by an interconnection of unit-delay elements and an interconnection of blocks. The following results are directly analogous to those derived in Section 6.6 for linear time-invariant continuous-time systems.

Interconnections of Unit-Delay Elements.   The *unit-delay element* is a system whose input/output relationship is given by

$$y[n] = x[n - 1] \tag{7.104}$$

The system given by (7.104) is referred to as the *unit-delay element*, since the output $y[n]$ is equal to a one-step time delay of the input $x[n]$. As shown in Figure 7.4, the unit-delay element is represented by a box with a D that stands for "delay."

Taking the z-transform of both sides of (7.104) with $x[-1] = 0$ yields the transfer function representation of the unit-delay element:

$$Y(z) = z^{-1}X(z) \tag{7.105}$$

From (7.105) it is seen that the transfer function of the unit-delay element is equal to $1/z$. Note that the unit delay is the discrete-time counterpart to the integrator in the sense that the unit delay has transfer function $1/z$ and the integrator has transfer function $1/s$.

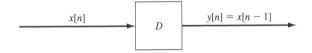

FIGURE 7.4
Unit-delay element.

Now suppose that a discrete-time system is given by an interconnection of unit delays, adders, subtracters, and scalar multipliers. We can compute the transfer function for any such interconnection by working in the z-domain with the unit delays represented by their transfer function $1/z$. The procedure is very similar to that considered in Section 6.6 for continuous-time systems consisting of interconnections of integrators.

Example 7.23   Computation of Transfer Function

Consider the discrete-time system given by the interconnection in Figure 7.5. Note that the outputs of the two unit-delay elements in Figure 7.4 are denoted by $q_1[n]$ and $q_2[n]$. The z-domain representation of the system with all initial conditions equal to zero is shown in Figure 7.6. From this figure it is clear that

$$zQ_1(z) = Q_2(z) + X(z) \tag{7.106}$$

$$zQ_2(z) = Q_1(z) - 3Y(z) \tag{7.107}$$

$$Y(z) = 2Q_1(z) + Q_2(z) \tag{7.108}$$

Solving (7.106) for $Q_1(z)$ and inserting the result into (7.107) and (7.108) yield

$$zQ_2(z) = z^{-1}Q_2(z) + z^{-1}X(z) - 3Y(z) \tag{7.109}$$

$$Y(z) = 2z^{-1}Q_2(z) + 2z^{-1}X(z) + Q_2(z) \tag{7.110}$$

Solving (7.109) for $Q_2(z)$ and inserting the result into (7.110) give

$$Y(z) = \frac{2z^{-1} + 1}{z - z^{-1}}[z^{-1}X(z) - 3Y(z)] + 2z^{-1}X(z)$$

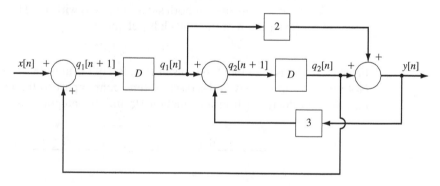

FIGURE 7.5
System in Example 7.23.

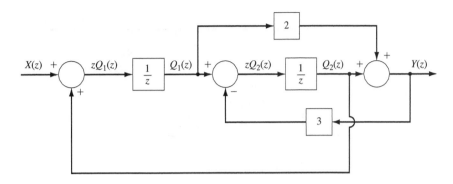

FIGURE 7.6
Representation of system in the z-domain.

Then,

$$\left[1 + \frac{3(2z^{-1} + 1)}{z - z^{-1}}\right]Y(z) = \frac{z^{-1}(2z^{-1} + 1)}{z - z^{-1}}X(z) + 2z^{-1}X(z)$$

$$\frac{z + 5z^{-1} + 3}{z - z^{-1}}Y(z) = \left[\frac{z^{-1}(2z^{-1} + 1)}{z - z^{-1}} + 2z^{-1}\right]X(z)$$

$$= \frac{z^{-1} + 2}{z - z^{-1}}X(z)$$

Thus,

$$Y(z) = \frac{z^{-1} + 2}{z + 5z^{-1} + 3}X(z) = \frac{2z + 1}{z^2 + 3z + 5}X(z)$$

So, the transfer function is

$$H(z) = \frac{2z + 1}{z^2 + 3z + 5}$$

Transfer Function of Basic Interconnections.   The transfer function of series, parallel, and feedback connections have exactly the same form as in the continuous-time case. The results are displayed in Figure 7.7.

7.5   SYSTEM ANALYSIS USING THE TRANSFER FUNCTION REPRESENTATION

Consider the linear time-invariant discrete-time system given by the transfer function

$$H(z) = \frac{B(z)}{A(z)} = \frac{b_M z^M + b_{M-1}z^{M-1} + \cdots + b_1 z + b_0}{a_N z^N + a_{N-1}z^{N-1} + \cdots + a_1 z + a_0} \tag{7.111}$$

The system is assumed to be causal, and thus $M \leq N$. It is also assumed that the polynomials $B(z)$ and $A(z)$ do not have any common factors. If there are common factors, they should be canceled.

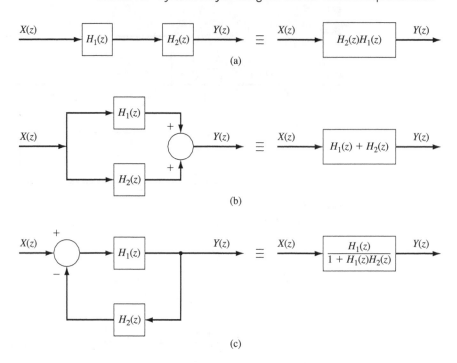

FIGURE 7.7

Transfer functions of basic interconnections: (a) series connection; (b) parallel connection; (c) feedback connection.

As observed in Section 7.4, the transfer function $H(z)$ is the z-transform of the system's unit-pulse response $h[n]$. From the development in Section 7.3, it follows that the form of the time variation of $h[n]$ is directly determined by the poles of the system, which are the roots of $A(z) = 0$. In particular, if $H(z)$ has a real pole p, then $h[n]$ contains a term of the form $c(p)^n$, and if $H(z)$ has a complex pair $a \pm jb$ of poles with magnitude σ and angles $\pm\Omega$, then $h[n]$ contains a term of the form $c(\sigma)^n \cos(\Omega n + \theta)$. If $H(z)$ has repeated poles, it will contain terms of the form $cn(n + 1)\cdots(n + i)(p)^n$ and/or $cn(n + 1)\cdots(n + i)\sigma^n \cos(\Omega n + \theta)$.

It follows from the relationship between the form of $h[n]$ and the poles of $H(z)$ that $h[n]$ converges to zero as $n \to \infty$ if and only if

$$|p_i| < 1 \qquad \text{for } i = 1, 2, \ldots, N \tag{7.112}$$

where $p_1, p_2, \ldots, p_N$ are the poles of $H(z)$. The condition (7.112) is equivalent to requiring that all the poles be located in the *open unit disk* of the complex plane. The open unit disk is that part of the complex plane consisting of all complex numbers whose magnitude is strictly less than 1. The open unit disk is the hatched region shown in Figure 7.8.

A discrete-time system with transfer function $H(z)$ given by (7.111) is said to be *stable* if its unit-pulse response $h[n]$ converges to zero as $n \to \infty$. Thus, stability is

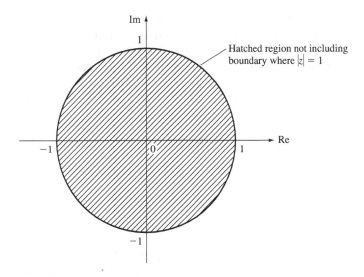

FIGURE 7.8
Open unit disk.

equivalent to requiring that all the poles of the system lie in the open unit disk of the complex plane. The system is marginally stable if the unit-pulse response $h[n]$ is bounded; that is,

$$|h[n]| < c \quad \text{for all } n \tag{7.113}$$

where c is a finite positive constant. It also follows from the relationship between the form of $h[n]$ and the poles of $H(z)$ that a system is marginally stable if and only if $|p_i| \leq 1$ for all nonrepeated poles of $H(z)$, and $|p_i| < 1$ for all repeated poles. This is equivalent to requiring that all poles lie in the open unit disk, except that nonrepeated poles can be located on the *unit circle* (i.e., all those complex numbers z such that $|z| = 1$).

Finally, a system is unstable if the magnitude of $h[n]$ grows without bound as $n \rightarrow \infty$; that is,

$$|h[n]| \rightarrow \infty \quad \text{as } n \rightarrow \infty \tag{7.114}$$

The instability condition (7.114) is equivalent to having one or more poles located outside the closed unit disk (all complex numbers z such that $|z| > 1$) or having one or more repeated poles located on the unit circle.

From the foregoing results, it is seen that the stability boundary in the discrete-time case is the unit circle of the complex plane.

The convergence of $h[n]$ to zero turns out to be equivalent (when $H(z)$ is rational) to absolute summability of $h[n]$; that is,

$$\sum_{n=0}^{\infty} |h[n]| < \infty \tag{7.115}$$

Thus, stability is equivalent to absolute summability of the unit-pulse response $h[n]$. In addition, convergence of $h[n]$ to zero is equivalent to bounded-input bounded-output

(BIBO) stability, which implies that the output $y[n]$ is a bounded signal whenever the input $x[n]$ is a bounded signal (assuming zero initial conditions). In mathematical terms, BIBO stability means that whenever $|x[n]| \leq c_1$ for all n and for some finite positive constant c_1, then $|y[n]| \leq c_2$ for all n and for some finite positive constant c_2, where $y[n]$ is the response to $x[n]$ with zero initial conditions. Hence, stability of a system (i.e., convergence of $h[n]$ to zero) is equivalent to BIBO stability.

7.5.1   Response to a Sinusoidal Input

Again, consider the linear time-invariant finite-dimensional discrete-time system with the transfer function

$$H(z) = \frac{B(z)}{A(z)} = \frac{b_M z^M + b_{M-1} z^{M-1} + \cdots + b_1 z + b_0}{a_N z^N + a_{N-1} z^{N-1} + \cdots + a_1 z + a_0} \qquad (7.116)$$

Throughout this section it is assumed that the system is stable, and thus all the poles of $H(z)$ are located in the open unit disk of the complex plane. We can determine the frequency response characteristics of the system by examining the response to a sinusoidal input. In particular, let the system input $x[n]$ be given by

$$x[n] = C\cos(\Omega_0 n), n = 0, 1, 2, \ldots \qquad (7.117)$$

where C and Ω_0 are real numbers. From Table 7.3, we see that the z-transform of the sinusoidal input (7.117) is

$$X(z) = \frac{C[z^2 - (\cos \Omega_0)z]}{z^2 - (2\cos \Omega_0)z + 1}$$

If the initial conditions are zero, the z-transform of the resulting output response is

$$Y(z) = \frac{CB(z)[z^2 - (\cos \Omega_0)z]}{A(z)[z^2 - (2\cos \Omega_0)z + 1]}$$

Now,

$$z^2 - (2\cos \Omega_0)z + 1 = (z - \cos \Omega_0 - j\sin \Omega_0)(z - \cos \Omega_0 + j\sin \Omega_0)$$

$$= (z - e^{j\Omega_0})(z - e^{-j\Omega_0})$$

and thus,

$$Y(z) = \frac{CB(z)[z^2 - (\cos \Omega_0)z]}{A(z)(z - e^{j\Omega_0})(z - e^{-j\Omega_0})}$$

Dividing $Y(z)$ by z gives

$$\frac{Y(z)}{z} = \frac{CB(z)(z - \cos \Omega_0)}{A(z)(z - e^{j\Omega_0})(z - e^{-j\Omega_0})}$$

Pulling out the terms $z - e^{j\Omega_0}$ and $z - e^{-j\Omega_0}$ yields

$$\frac{Y(z)}{z} = \frac{\eta(z)}{A(z)} + \frac{c}{z - e^{j\Omega_0}} + \frac{\bar{c}}{z - e^{-j\Omega_0}}$$

where $\eta(z)$ is a polynomial in z with the degree of $\eta(z)$ less than N. The constant c is given by

$$c = \left[(z - e^{j\Omega_0}) \frac{Y(z)}{z} \right]_{z = e^{j\Omega_0}}$$

$$= \left[\frac{CB(z)(z - \cos \Omega_0)}{A(z)(z - e^{-j\Omega_0})} \right]_{z = e^{j\Omega_0}}$$

$$= \frac{CB(e^{j\Omega_0})(e^{j\Omega_0} - \cos \Omega_0)}{A(e^{j\Omega_0})(e^{j\Omega_0} - e^{-j\Omega_0})}$$

$$= \frac{CB(e^{j\Omega_0})(j \sin \Omega_0)}{A(e^{j\Omega_0})(j2 \sin \Omega_0)}$$

$$= \frac{CB(e^{j\Omega_0})}{2A(e^{j\Omega_0})} = \frac{C}{2} H(e^{j\Omega_0})$$

Multiplying the expression for $Y(z)/z$ by z gives

$$Y(z) = \frac{z\eta(z)}{A(z)} + \frac{(C/2)H(e^{j\Omega_0})z}{z - e^{j\Omega_0}} + \frac{(C/2)\overline{H(e^{j\Omega_0})}z}{z - e^{-j\Omega_0}} \tag{7.118}$$

Let $y_{\text{tr}}[n]$ denote the inverse z-transform of $z\eta(z)/A(z)$. Since the system is stable, the roots of $A(z) = 0$ are within the open unit disk of the complex plane, and thus $y_{\text{tr}}[n]$ converges to zero as $n \to \infty$. Hence, $y_{\text{tr}}[n]$ is the *transient part* of the output response resulting from the sinusoidal input $x[n] = C \cos(\Omega_0 n)$.

Now, let $y_{\text{ss}}[n]$ denote the inverse z-transform of the second and third terms on the right-hand side of (7.118). By using the trigonometric identity (7.68), we can write $y_{\text{ss}}[n]$ in the form

$$y_{\text{ss}}[n] = C|H(e^{j\Omega_0})| \cos[\Omega_0 n + \angle H(e^{j\Omega_0})], \quad n = 0, 1, 2, \ldots \tag{7.119}$$

The response $y_{\text{ss}}[n]$ clearly does not converge to zero as $n \to \infty$, and thus it is the *steady-state part* of the output response resulting from the input $x[n] = C \cos(\Omega_0 n)$.

Note that the steady-state response to a sinusoidal input is also a sinusoid with the same frequency, but is amplitude-scaled by the amount $|H(e^{j\Omega_0})|$ and is phase-shifted by the amount $\angle H(e^{j\Omega_0})$. This result corresponds to the development given in Chapter 5 in terms of the discrete-time Fourier transform (DTFT). More precisely, it follows directly from the formulation in Section 5.5 that the output response $y[n]$ resulting from the input

$$x[n] = C \cos(\Omega_0 n), n = 0, \pm 1, \pm 2, \ldots$$

is given by

$$y[n] = C|H(\Omega_0)| \cos[\Omega_0 n + \angle H(\Omega_0)], \quad n = 0, \pm 1, \pm 2, \dots \qquad (7.120)$$

where $H(\Omega_0)$ is the value at $\Omega = \Omega_0$ of the DTFT $H(\Omega)$ of the unit-pulse response $h[n]$; that is,

$$H(\Omega_0) = H(e^{j\Omega_0}) = \sum_{n=0}^{\infty} h[n]e^{-j\Omega_0 n}$$

Now, since the system is stable, the unit-pulse response $h[n]$ is absolutely summable, and thus the DTFT of $h[n]$ is equal to the transfer function $H(z)$ evaluated at $z = e^{j\Omega}$; that is,

$$H(\Omega) = H(z)|_{z=e^{j\Omega}} \qquad (7.121)$$

It then follows that the expressions (7.119) and (7.120) for the output response $y[n]$ are identical for $n \geq 0$, and thus the preceding transfer function analysis directly corresponds to the Fourier analysis given in Chapter 5.

As first defined in Section 5.5, the DTFT $H(\Omega)$ of $h[n]$ is the frequency response function of the system, and the plots of $|H(\Omega)|$ and $\angle H(\Omega)$ versus Ω are the magnitude and phase plots of the system. Since

$$|H(e^{j\Omega_0})| = |H(\Omega)|_{\Omega=\Omega_0} \quad \text{and} \quad \angle H(e^{j\Omega_0}) = \angle H(\Omega)|_{\Omega=\Omega_0}$$

from (7.119) it is seen that the steady-state response resulting from the sinusoidal input $x[n] = C \cos(\Omega_0 n)$ can be determined directly from the magnitude and phase plots.

7.5.2   Computation of the Frequency Response Curves from the Transfer Function

MATLAB can be used to compute the frequency response function $H(\Omega)$ directly from the transfer function $H(z)$ given by (7.116) with $M \leq N$. $H(z)$ can be defined in MATLAB as a function of z where $z = e^{j\Omega}$, and Ω is a vector ranging from $\Omega = 0$ to $\Omega = \pi$. The resulting $H(z)$ is a vector that contains the values of the frequency response function whose magnitude and angle can be plotted versus Ω. The use of the MATLAB software is illustrated in the following application:

7.5.3   Filtering Signals with Noise

Again, consider a signal $x[n]$ given by the additive form $x[n] = s[n] + e[n]$, where $s[n]$ is the smooth part of $x[n]$ and $e[n]$ is the noisy part of $x[n]$. The goal is to filter $x[n]$ so that $e[n]$ is reduced in magnitude as much as possible and $s[n]$ is passed with as little distortion or time delay as possible. As discussed in Section 5.6, this can be accomplished by the use of a causal lowpass digital filer. The N-point MA, EWMA, and WMA filters considered in Sections 2.1 and 5.6 are all examples of causal lowpass digital filters whose unit-pulse response $h[n]$ is of finite duration in time. In digital signal processing (DSP), such filters are referred to as FIR digital filters, where FIR stands for "finite impulse response." (In DSP, the unit-pulse response $h[n]$ is called the impulse response.). In an FIR digital filter, the output $y[n]$ at time n depends only on the input

$x[i]$ at times $i = n, n - 1, n - 2, \ldots, n - N + 1$, for some positive integer N. As noted in Section 2.3, such filters are said to be nonrecursive.

A recursive digital filter (see Section 2.3) has an impulse response (or unit-pulse response) $h[n]$ that is of infinite duration in time. In DSP, these filters are referred to as IIR (infinite impulse response) digital filters. It turns out that we can extend the N-point EWMA filter to an IIR filter by taking $N = \infty$. Hence, the impulse response $h[n]$ of the IIR EWMA filter is given by

$$h[n] = ab^n, n = 0, 1, 2, \ldots \qquad (7.122)$$

where $0 < b < 1$, and the constant a is chosen so that the value of the frequency response function $H(\Omega)$ at $\Omega = 0$ is equal to 1. Taking the z-transform of both sides of (7.122) results in the following transfer function for the IIR EWMA filter:

$$H(z) = \frac{az}{z - b} \qquad (7.123)$$

Then, using 7.121 and setting $z = 1$ in the right-hand side of (7.123) give $H(0) = a/(1 - b) = 1$, and thus $a = 1 - b$.

From (7.123) with $a = 1 - b$, it follows that the IIR EWMA filter is given by the following recursion:

$$y[n + 1] - by[n] = (1 - b)x[n + 1] \qquad (7.124)$$

Rearranging terms in (7.124) and replacing $-by[n]$ by $(1 - b)y[n] - y[n]$ yield

$$y[n + 1] = y[n] + (1 - b)(x[n + 1] - y[n]) \qquad (7.125)$$

From (7.125) it is seen that the next value $y[n + 1]$ of the output is equal to the previous value $y[n]$ of the output, plus an *update term* equal to $(1 - b)(x[n + 1] - y[n])$.

Setting $b = 0$ in (7.125) results in the input/output relationship $y[n + 1] = x[n + 1]$, and thus there is no filtering in this case. As b is increased from zero, the filter provides a sharper drop off in the magnitude of the frequency response function as Ω is increased from 0 to π. To reveal this, the magnitude function of the filter is displayed in Figure 7.9a for the two cases $b = 0.7$ and $b = 0.8$. The dotted curve in Figure 7.9a is the case $b = 0.8$. The phase function in the two cases is shown in Figure 7.9b, where, again, the case $b = 0.8$ is plotted by the use of dots. The magnitude and phase functions when $b = 0.8$ were generated by the MATLAB commands

```
W = 0:.01:1;
OMEGA = W*pi;
z = exp(j*OMEGA);
Hz = 0.2*z./(z-0.8);
magH = abs(Hz);
angH = 180/pi*unwrap(angle(Hz));
```

It should be noted that the purpose of the command unwrap is to smooth out the phase plot, since the angle command may result in jumps of $\pm 2\pi$. Also, the element-by-element operator "." is used, since z is a vector.

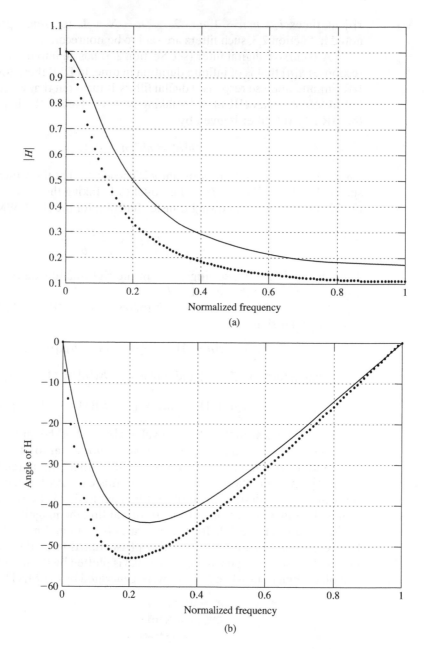

(a)

(b)

FIGURE 7.9
Magnitude (a) and phase (b) of the IIR EWMA filter with $b = 0.7$ and $b = 0.8$ (dotted line)

From Figure 7.9a it is seen that the magnitude function of the IIR EWMA filter drops off at a very fast rate as Ω is increased from zero, and thus the filter (with $b = 0.7$ or 0.8) has an excellent lowpass filter characteristic. However, the phase functions displayed in Figure 7.9b show that the filter introduces a significant time delay for the low-frequency components (the components of the smooth part $s[n]$) of the filter input $x[n]$. To be precise, in the case $b = 0.8$, the phase is approximately equal to $-45° = -0.25\pi$ radians when $\Omega = 0.1\pi$. Hence, the slope of the phase curve is approximately equal to $-0.25\pi/0.1\pi = -2.5$ as Ω is varied from $\Omega = 0$ to $\Omega = 0.1\pi$. This means that the IIR EWMA filter in the case $b = 0.8$ delays the low-frequency components of the input by 2.5 time units. Thus, in the application to filtering stock price data, the filter delays the smooth part of the price data by 2.5 days. To verify this, the filter was applied to the closing price of QQQQ for the 50-business-day period from March 1, 2004, up to May 10, 2004. The output $y[n]$ of the filter and the closing price $c[n]$ of QQQQ are plotted for $n \geq 2$ in Figure 7.10. In the figure, the values of $c[n]$ are displayed by o's and the values of $y[n]$ are displayed by *'s. Note that the filter response shown in Figure 7.10 is almost as smooth as the response of the 11-day MA filter considered in Section 1.4, which has a time delay of 5 days. The MATLAB commands for generating the plots in Figure 7.10 are as follows:

```
c=csvread('QQQQdata2.csv',1,4,[1 4 50 4]);
y(1)=c(1);
for i=2:50,
    y(i)=y(i-1)+.2*(c(i)-y(i-1));
end
n=2:50;
plot(n,c(n),n,c(n),'o',n,y(n),n,y(n),'*')
```

The for loop in the code implements the recursion. The recursion can also be performed by the command recur, which was generated in Chapter 2.

Application to Trading Stocks.   In the application to trading stocks, the IIR EWMA filter is referred to as the EMA filter, where EMA stands for "exponential moving average." The "N-day EMA filter" that is used in trading is the same as the IIR EWMA filter defined previously, with $b = (N - 1)/(N + 1)$. This filter should not be confused with the N-point EWMA filter, which was defined in Section 2.1. Recall that the N-point EWMA filter is an FIR filter, whereas the N-day EMA filter is a recursive filter given by the difference equation (7.125), with $b = (N - 1)/(N + 1)$. In the trading strategy given subsequently, the filter defined by (7.125) will be referred to as the EMA filter. This particular approach to trading is called the MACD method, where MACD stands for "moving average convergence divergence." Information on the MACD method is available on the Web, simply type "MACD" into a search engine.

Given two values $b1$ and $b2$ of b, with $b1 < b2$, let $y1[n]$ denote the output of the EMA filter with $b = b1$, and let $y2[n]$ denote the output of the EMA filter with $b = b2$. The input to the filters is the closing price of a stock such as QQQQ. Note that, since $b1 < b2$, the output $y1[n]$ of the EMA filter with $b = b1$ will be "faster" than the output $y2[n]$ of the EMA filter with $b = b2$. Hence, when $y1[n]$ crosses above $y2[n]$,

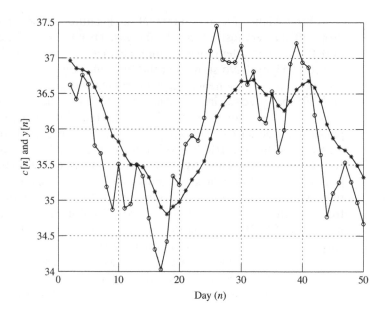

FIGURE 7.10
Plot of $c[n]$ and the output $y[n]$ of the IIR EWMA filter, with $b = 0.8$.

this is an indication of "bullishness," which means that the trend is in the upward direction and the stock should be bought. On the other hand, when $y1[n]$ falls below $y2[n]$, this is an indication of "bearishness," which means that the trend is in the downward direction and the stock should be sold.

The extent to which this scheme works (in terms of producing profits) depends on the choice of values for $b1$ and $b2$. For a given stock, optimal values (in terms of maximizing profit) of $b1$ and $b2$ will vary from period to period, depending on the pattern and the rate of change of the price data. There is no exact science for selecting values for $b1$ and $b2$. Practice on selecting $b1$ and $b2$ by experimenting with historical data is a good way to gain experience in being able to make profitable trades.

To illustrate this approach to trading, the closing price of QQQQ for the 149-business-day period from July 1, 2004, through February 1, 2005, is applied to the EMA filters, with $b = 0.8$ and $b = 0.9$. The "MACD signal," which is the difference $D[n] = y1[n] - y2[n]$ between the outputs of the two filters, is shown in Figure 7.11a for $n = 2$ to 149, and in Figure 7.11b a plot of the closing price $c[n]$ is given for $n = 2$ to 149. The output of the EMA filter with $b = 0.8$ crosses above the output of the EMA filter with $b = 0.9$ when $D[n]$ becomes positive, and the output of the EMA filter with $b = 0.8$ falls below the output of the EMA filter with $b = 0.9$ when $D[n]$ becomes negative. From Figure 7.11a, we can see (although the values of $D[n]$ must be checked for us to be certain) that $D[n]$ becomes positive on day 40, and then becomes negative on day 130. Thus, QQQQ should have been bought at the close on day 40 and then sold at the close on day 130. This would have resulted in a gain of $c[130] - c[40] = 38.78 - 34.40 = \$4.38$ per share (not including commissions). The percent gain is $(4.38/34.40)(100) = 12.7\%$.

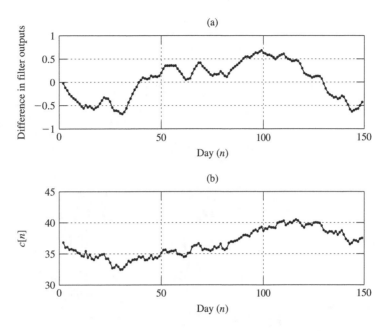

FIGURE 7.11
Plots of (a) difference in filter outputs and (b) closing price $c[n]$.

The MACD approach with $b = 0.8$ and 0.9 works well (in terms of producing a nice profit) for the range of closing prices of QQQQ shown in Figure 7.11b, because the pattern of the prices is a ramp (with noise) running from approximately day 30 to day 120. However, this approach with $b = 0.8$ and 0.9 will produce a loss, in general, if the closing prices of QQQQ are cycling; that is, the prices are moving up and down with a significant price difference between the peaks and valleys and with 100 or fewer days between the peaks. For example, consider the closing prices of QQQQ shown in Figure 7.12, which covers the 100-business-day period from March 1, 2004 to July 22, 2004. If the MACD approach with $b = 0.8$ and 0.9 is applied to the closing prices displayed in Figure 7.12, a loss would have resulted. (The reader is invited to check this.) The method does not work in this case, because the closing prices of QQQQ are cycling with a period of approximately 60 days. The EMA filters with $b = 0.8$ and 0.9 are simply too slow, given the variation of the prices.

To achieve a faster response, EMA filters with $b = 0.6$ and 0.8 can be used, which results in the MACD signal $D[n]$ shown in Figure 7.13, where $D[n]$ is now the difference between the output of the EMA filter with $b = 0.6$ and the output of the EMA filter with $b = 0.8$. Applying the trading strategy with the MACD signal shown in Figure 7.13 still does not give a good result (in terms of making a profit). However, the trading method can be modified as follows to yield a good result: Buy the stock at the close on day n if $D[n] > -\varepsilon$ and $D[n - 1] < -\varepsilon$, where ε is a small positive threshold. Then, hold the stock until day n when $D[n] < D[n - 1]$, in which case sell the stock at the close on that day. After selling the stock, wait until day n when $D[n] > -\varepsilon$ and

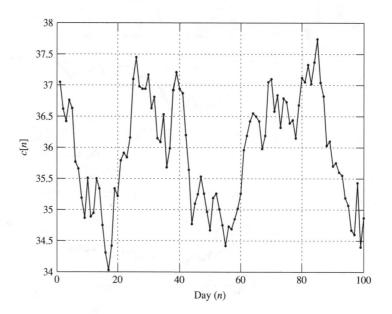

FIGURE 7.12
Closing price of QQQQ for 3/1/04 through 7/22/04.

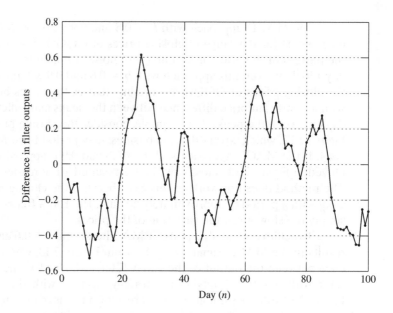

FIGURE 7.13
Plot of the difference in filter outputs with $b = 0.6$ and 0.8 for the data shown in Figure 7.12

$D[n-1] < -\varepsilon$, and then buy at the close on that day. This process is then repeated over and over again.

If this scheme with $\varepsilon = 0.05$ is applied to the signal $D[n]$ displayed in Figure 7.13, there are buys on days 20, 38, 59, and 79; and there are sells on days 27, 41, 65, and 83. This results in a net profit of \$3.53 per share, not including commissions. The reader is invited to verify this.

It should be stressed that the aforementioned modified MACD approach to trading does not always produce a profit, and therefore anyone who uses this strategy for actual trading should be aware of the risk. The authors and publisher are not responsible for any losses that may occur as a result of the use of this trading method or any other trading strategy discussed in this book.

It is worth noting that the MACD approach to trading can be modified by the use of different filters in place of the EMA filters, such as the two-pole IIR digital filters constructed in Section 10.2.

7.6   CHAPTER SUMMARY

The z-transform is a very useful tool for solving difference equations, finding the response of a discrete-time system, and analyzing the behavior of a discrete-time system. The two-sided z-transform is defined as

$$X(z) = \sum_{n=-\infty}^{\infty} x[n]z^{-n}$$

We obtain the one-sided z-transform by starting the sum at $n = 0$. The two forms are the same when $x[n] = 0$ for $n < 0$. There are common transform pairs for basic functions such as the unit pulse, step, ramp, exponential with the form $a^n u[n]$, and sinusoids. The properties of the z-transform include linearity, right and left shifts, convolution, and final-value theorem. The inverse z-transform is determined most easily by use of partial fraction expansion. However, partial fraction expansion requires that the expression for $X(z)$ be strictly proper; that is, the degree of the numerator must be less than that of the denominator. Since $X(z)$ is generally not strictly proper, the partial fraction expansion is performed on $X(z)/z$, and then the resulting terms are multiplied by z to reform an expression for $X(z)$ that is a sum of terms found in a table of common transform pairs.

The transfer function of a discrete-time system is defined as $H(z) = Y(z)/X(z)$, where $X(z)$ is the input z-transform and $Y(z)$ is the corresponding output z-transform with zero initial conditions. The convolution property of z-transforms is used to relate this input/output relationship to the time-domain relationship defined by the convolution representation given in Chapter 2, where $H(z)$ is the z-transform of the unit pulse response $h[n]$. Interconnections of systems can be represented by block diagrams where each block contains a transfer function. Common block diagram connections include series, parallel, and feedback connections. Each of these types of connections can be reduced to a single block that contains the transfer function of the combined system. A more complex block diagram can be similarly reduced to one block by successive reduction of any series, parallel or feedback connections in the block diagram to one block.

A transfer function characterizes the transient response of a system. In particular, the poles of the transfer function appear in the time-domain representation of the system response. For example, a pole at p results in a term in the response of p^n. Clearly, if the pole has magnitude greater than 1, the response will grow over time. A system is stable if all of its poles are inside of the unit circle, that is, have magnitude less than 1. A system is unstable if at least one of the poles is outside of the unit circle, and a system is marginally stable if it is not unstable and if it has at least one pole on the unit circle.

The transfer function of a system $H(z)$ is related to the frequency response of the system $H(\Omega)$ by the relationship $z = e^{j\Omega}$. In Chapter 5, it was seen that the frequency response is used to determine the response of a system to signals of different frequencies. Moreover, the shape of the frequency response determines the type of filtering present in the system, that is, which frequency ranges are attenuated. An example of a lowpass filter given in this chapter is the IIR EWMA filter.

PROBLEMS

7.1. Consider the discrete-time signal $x[n]$, where

$$x[n] = \begin{cases} b^n, & \text{for } n = 0, 1, 2, \ldots, N - 1 \\ 0, & \text{for all other } N \end{cases}$$

Here, b is an arbitrary real number and N is a positive integer.
   (a) For what real values of b does $x[n]$ possess a z-transform?
   (b) For the values of b determined in part (a), find the z-transform of $x[n]$. Express your answer as a ratio of polynomials.

7.2. A discrete-time signal $x[n]$ has z-transform

$$X(z) = \frac{z}{8z^2 - 2z - 1}$$

Determine the z-transform $V(z)$ of the following signals:
   (a) $v[n] = x[n - 4]u[n - 4]$
   (b) $v[n] = x[n + 2]u[n + 2]$
   (c) $v[n] = \cos(2n)x[n]$
   (d) $v[n] = e^{3n}x[n]$
   (e) $v[n] = n^2 x[n]$
   (f) $v[n] = x[n] * x[n]$
   (g) $v[n] = x[0] + x[1] + x[2] + \cdots + x[n]$

7.3. Compute the z-transform of the discrete-time signals (a)–(j). Express your answer as a ratio of polynomials in z whenever possible. Check your answer by using the Symbolic Math Toolbox command `ztrans`.
   (a) $x[n] = \delta[n] + 2\delta[n - 1]$
   (b) $x[n] = 1$ for $n = 0, 1$ and $x[n] = 2$ for all $n \geq 2$ (i.e., $n = 2, 3, \ldots$)
   (c) $x[n] = e^{0.5n}u[n] + u[n - 2]$
   (d) $x[n] = e^{0.5n}$ for $n = 0, 1$, and $x[n] = 1$ for all $n \geq 2$
   (e) $x[n] = \sin(\pi n/2)u[n - 2]$
   (f) $x[n] = (0.5)^n nu[n]$
   (g) $x[n] = u[n] - nu[n - 1] + (1/3)^n u[n - 2]$

(h) $x[n] = n$ for $n = 0, 1, 2$ and $x[n] = -n$ for all $n \geq 3$

(i) $x[n] = (n - 1)u[n] - nu[n - 3]$

(j) $x[n] = (0.25)^{-n}u[n - 2]$

7.4. Using the transform pairs in Table 7.3 and the properties of the z-transform in Table 7.2, determine the z-transform of the following discrete-time signals:

(a) $x[n] = (\cos^2 \omega n)u[n]$

(b) $x[n] = (\sin^2 \omega n)u[n]$

(c) $x[n] = n(\cos \omega n)u[n]$

(d) $x[n] = n(\sin \omega n)u[n]$

(e) $x[n] = ne^{-bn}(\cos \omega n)u[n]$

(f) $x[n] = ne^{-bn}(\sin \omega n)u[n]$

(g) $x[n] = e^{-bn}(\cos^2 \omega n)u[n]$

(h) $x[n] = e^{-bn}(\sin^2 \omega n)u[n]$

7.5. Let $x[n]$ be a discrete-time signal with $x[n] = 0$ for $n = -1, -2, \ldots$. The signal $x[n]$ is said to be summable if

$$\sum_{n=0}^{\infty} x[n] < \infty$$

If $x[n]$ is summable, the sum x_{sum} of $x[n]$ is defined by

$$x_{\text{sum}} = \sum_{n=0}^{\infty} x[n]$$

Now, suppose that the z-transform $X(z)$ of $x[n]$ can be expressed in the form

$$X(z) = \frac{B(z)}{a_N(z - p_1)(z - p_2) \cdots (z - p_N)}$$

where $B(z)$ is a polynomial in z. By using the final-value theorem, show that, if $|p_i| < 1$ for $i = 1, 2, \ldots, N$, $x[n]$ is summable and

$$x_{\text{sum}} = \lim_{z \to 1} X(z)$$

7.6. Using the results of Problem 7.5, compute x_{sum} for the signals (a)–(d). In each case, assume that $x[n] = 0$ for all $n < 0$.

(a) $x[n] = a^n, |a| < 1$

(b) $x[n] = n(a^n), |a| < 1$

(c) $x[n] = a^n \cos \pi n, |a| < 1$

(d) $x[n] = a^n \sin(\pi n/2), |a| < 1$

7.7. Let p and c be complex numbers defined in polar coordinates as $p = \sigma e^{j\Omega}$ and $c = |c|e^{j\angle c}$. Prove the following relationship:

$$cp^n + \bar{c}\,\bar{p}^n = 2|c|\sigma^n \cos(\Omega n + \angle c)$$

7.8. A discrete-time signal $x[n]$ has z-transform

$$X(z) = \frac{z + 1}{z(z - 1)}$$

Compute $x[0], x[1]$, and $x[10{,}000]$.

7.9. Compute the inverse z-transform $x[n]$ of the transforms (a)–(h). Determine $x[n]$ for all integers $n \geq 0$. Check your answer by using the Symbolic Math Toolbox command `iztrans`.

(a) $X(z) = \dfrac{z}{z^2 + 1}$

(b) $X(z) = \dfrac{z^2}{z^2 + 1}$

(c) $X(z) = \dfrac{1}{z^2 + 1} + \dfrac{1}{z^2 - 1}$

(d) $X(z) = \dfrac{z^2}{z^2 + 1} + \dfrac{z}{z^2 - 1}$

(e) $X(z) = \dfrac{z^2 - 1}{z^2 + 1}$

(f) $X(z) = \dfrac{z + 2}{(z - 1)(z^2 + 1)}$

(g) $X(z) = \dfrac{z^2 + 2}{(z - 1)(z^2 + 1)}$

(h) $X(z) = \ln\left(\dfrac{2z - 1}{2z}\right)$

7.10. For the transforms given in Problem 7.9 (a) to (g), compute the inverse z-transform numerically by using the command `filter`. Compare these results with the answers obtained analytically for $n = 0$ to $n = 5$.

7.11. Find the inverse z-transform $x[n]$ of the transforms that follow. Determine $x[n]$ for all n.

(a) $X(z) = \dfrac{z + 0.3}{z^2 + 0.75z + 0.125}$

(b) $X(z) = \dfrac{5z + 1}{4z^2 + 4z + 1}$

(c) $X(z) = \dfrac{4z + 1}{z^2 - z + 0.5}$

(d) $X(z) = \dfrac{z}{16z^2 + 1}$

(e) $X(z) = \dfrac{2z + 1}{z(10z^2 - z - 2)}$

(f) $X(z) = \dfrac{z + 1}{(z - 0.5)(z^2 - 0.5z + 0.25)}$

(g) $X(z) = \dfrac{z^3 + 1}{(z - 0.5)(z^2 - 0.5z + 0.25)}$

(h) $X(z) = \dfrac{z + 1}{z(z - 0.5)(z^2 - 0.5z + 0.25)}$

7.12. For each of the transforms given in Problem 7.11, compute the inverse z-transform numerically by using the command `filter`. Compare these results with the answers obtained analytically.

7.13. By using the z-transform, compute the convolution $x[n] * v[n]$ for all $n \geq 0$, where

(a) $x[n] = u[n] + 3\delta[n - 1]$, $v[n] = u[n - 2]$

(b) $x[n] = u[n]$, $v[n] = nu[n]$

(c) $x[n] = \sin(\pi n/2)u[n]$, $v[n] = e^{-n}u[n - 2]$

(d) $x[n] = u[n - 1] + \delta[n]$, $v[n] = e^{-n}u[n] - 2e^{-2n}u[n - 2]$

7.14. A linear time-invariant discrete-time system has unit-pulse response

$$h[n] = \begin{cases} \dfrac{1}{n}, & \text{for } n = 1, 2, 3 \\ n - 2, & \text{for } n = 4, 5 \\ 0, & \text{for all other } n \end{cases}$$

Compute the transfer function $H(z)$.

7.15. The input $x[n] = (-1)^n u[n]$ is applied to a linear time-invariant discrete-time system. The resulting output response $y[n]$ with zero initial conditions is given by

$$y[n] = \begin{cases} 0, & \text{for } n < 0 \\ n + 1, & \text{for } n = 0, 1, 2, 3 \\ 0, & \text{for } n \geq 4 \end{cases}$$

Determine the transfer function $H(z)$ of the system.

7.16. For the system defined in Problem 7.15, compute the output response $y[n]$ resulting from the input $x[n] = (1/n)(u[n - 1] - u[n - 3])$ with zero initial conditions.

7.17. A system is described by the difference equation

$$y[n] + 0.7y[n - 1] = u[n]; \quad y[-1] = 1$$

(a) Find an analytical expression for $y[n]$.

(b) Verify your result by simulating the system by using MATLAB.

7.18. Repeat Problem 7.17 for the system described by the following difference equation:

$$y[n] - 0.2y[n - 1] - 0.8y[n - 2] = 0; \quad y[-1] = 1, y[-2] = 1$$

7.19. A linear time-invariant discrete-time system is described by the input/output difference equation

$$y[n + 2] + y[n] = 2x[n + 1] - x[n]$$

(a) Compute the unit-pulse response $h[n]$.

(b) Compute the step response $y[n]$.

(c) Compute $y[n]$ for all $n \geq 0$ when $x[n] = 2^n u[n]$ with $y[-1] = 3$ and $y[-2] = 2$.

(d) An input $x[n]$ with $x[-2] = x[-1] = 0$ produces the output response $y[n] = (\sin \pi n)u[n]$ with zero initial conditions. Determine $x[n]$.

(e) An input $x[n]$ with $x[-2] = x[-1] = 0$ produces the output response $y[n] = \delta[n - 1]$. Compute $x[n]$.

(f) Verify the results of parts (a) to (e) via computer simulation.

7.20. A linear time-invariant discrete-time system is given by the input/output difference equation

$$y[n] + y[n - 1] - 2y[n - 2] = 2x[n] - x[n - 1]$$

Find an input $x[n]$ with $x[n] = 0$ for $n < 0$ that gives the output response $y[n] = 2(u[n] - u[n - 3])$ with initial conditions $y[-2] = 2$, $y[-1] = 0$.

7.21. The input $x[n] = u[n] - 2u[n - 2] + u[n - 4]$ is applied to a linear time-invariant discrete-time system. The resulting response with zero initial conditions is $y[n] = nu[n] - nu[n - 4]$. Compute the transfer function $H(z)$.

7.22. A system has the transfer function

$$H(z) = \frac{-0.4z^{-1} - 0.5z^{-2}}{(1 - 0.5z^{-1})(1 - 0.8z^{-1})}$$

(a) Compute an analytical expression for the step response.

(b) Verify your result by simulating the step response by using MATLAB.

7.23. Repeat Problem 7.22 for the system with the transfer function

$$H(z) = \frac{z^2 - 0.1}{z^2 - 0.6484z + 0.36}$$

7.24. A linear time-invariant discrete-time system has transfer function

$$H(z) = \frac{z^2 - z - 2}{z^2 + 1.5z - 1}$$

(a) Compute the unit-pulse response $h[n]$ for all $n \geq 0$.

(b) Compute the step response $y[n]$ for all $n \geq 0$.

(c) Compute the output values $y[0]$, $y[1]$, $y[2]$ resulting from the input $x[n] = 2^n \sin(\pi n/4) + \tan(\pi n/3)$, $n = 0, 1, 2, \ldots$, with zero initial conditions.

(d) If possible, find an input $x[n]$ with $x[n] = 0$ for all $n < 0$ such that the output response $y[n]$ resulting from $x[n]$ is given by $y[0] = 2$, $y[1] = -3$, and $y[n] = 0$ for all $n \geq 2$. Assume that the initial conditions are equal to zero.

(e) Verify the results of parts (a) to (d) by computer simulation.

7.25. A linear time-invariant discrete-time system has transfer function

$$H(z) = \frac{z}{(z - 0.5)^2(z^2 + 0.25)}$$

(a) Find the unit-pulse response $h[n]$ for all $n \geq 0$.

(b) Simulate the unit-pulse response by using MATLAB, and compare this result with the result for $h[n]$ obtained analytically in part (a).

7.26. The input $x[n] = (0.5)^n u[n]$ is applied to a linear time-invariant discrete-time system with the initial conditions $y[-1] = 8$ and $y[-2] = 4$. The resulting output response is

$$y[n] = 4(0.5)^n u[n] - n(0.5)^n u[n] - (-0.5)^n u[n]$$

Find the transfer function $H(z)$.

7.27. A linear time-invariant discrete-time system has transfer function

$$H(z) = \frac{3z}{(z + 0.5)(z - 0.5)}$$

The output response resulting from the input $x[n] = u[n]$ and initial conditions $y[-1]$ and $y[-2]$ is

$$y[n] = [(0.5)^n - 3(-0.5)^n + 4]u[n]$$

Determine the initial conditions $y[-1]$, $y[-2]$, and the part of the output response due to the initial conditions.

7.28. A linear time-invariant discrete-time system has unit-pulse response $h[n]$ equal to the Fibonacci sequence; that is, $h[0] = 0$, $h[1] = 1$, and $h[n] = h[n - 2] + h[n - 1]$ for $n \geq 2$. Show that the system's transfer function $H(z)$ is rational in z. Express $H(z)$ as a ratio of polynomials in positive powers of z.

7.29. Consider each of the following transfer functions:

    (i) $H(z) = z/(z - 0.5)$

    (ii) $H(z) = z/(z + 0.5)$

    (iii) $H(z) = z/(z - 1)$

    (iv) $H(z) = z/(z + 1)$

    (v) $H(z) = z/(z - 2)$

    (vi) $H(z) = z/(z + 2)$

        (a) Compute the pole of the system. From this pole position, describe the type of behavior that you would expect in the transient response.

        (b) Verify your prediction in part (a) by determining an analytical expression for the unit-pulse response.

        (c) Simulate the unit-pulse response, and compare with the answer obtained analytically in part (b).

7.30. Consider the following transfer functions:

    (i) $H(z) = (z^2 - 0.75z)/(z^2 - 1.5z + 2.25)$

    (ii) $H(z) = (z^2 - 0.5z)/(z^2 - z + 1)$

    (iii) $H(z) = (z^2 - 0.25z)/(z^2 - 0.5z + 0.25)$

        (a) Compute the pole positions. From knowledge of the pole positions, describe the type of behavior you would expect in the transient response.

        (b) Without computing the actual response, give a general expression for the step response.

        (c) Verify your prediction by simulating the system for a step input.

7.31. By using the z-domain representation, determine the transfer functions of the discrete-time systems shown in Figure P7.31.

7.32. Consider the discrete-time system shown in Figure P7.32.

    (a) Determine the transfer function $H(z)$ of the system.

    (b) Determine the system's input/output difference equation.

    (c) Compute the output response $y[n]$ when $x[n] = 4u[n]$ with zero initial conditions.

7.33. Consider the cascade connection shown in Figure P7.33. Determine the unit-pulse response $h_2[n]$ of the system with transfer function $H_2(z)$ so that, when $x[n] = \delta[n]$ with zero initial conditions, the response $y[n]$ is equal to $\delta[n]$.

7.34. A linear time-invariant discrete-time system is given by the feedback connection shown in Figure P7.34. In Figure P7.34, $X(z)$ is the z-transform of the system's input $x[n]$, $Y(z)$ is the z-transform of the system's output $y[n]$, and $H_1(z)$, $H_2(z)$ are the transfer functions of the subsystems given by

$$H_1(z) = \frac{z}{z + 1}, \qquad H_2(z) = \frac{9}{z - 8}$$

    (a) Determine the unit-pulse response of the overall system.

    (b) Compute the step response of the overall system.

    (c) Compute $y[n]$ when $x[n] = (0.5)^n u[n]$ with $y[-1] = -3$, $y[-2] = 4$.

    (d) Compute $y[n]$ when $x[n] = (0.5)^n u[n]$ with $y[-2] = 1$, $w[-1] = 2$, where $w[n]$ is the output of the feedback system in Figure P7.34.

    (e) Verify the results of parts (a) to (d) via computer simulation.

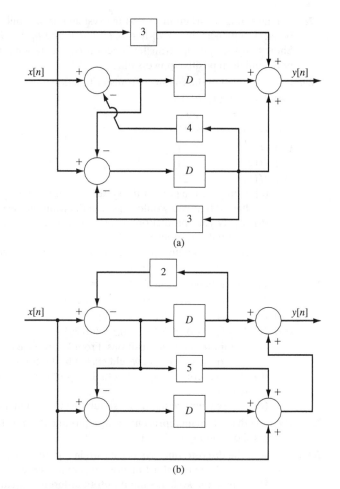

(a)

(b)

FIGURE P7.31

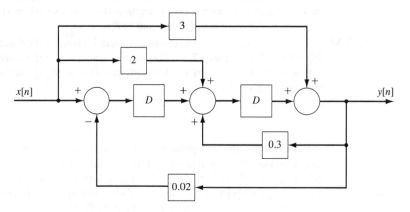

FIGURE P7.32

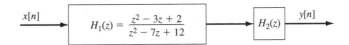

FIGURE P7.33

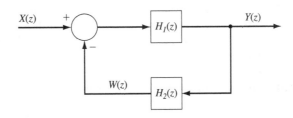

FIGURE P7.34

7.35. A model for the generation of echoes is shown in Figure P7.35. As shown, each successive echo is represented by a delayed and scaled version of the output, which is fed back to the input.

    (a) Determine the transfer function $H(z)$ of the echo system.

    (b) Suppose that we would like to recover the original signal $x[n]$ from the output $y[n]$ by using a system with transfer function $W(z)$ [and with input $y[n]$ and output $x[n]$]. Determine $W(z)$.

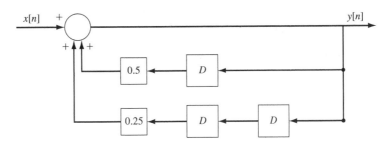

FIGURE P7.35

7.36. A linear time-invariant discrete-time system is given by the cascade connection shown in Figure P7.36.

    (a) Compute the unit-pulse response of the overall system.

    (b) Compute the input/output difference equation of the overall system.

    (c) Compute the step response of the overall system.

    (d) Compute $y[n]$ when $x[n] = u[n]$ with $y[-1] = 3$, $q[-1] = 2$.

    (e) Compute $y[n]$ when $x[n] = (0.5)^n u[n]$ with $y[-2] = 2$, $q[-2] = 3$.

    (f) Verify the results of part (a) and parts (c) to (e) via computer simulation.

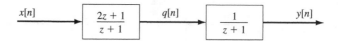

FIGURE P7.36

7.37. A linear time-invariant discrete-time system is excited by the input $x[n] = \delta[n] + 2u[n - 1]$. The resulting output response with zero initial conditions is $y[n] = (0.5)^n u[n]$. Determine if the system is stable, marginally stable, or unstable. Justify your answer.

7.38. Determine if the system in Problem 7.32 is stable, marginally stable, or unstable. Justify your answer.

7.39. A discrete-time system is given by the input/output difference equation

$$y[n + 2] - y[n + 1] + y[n] = x[n + 2] - x[n + 1]$$

Is the system stable, marginally stable, or unstable? Justify your answer.

7.40. For the following linear time-invariant discrete-time systems with unit-pulse response $h[n]$, determine if the system is BIBO stable.
  (a) $h[n] = \sin(\pi n/6)(u[n] - u[n - 10])$
  (b) $h[n] = (1/n)u[n - 1]$
  (c) $h[n] = (1/n^2)u[n - 1]$
  (d) $h[n] = e^{-n}\sin(\pi n/6)u[n]$

7.41. Determine whether or not the following linear time-invariant discrete-time systems are stable:
  (a) $H(z) = \dfrac{z - 4}{z^2 + 1.5z + 0.5}$
  (b) $H(z) = \dfrac{z^2 - 3z + 1}{z^3 + z^2 - 0.5z + 0.5}$
  (c) $H(z) = \dfrac{1}{z^3 + 0.5z + 0.1}$

7.42. A linear time-invariant discrete-time system has transfer function

$$H(z) = \frac{z}{z + 0.5}$$

  (a) Find the transient response and steady-state response resulting from the input $x[n] = 5\cos 3n$, $n = 0, 1, 2, \ldots$, with zero initial conditions.
  (b) Sketch the frequency response curves.
  (c) Plot the actual frequency response curves by using MATLAB.

7.43. A linear time-invariant discrete-time system has transfer function

$$H(z) = \frac{-az + 1}{z - a}$$

where $-1 < a < 1$.
  (a) Compute the transient and steady-state responses when the input $x[n] = \cos(\pi n/2)u[n]$ with zero initial conditions.
  (b) When $a = 0.5$, sketch the frequency response curves.
  (c) Show that $|H(\Omega)| = C$ for $0 \le \Omega \le 2\pi$, where C is a constant. Derive an expression for C in terms of a. Since $|H(\Omega)|$ is constant, the system is called an *allpass filter*.

7.44. Consider the discrete-time system shown in Figure P7.44. Compute the steady-state output response $y_{ss}[n]$ and the transient output response $y_{tr}[n]$ when $y[-1] = x[-1] = 0$ and $x[n] = 2\cos(\pi n/2)u[n]$.

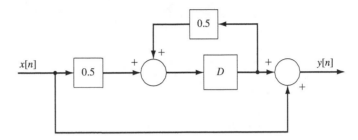

FIGURE P7.44

7.45. A discrete-time system is given by the input/output difference equation

$$y[n + 2] + 0.3y[n + 1] + 0.02y[n] = x[n + 1] + 3x[n]$$

    (a) Compute the transient and steady-state output responses when $x[n] = (\cos \pi n)u[n]$ with zero initial conditions.

    (b) Use MATLAB to plot the frequency response curves of the system.

7.46. A linear time-invariant discrete-time system has transfer function

$$H(z) = \frac{z}{(z^2 + 0.25)(z - 0.5)^2}$$

    (a) Assuming that the initial conditions are zero, find the transient and steady-state response to $x[n] = 12\cos(\pi n/2)u[n]$.

    (b) Use MATLAB to plot the frequency response curves of the system.

7.47. The *differencer* is the discrete-time system with the input/output difference equation

$$y[n] = x[n] - x[n - 1]$$

Use MATLAB to plot the frequency response curves of the differencer.

7.48. As discussed in Section 7.5, the IIR EWMA filter is given by the transfer function $H(z) = (1 - b)z/(z - b)$, where $0 < b < 1$. Note that the pole of the filter is located at $z = b$, and since b is positive, the pole is on the positive real axis of the complex plane.

    (a) Consider the IIR filter with transfer function $H(z) = (1 - b)z/(z + b)$, where $0 < b < 1$. Note that the pole of this system is on the negative real axis of the complex plane. Use MATLAB to plot the frequency response curves of the system when $b = 0.8$. What type of filter is this? Explain.

    (b) Now consider the series connection of the IIR EWMA filter and the filter in part (b). The transfer function of the filter is

$$H(z) = \frac{(1 - b)^2 z^2}{(z - b)(z + b)}$$

Use MATLAB to plot the frequency response curves of the system when $b = 0.8$. What type of filter is this? Explain.

7.49. For various values of b in the EMA filter, apply the MACD trading strategy and the modified MACD strategy to the closing price of QQQQ for the period from February 1, 2005, through September 1, 2005. For the values of b that are used, what is the net gain or loss over the time period?

Analysis of Continuous-Time Systems by Use of the Transfer Function Representation

In Section 7.5 of the previous chapter, the transfer function representation was used to analyze the behavior of discrete-time systems. In this chapter the transfer function is applied to the analysis of causal linear time-invariant continuous-time systems. The presentation begins in Sections 8.1 and 8.2 with the study of stability of a continuous-time system. Then in Section 8.3 the transfer function representation is used to study the basic characteristics of the output response resulting from an input, with the focus on the case of a step input. Here it is shown that the poles of a system determine the basic features of the transient response resulting from a step input. In Section 8.4 the transfer function representation is utilized to study the steady-state response resulting from a sinusoidal input. This leads to the concept of a system's frequency response function that was first considered in Chapter 5. A detailed development of the frequency response function is given in Section 8.5, including the description of frequency response data in terms of Bode diagrams. In Section 8.6 the frequency function analysis is applied to the study of causal filters, and in Section 8.7 a summary of the chapter is given.

8.1 STABILITY AND THE IMPULSE RESPONSE

Consider a causal linear time-invariant continuous-time system with input $x(t)$ and output $y(t)$. Throughout this chapter it is assumed that the system is finite dimensional, and thus the system's transfer function $H(s)$ is rational in s; that is,

$$H(s) = \frac{b_M s^M + b_{M-1} s^{M-1} + \cdots + b_1 s + b_0}{s^N + a_{N-1} s^{N-1} + \cdots + a_1 s + a_0} \tag{8.1}$$

In the following development it is assumed that the order N of the system is greater than or equal to M and that the transfer function $H(s)$ does not have any common poles and zeros. If there are common poles and zeros, they should be canceled.

As first observed in Section 6.5, the transfer function $H(s)$ is the Laplace transform of the system's impulse response $h(t)$. That is, if the input $x(t)$ applied to the system is the impulse $\delta(t)$, the transform of the resulting output response is $H(s)$ (assuming zero initial conditions). Since $H(s)$ is the Laplace transform of the impulse response $h(t)$, it follows directly from the results in Section 6.3 that the form of the impulse response is determined directly by the poles of the system [i.e., the poles of $H(s)$]. In particular, if $H(s)$ has a real pole p, then $h(t)$ contains a term of the form ce^{pt}; and if $H(s)$ has a complex pair

$\sigma \pm j\omega$ of poles, then $h(t)$ contains a term of the form $ce^{\sigma t}\cos(\omega t + \theta)$. If $H(s)$ has repeated poles, $h(t)$ will contain terms of the form $ct^i e^{pt}$ and/or $ct^i e^{\sigma t}\cos(\omega t + \theta)$.

It follows from the relationship between the form of $h(t)$ and the poles of $H(s)$ that the impulse response $h(t)$ converges to zero as $t \to \infty$ if and only if

$$\operatorname{Re}(p_i) < 0 \text{ for } i = 1, 2, \ldots, N \tag{8.2}$$

where $p_1, p_2, \ldots, p_N$ are the poles of $H(s)$. The condition (8.2) is equivalent to requiring that all the poles of the system are located in the open left-half plane, where the open left-half plane is the region of the complex plane consisting of all points to the left of the $j\omega$-axis, but not including the $j\omega$-axis. The open left-half plane is indicated by the hatched region shown in Figure 8.1.

A system with transfer function $H(s)$ given by (8.1) is said to be *stable* if its impulse response $h(t)$ converges to zero as $t \to \infty$. Hence, a system is stable if and only if all the poles are located in the open left-half plane.

A system with transfer function $H(s)$ is said to be *marginally stable* if its impulse response $h(t)$ is bounded; that is, there exists a finite positive constant c such that

$$|h(t)| \leq c \quad \text{for all } t \tag{8.3}$$

Again, from the relationship between the form of $h(t)$ and the poles of $H(s)$, it follows that a system is marginally stable if and only if $\operatorname{Re}(p_i) \leq 0$ for all nonrepeated poles of $H(s)$, and $\operatorname{Re}(p_i) < 0$ for all repeated poles. Thus, a system is marginally stable if and only if all the poles are in the open left-half plane, except that there can be nonrepeated poles on the $j\omega$-axis.

Finally, a system is *unstable* if the impulse response $h(t)$ grows without bound as $t \to \infty$. In mathematical terms, the system is unstable if and only if

$$|h(t)| \to \infty \qquad \text{as } t \to \infty \tag{8.4}$$

The relationship between the form of $h(t)$ and the poles reveals that a system is unstable if and only if there is at least one pole p_i with $\operatorname{Re}(p_i) > 0$ or if there is at least one

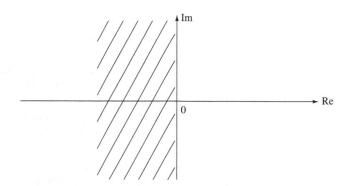

FIGURE 8.1
Open left-half plane.

repeated pole p_i with $\text{Re}(p_i) = 0$. Hence, a system is unstable if and only if there are one or more poles in the open right-half plane (the region to the right of the $j\omega$-axis) or if there are repeated poles on the $j\omega$-axis.

Example 8.1   Series *RLC* Circuit

Consider the series *RLC* circuit that was studied in Example 6.37. The circuit is redrawn in Figure 8.2. In the following analysis it is assumed that $R > 0$, $L > 0$, and $C > 0$. As computed in Example 6.37, the transfer function $H(s)$ of the circuit is

$$H(s) = \frac{1/LC}{s^2 + (R/L)s + 1/LC}$$

From the quadratic formula, the poles of the system are

$$p_1, p_2 = -\frac{R}{2L} \pm \sqrt{b}$$

where

$$b = \left(\frac{R}{2L}\right)^2 - \frac{1}{LC}$$

Now if $b < 0$, both poles are complex with real part equal to $-R/2L$, and thus in this case the circuit is stable. If $b \geq 0$, both poles are real. In this case,

$$-\frac{R}{2L} - \sqrt{b} < 0$$

In addition, $b \geq 0$ implies that

$$b < \left(\frac{R}{2L}\right)^2$$

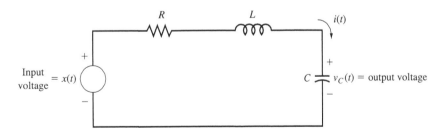

FIGURE 8.2
Series *RLC* circuit in Example 8.1.

and thus

$$\sqrt{b} < \frac{R}{2L}$$

Therefore,

$$-\frac{R}{2L} + \sqrt{b} < 0$$

and thus the circuit is still stable. So the circuit is stable for any values of $R, L, C > 0$. This means that if an impulsive input voltage $x(t)$ is applied to the circuit with zero initial conditions, the voltage $v_c(t)$ across the capacitor decays to zero as $t \to \infty$.

8.1.1   Alternative Characterizations of Stability

Again consider the system with the rational transfer function $H(s)$ given by (8.1). It turns out that the system's impulse response $h(t)$ will converge to zero as $t \to \infty$ if and only if $h(t)$ is absolutely integrable; that is,

$$\int_0^\infty |h(t)| \, dt < \infty \tag{8.5}$$

Hence, stability of the system is equivalent to absolute integrability of the system's impulse response.

Now let $y(t)$ denote the output response of the system resulting from input $x(t)$ applied for $t \geq 0$ with zero initial conditions. The system is said to be *bounded-input bounded-output (BIBO) stable* if $y(t)$ is bounded whenever the input $x(t)$ is bounded. In mathematical terms, this means that if $|x(t)| \leq c_1$ for all t where c_1 is a finite positive constant, then the resulting output response (with zero initial conditions) satisfies the condition

$$|y(t)| \leq c_2 \qquad \text{for all } t$$

where c_2 is a finite positive constant [depending in general on $x(t)$]. It turns out that BIBO stability is equivalent to the integrability condition (8.5). Then, since (8.5) is equivalent to the condition that $h(t) \to 0$ as $t \to \infty$, it is seen that BIBO stability is equivalent to stability as previously defined. It should be stressed that the validity of this result is based on the assumption that the transfer function $H(s)$ is a rational function of s and does not have any common poles and zeros.

8.2   ROUTH–HURWITZ STABILITY TEST

By the results in Section 8.1, we can check the stability of a system with rational transfer function $H(s) = B(s)/A(s)$ by first determining the poles of $H(s)$, which are the roots of $A(s) = 0$. The poles of $H(s)$ can be computed by the MATLAB command roots. The use of this command was illustrated in Chapter 6.

It turns out that there are procedures for testing for stability that do not require the computation of the poles of the system. One such procedure is the Routh–Hurwitz stability test, which is based on simple computations involving the coefficients of the polynomial $A(s)$. The details are as follows.

Suppose that

$$A(s) = a_N s^N + a_{N-1} s^{N-1} + \cdots + a_1 s + a_0, \, a_N > 0 \tag{8.6}$$

Note that the leading coefficient a_N of $A(s)$ may be any nonzero positive number. By the results in Section 8.1, the system is stable if and only if all the zeros of $A(s)$ are in the open left-half plane (OLHP). A necessary (but, in general, insufficient) condition for this to be the case is that all the coefficients of $A(s)$ must be strictly positive; that is,

$$a_i > 0 \text{ for } i = 0, 1, 2, \ldots, N - 1 \tag{8.7}$$

Thus, if $A(s)$ has one or more coefficients that are zero or negative, there is at least one pole not in the OLHP, and the system is not stable. Here, the expression "pole not in the OLHP" means a pole located on the $j\omega$-axis or located in the open right-half plane (ORHP).

It should be stressed that the condition (8.7) is not a sufficient condition for stability, in general. In other words, there are unstable systems for which (8.7) is satisfied.

Now the Routh–Hurwitz stability test will be stated, which gives necessary and sufficient conditions for stability. Given the polynomial $A(s)$ defined by (8.6), the first step is to construct the Routh array shown in Table 8.1.

As seen from Table 8.1, the Routh array has $N + 1$ rows, with the rows indexed by the powers of s. The number of columns of the array is $(N/2) + 1$ if N is even or $(N + 1)/2$ if N is odd. The first two rows of the Routh array are filled by the coefficients of $A(s)$, starting with the leading coefficient a_N. The elements in the third row are given by

$$b_{N-2} = \frac{a_{N-1} a_{N-2} - a_N a_{N-3}}{a_{N-1}} = a_{N-2} - \frac{a_N a_{N-3}}{a_{N-1}}$$

$$b_{N-4} = \frac{a_{N-1} a_{N-4} - a_N a_{N-5}}{a_{N-1}} = a_{N-4} - \frac{a_N a_{N-5}}{a_{N-1}}$$

$$\vdots$$

TABLE 8.1   Routh Array

s^N	a_N	a_{N-2}	a_{N-4}	$\cdots$
s^{N-1}	a_{N-1}	a_{N-3}	a_{N-5}	$\cdots$
s^{N-2}	b_{N-2}	b_{N-4}	b_{N-6}	$\cdots$
s^{N-3}	c_{N-3}	c_{N-5}	c_{N-7}	$\cdots$
.	.	.	.	
.	.	.	.	
.	.	.	.	
s^2	d_2	d_0	0	$\cdots$
s^1	e_1	0	0	$\cdots$
s^0	f_0	0	0	$\cdots$

The elements in the fourth row are given by

$$c_{N-3} = \frac{b_{N-2}a_{N-3} - a_{N-1}b_{N-4}}{b_{N-2}} = a_{N-3} - \frac{a_{N-1}b_{N-4}}{b_{N-2}}$$

$$c_{N-5} = \frac{b_{N-2}a_{N-5} - a_{N-1}b_{N-6}}{b_{N-2}} = a_{N-5} - \frac{a_{N-1}b_{N-6}}{b_{N-2}}$$

$$\vdots$$

The other rows (if there are any) are computed in a similar fashion. As a check on the computations, it should turn out that the last nonzero element in each column of the array is equal to the coefficient a_0 of $A(s)$.

The Routh–Hurwitz stability test states that the system is stable (all poles in the OLHP) if and only if all the elements in the first column of the Routh array are strictly positive (>0). In addition, the number of poles in the ORHP is equal to the number of sign changes in the first column.

In calculating the Routh array, it may happen that an element in the first column is zero, in which case it is not possible to perform the division in computing the elements in the subsequent row. It is clear from the Routh-Hurwitz stability test that the system is not stable. In order to determine if the system is marginally stable or unstable, the array must be completed. To complete the array, replace the zero element with ε (a very small positive number), and then continue. (Note that if any zero elements are set equal to small positive numbers, the last nonzero element of the columns of the Routh array will, in general, not be equal to a_0.) The number of sign changes in the first column of the array is equal to the number of poles in the ORHP, and so no sign changes would indicate poles on the $j\omega$ axis.

The application of the Routh–Hurwitz stability test is illustrated in the examples that follow. The proof of the Routh–Hurwitz stability test is well beyond the scope of this book. As shown in the following examples, when the degree N of $A(s)$ is less than or equal to 3, the Routh–Hurwitz test can be used to derive simple conditions for stability given directly in terms of the coefficients of $A(s)$:

Example 8.2   Second-Order Case

Let $N = 2$ and $a_2 = 1$, so that

$$A(s) = s^2 + a_1 s + a_0$$

The Routh array for this case is given in Table 8.2. The elements in the first column of the Routh array are $1, a_1,$ and a_0; and thus the poles are in the OLHP if and only if the coefficients a_1 and a_0

TABLE 8.2	Routh Array in the $N = 2$ Case	
s^2	1	a_0
s^1	a_1	0
s^0	$\dfrac{a_1 a_0 - (1)(0)}{a_1} = a_0$	0

are both positive. So in this case, the positive-coefficient condition (8.7) is necessary and sufficient for stability. Now, suppose that $a_1 > 0$ and $a_0 < 0$. Then there is one sign change in the first column of the Routh array, which means that there is one pole in the ORHP. If $a_1 < 0$ and $a_0 < 0$, there still is one sign change and thus there is one pole in the ORHP. If $a_1 < 0$ and $a_0 > 0$, there are two sign changes in the first column, and therefore both poles are in the ORHP.

Example 8.3   Third-Order Case

Consider the third-order case

$$A(s) = s^3 + a_2 s^2 + a_1 s + a_0$$

The Routh array is displayed in Table 8.3. Since

$$a_1 - \frac{a_0}{a_2} > 0$$

if and only if

$$a_1 > \frac{a_0}{a_2}$$

then all three poles are in the OLHP if and only if

$$a_2 > 0, \quad a_1 > \frac{a_0}{a_2}, \quad a_0 > 0$$

This result shows that when $N = 3$, it is not true, in general, that positivity of a_2, a_1, and a_0 implies that the system is stable. Note that if $a_2 < 0$ and $a_0 > 0$, there are two sign changes in the first column of the Routh array, and thus there are two poles in the ORHP. If $a_2 < 0$, $a_1 > a_0/a_2$, and $a_0 < 0$, there are three sign changes, and therefore all three poles are in the ORHP. If $a_2 < 0$, $a_1 < a_0/a_2$, and $a_0 < 0$, there is one sign change, which means that there is one pole in the ORHP.

TABLE 8.3   The $N = 3$ Case

s^3	1	a_1
s^2	a_2	a_0
s^1	$\dfrac{a_2 a_1 - (1)a_0}{a_2} = a_1 - \dfrac{a_0}{a_2}$	0
s^0	a_0	0

As N is increased above the value $N = 3$, the conditions for stability in terms of the coefficients of $A(s)$ get rather complicated. For $N \geq 4$, the Routh–Hurwitz test can still be applied on a case-by-case basis.

Example 8.4   Higher-Order Case

Suppose that $H(s) = B(s)/A(s)$, where

$$A(s) = 6s^5 + 5s^4 + 4s^3 + 3s^2 + 2s + 1$$

Then, $N = 5$ and

$$a_0 = 1, a_1 = 2, a_2 = 3, a_3 = 4, a_4 = 5, a_5 = 6$$

The Routh array for this example is shown in Table 8.4. There are two sign changes in the first column of the Routh array, and thus two of the five poles are located in the open right-half plane. The system is therefore not stable.

TABLE 8.4   Routh Array for Example 8.4

s^5	$a_5 = 6$	$a_3 = 4$	$a_1 = 2$
s^4	$a_4 = 5$	$a_2 = 3$	$a_0 = 1$
s^3	$\dfrac{(5)(4) - (6)(3)}{5} = 0.4$	$\dfrac{(5)(2) - (6)(1)}{5} = 0.8$	0
s^2	$\dfrac{(0.4)(3) - (5)(0.8)}{0.4} = -7$	$a_0 = 1$	0
s^1	$\dfrac{(-7)(0.8) - (0.4)(1)}{-7} = 6/7$	0	0
s^0	$a_0 = 1$	0	0

The Routh–Hurwitz test can also be used to determine the values of complex poles located on the $j\omega$-axis: Given a rational transfer function $H(s) = B(s)/A(s)$, there is a pair of poles on the $j\omega$-axis with all other poles in the open left-half plane if and only if all the entries in the first column of the Routh array are strictly positive, except for the entry in the row indexed by s^1, which is zero. If this is the case, there is a pair of poles at $\pm j\sqrt{a_0/\gamma_2}$, where γ_2 is the entry in the first row of the Routh array indexed by s^2 and a_0 is the constant term of $A(s)$.

Example 8.5   Fourth-Order Case

Suppose that

$$A(s) = s^4 + s^3 + 3s^2 + 2s + 2$$

The Routh array for this example is shown in Table 8.5. Note that all the entries in the first column of the array are strictly positive except for the entry in the row indexed by s^1, which is zero. As a result, two of the poles are in the open left-half plane, and the other two poles are on the $j\omega$-axis, located at $\pm j\sqrt{2/1} = \pm j\sqrt{2}$.

TABLE 8.5   Routh Array for Example 8.5

s^4	1	3	2
s^3	1	2	0
s^2	$\dfrac{3-2}{1} = 1$	2	0
s^1	$\dfrac{2-2}{1} \approx \varepsilon$	0	0
s^0	$\dfrac{2\varepsilon - 0}{\varepsilon} = 2$	0	0

8.3   ANALYSIS OF THE STEP RESPONSE

Again consider the system with the rational transfer function $H(s) = B(s)/A(s)$, where the degree of $B(s)$ is less than or equal to the degree of $A(s)$. If an input $x(t)$ is applied to the system for $t \geq 0$ with zero initial conditions, then from the results in Section 6.5 we see that the transform $Y(s)$ of the resulting output response is given by

$$Y(s) = \frac{B(s)}{A(s)} X(s) \tag{8.8}$$

Now, suppose that $x(t)$ is the unit-step function $u(t)$, so that $X(s) = 1/s$. Then, inserting $X(s) = 1/s$ into (8.8) results in the transform of the step response

$$Y(s) = \frac{B(s)}{A(s)s} \tag{8.9}$$

Then, if $A(0) \neq 0$, "pulling out" the $1/s$ term from $Y(s)$ and using the residue formula (6.67) yield

$$Y(s) = \frac{E(s)}{A(s)} + \frac{c}{s} \tag{8.10}$$

where $E(s)$ is a polynomial in s and c is the constant given by

$$c = [sY(s)]_{s=0} = H(0)$$

Then, taking the inverse Laplace transform of $Y(s)$ results in the step response

$$y(t) = y_1(t) + H(0), t \geq 0 \tag{8.11}$$

where $y_1(t)$ is the inverse Laplace transform of $E(s)/A(s)$.

Note that if the system is stable so that all the roots of $A(s) = 0$ are in the open left-half plane, the term $y_1(t)$ in (8.11) converges to zero as $t \to \infty$, in which case $y_1(t)$ is the *transient part of the response*. So if the system is stable, the step response contains a transient that decays to zero, and it contains a constant with value $H(0)$. The constant $H(0)$ is the *steady-state* value of the step response.

It is very important to note that, since the transform of the term $y_1(t)$ in (8.11) is equal to $E(s)/A(s)$, the form of $y_1(t)$ depends on the system's poles [the poles of $H(s)$]. This is examined in detail next, beginning with first-order systems.

8.3.1   First-Order Systems

Consider the system given by the first-order transfer function

$$H(s) = \frac{k}{s - p} \tag{8.12}$$

where k is a real constant and p is the pole (which is real). Then with $x(t) = u(t)$, the transform of $y(t)$ is equal to $H(s)/s$ and the partial fraction expansion for $Y(s)$ is

$$Y(s) = \frac{-k/p}{s} + \frac{k/p}{s - p}$$

Taking the inverse Laplace transform of $Y(s)$ yields the following step response:

$$y(t) = -\frac{k}{p}(1 - e^{pt}), \qquad t \geq 0 \tag{8.13}$$

In this case, the step response $y(t)$ can be expressed in the form (8.11) with

$$y_1(t) = \frac{k}{p}e^{pt}, \quad t \geq 0$$

$$H(0) = -\frac{k}{p}$$

Note that the time behavior of the term $y_1(t) = (k/p)\exp(pt)$ depends directly on the location of the pole p in the complex plane. In particular, if p is in the right-half plane, the system is unstable and $y_1(t)$ grows without bound. Furthermore, the farther the pole is to the right in the right-half plane, the faster the rate of growth of $y_1(t)$. On the other hand, if the system is stable, so that p lies in the open left-half plane, then $y_1(t)$ decays to zero and thus $y_1(t)$ is the transient part of the response. Note that the rate at which the transient decays to zero depends on how far over to the left the pole is in the open left-half plane. Also, since the step response is equal to $y_1(t) - k/p$, the rate at which the step response converges to the constant $-k/p$ is equal to the rate at which the transient decays to zero. These properties of the step response are verified in the next example, by MATLAB. There are several methods to compute and plot a step response in MATLAB: the Symbolic Math Toolbox, Simulink, and the command `step` available from the Control System Toolbox. All of these methods were introduced in Examples 6.34–6.36.

Example 8.6   First-Order System

Consider the first-order system given by the transfer function (8.12) with $k = 1$. Given any specific value for p, the MATLAB commands to generate the step response are

```
num = 1; den = [1 -p];
t = 0:0.05:10;
y = step(num,den,t);
```

The step responses for $p = 1, 2$, and 3 are displayed in Figure 8.3. Note that in all three cases the step response grows without bound, which shows unstable behavior. Note also that the response for $p = 3$ grows with the fastest rate, since the pole $p = 3$ is farther to the right in the right-half plane (in comparison with $p = 1$ or $p = 2$).

If the pole p is negative, then the system is stable and $y_1(t) = (k/p)\exp(pt)$ is the transient part of the step response. In this case the step response will converge to the constant value $H(0) = -k/p$. For $k = -p$ [which yields $H(0) = 1$] and $p = -1, -2, -5$, the step response is plotted in Figure 8.4. As seen in the figure, the step response approaches the steady-state value of 1 at a faster rate as p becomes more negative, that is, as the pole moves farther to the left in the open left-half plane. This also corresponds to the fact that the rate of decay to zero of the transient is fastest for the case $p = -5$.

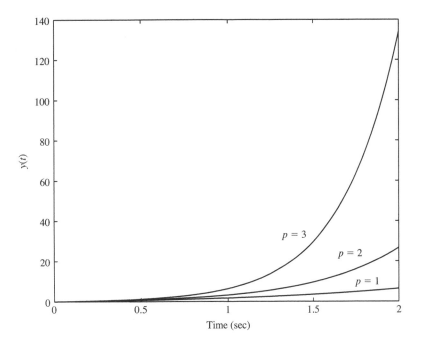

FIGURE 8.3
Step response when $p = 1, 2, 3$.

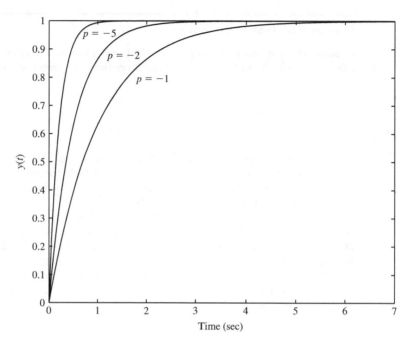

FIGURE 8.4
Step response when $p = -1, -2, -5$.

The MATLAB commands for using the `step` command are given next, followed by the commands for using the Symbolic Math Toolbox for the case $k = -p$ and $p = -2$. The Simulink model method follows the procedure described in Example 6.36.

```
num = 2; den = [1 2];
H = tf(num,den);  % forms a transfer function object
step(H);  % computes and plots the step response
syms X H y s
X = 1/s;  H = 2/(s+2);
y = ilaplace(H*X);
ezplot(y,[0 1])  % plots y from t = 0 to 1
axis([0 7 0 1])  % resets the axes
```

The Symbolic Math approach results in the expression `y = 2*exp(-t)*sinh(t)`, where `sinh` is the hyperbolic sine defined as `sinh(t)=0.5(e`^{t}`-e`$^{-t}$`)`. This expression is equal to the solution found analytically and given in (8.13).

An important quantity that characterizes the rate of decay to zero of the transient part of a response is the *time constant* τ, which is defined to be the amount of time that it takes for the transient to decay to $1/e$ ($\approx 37\%$) of its initial value. Since the transient for the first-order system (8.12) is equal to $(k/p)\exp(pt)$, we see that the time constant τ

is equal to $-1/p$, assuming that $p < 0$. To verify that τ is equal to $-1/p$, first let $y_{tr}(t)$ denote the transient so that $y_{tr}(t) = (k/p) \exp(pt)$. Then setting $t = \tau = -1/p$ in $y_{tr}(t)$ gives

$$y_{tr}(\tau) = \frac{k}{p}e^{p(-1/p)} = \frac{k}{p}e^{-1} = \frac{1}{e}y_{tr}(0)$$

In Example 8.6 the time constants for $p = -1, -2,$ and -5 are $\tau = 1, 0.5,$ and 0.2 second, respectively. Note that the smaller the value of τ, the faster the rate of decay of the transient.

Example 8.7   Determining the Pole Location from the Step Response

Consider a first-order system $H(s) = k/(s - p)$ with the step response shown in Figure 8.5. From the plot, it is possible to determine both k and the pole position (i.e., the value of p). First, since the step response displayed in Figure 8.5 is bounded, the system must be stable, and thus p must be negative. From the plot, it is seen that the steady-state value of the step response is equal to 2. Hence, $H(0) = -k/p = 2$, and from (8.13) the step response is

$$y(t) = 2(1 - e^{pt}) \tag{8.14}$$

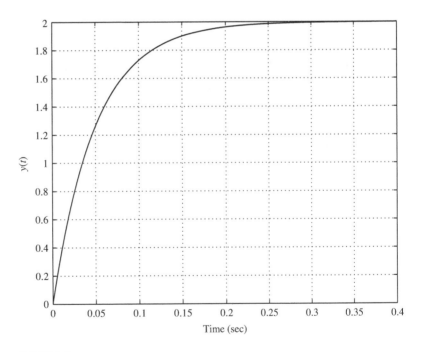

FIGURE 8.5
Step response in Example 8.7.

Now, from the plot in Figure 8.5, we see that $y(0.1) = 1.73$, and thus evaluating both sides of (8.14) at $t = 0.1$ gives

$$y(0.1) = 1.73 = 2[1 - e^{p(0.1)}] \tag{8.15}$$

Solving (8.15) for p gives $p = -20$.

8.3.2   Second-Order Systems

Now consider the second-order system given by the transfer function

$$H(s) = \frac{k}{s^2 + 2\zeta\omega_n s + \omega_n^2} \tag{8.16}$$

The real parameter ζ in (8.16) is called the *damping ratio*, and the real parameter ω_n is called the *natural frequency*. The reason for this terminology will become clear from the results given subsequently. In the following development it is assumed that $\zeta > 0$ and $\omega_n > 0$, and thus by the Routh–Hurwitz criterion, the system is stable.

Using the quadratic formula reveals that the poles of $H(s)$ are

$$p_1 = -\zeta\omega_n + \omega_n\sqrt{\zeta^2 - 1} \tag{8.17}$$

$$p_2 = -\zeta\omega_n - \omega_n\sqrt{\zeta^2 - 1} \tag{8.18}$$

From (8.17) and (8.18), it is seen that both the poles are real when $\zeta > 1$, the poles are real and repeated when $\zeta = 1$, and the poles are a complex-conjugate pair when $0 < \zeta < 1$. The step response for these three cases is considered as follows.

Case when both poles are real.   When $\zeta > 1$, the poles p_1 and p_2 given by (8.17) and (8.18) are both real and nonrepeated, in which case $H(s)$ can be expressed in the factored form

$$H(s) = \frac{k}{(s - p_1)(s - p_2)} \tag{8.19}$$

The transform $Y(s)$ of the step response is then given by

$$Y(s) = \frac{k}{(s - p_1)(s - p_2)s}$$

Performing a partial fraction expansion on $Y(s)$ yields the step response

$$y(t) = \frac{k}{p_1 p_2}(k_1 e^{p_1 t} + k_2 e^{p_2 t} + 1), \quad t \geq 0 \tag{8.20}$$

where k_1 and k_2 are real constants whose values depend on the poles p_1 and p_2. Hence, in this case the transient part $y_{tr}(t)$ of the step response is a sum of two exponentials given by

$$y_{tr}(t) = \frac{k}{p_1 p_2}(k_1 e^{p_1 t} + k_2 e^{p_2 t}), \quad t \geq 0$$

and the steady-state value of the step response is

$$H(0) = \frac{k}{\omega_n^2} = \frac{k}{p_1 p_2}$$

One of the exponential terms in (8.20) usually dominates the other exponential term; that is, the magnitude of one of the exponential terms is often much larger than the other. In this case the pole corresponding to the dominant exponential term is called the *dominant pole*. The dominant pole is usually the one nearest the imaginary axis, since it has the largest time constant (equal to $-1/p$, where p is the dominant pole). If one of the poles is dominant, the transient part of the step response (8.20) looks similar to the transient part of the step response in the first-order case considered previously.

Example 8.8   Case when Both Poles Are Real

In (8.19), let $k = 2$, $p_1 = -1$, and $p_2 = -2$. Then, expanding $H(s)/s$ via partial fractions gives

Pole Positions and Step Response

$$Y(s) = \frac{-2}{s+1} + \frac{1}{s+2} + \frac{1}{s}$$

Thus, the step response is

$$y(t) = -2e^{-t} + e^{-2t} + 1, \quad t \geq 0$$

and the transient response is

$$y_{tr}(t) = -2e^{-t} + e^{-2t}, \quad t \geq 0$$

The step response obtained from MATLAB is shown in Figure 8.6. In this example it turns out that the transient response is dominated by p_1, since the term in the transient due to p_2 decays faster. The response displayed in Figure 8.6 is similar to the response of the first-order system, with $p = -1$ shown in Figure 8.4.

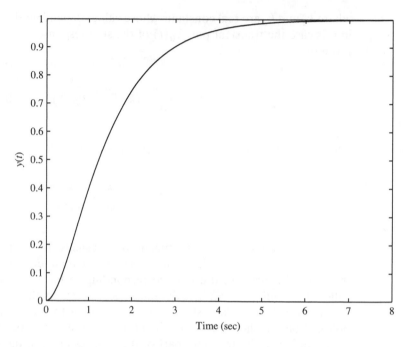

FIGURE 8.6
Step response of system in Example 8.8.

Case when poles are real and repeated. When $\zeta = 1$, the poles p_1 and p_2 given by (8.17) and (8.18) are real and are both equal to $-\omega_n$. In this case the transfer function $H(s)$ given by (8.16) has the factored form

$$H(s) = \frac{k}{(s + \omega_n)^2} \qquad (8.21)$$

Then, expanding $H(s)/s$ via partial fractions and taking the inverse transform yield the following step response:

$$y(t) = \frac{k}{\omega_n^2}[1 - (1 + \omega_n t)e^{-\omega_n t}], \quad t \geq 0 \qquad (8.22)$$

Hence, in this case the transient response is

$$y_{\text{tr}}(t) = -\frac{k}{\omega_n^2}(1 + \omega_n t)e^{-\omega_n t}, \quad t \geq 0$$

Example 8.9   Both Poles Real and Repeated

In (8.21), let $k = 4$ and $\omega_n = 2$. Then, both poles are equal to -2, and from (8.22) the step response is

$$y(t) = 1 - (1 + 2t)e^{-2t}, t \geq 0$$

A MATLAB plot of the step response is given in Figure 8.7.

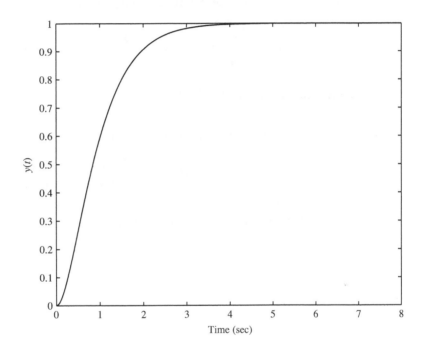

FIGURE 8.7
Step response in Example 8.9.

Case when poles are a complex pair.   Now, suppose that $0 < \zeta < 1$, so that the poles p_1 and p_2 are a complex pair. With $\omega_d = \omega_n\sqrt{1 - \zeta^2}$, the poles are $p_1, p_2 = -\zeta\omega_n \pm j\omega_d$. Note that the real part of the poles is equal to $-\zeta\omega_n$ and the imaginary part of the poles is equal to $\pm\omega_d$.

Then, given the transfer function (8.16), completing the square in the denominator of $H(s)$ yields

$$H(s) = \frac{k}{(s + \zeta\omega_n)^2 + \omega_d^2} \tag{8.23}$$

Expanding $Y(s) = H(s)/s$ gives

$$Y(s) = \frac{-(k/\omega_n^2)s - 2k\zeta/\omega_n}{(s + \zeta\omega_n)^2 + \omega_d^2} + \frac{k/\omega_n^2}{s}$$

$$= \frac{-(k/\omega_n^2)(s + \zeta\omega_n)}{(s + \zeta\omega_n)^2 + \omega_d^2} - \frac{(k\zeta/\omega_n)}{(s + \zeta\omega_n)^2 + \omega_d^2} + \frac{k/\omega_n^2}{s}$$

Thus, from Table 6.2 we see that the step response is

$$y(t) = -\frac{k}{\omega_n^2}e^{-\zeta\omega_n t}\cos \omega_d t - \frac{k\zeta}{\omega_n \omega_d}e^{-\zeta\omega_n t}\sin \omega_d t + \frac{k}{\omega_n^2}, \quad t \geq 0$$

Finally, using the trigonometric identity,

$$C \cos \beta + D \sin \beta = \sqrt{C^2 + D^2}\sin(\beta + \theta),$$

$$\text{where } \theta = \begin{cases} \tan^{-1}(C/D), C \geq 0 \\ \pi + \tan^{-1}(C/D), C < 0 \end{cases}$$

results in the following form for the step response:

$$y(t) = \frac{k}{\omega_n^2}\left[1 - \frac{\omega_n}{\omega_d}e^{-\zeta\omega_n t}\sin(\omega_d t + \phi)\right], \quad t \geq 0 \tag{8.24}$$

Here, $\phi = \tan^{-1}(\omega_d/\zeta\omega_n)$. Note that the steady-state value is equal to k/ω_n^2 and the transient response is an exponentially decaying sinusoid with frequency ω_d rad/sec. Thus, second-order systems with complex poles have an oscillatory step response with the frequency of the oscillation equal to ω_d.

Example 8.10   Poles Are a Complex Pair

Consider the second-order system given by the transfer function

Pole Positions and Step Response

$$H(s) = \frac{17}{s^2 + 2s + 17}$$

Writing $H(s)$ in the form (8.23) reveals that $k = 17$, $\zeta = 0.242$, and $\omega_n = \sqrt{17}$. Also, $\omega_d = 4$, and the poles of the system are $-1 \pm j4$. The step response is found from (8.24) to be given by

$$y(t) = 1 - \frac{\sqrt{17}}{4}e^{-t}\sin(4t + 1.326)$$

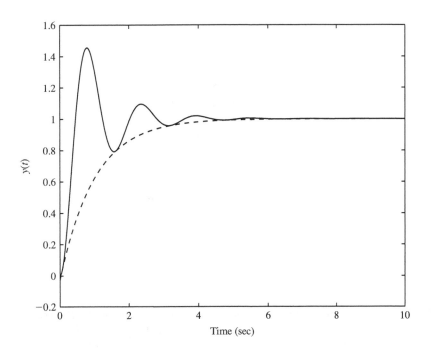

FIGURE 8.8
Step response in Example 8.10.

The step response can be obtained numerically from MATLAB by use of the following commands from the Control System Toolbox:

```
num = 17; den = [1 2 17];
H = tf(num,den);
step(H);
```

The commands from the Symbolic Math Toolbox are

```
syms Y s y
Y = 17/(s^3+2*s^2+17*s);
y = ilaplace(Y);
ezplot(y,[0, 10])
axis([0 10 0 2])
```

The step response is shown in Figure 8.8. Notice that the response oscillates with frequency $\omega_d = 4$ rad/sec and that the oscillations decay exponentially. An envelope corresponding to the decay of the transient part of the step response is shown as a dashed line in Figure 8.8. As seen from (8.24), the rate of decay of the transient is determined by the real part of the poles, $\zeta\omega_n = -1$. The time constant corresponding to the poles is equal to $1/\zeta\omega_n = 1$ sec.

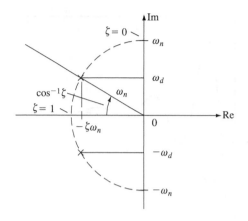

FIGURE 8.9
Location of poles in the complex plane.

As noted, when $0 < \zeta < 1$, the poles are given by the complex pair $-\zeta\omega_n \pm j\omega_d$, where $\omega_d = \omega_n\sqrt{1 - \zeta^2}$. The location of the poles in the complex plane is illustrated in Figure 8.9. As shown in the figure, ω_n is equal to the distance from the origin to the poles and ζ is equal to the cosine of the angle formed from the negative real axis. If ω_n is held constant (at some strictly positive value) and ζ is varied from one to zero, the pole positions trace a circular arc in the left-half plane starting on the negative real axis (when $\zeta = 1$) and ending on the imaginary axis (when $\zeta = 0$). This is illustrated by the dashed line in Figure 8.9. As a general rule, the closer the poles are to the $j\omega$-axis, the more oscillatory the response is. Hence, as ζ is decreased from 1 to 0 (with ω_n held constant), the step response becomes more oscillatory. This is verified by the following example:

Example 8.11   Effect of Damping Ratio on the Step Response

Consider the transfer function (8.23) with $\omega_n = 1$ and $k = 1$. The step responses for $\zeta = 0.1$, $\zeta = 0.25$, and $\zeta = 0.7$ are shown in Figure 8.10. Note that the smaller the value of ζ is, the more pronounced the oscillation is.

Again consider the step response given by (8.24). In addition to the parameter ζ, the value of the natural frequency ω_n also has a substantial effect on the response. To see this, suppose that ζ is held constant and ω_n is varied. Since ζ determines the angle of the pole in polar coordinates (see Figure 8.9), keeping ζ constant will keep the angle constant, and thus in the plot of the poles, increasing the value of ω_n will generate a radial line starting from the origin and continuing outward into the left-half plane (see Figure 8.9). It follows that the transient response should decay faster and the frequency of the oscillation should increase as ω_n is increased. This is investigated in the next example.

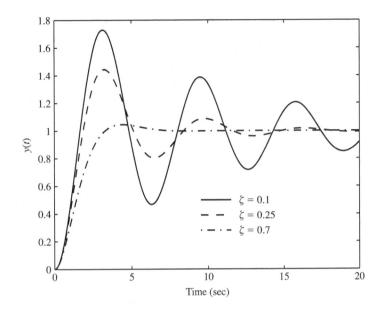

FIGURE 8.10
Step response in Example 8.11.

Example 8.12   Effect of ω_n on Step Response

Consider the transfer function (8.23) with $\zeta = 0.4$ and $k = \omega_n^2$ (so that the steady-state value is equal to 1). The step responses for $\omega_n = 0.5$, 1, and 2 rad/sec are shown in Figure 8.11. Note that the larger the value of ω_n is, the smaller the time constant and the higher the frequency of oscillations are. Note also that, since ζ is kept constant, the peak values of the oscillations are the same for each value of ω_n.

Comparison of cases.   Again, consider the system with transfer function

$$H(s) = \frac{k}{s^2 + 2\zeta\omega_n s + \omega_n^2} \tag{8.25}$$

From the foregoing results, it is seen that when $0 < \zeta < 1$, the transient part of the step response is oscillatory with "damped natural frequency" equal to $\omega_n \sqrt{1 - \zeta^2}$, and the oscillation is more pronounced as ζ is decreased to zero. For $\zeta \geq 1$, there is no oscillation in the transient. The existence of the oscillation implies a lack of "damping" in the system, and thus ζ does give a measure of the degree of damping in the system. When $0 < \zeta < 1$, the system is said to be *underdamped*, since in this case the damping ratio ζ is not large enough to prevent an oscillation in the transient resulting from a

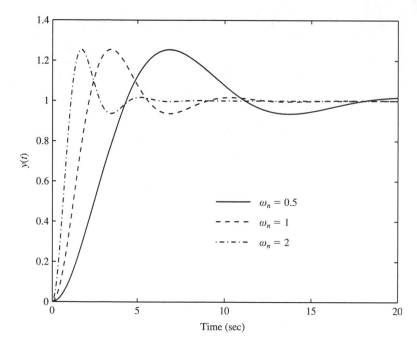

FIGURE 8.11
Step response in Example 8.12.

step input. When $\zeta > 1$, the system is said to be *overdamped,* since in this case ζ is larger than necessary to prevent an oscillation in the transient. When $\zeta = 1$, the system is said to be *critically damped,* since this is the smallest value of ζ for which there is no oscillation in the transient response to a step input.

Example 8.13   Comparison of Cases

Consider the transfer function (8.25) with $k = 4$ and $\omega_n = 2$. To compare the underdamped, critically damped, and overdamped cases, the step response of the system will be computed for $\zeta = 0.5, 1,$ and 1.5. The results are shown in Figure 8.12. Note that, if the overshoot can be tolerated, the fastest response displayed in Figure 8.12 is the one for which $\zeta = 0.5$. Here, "the fastest" refers to the step response that reaches the steady-state value (equal to 1 here) in the fastest time of all three of the responses shown in Figure 8.12.

Returning to the system with transfer function $H(s)$ given by (8.25), it is worth noting that if $\zeta < 0$ and $\omega_n > 0$, both of the poles are in the open right-half plane, and thus the system is unstable. In this case the "transient part" of the step response will grow without bound as $t \to \infty$. Hence, the transient is not actually a transient, since it does not decay to zero as $t \to \infty$. The transient will decay to zero if and only if the system is stable. This follows directly from the analysis of stability given in Section 8.1.

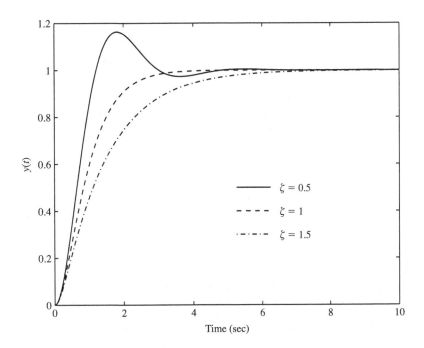

FIGURE 8.12
Step response in Example 8.13.

Example 8.14   Mass–Spring–Damper System

Mass–
Spring–
Damper
System

Consider the mass–spring–damper system (see Example 6.32) with the input/output differential equation

$$M\frac{d^2y(t)}{dt^2} + D\frac{dy(t)}{dt} + Ky(t) = x(t)$$

where M is the mass, D is the damping constant, K is the stiffness constant, $x(t)$ is the force applied to the mass, and $y(t)$ is the displacement of the mass relative to the equilibrium position. It is assumed that M, D, and K are strictly positive real numbers (greater than 0), which is the condition for stability of the system. The transfer function of the system is given by

$$H(s) = \frac{1}{Ms^2 + Ds + K} = \frac{1/M}{s^2 + (D/M)s + (K/M)} \qquad (8.26)$$

Equating the coefficients of the polynomials in the denominators of (8.25) and (8.26) results in the relationships

$$2\zeta\omega_n = \frac{D}{M} \text{ and } \omega_n^2 = \frac{K}{M}$$

Solving for the damping ratio ζ and the natural frequency ω_n yields

$$\zeta = \frac{D}{2\sqrt{MK}}, \ \omega_n = \sqrt{\frac{K}{M}}$$

Note that the damping ratio ζ is directly proportional to the damping constant D, and thus the damping in the system is a result of the term $Ddy(t)/dt$ in the input/output differential equation. In particular, there is no damping in the system if $D = 0$. The system is underdamped when

$$0 < \frac{D}{2\sqrt{MK}} < 1$$

which is equivalent to the following condition on the damping constant D:

$$0 < D < 2\sqrt{MK}$$

The system is critically damped when $D = 2\sqrt{MK}$, and the system is overdamped when $D > 2\sqrt{MK}$. The reader is invited to check out animations of the step response for these three cases by using the online demo on the website. To generate the three cases, values of M, D, and K need to be selected on the basis of the ranges for D given previously.

8.3.3   Higher-Order Systems

Higher-order systems can sometimes be approximated as first- or second-order systems, since one or two of the poles are usually more dominant than the other poles, and thus these other poles can simply be neglected. An example where one pole dominates over another pole occurred in Example 8.8. In this example, the two-pole system behaves similarly to the one-pole system with the dominant pole, and thus the system can be approximated by the dominant pole.

There are two situations when care must be exercised in carrying out system approximation based on the concept of dominant poles. First, if the dominant poles are not significantly different from the other poles, the approximation made by neglecting the faster poles (i.e., the poles with the smaller time constants) may not be very accurate. Second, a zero near a pole causes the residue of the pole to be small, making the magnitude of the corresponding term in the transient response small. Hence, although such a pole may appear to be dominant, it in fact is not. This is illustrated in the following example:

Example 8.15   Third-Order System

Consider the third-order system with the following transfer function:

$$H(s) = \frac{25}{(s^2 + 7s + 25)(s + 1)} \tag{8.27}$$

The plot of the poles is given in Figure 8.13. From the plot, it is seen that the pole at $s = -1$ is the most dominant, since it is closest to the imaginary axis. Thus, it should be possible to neglect the

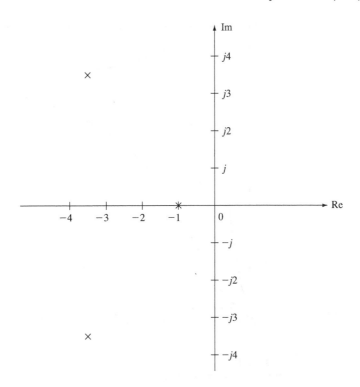

FIGURE 8.13
Location of poles in Example 8.15.

other two complex poles. To verify that this can be done, we will calculate the step response of the system by first expanding $Y(s) = H(s)/s$ in the partial fraction expansion:

$$Y(s) = \frac{1}{s} - 1.316\frac{1}{s+1} + 0.05263\frac{6s+17}{s^2+7s+25}$$

Taking the inverse transform gives the step response:

$$y(t) = 1 - 1.316e^{-t} + 0.321e^{-3.5t}\sin(3.57t + 1.754), \qquad t \geq 0 \tag{8.28}$$

Note that the second term on the right-hand side of (8.28) is larger than the third term and will decay more slowly, making it a dominant term. So, this corresponds to the observation made previously that the pole at $s = -1$ is dominant. To check this out further, we will obtain the step response from the Control System Toolbox in MATLAB by using the commands

```
num = 25;
den = conv([1 7 25],[1 1]); % this multiplies the polynomials
H = tf(num,den);
t = 0:0.01:4;
step(H,t);
```

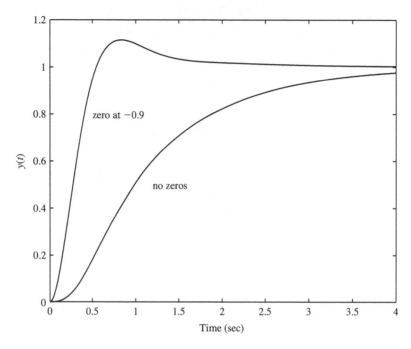

FIGURE 8.14
Step response in Example 8.15.

The resulting step response, shown in Figure 8.14, is very similar to a simple exponential first-order response with a pole at $s = -1$. Hence, this again confirms the observation that the pole at $s = -1$ is dominant, and thus the other two poles can be neglected.

Now, suppose that a zero at $s = -0.9$ is added to the transfer function (8.27) and the constant in the numerator is adjusted so that the steady-state value is still 1. The resulting transfer function is

$$H(s) = \frac{(25/0.9)s + 25}{(s^2 + 7s + 25)(s + 1)}$$

Note that the zero at $s = -0.9$ very nearly cancels the pole at $s = -1$. The partial fraction expansion of $H(s)/s$ is then given by

$$Y(s) = \frac{1}{s} + 0.1462\frac{1}{s + 1} + \frac{-1.146s - 7.87}{s^2 + 7s + 25}$$

Note that the residue corresponding to the pole -1 is now very small. This is a result of the zero being near the pole. Taking the inverse Laplace transform of $Y(s)$ then gives

$$y(t) = 1 + 0.1462e^{-t} + 1.575e^{-3.5t}\sin(3.57t - 2.327), t \geq 0$$

Since the coefficient multiplying e^{-t} is much smaller in this case, the pole at $s = -1$ is less significant than the other two poles, and thus the dominant poles are now the complex-conjugate pair (at $s = -3.5 \pm 3.57j$), even though they are farther from the imaginary axis than the pole -1. The resulting step response obtained from MATLAB is shown in Figure 8.14. The reader is invited to compare this response with a second-order system with poles at $s = -3.5 \pm 3.57j$.

8.4   RESPONSE TO SINUSOIDS AND ARBITRARY INPUTS

Again, consider the system with the rational transfer function $H(s) = B(s)/A(s)$ with the degree of $B(s)$ less than the degree of $A(s)$. In the first part of this section, the objective is to determine the output response of the system when $x(t)$ is the sinusoid

$$x(t) = C \cos \omega_0 t, \, t \geq 0$$

where the magnitude C and the frequency ω_0 (in rad/sec) are arbitrary constants. From Table 6.2, we see that the Laplace transform of the input is

$$X(s) = \frac{Cs}{s^2 + \omega_0^2} = \frac{Cs}{(s + j\omega_0)(s - j\omega_0)}$$

Hence, in this case the transform $X(s)$ of the input has a zero at $s = 0$ and two poles at $s = \pm j\omega_0$.

  If the system has zero initial conditions, the transform $Y(s)$ of the resulting output is given by

$$Y(s) = \frac{CsB(s)}{A(s)(s + j\omega_0)(s - j\omega_0)} \tag{8.29}$$

The computation of the output response $y(t)$ from (8.29) will be carried out in a manner similar to that done in Section 8.3 for the case of a step input. Here, we "pull out" the terms $s + j\omega_0$ and $s - j\omega_0$ in (8.29) by using the partial fraction expansion, assuming that $A(\pm j\omega_0) \neq 0$. This gives

$$Y(s) = \frac{\gamma(s)}{A(s)} + \frac{c}{s - j\omega_0} + \frac{\bar{c}}{s + j\omega_0} \tag{8.30}$$

where $\gamma(s)$ is a polynomial in s, c is a complex constant, and $\bar{c}$ is the complex conjugate of c. From the residue formula (6.67), c is given by

$$c = [(s - j\omega_0)Y(s)]_{s=j\omega_0} = \left[\frac{CsB(s)}{A(s)(s + j\omega_0)} \right]_{s=j\omega_0}$$

$$= \frac{jC\omega_0 B(j\omega_0)}{A(j\omega_0)(j2\omega_0)} = \frac{C}{2} H(j\omega_0)$$

Then, inserting the values for c and $\bar{c}$ into (8.30) gives

$$Y(s) = \frac{\gamma(s)}{A(s)} + \frac{(C/2)H(j\omega_0)}{s - j\omega_0} + \frac{(C/2)\overline{H(j\omega_0)}}{s + j\omega_0} \qquad (8.31)$$

where $\overline{H(j\omega_0)}$ is the complex conjugate of $H(j\omega_0)$.

Now, let $y_1(t)$ denote the inverse Laplace transform of $\gamma(s)/A(s)$. Then, taking the inverse Laplace transform of both sides of (8.31) yields

$$y(t) = y_1(t) + \frac{C}{2}[H(j\omega_0)e^{j\omega_0 t} + \overline{H(j\omega_0)}e^{-j\omega_0 t}] \qquad (8.32)$$

Finally, using the identity [see (6.71)]

$$\beta e^{j\omega_0 t} + \bar{\beta} e^{-j\omega_0 t} = 2|\beta| \cos(\omega_0 t + \angle\beta)$$

expression (8.32) for $y(t)$ can be written in the form

$$y(t) = y_1(t) + C|H(j\omega_0)| \cos(\omega_0 t + \angle H(j\omega_0)), \quad t \geq 0 \qquad (8.33)$$

When the system is stable [all poles of $H(s)$ are in the open left-half plane], the term $y_1(t)$ in (8.33) decays to zero as $t \to \infty$, and thus $y_1(t)$ is the transient part of the response. The sinusoidal term in the right-hand side of (8.33) is the steady-state part of the response, which is denoted by $y_{ss}(t)$; that is,

$$y_{ss}(t) = C|H(j\omega_0)| \cos(\omega_0 t + \angle H(j\omega_0)), \quad t \geq 0 \qquad (8.34)$$

From (8.34) it is seen that the steady-state response $y_{ss}(t)$ to the sinusoidal input $x(t) = C \cos(\omega_0 t), t \geq 0$, has the same frequency as the input, but it is scaled in magnitude by the amount $|H(j\omega_0)|$, and it is phase-shifted by the amount $\angle H(j\omega_0)$. This result resembles the development given in Section 5.1, where it was shown that the response to the input

$$x(t) = C \cos \omega_0 t, \ -\infty < t < \infty$$

is given by

$$y(t) = C|H(\omega_0)| \cos(\omega_0 t + \angle H(\omega_0)), \quad -\infty < t < \infty \qquad (8.35)$$

where $H(\omega_0)$ is the Fourier transform $H(\omega)$ of the impulse response $h(t)$ with $H(\omega)$ evaluated at $\omega = \omega_0$. Note that in the expression (8.35) for the output, there is no transient, since the input is first applied at time $t = -\infty$. A transient response is generated only when the input is applied at some finite value of time (not $t = -\infty$).

If the given system is causal and stable, there is a direct correspondence between the previously given derivation of $y_{ss}(t)$ and the result given by (8.35). To see this, first recall from Section 8.1 that stability implies the integrability condition

$$\int_0^\infty |h(t)|\, dt < \infty$$

It then follows from the discussion given in Section 6.1 that the Fourier transform $H(\omega)$ of $h(t)$ is equal to the Laplace transform $H(s)$ evaluated at $s = j\omega$, that is,

$$H(\omega) = H(j\omega) = H(s)|_{s=j\omega} \qquad (8.36)$$

Note that $H(j\omega)$ is denoted by $H(\omega)$. This notation will be followed from here on.

As a consequence of (8.36), the expressions (8.34) and (8.35) are identical for $t \geq 0$, and thus there is a direct correspondence between the two results. In addition, by (8.36) the frequency response function of the system [which was first defined in Section 5.1 to be the Fourier transform $H(\omega)$ of $h(t)$] is equal to the transfer function $H(s)$ evaluated at $s = j\omega$. Hence, the frequency response behavior of a stable system can be determined directly from the transfer function $H(s)$. In particular, the magnitude function $|H(\omega)|$ and the phase function $\angle H(\omega)$ can both be directly generated from the transfer function $H(s)$.

Example 8.16   First-Order System

Consider the first-order system with the transfer function

$$H(s) = \frac{k}{s - p} \qquad (8.37)$$

It is assumed that $k > 0$ and $p < 0$, so that the system is stable. With $k = -p = 1/RC$, the system with transfer function (8.37) could be the RC circuit shown in Figure 8.15.

Now, setting $s = j\omega$ in $H(s)$ gives

$$H(\omega) = \frac{k}{j\omega - p}$$

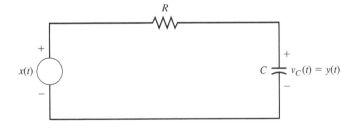

FIGURE 8.15
RC circuit with transfer function $H(s) = k/(s - p)$, where $k = -p = 1/RC$.

and taking the magnitude and angle of $H(\omega)$ yields

$$|H(\omega)| = \frac{|k|}{|j\omega - p|} = \frac{k}{\sqrt{\omega^2 + p^2}} \tag{8.38}$$

$$\angle H(\omega) = -\angle(j\omega - p) = -\tan^{-1}\frac{-\omega}{p} \tag{8.39}$$

The output response resulting from the sinusoidal input $x(t) = C \cos \omega_0 t, t \geq 0$ (with zero initial conditions) can be computed using (8.38) and (8.39) as follows. First, the Laplace transform of the output response is given by

$$Y(s) = H(s)X(s) = \frac{kCs}{(s - p)(s^2 + \omega_0^2)}$$

In this case, the partial fraction expression (8.30) for $Y(s)$ becomes

$$Y(s) = \frac{\gamma}{s - p} + \frac{c}{s - j\omega_0} + \frac{\bar{c}}{s + j\omega_0}$$

where

$$\gamma = [(s - p)Y(s)]|_{s=p} = \frac{kCp}{p^2 + \omega_0^2}$$

Then the transient part of the output response is given by

$$y_{\text{tr}}(t) = \gamma e^{pt} = \frac{kCp}{p^2 + \omega_0^2}e^{pt}, \quad t \geq 0$$

and from (8.33), the complete output response is

$$y(t) = \frac{kCp}{p^2 + \omega_0^2}e^{pt} + C|H(\omega_0)| \cos(\omega_0 t + \angle H(\omega_0)), \quad t \geq 0 \tag{8.40}$$

Finally, inserting (8.38) and (8.39) into (8.40) yields the output response

$$y(t) = \frac{kCp}{p^2 + \omega_0^2}e^{pt} + \frac{Ck}{\sqrt{\omega_0^2 + p^2}}\cos\left[\omega_0 t - \tan^{-1}\left(-\frac{\omega_0}{p}\right)\right], \quad t \geq 0 \tag{8.41}$$

Equation (8.41) is the complete response resulting from the input $x(t) = C \cos \omega_0 t$ applied for $t \geq 0$. Note that the transient part of the response is a decaying exponential, since $p < 0$, with the rate of decay depending on the value of the pole p.

Now suppose that $k = 1, p = -1$, and the input is $x(t) = 10 \cos(1.5t), t \geq 0$, so that $C = 10$ and $\omega_0 = 1.5$ rad/sec. Then,

$$\gamma = \frac{(1)(10)(-1)}{1 + (1.5)^2} = -3.08$$

$$|H(1.5)| = 5.55$$

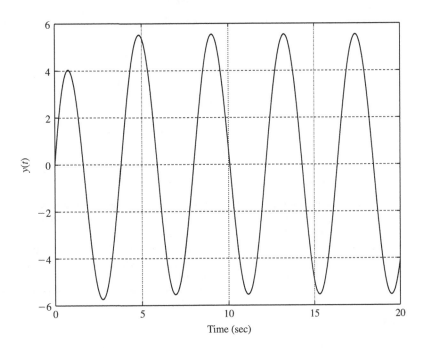

FIGURE 8.16
Output response in Example 8.16.

$$\angle H(1.5) = -56.31°$$

and thus from (8.41), we see that the output response is

$$y(t) = -3.08e^{-t} + 5.55 \cos(1.5t - 56.31°), t \geq 0 \qquad (8.42)$$

The response is shown in Figure 8.16. Note that the transient $-3.08e^{-t}$ can be seen for small values of t, but then disappears. Since the time constant τ associated with the pole at -1 is $\tau = 1$, the transient decays to $1/e = 37\%$ of its initial value at $t = 1$ second. It can also be observed from the plot in Figure 8.16 that the amplitude and the phase of the steady-state part of the response matches the values previously obtained analytically. [See (8.42).] It should be noted that when the transient has died out, the resulting (steady-state) response previously obtained is identical to the solution obtained by use of the Fourier theory in Example 5.2 for the RC circuit (with $RC = 1$, $A = C = 10$, and $\omega_0 = 1.5$). As remarked before, the Fourier setup considered in Chapter 5 has no transient response, since the input is applied at time $t = -\infty$. The reader should check that the two solutions do in fact correspond. The response shown in Figure 8.16 was obtained by the use of the following MATLAB commands from the Control System Toolbox:

```
t = 0:0.05:20;
num = 1; den = [1 1];
H = tf(num,den);
x = 10*cos(1.5*t);
y = lsim(H,x,t);
```

The Symbolic Math Toolbox could also be used to compute the response by the following commands:

```
syms X H y s
X = 10*s/(s^2+1.5^2);  H = 1/(s+1);
y = ilaplace(H*X);
ezplot(y,[0 20])
axis([0 20 -6 6])
```

Example 8.17   Mass–Spring–Damper System

Mass–
Spring–
Damper
System

For the mass–spring–damper system (see Example 8.14), we can generate an animation of the output response resulting from the input $x(t) = \cos \omega_0 t, t \geq 0$ by running the demo on the website. This provides a nice visualization of the transient response and the eventual convergence of the output response to the steady-state behavior. The reader is invited to run the demo for various values of M, D, and K and to compare the results with an analytical computation of the output response by using the Laplace transform.

8.4.1   Response to Arbitrary Inputs

The aforementioned analysis for a sinusoidal input, and the one given in Section 8.3 for a step input, generalize to arbitrary inputs as follows. Suppose that the transform $X(s)$ of the input $x(t)$ is a rational function; that is, $X(s) = C(s)/D(s)$, where $C(s)$ and $D(s)$ are polynomials in s with the degree of $C(s)$ less than the degree of $D(s)$. In terms of this notation, the poles of $X(s)$ are the roots of $D(s) = 0$.

Now if $x(t)$ is applied to a system with transfer function $H(s) = B(s)/A(s)$, the transform of the resulting response (with zero initial conditions) is

$$Y(s) = \frac{B(s)C(s)}{A(s)D(s)}$$

If there are no common poles between $H(s)$ and $X(s)$, $Y(s)$ can be expressed in the form

$$Y(s) = \frac{E(s)}{A(s)} + \frac{F(s)}{D(s)} \tag{8.43}$$

where $E(s)$ and $F(s)$ are polynomials in s. Then, taking the inverse transform of both sides of (8.43) gives

$$y(t) = y_1(t) + y_2(t) \tag{8.44}$$

where $y_1(t)$ is the inverse transform of $E(s)/A(s)$ and $y_2(t)$ is the inverse transform of $F(s)/D(s)$.

It is very important to note that the form of $y_1(t)$ depends directly on the poles of $H(s)$ and the form of $y_2(t)$ depends directly on the poles of $X(s)$. When the system is stable, $y_1(t)$ converges to zero as $t \to \infty$, in which case $y_1(t)$ is identified as the transient part of the response [although there may be terms in $y_2(t)$ that are also converging to zero]. A key point here is that the form of $y_1(t)$ (i.e., the transient) depends only on the poles of the system, regardless of the particular form of the input signal $x(t)$.

If $X(s)$ has poles on the $j\omega$-axis, these poles appear in the transform of $y_2(t)$, and thus $y_2(t)$ will not converge to zero. Hence, $y_2(t)$ is identified as the steady-state part of the response. It should be stressed that the form of $y_2(t)$ (i.e., the steady-state response) depends only on the poles of the input transform $X(s)$, regardless of what the system transfer function $H(s)$ is.

Example 8.18   Form of Output Response

Suppose that a system has transfer function $H(s)$ with two real poles a, b and a complex pair of poles $\sigma \pm jc$, where $a < 0, b < 0$, and $\sigma < 0$, so that the system is stable. Let $x(t)$ be any input whose transform $X(s)$ is rational in s and whose poles are different from those of $H(s)$. Then the form of the transient response is

$$y_{tr}(t) = k_1 e^{at} + k_2 e^{bt} + k_3 e^{\sigma t} \cos(ct + \theta), t \geq 0$$

where k_1, k_2, k_3, and θ are all constants that depend on the specific input and the zeros of the system.

If $x(t)$ is the step function, the form of the complete response is

$$y(t) = k_1 e^{at} + k_2 e^{bt} + k_3 e^{\sigma t} \cos(ct + \theta) + A, t \geq 0$$

where A is a constant. When $x(t)$ is the ramp $x(t) = tu(t)$, the form of the complete response is

$$y(t) = k_1 e^{at} + k_2 e^{bt} + k_3 e^{\sigma t} \cos(ct + \theta) + A + Bt, t \geq 0$$

where A and B are constants. When $x(t)$ is a sinusoid with frequency ω_0, the form of the complete response is

$$y(t) = k_1 e^{at} + k_2 e^{bt} + k_3 e^{\sigma t} \cos(ct + \theta) + B \cos(\omega_0 t + \phi), t \geq 0$$

for some constants B and ϕ. It should be noted that the values of the k_i in the prior expressions are not the same.

8.5   FREQUENCY RESPONSE FUNCTION

Given a stable system with rational transfer function $H(s) = B(s)/A(s)$, in Section 8.4 we showed that the steady-state response to the sinusoid $x(t) = C \cos \omega_0 t, t \geq 0$, with zero initial conditions is given by

$$y_{ss}(t) = C|H(\omega_0)| \cos(\omega_0 t + \angle H(\omega_0)), t \geq 0 \tag{8.45}$$

where $H(\omega)$ is the frequency response function [which is equal to $H(s)$ with $s = j\omega$]. As a result of the fundamental relationship (8.45), the behavior of the system as it relates to the response to sinusoidal inputs can be studied in terms of the frequency response curves given by the plots of the magnitude function $|H(\omega)|$ and the phase function $\angle H(\omega)$.

The magnitude function $|H(\omega)|$ is sometimes given in *decibels*, denoted by $|H(\omega)|_{dB}$ and defined by

$$|H(\omega)|_{dB} = 20 \log_{10} |H(\omega)|$$

The term *decibel* (denoted by dB) was first defined as a unit of power gain in an electrical circuit. Specifically, the power gain through a circuit is defined to be 10 times the logarithm (to the base 10) of the output power divided by the input power. Since power in an electrical circuit is proportional to the square of the voltage or current and for any constant K,

$$10 \log_{10}(K^2) = 20 \log_{10} K$$

the foregoing definition of $|H(\omega)|_{dB}$ can be viewed as a generalization of the original meaning of the term *decibel*.

Note that

$$|H(\omega)|_{dB} < 0 \, dB \quad \text{when } |H(\omega)| < 1$$

$$|H(\omega)|_{dB} = 0 \, dB \quad \text{when } |H(\omega)| = 1$$

$$|H(\omega)|_{dB} > 0 \, dB \quad \text{when } |H(\omega)| > 1$$

Thus, it follows from (8.45) that when $|H(\omega_0)|_{dB} < 0$ dB, the system attenuates the sinusoidal input $x(t) = C \cos \omega_0 t$; when $|H(\omega_0)|_{dB} = 0$ dB, the system passes $x(t)$ with no attenuation; and when $|H(\omega_0)|_{dB} > 0$ dB, the system amplifies $x(t)$.

The plots of $|H(\omega)|$ (or $|H(\omega)|_{dB}$) versus ω and $\angle H(\omega)$ versus ω with ω on a logarithmic scale are called the *Bode diagrams* of the system. A technique for generating the Bode diagrams by the use of asymptotes is given in the development that follows.

We can often determine the frequency response curves (or the Bode diagrams) experimentally by measuring the steady-state response resulting from the sinusoidal input $x(t) = C \cos \omega_0 t$. By performing this experiment for various values of ω_0, it is possible to extrapolate the results to obtain the magnitude function $|H(\omega)|$ and phase function $\angle H(\omega)$ for all values of ω ($\omega \geq 0$). This then determines the frequency function $H(\omega)$, since

$$H(\omega) = |H(\omega)| \exp[j\angle H(\omega)]$$

The frequency response curves can be generated directly from the transfer function $H(s)$ by the MATLAB command bode. The use of this command is illustrated in the examples that follow. It will also be shown that if the number of poles and zeros of the system is not large, the general shape of the frequency response curves can be determined from vector representations in the complex plane of the factors included in $H(s)$. The development begins with the first-order case.

8.5.1   First-Order Case

Consider the first-order system given by the transfer function

$$H(s) = \frac{k}{s + B} \qquad (8.46)$$

where $k > 0$ and $B > 0$. The frequency response function is $H(\omega) = k/(j\omega + B)$, and the magnitude and phase functions are given by

$$|H(\omega)| = \frac{k}{\sqrt{\omega^2 + B^2}} \qquad (8.47)$$

$$\angle H(\omega) = -\tan^{-1}\frac{\omega}{B} \qquad (8.48)$$

We can generate the frequency response curves by evaluating (8.47) and (8.48) for various values of ω. Instead of doing this, we will show that the shape of the frequency response curves can be determined from the vector representation of the factor $j\omega + B$ making up $H(\omega)$. The vector representation of $j\omega + B$ is shown in Figure 8.17.

The magnitude $|j\omega + B|$ and the angle $\angle(j\omega + B)$ can be computed from the vector representation of $j\omega + B$ shown in Figure 8.17. Here, the magnitude $|j\omega + B|$ is the length of the vector from the pole $s = -B$ to the point $s = j\omega$ on the imaginary axis, and the angle $\angle(j\omega + B)$ is the angle between this vector and the real axis of the complex plane. From Figure 8.17, it is clear that $|j\omega + B|$ becomes infinite as $\omega \to \infty$ and $\angle(j\omega + B)$ approaches $90°$ as $\omega \to \infty$. Then, from (8.47) and (8.48), it is seen that the magnitude function $|H(\omega)|$ starts with value k/B when $\omega = 0$ and approaches zero as $\omega \to \infty$, while the phase $\angle H(\omega)$ starts with value $0°$ when $\omega = 0$ and approaches $-90°$ as $\omega \to \infty$. This provides a good indication as to the shape of the frequency response curves.

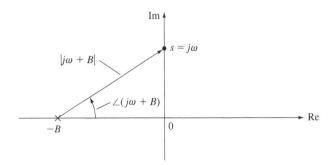

FIGURE 8.17
Vector representation of $j\omega + B$.

To generate accurate plots, MATLAB can be used to compute the complex vector H for a range of frequency values. For example, when $k = B = 2$, the curves can be generated by the following commands:

```
w = 0:.05:10; s = j*w;
H = 2./(s+2);
subplot(211),plot(w,abs(H));
subplot(212),plot(w,angle(H)*180/pi);
```

The results are displayed in Figure 8.18. The magnitude plot in Figure 8.18a reveals that the system is a lowpass filter, since it passes sinusoids whose frequency is less than 2 rad/sec, while it attenuates sinusoids whose frequency is above 2 rad/sec. Recall that the lowpass frequency response characteristic was first encountered in Example 5.2, which was given in terms of the Fourier analysis.

For an arbitrary value of $B > 0$, when $k = B$ the system with transfer function $H(s) = k/(s + B)$ is a lowpass filter, since the magnitude function $|H(\omega)|$ starts with value $H(0) = k/B = 1$ and then rolls off to zero as $\omega \to \infty$. The point $\omega = B$ is called the 3-dB point of the filter, since this is the value of ω for which $|H(\omega)|_{dB}$ is down by 3 dB from the peak value of $|H(0)|_{dB} = 0$ dB. This lowpass filter is said to have a *3-dB bandwidth* of B rad/sec, since it passes (with less than 3 dB of attenuation) sinusoids whose frequency is less than B rad/sec. The *passband* of the filter is the frequency range from 0 rad/sec to B rad/sec. The *stopband* of the filter is the frequency range from B rad/sec to ∞. As seen from the magnitude plot in Figure 8.18a, for this filter, the cutoff

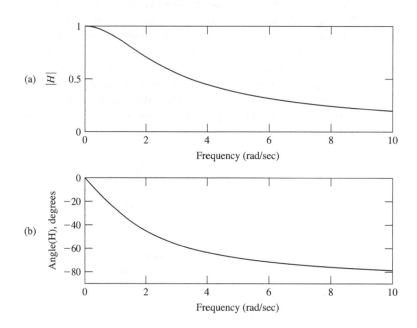

FIGURE 8.18
Frequency response curves for $H(s) = 2/(s + 2)$: (a) magnitude curve; (b) phase curve.

between the passband and the stopband is not very sharp. It will be shown in the next section that a much sharper cutoff can be obtained by increasing the number of poles of the system.

Single-pole systems with a zero.   From the results derived previously, it was discovered that a single-pole system with no zero is a lowpass filter. We can change this frequency response characteristic by adding a zero to the system. In particular, consider the single-pole system with the transfer function

$$H(s) = \frac{s + C}{s + B}$$

It is assumed that $B > 0$ and $C > 0$. Setting $s = j\omega$ in $H(s)$ gives

$$H(\omega) = \frac{j\omega + C}{j\omega + B}$$

Then, the magnitude and phase functions are given by

$$|H(\omega)| = \frac{|j\omega + C|}{|j\omega + B|} = \sqrt{\frac{\omega^2 + C^2}{\omega^2 + B^2}}$$

$$\angle H(\omega) = \angle(j\omega + C) - \angle(j\omega + B)$$

$$= \tan^{-1}\frac{\omega}{C} - \tan^{-1}\frac{\omega}{B}$$

The frequency response curves will be determined in the case when $0 < C < B$. First, consider the vector representations of $j\omega + B$ and $j\omega + C$ shown in Figure 8.19. As seen from the figure, both $|j\omega + B|$ and $|j\omega + C|$ increase as ω is increased from zero; however, the percent increase in $|j\omega + C|$ is larger. Thus, $|H(\omega)|$ starts with value

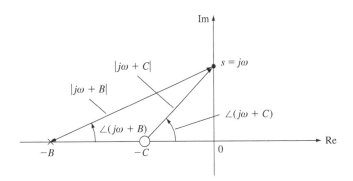

FIGURE 8.19
Vector representations of $j\omega + B$ and $j\omega + C$ when $0 < C < B$.

$|H(0)| = C/B$, and then $|H(\omega)|$ increases as ω is increased from zero. For large values of ω, the difference in $|j\omega + B|$ and $|j\omega + C|$ is very small, and thus $|H(\omega)| \to 1$ as $\omega \to \infty$.

From Figure 8.19, it is seen that the angles $\angle(j\omega + B)$ and $\angle(j\omega + C)$ both increase as ω is increased from zero; however, at first the increase in $\angle(j\omega + C)$ is larger. Hence, $\angle H(\omega)$ starts with value $\angle H(0) = 0°$, and then $\angle H(\omega)$ increases as ω is increased from zero. For $\omega > B$, the percent increase in $\angle(j\omega + B)$ is greater than that of $\angle(j\omega + C)$, and thus $\angle H(\omega)$ decreases as ω is increased from the value $\omega = B$. Hence, $\angle H(\omega)$ will have a maximum value at some point ω between $\omega = C$ and $\omega = B$. As $\omega \to \infty$, both angles $\angle(j\omega + B)$ and $\angle(j\omega + C)$ approach $90°$, and therefore $\angle H(\omega) \to 0°$ as $\omega \to \infty$.

In the case $C = 1$ and $B = 20$, the exact frequency response curves were computed by the MATLAB command bode. The results are shown in Figure 8.20. From 8.20a it is seen that the system is a *highpass filter*, since it passes with little attenuation all frequencies above $B = 20$ rad/sec. Although highpass filters exist in theory, they do not exist in practice, since actual systems cannot pass sinusoids with arbitrarily large frequencies. In other words, no actual system can have an infinite bandwidth. Thus, any implementation of the transfer function $H(s) = (s + C)/(s + B)$ would only be an approximation of a highpass filter (in the case $0 < C < B$).

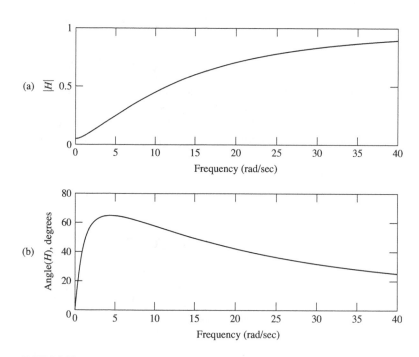

FIGURE 8.20

Frequency response curves for $H(s) = (s + 1)/(s + 20)$: (a) magnitude curve; (b) phase curve.

8.5.2   Second-Order Systems

Now consider the system with the transfer function

$$H(s) = \frac{k}{s^2 + 2\zeta\omega_n s + \omega_n^2}$$

where $k > 0$, $\zeta > 0$, and $\omega_n > 0$, so that the system is stable. As discussed in Section 8.3, the poles of the system are

$$p_1 = -\zeta\omega_n + \omega_n\sqrt{\zeta^2 - 1}$$

$$p_2 = -\zeta\omega_n - \omega_n\sqrt{\zeta^2 - 1}$$

Expressing $H(s)$ in terms of p_1 and p_2 gives

$$H(s) = \frac{k}{(s - p_1)(s - p_2)}$$

and thus the magnitude and phase functions are given by

$$|H(\omega)| = \frac{k}{|j\omega - p_1||j\omega - p_2|} \tag{8.49}$$

$$\angle H(\omega) = -\angle(j\omega - p_1) - \angle(j\omega - p_2) \tag{8.50}$$

As noted in Section 8.3, the poles p_1 and p_2 are real if and only if $\zeta \geq 1$. In this case the shape of the frequency response curves can be determined by a consideration of the vector representations of $j\omega - p_1$ and $j\omega - p_2$, shown in Figure 8.21. Here the magnitudes $|j\omega - p_1|$ and $|j\omega - p_2|$ become infinite as $\omega \to \infty$, and the angles $\angle(j\omega - p_1)$ and $\angle(j\omega - p_2)$ approach $90°$ as $\omega \to \infty$. Then, from (8.49) and (8.50), it is seen that the magnitude $|H(\omega)|$ starts with value $|k/p_1 p_2| = k/\omega_n^2$ at $\omega = 0$ and approaches zero as $\omega \to \infty$. The phase $\angle H(\omega)$ starts with value $0°$ when $\omega = 0$ and approaches $-180°$ as

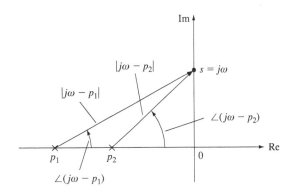

FIGURE 8.21
Vector representations of $j\omega - p_1$ and $j\omega - p_2$ when p_1 and p_2 are real.

$\omega \to \infty$. Thus, when $k = \omega_n^2$, $H(0) = k/\omega_n^2 = 1$ and the system is a lowpass filter whose 3-dB bandwidth depends on the values of ζ and ω_n. When $\zeta = 1$, the 3-dB bandwidth is equal to $\left(\sqrt{\sqrt{2} - 1}\right)\omega_n^2$. This can be verified by a consideration of the vector representations in Figure 8.21. The details are left to the reader.

With $\zeta = 1$, $\omega_n = 3.1$ rad/sec, and $k = \omega_n^2$, the 3-dB bandwidth of the lowpass filter is approximately equal to 2 rad/sec., The frequency response curves for this case were computed by the following MATLAB commands:

```
w = 0:.05:10; s = j*w;
zeta = 1; wn = 3.1;
H = wn^2./(s.^2+2*zeta*wn*s+wn^2);
subplot(211),plot(w,abs(H));
subplot(212),plot(w,angle(H)*180/pi);
```

The results are shown in Figure 8.22. Also displayed in Figure 8.22 are the frequency response curves of the one-pole lowpass filter with transfer function $H(s) = 2/(s + 2)$. Note that the two-pole filter has a sharper cutoff than the one-pole filter.

Complex pole case.   Now it is assumed that $0 < \zeta < 1$, so that the poles p_1 and p_2 are complex. With $\omega_d = \omega_n\sqrt{1 - \zeta^2}$ (as defined in Section 8.3), the poles are $p_1, p_2 = -\zeta\omega_n \pm j\omega_d$. Then $H(\omega)$ can be expressed in the form

$$H(\omega) = \frac{k}{(j\omega + \zeta\omega_n + j\omega_d)(j\omega + \zeta\omega_n - j\omega_d)}$$

The vector representations of $j\omega + \zeta\omega_n + j\omega_d$ and $j\omega + \zeta\omega_n - j\omega_d$ are shown in Figure 8.23. Note that as ω increases from $\omega = 0$, the magnitude $|j\omega + \zeta\omega_n - j\omega_d|$ decreases, while the magnitude $|j\omega + \zeta\omega_n + j\omega_d|$ increases. For $\omega > \omega_d$, both these magnitudes grow until they become infinite, and thus $|H(\omega)| \to 0$ as $\omega \to \infty$. However, it is not clear if $|H(\omega)|$ first increases or decreases as ω is increased from $\omega = 0$. It turns out that when $\zeta < 1/\sqrt{2}$, the magnitude $|H(\omega)|$ increases as ω is increased from 0; and when $\zeta \geq 1/\sqrt{2}$, the magnitude $|H(\omega)|$ decreases as ω is increased from 0. We can prove this by taking the derivative of $|H(\omega)|$ with respect to ω. The details are left to a homework problem. (See Problem 8.23.)

Since the magnitude function $|H(\omega)|$ has a peak when $\zeta < 1/\sqrt{2}$, the system is said to have a *resonance* when $\zeta < 1/\sqrt{2}$. In addition, it can be shown that the peak occurs when $\omega = \omega_r = \omega_n\sqrt{1 - 2\zeta^2}$, and thus ω_r is called the *resonant frequency* of the system. The magnitude of the resonance (i.e., the peak value of $|H(\omega)|$) increases as $\zeta \to 0$, which corresponds to the poles approaching the $j\omega$-axis. (See Figure 8.23.) When $\zeta \geq 1/\sqrt{2}$, the system does not have a resonance, and there is no resonant frequency.

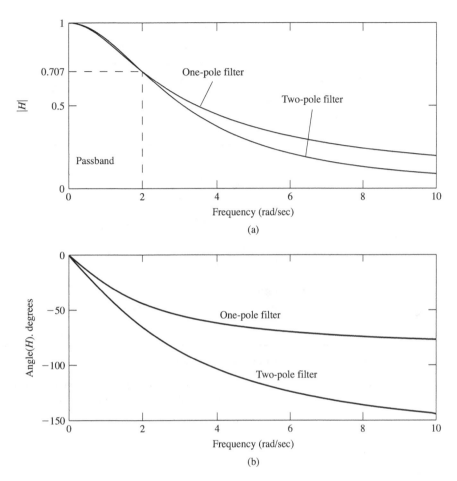

FIGURE 8.22
Frequency response curves of two-pole lowpass filter with $\zeta = 1$ and one-pole lowpass filter:
(a) magnitude curve; (b) phase curve.

When $\zeta < 1/\sqrt{2}$ and the peak value of $|H(\omega)|$ is equal to 1 (i.e., $|H(\omega_r)| = 1$), the system behaves like a *bandpass filter* since it will pass input sinusoids whose frequencies are in a neighborhood of the resonant frequency ω_r. The *center frequency* of the filter is equal to ω_r. The 3-dB bandwidth of the filter is defined to be all those frequencies ω for which the magnitude $|H(\omega)|$ is greater than or equal to $M_p/\sqrt{2}$, where $M_p = |H(\omega_r)|$ is the peak value of $|H(\omega)|$. It follows from the vector representations in Figure 8.23 that the 3-dB bandwidth is approximately equal to $2\zeta\omega_n$. This bandpass filter characteristic is illustrated in the next example. An in-depth treatment of filtering is given in Section 8.6.

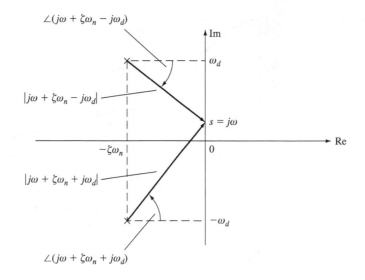

FIGURE 8.23
Vector representations in the complex-pole case.

Example 8.19   Two-Pole Bandpass Filter

Suppose that the objective is to design a two-pole bandpass filter with center frequency $\omega_r = 10$ rad/sec, and with 3-dB bandwidth equal to 2 rad/sec. Then,

$$10 = \omega_r = \omega_n \sqrt{1 - 2\zeta^2}$$

and

$$2 = 2\zeta\omega_n$$

Solving the second equation for ω_n and inserting the result into the first equation give

$$10 = \frac{\sqrt{1 - 2\zeta^2}}{\zeta}$$

and thus

$$\frac{1 - 2\zeta^2}{\zeta^2} = 100$$

Solving for ζ gives

$$\zeta = \frac{1}{\sqrt{102}} \approx 0.099$$

Then,

$$\omega_n = \frac{1}{\zeta} = 10.1$$

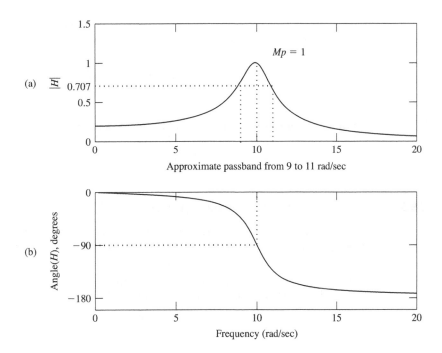

FIGURE 8.24
Frequency response curves for Example 8.19: (a) magnitude curve; (b) phase curve.

and the transfer function of the desired filter is

$$H(s) = \frac{k}{s^2 + 2s + 102}$$

Now the constant k should be chosen so that the peak value of $|H(\omega)|$ is equal to 1. Since the center frequency ω_r of the filter is equal to 10 rad/sec, the peak occurs at $\omega = 10$, and thus k must be chosen so that $|H(10)| = 1$. Then setting $s = j10$ in $H(s)$ and taking the magnitude give

$$|H(10)| = \frac{k}{|-100 + j20 + 102|} = \frac{k}{20.1}$$

Hence, $k = 20.1$. Using MATLAB results in the frequency response curves shown in Figure 8.24. From the plot it can be seen that the desired center frequency and 3-dB bandwidth have been obtained.

Example 8.20   Mass–Spring–Damper System

Mass–
Spring–
Damper
System

For the mass–spring–damper system (see Example 8.14), recall that the damping ratio ζ and the natural frequency ω_n are given by

$$\zeta = \frac{D}{2\sqrt{MK}}, \omega_n = \sqrt{\frac{K}{M}}$$

When $\zeta < 1/\sqrt{2}$, which implies that $D < \sqrt{2MK}$, the system has a resonance with the resonance frequency ω_r given by

$$\omega_r = \omega_n \sqrt{1 - 2\zeta^2} = \sqrt{\frac{K}{M}} \sqrt{1 - \frac{D^2}{2MK}} = \frac{\sqrt{2MK - D^2}}{\sqrt{2M}} \tag{8.51}$$

For various positive values of M, D, and K satisfying the condition $D < \sqrt{2MK}$, the reader is invited to run the online demo with the input equal to the sine sweep. Verify that the resonance frequency observed in the demo is the same as the value computed by the use of (8.51).

8.5.3   Construction of Bode Plots via Asymptotes

Given a system with transfer function $H(s)$, recall that the Bode diagrams are the plots of the magnitude function $|H(\omega)|_{dB} = 20 \log |H(\omega)|$ and the phase function $\angle H(\omega)$, where the scale for the frequency variable ω is logarithmic. The use of the log function in the definition of the magnitude $|H(\omega)|_{dB}$ and the log scale for ω enable the Bode plots to be approximated by straight lines, referred to as *asymptotes*, which can be drawn easily. To see this construction, first consider the system with the transfer function

$$H(s) = \frac{A(s + C_1)(s + C_2) \cdots (s + C_M)}{s(s + B_1)(s + B_2) \cdots (s + B_{N-1})} \tag{8.52}$$

In (8.52), A is a real constant, the zeros $-C_1, -C_2, \ldots, -C_M$ are real numbers, and the poles $-B_1, -B_2, \ldots, -B_{N-1}$ are real numbers. (The case of complex poles and/or zeros will be considered later.) Then, setting $s = j\omega$ in (8.52) gives

$$H(\omega) = \frac{A(j\omega + C_1)(j\omega + C_2) \cdots (j\omega + C_M)}{j\omega(j\omega + B_1)(j\omega + B_2) \cdots (j\omega + B_{N-1})}$$

Dividing each factor $j\omega + C_i$ in the numerator by C_i and dividing each factor $j\omega + B_i$ in the denominator by B_i yield

$$H(\omega) = \frac{K\left(j\dfrac{\omega}{C_1} + 1\right)\left(j\dfrac{\omega}{C_2} + 1\right) \cdots \left(j\dfrac{\omega}{C_M} + 1\right)}{j\omega\left(j\dfrac{\omega}{B_1} + 1\right)\left(j\dfrac{\omega}{B_2} + 1\right) \cdots \left(j\dfrac{\omega}{B_{N-1}} + 1\right)} \tag{8.53}$$

where K is the real constant given by

$$K = \frac{AC_1 C_2 \cdots C_M}{B_1 B_2 \cdots B_{N-1}}$$

Now, since $\log(AB) = \log(A) + \log(B)$ and $\log(A/B) = \log(A) - \log(B)$, from (8.53) the magnitude in dB of $H(\omega)$ is given by

$$|H(\omega)|_{\text{dB}} = 20 \log|K| + 20 \log\left|j\frac{\omega}{C_1} + 1\right| + \cdots + 20 \log\left|j\frac{\omega}{C_M} + 1\right|$$
$$-20 \log|j\omega| - 20 \log\left|j\frac{\omega}{B_1} + 1\right| - \cdots - 20 \log\left|j\frac{\omega}{B_{N-1}} + 1\right|$$

The phase of $H(\omega)$ is given by

$$\angle H(\omega) = \angle K + \angle\left(j\frac{\omega}{C_1} + 1\right) + \cdots + \angle\left(j\frac{\omega}{C_M} + 1\right)$$
$$- \angle(j\omega) - \angle\left(j\frac{\omega}{B_1} + 1\right) - \cdots - \angle\left(j\frac{\omega}{B_{N-1}} + 1\right)$$

Thus, the magnitude and phase functions can be decomposed into a sum of individual factors. The Bode diagrams can be computed for each factor and then added graphically to obtain the Bode diagrams for $H(\omega)$. To carry out this procedure, it is first necessary to determine the Bode plots for the three types of factors in $H(\omega)$: a constant K, the factor $j\omega$, and factors of the form $j\omega T + 1$, where T is a real number. Straight-line approximations (asymptotes) to the Bode plots are derived next for each factor, and from this the actual curves can be sketched. In this development, it is assumed that $T > 0$ in the factor $j\omega T + 1$.

Constant factors. The magnitude plot for the constant factor K is a constant line versus ω given by

$$|K|_{\text{dB}} = 20 \log|K|$$

Similarly, the phase of the factor K is a constant line versus ω:

$$\angle K = \begin{cases} 0° & \text{for } K > 0 \\ -180° & \text{for } K < 0 \end{cases}$$

$(j\omega T + 1)$ factors. The magnitude of $(j\omega T + 1)$ in dB is given by

$$|j\omega T + 1|_{\text{dB}} = 20 \log\sqrt{\omega^2 T^2 + 1}$$

Define the corner frequency ω_{cf} to be the value of ω for which $\omega T = 1$; that is, $\omega_{\text{cf}} = 1/T$. Then, for $\omega < \omega_{\text{cf}}$, ωT is less than 1, and hence the magnitude can be approximated by

$$|j\omega T + 1|_{\text{dB}} \approx 20 \log(1) = 0 \text{ dB}$$

For frequencies $\omega > \omega_{\text{cf}}$, ωT is greater than 1 and the magnitude can be approximated by

$$|j\omega T + 1|_{\text{dB}} \approx 20 \log(\omega T)$$

When plotted on a logarithmic scale for ω, the term $20 \log(\omega T)$ is a straight line with slope equal to 20 dB/decade, where a decade is a factor of 10 in frequency.

The plot of the constant 0 dB for $\omega < \omega_{\text{cf}}$ and the plot of the line $20 \log(\omega T)$ for $\omega > \omega_{\text{cf}}$ are the asymptotes for the magnitude term $|j\omega T + 1|_{\text{dB}}$. The asymptotes are plotted in Figure 8.25a, along with the exact Bode magnitude function for the factor $j\omega T + 1$. As seen from the figure, the asymptotes provide a good approximation for frequencies away from the corner frequency. At the corner frequency, the asymptote approximation is off by approximately 3 dB.

The angle of the factor $(j\omega T + 1)$ is given by

$$\angle(j\omega T + 1) = \tan^{-1} \omega T$$

For very small frequencies, $\angle(j\omega T + 1) \approx 0°$, and for very large frequencies, $\angle(j\omega T + 1) \approx 90°$. A straight-line (asymptote) approximation of $\angle(j\omega T + 1)$ for $\omega \leq \omega_{\text{cf}}/10$ is $\angle(j\omega T + 1) = 0°$, and for $\omega \geq 10\omega_{\text{cf}}$, $\angle(j\omega T + 1) = 90°$. The transition from 0° to 90° can be approximated by a straight line with slope 45°/decade drawn over a two-decade range from $\omega_{\text{cf}}/10$ to $10\omega_{\text{cf}}$. A plot of the asymptote approximations as well as the exact angle plot are shown in Figure 8.25b. The approximations are fairly accurate with errors of about 5° at the corners.

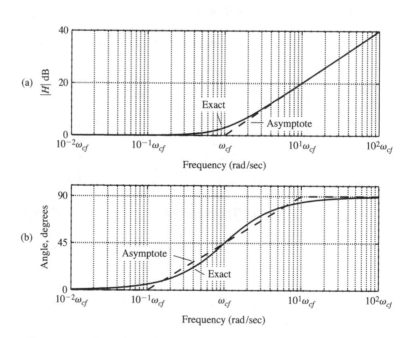

FIGURE 8.25
Magnitude (a) and phase (b) plots for the factor $j\omega T + 1$.

When the factor $(Tj\omega + 1)$ is in the numerator of $H(\omega)$, it represents a zero at $s = -1/T$ in the transfer function $H(s)$. From the prior analysis, it is seen that each (real) zero in the transfer function contributes a phase angle of approximately $+90°$ at high frequencies and a slope of $+20$ dB/decade in magnitude at high frequencies. When the factor $j\omega T + 1$ is in the denominator of $H(\omega)$, it corresponds to a pole of $H(s)$ at $-1/T$. Since

$$|(j\omega T + 1)^{-1}|_{dB} = -|j\omega T + 1|_{dB}$$

and

$$\angle(j\omega T + 1)^{-1} = -\angle(j\omega T + 1)$$

a pole factor $j\omega T + 1$ contributes a phase angle of approximately $-90°$ at high frequencies and a slope of -20 dB/decade in magnitude at high frequencies. The magnitude and phase curves for a pole factor $j\omega T + 1$ are the negative of the magnitude and phase curves for a zero factor $j\omega T + 1$ given in Figure 8.25.

$j\omega$ factors.   The magnitude of $j\omega$ is given by

$$|j\omega|_{dB} = 20\log(\omega)$$

This is a straight line with slope 20 dB/decade when plotted on a logarithmic scale for ω. The line crosses the 0-dB line at $\omega = 1$. To see this, note that when $\omega = 1$,

$$|j\omega|_{dB} = 20\log(1) = 0 \text{ dB}$$

The plot of $|j\omega|_{dB}$ is given in Figure 8.26a. In this case, an approximation is not needed, since the exact plot is already a straight line. The phase plot is also a straight line with a

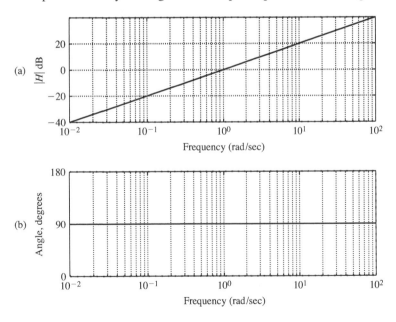

FIGURE 8.26
(a) Magnitude and (b) phase plots for the factor $j\omega$.

constant value of $\angle j\omega = 90°$, as shown in Figure 8.26b. Clearly, a $j\omega$ factor in the numerator of $H(\omega)$ corresponds to a zero of $H(s)$ at $s = 0$. When the $j\omega$ factor is in the denominator of $H(\omega)$, the plots in Figure 8.26 are negated, resulting in a slope of -20 dB/decade in the magnitude plot and $-90°$ in the phase plot.

Plotting Bode diagrams.   Now, to compute the Bode diagrams for a system with $H(\omega)$ given by (8.53), the Bode diagrams for the various factors can simply be added together. The procedure is illustrated in the following example:

Example 8.21   Bode Diagrams

Consider the system with transfer function

$$H(s) = \frac{1000(s + 2)}{(s + 10)(s + 50)}$$

Writing $H(\omega)$ in the form (8.53) yields

$$H(\omega) = \frac{1000(j\omega + 2)}{(j\omega + 10)(j\omega + 50)} = \frac{4(j\omega(0.5) + 1)}{(j\omega(0.1) + 1)(j\omega(0.02) + 1)}$$

The factors making up $H(\omega)$ are 4, $[j\omega(0.5) + 1]$, $[j\omega(0.1) + 1]^{-1}$, and $[j\omega(0.02) + 1]^{-1}$. The constant factor has a magnitude in decibels of $20 \log(4) = 12.04$ dB and an angle of $0°$. The other factors have corner frequencies of $\omega_{cf} = 2$, 10, and 50, respectively. The asymptote approximations of the magnitude and phase for each factor (numbered 1 through 4) are shown in Figure 8.27 by dashed lines. The addition of all the asymptotes is also shown via a solid line in Figure 8.27.

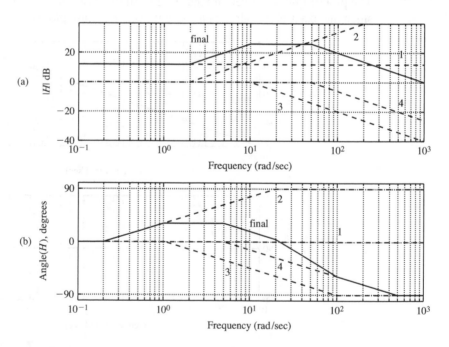

FIGURE 8.27
(a) Magnitude and (b) phase asymptotes for Example 8.21.

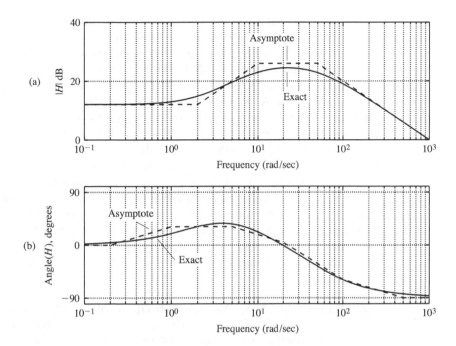

FIGURE 8.28
Summed asymptotes and exact Bode diagrams for Example 8.21.

Note that when adding plots, it is easiest to add the slopes algebraically. For example, the slope of the magnitude curve for $\omega < 2$ is 0 dB/decade; the slope between $\omega = 2$ and $\omega = 10$ is 20 dB/decade; the slope between $\omega = 10$ and $\omega = 50$ is 0 dB/decade; and the slope for $\omega > 50$ is -20 dB/decade. The slopes of the angle plot add similarly. The exact plot obtained by MATLAB is shown in Figure 8.28, along with the summed asymptotes. The MATLAB commands used to generate the exact frequency response are

```
w = logspace(-1,3,300); s = j*w;
H = 1000*(s+2)./(s+10)./(s+50);
magdb = 20*log10(abs(H));
phase = angle(H)*180/pi;
```

There is a command called bode in the Control System Toolbox that can also be used for the exact Bode plot. Its use is demonstrated with the following commands:

```
num = [1000 2000];
den = conv([1 10],[1 50]); % conv multiplies the polynomials
bode(num,den);
```

The addition of slopes to compute the final curve, as shown in Example 8.21, is the foundation for a shorter way of constructing magnitude plots: First, determine the lowest corner frequency. Below that frequency, the only nonzero factors are the constant and $j\omega$ factors. Each $j\omega$ factor in the numerator results in a slope of 20 dB/decade with a 0-dB intercept at $\omega = 1$ rad/sec, while each $j\omega$ factor in the denominator results in a slope of -20 dB/decade with a 0-dB intercept of $\omega = 1$ rad/sec. Therefore, a factor of $(j\omega)^q$ in the numerator (where q is a positive or negative integer) corresponds to a line with slope $20q$ dB/decade and a 0-dB intercept at $\omega = 1$ rad/sec. This line is then offset by the magnitude in decibels of the constant factor to yield the low-frequency asymptote. This asymptote extends until the lowest corner frequency at which a change in slope will occur. A change in slope will occur also at each of the other corner frequencies. The slope change at a corner frequency ω_{cf} is -20 dB/decade if the corner frequency corresponds to a single pole, and $+20$ dB/decade if ω_{cf} corresponds to a single zero. The slope at high frequency should be $-20(N - M)$ dB/decade, where N is the number of poles and M is the number of zeros.

Similarly, we can graph the angle plot by using a similar shortcut, since every pole adds $-90°$ at high frequencies with a transitional slope of $-45°$/decade, and every zero adds $+90°$ at high frequencies with a transitional slope of $+45°$/decade. Since there are two slope changes with each pole or zero, it is easier to plot the individual terms and then add them graphically. A good way to check the final plot is to verify that the angle at high frequencies is equal to $-90(N - M)$, where N is the number of poles and M is the number of zeros.

Complex poles or zeros.   Suppose that $H(s)$ contains a quadratic factor of the form $s^2 + 2\zeta\omega_n s + \omega_n^2$, with $0 < \zeta < 1$ and $\omega_n > 0$ (so that the zeros are complex). Setting $s = j\omega$ and dividing by ω_n^2 result in the factor $(j\omega/\omega_n)^2 + (2\zeta/\omega_n)j\omega + 1$. The magnitude in decibels for this quadratic term is

$$\left|\left(\frac{j\omega}{\omega_n}\right)^2 + \frac{2\zeta}{\omega_n}j\omega + 1\right|_{dB} = 20\log\sqrt{\left(1 - \frac{\omega^2}{\omega_n^2}\right)^2 + \left(\frac{2\zeta\omega}{\omega_n}\right)^2}$$

Define the corner frequency to be the frequency for which $\omega/\omega_n = 1$; that is, $\omega_{cf} = \omega_n$. Then we can carry out an asymptote construction by making the following approximation for low frequencies $\omega < \omega_n$:

$$\left|\left(\frac{j\omega}{\omega_n}\right)^2 + \left(\frac{2\zeta}{\omega_n}\right)j\omega + 1\right|_{dB} \approx 20\log(1) = 0 \text{ dB}$$

And for large frequencies $\omega > \omega_n$,

$$\left|\left(\frac{j\omega}{\omega_n}\right)^2 + \frac{2\zeta}{\omega_n}j\omega + 1\right|_{dB} \approx 20\log\left(\frac{\omega^2}{\omega_n^2}\right) = 40\log\left(\frac{\omega}{\omega_n}\right)$$

Thus, the high-frequency asymptote is a straight line with slope of 40 dB/decade. The asymptote approximation of the magnitude of the quadratic term is shown in Figure 8.29a. In this case the difference between the asymptote approximation and the exact plot depends on the value of ζ, as will be shown later.

The angle for low frequencies $\omega < \omega_n$ is approximated by

$$\angle[(j\omega/\omega_n)^2 + (2\zeta/\omega_n)j\omega + 1] = \tan^{-1}\frac{2\zeta\omega/\omega_n}{1 - (\omega/\omega_n)^2} \approx \tan^{-1}\frac{0}{1} = 0°$$

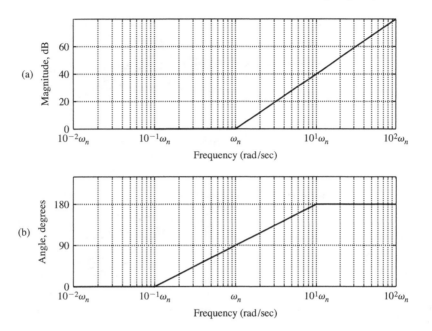

FIGURE 8.29
(a) Magnitude and (b) phase asymptote approximations for the quadratic term.

For high frequencies $\omega > \omega_n$, the angle is approximated by

$$\angle[(j\omega/\omega_n)^2 + (2\zeta/\omega_n)j\omega + 1] = \tan^{-1}\frac{2\zeta\omega/\omega_n}{1 - (\omega/\omega_n)^2} \approx \tan^{-1}\frac{2\zeta\omega_n}{-\omega} \approx 180°$$

The transition between the low- and high-frequency asymptotes is a line spanning two decades from $(0.1)\omega_n$ to $10\omega_n$, with a slope of 90°/decade, as shown in Figure 8.29b.

When the quadratic term is in the numerator of $H(\omega)$, it corresponds to a complex pair of zeros in $H(s)$. Thus, by the preceding analysis, it is seen that a pair of complex zeros contributes a phase angle of approximately 180° and a slope of +40 dB/decade in magnitude at high frequencies. This corresponds to the comment made previously that each real zero of a transfer function contributes a phase angle of +90° and a slope of +20 dB/decade in magnitude to the Bode diagrams at high frequencies.

When the quadratic term is in the denominator, it corresponds to a complex pair of poles of $H(s)$. In this case the asymptote approximations shown in Figure 8.29 are negated to yield the Bode plots for $[(j\omega/\omega_n)^2 + (2\zeta/\omega_n)j\omega + 1]^{-1}$. Hence, a pair of complex poles contributes to the Bode diagrams a phase angle of −180° and a slope in the magnitude plot of −40 dB/decade at high frequencies. As discussed previously, every real pole contributes −90° and −20 dB/decade at high frequencies.

As mentioned before, the exact Bode plot in the quadratic case depends on the value of ζ. The exact Bode diagrams are plotted in Figure 8.30 for various values of ζ. Note from the figure that as ζ approaches 0, the exact magnitude curve has a peak at

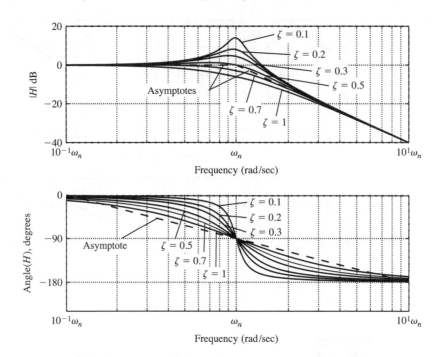

FIGURE 8.30
Exact Bode diagrams for a quadratic term.

ω_n that grows in magnitude while the angle transition in the phase curve becomes sharper. Therefore, the asymptote approximation is not very accurate for small values of ζ. As ζ approaches 1, the quadratic term approaches a term containing two repeated real poles. The resulting Bode plot is equal to the sum of two single-pole plots. In this case the error in the magnitude approximation is 6 dB at the corners (3 dB at each corner frequency corresponding to the real poles).

Example 8.22   Quadratic Term in Denominator

Consider the system with transfer function

$$H(s) = \frac{63(s + 1)}{s(s^2 + 6s + 100)}$$

First, rewrite $H(\omega)$ in the standard form

$$H(\omega) = \frac{0.63(j\omega + 1)}{j\omega[(j\omega/10)^2 + 0.06j\omega + 1]}$$

The factors of $H(\omega)$ are 0.63, $j\omega + 1$, $(j\omega)^{-1}$, and $[(j\omega/10)^2 + 0.06j\omega + 1]^{-1}$. The corner frequencies are $\omega_{\text{cf}} = 1$ rad/sec for the zero and $\omega_n = 10$ rad/sec for the quadratic in the denominator. To obtain the magnitude plot, note that for $\omega < 1$ rad/sec, the only nonzero factors are the constant with magnitude $20 \log(0.63) = -4$ dB and $(j\omega)^{-1}$, which is a line with slope -20 dB/decade and

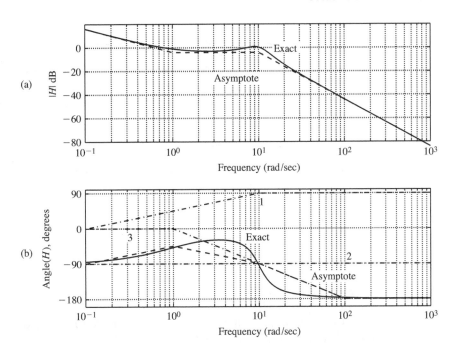

FIGURE 8.31
Asymptotes and exact Bode plots for Example 8.22: (a) magnitude; (b) phase.

an intercept of 0 dB at 1 rad/sec. Combining these factors for $\omega < 1$ simply offsets the -20 dB/decade line by -4 dB. This low-frequency asymptote is plotted in Figure 8.31a. The corner frequency at $\omega_{cf} = 1$ corresponds to a zero, which means that there is a change in slope of $+20$ dB/decade. Therefore, the slope for $1 < \omega < 10$ is 0 dB/decade. The corner frequency at $\omega = 10$ corresponds to a quadratic in the denominator, which means that the slope changes by -40 dB/decade for $\omega > 10$. The final magnitude plot is shown in Figure 8.31a, along with the exact Bode plot obtained from MATLAB. The phase plots for the individual factors are shown in Figure 8.31b, along with the final asymptote plot and the exact plot. The MATLAB commands used to generate the exact magnitude and phase are

```
w = logspace(-1,3,300); s = w*j;
H = 63*(s+1)./(s.^2+6*s+100)./s;
magdb = 20*log10(abs(H));
phase = angle(H)*180/pi;
```

8.6   CAUSAL FILTERS

In real-time filtering applications, it is not possible to utilize ideal filters, since they are noncausal. (See Section 5.3.) In such applications, it is necessary to use causal filters which are nonideal; that is, the transition from the passband to the stopband (and vice versa) is gradual. In particular, the magnitude functions of causal versions of lowpass, highpass, bandpass, and bandstop filters have gradual transitions from the passband to

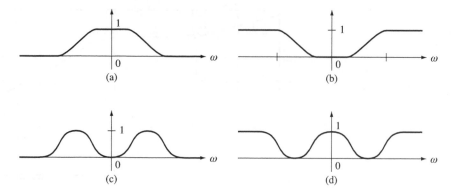

FIGURE 8.32
Causal filter magnitude functions: (a) lowpass; (b) highpass; (c) bandpass; (d) bandstop.

the stopband. Examples of magnitude functions for these basic types of filters are shown in Figure 8.32.

Consider a causal filter with frequency function $H(\omega)$ and with the peak value of $|H(\omega)|$ equal to 1. Then the passband is defined as the set of all frequencies ω for which

$$|H(\omega)| \geq \frac{1}{\sqrt{2}} = 0.707 \tag{8.54}$$

Note that (8.54) is equivalent to the condition that $|H(\omega)|_{dB}$ is less than 3 dB down from the peak value of 0 dB. For lowpass or bandpass filters, the width of the passband is called the 3-dB bandwidth.

A stopband in a causal filter is a set of frequencies ω for which $|H(\omega)|_{dB}$ is down some desired amount (e.g., 40 or 50 dB) from the peak value of 0 dB. The range of frequencies between a passband and a stopband is called a *transition region*. In causal filter design, a key objective is to have the transition regions be suitably small in extent. Later in this section, we will see that we can achieve the sharpest transitions by allowing for ripple in the passband and/or stopband (as opposed to the monotone characteristics shown in Figure 8.32).

To be able to build a causal filter from circuit components, it is necessary that the filter transfer function $H(s)$ be rational in s. For ease of implementation, the order of $H(s)$ (i.e., the degree of the denominator) should be as small as possible. However, there is always a trade-off between the magnitude of the order and desired filter characteristics such as the amount of attenuation in the stopbands and the width of the transition regions.

As in the case of ideal filters, to avoid phase distortion in the output of a causal filter, the phase function should be linear over the passband of the filter. However, the phase function of a causal filter with a rational transfer function cannot be exactly linear over the passband, and thus there will always be some phase distortion. The amount of phase distortion that can be tolerated is often included in the list of filter specifications in the design process.

8.6.1   Butterworth Filters

For the two-pole system with the transfer function

$$H(s) = \frac{\omega_n^2}{s^2 + 2\zeta\omega_n s + \omega_n^2}$$

it follows from the results in Section 8.5 that the system is a lowpass filter when $\zeta \geq 1/\sqrt{2}$. If $\zeta = 1/\sqrt{2}$, the resulting lowpass filter is said to be *maximally flat*, since the variation in the magnitude $|H(\omega)|$ is as small as possible across the passband of the filter. This filter is called the two-pole *Butterworth filter*.

The transfer function of the two-pole Butterworth filter is

$$H(s) = \frac{\omega_n^2}{s^2 + \sqrt{2}\omega_n s + \omega_n^2}$$

Factoring the denominator of $H(s)$ reveals that the poles are located at

$$s = -\frac{\omega_n}{\sqrt{2}} \pm j\frac{\omega_n}{\sqrt{2}}$$

Note that the magnitude of each of the poles is equal to ω_n.

Setting $s = j\omega$ in $H(s)$ yields the magnitude function of the two-pole Butterworth filter:

$$|H(\omega)| = \frac{\omega_n^2}{\sqrt{(\omega_n^2 - \omega^2)^2 + 2\omega_n^2\omega^2}}$$

$$= \frac{\omega_n^2}{\sqrt{\omega_n^4 - 2\omega_n^2\omega^2 + \omega^4 + 2\omega_n^2\omega^2}}$$

$$= \frac{\omega_n^2}{\sqrt{\omega_n^4 + \omega^4}}$$

$$= \frac{1}{\sqrt{1 + (\omega/\omega_n)^4}} \tag{8.55}$$

From (8.55) it is seen that the 3-dB bandwidth of the Butterworth filter is equal to ω_n; that is, $|H(\omega_n)|_{dB} = -3$ dB. For a lowpass filter, the point where $|H(\omega)|_{dB}$ is down by 3 dB is often referred to as the *cutoff frequency*. Hence, ω_n is the cutoff frequency of the lowpass filter with magnitude function given by (8.55).

For the case $\omega_n = 2$ rad/sec, the frequency response curves of the Butterworth filter are plotted in Figure 8.33. Also displayed are the frequency response curves for the one-pole lowpass filter with transfer function $H(s) = 2/(s + 2)$, and the two-pole

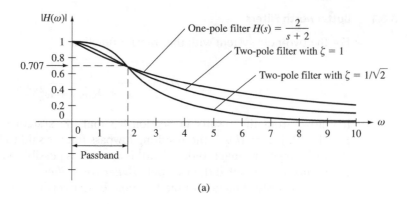

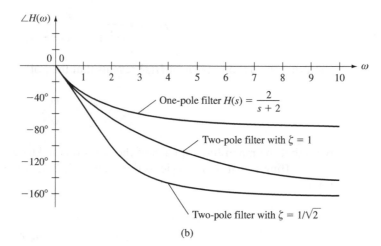

FIGURE 8.33
Frequency curves of one- and two-pole lowpass filters: (a) magnitude curves; (b) phase curves.

lowpass filter with $\zeta = 1$ and with cutoff frequency equal to 2 rad/sec. Note that the Butterworth filter has the sharpest transition of all three filters.

N-pole Butterworth filter.   For any positive integer N, the *N-pole Butterworth filter* is the lowpass filter of order N with a maximally flat frequency response across the passband. The distinguishing characteristic of the Butterworth filter is that the poles lie on a semicircle in the open left-half plane. The radius of the semicircle is equal to ω_c, where ω_c is the cutoff frequency of the filter. In the third-order case, the poles are as displayed in Figure 8.34.

The transfer function of the three-pole Butterworth filter is

$$H(s) = \frac{\omega_c^3}{(s + \omega_c)(s^2 + \omega_c s + \omega_c^2)} = \frac{\omega_c^3}{s^3 + 2\omega_c s^2 + 2\omega_c^2 s + \omega_c^3}$$

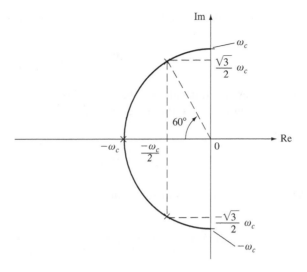

FIGURE 8.34
Pole locations for the three-pole Butterworth filter.

Setting $s = j\omega$ in $H(s)$ and taking the magnitude result in the magnitude function of the three-pole filter:

$$|H(\omega)| = \frac{1}{\sqrt{1 + (\omega/\omega_c)^6}}$$

The magnitude function is plotted in Figure 8.35 for the case $\omega_c = 2$. Also plotted is the magnitude function of the two-pole Butterworth filter with cutoff frequency equal to 2. Clearly, the three-pole filter has a sharper transition than the two-pole filter.

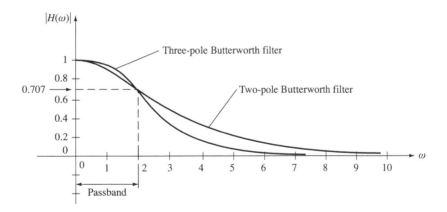

FIGURE 8.35
Magnitude curves of two- and three-pole Butterworth filters.

In the general case, the magnitude function of the N-pole Butterworth filter is

$$|H(\omega)| = \frac{1}{\sqrt{1 + (\omega/\omega_c)^{2N}}}$$

The transfer function can be determined from a table for Butterworth polynomials. For example, when $N = 4$, the transfer function is

$$H(s) = \frac{\omega_c^4}{(s^2 + 0.765\omega_c s + \omega_c^2)(s^2 + 1.85\omega_c s + \omega_c^2)}$$

When $N = 5$, the transfer function is

$$H(s) = \frac{\omega_c^5}{(s + \omega_c)(s^2 + 0.618\omega_c s + \omega_c^2)(s^2 + 1.62\omega_c s + \omega_c^2)}$$

The Signal Processing Toolbox in MATLAB contains a command for designing Butterworth filters with the cutoff frequency normalized to 1 rad/sec. For example, the following commands may be used to create the two-pole Butterworth filter and to obtain the magnitude and phase functions of the filter:

```
[z,p,k] = buttap(2);    % 2 pole filter
[b,a] = zp2tf(z,p,k);
w = 0:.01:4;
H = tf(b,a);     % convert to a transfer function object
bode(H,w);
```

Executing the foregoing commands yields

$$b = 1$$
$$a = [1 \; 1.414 \; 1]$$

which are the coefficients of the numerator and denominator polynomials of the filter transfer function. Running the MATLAB software for the two-, five-, and 10-pole Butterworth filters results in the frequency response curves shown in Figure 8.36. Note that the higher the order of the filter is, the sharper the transition is from the passband to the stopband.

8.6.2   Chebyshev Filters

The magnitude function of the N-pole Butterworth filter has a monotone characteristic in both the passband and stopband of the filter. Here, *monotone* means that the magnitude curve is gradually decreasing over the passband and stopband. In contrast to the Butterworth filter, the magnitude function of a type 1 Chebyshev filter has ripple in the passband and is monotone decreasing in the stopband. (A type 2 Chebyshev filter has the opposite characteristic.) By allowing ripple in the passband or stopband,

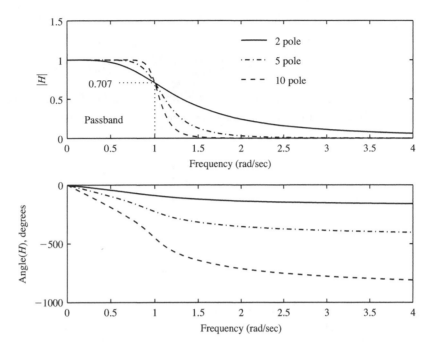

FIGURE 8.36
Frequency response curves for the 2-, 5-, and 10-pole Butterworth filters.

it is possible to achieve a sharper transition between the passband and stopband in comparison with the Butterworth filter.

The N-pole type 1 Chebyshev filter is given by the frequency function

$$|H(\omega)| = \frac{1}{\sqrt{1 + \varepsilon^2 T_N^2(\omega/\omega_1)}} \tag{8.56}$$

where $T_N(\omega/\omega_1)$ is the Nth-order Chebyshev polynomial and ε is a positive number. The Chebyshev polynomials can be generated from the recursion

$$T_N(x) = 2xT_{N-1}(x) - T_{N-2}(x)$$

where $T_0(x) = 1$ and $T_1(x) = x$. The polynomials for $N = 2, 3, 4, 5$ are

$$T_2(x) = 2x(x) - 1 = 2x^2 - 1 \tag{8.57}$$

$$T_3(x) = 2x(2x^2 - 1) - x = 4x^3 - 3x$$

$$T_4(x) = 2x(4x^3 - 3x) - (2x^2 - 1) = 8x^4 - 8x^2 + 1$$

$$T_5(x) = 2x(8x^4 - 8x^2 + 1) - (4x^3 - 3x) = 16x^5 - 20x^3 + 5x$$

Using (8.57) yields the two-pole type 1 Chebyshev filter with frequency function

$$|H(\omega)| = \frac{1}{\sqrt{1 + \varepsilon^2[2(\omega/\omega_1)^2 - 1]^2}}$$

For the N-pole filter defined by (8.56), it can be shown that the magnitude function $|H(\omega)|$ of the filter oscillates between the value 1 and the value $1/\sqrt{1 + \varepsilon^2}$ as ω is varied from 0 to ω_1, with

$$H(0) = \begin{cases} 1 & \text{if } N \text{ is odd} \\ \dfrac{1}{\sqrt{1 + \varepsilon^2}} & \text{if } N \text{ is even} \end{cases}$$

and

$$|H(\omega_1)| = \frac{1}{\sqrt{1 + \varepsilon^2}}$$

The magnitude function $|H(\omega)|$ is monotone decreasing for $\omega > \omega_1$, and thus the filter is a lowpass filter with ripple over the passband. In general, ω_1 is not equal to the cutoff frequency (the 3-dB point) of the filter; however, if

$$\frac{1}{\sqrt{1 + \varepsilon^2}} = \frac{1}{\sqrt{2}}$$

so that $\varepsilon = 1$, then $|H(\omega_1)| = 1/\sqrt{2}$, and in this case ω_1 is the cutoff frequency. When $\varepsilon = 1$, the ripple varies by 3 dB across the passband of the filter.

For the case of a 3-dB ripple ($\varepsilon = 1$), the transfer functions of the two- and three-pole type 1 Chebyshev filters are

$$H(s) = \frac{0.50\omega_c^2}{s^2 + 0.645\omega_c s + 0.708\omega_c^2}$$

$$H(s) = \frac{0.251\omega_c^3}{s^3 + 0.597\omega_c s^2 + 0.928\omega_c^2 s + 0.251\omega_c^3}$$

where ω_c is the cutoff frequency. The frequency response curves for these two filters are plotted in Figure 8.37 for the case $\omega_c = 2$ radians.

The magnitude response functions of the three-pole Butterworth filter and the three-pole type 1 Chebyshev filter are compared in Figure 8.38 with the cutoff frequency of both filters equal to 2 rad/sec. Note that the transition from passband to stopband is sharper in the Chebyshev filter; however, the Chebyshev filter does have the 3-dB ripple over the passband.

The transition from the passband to the stopband can be made sharper (in comparison with the Chebyshev filter) by the allowance of ripple in both the passband and stopband. *Elliptic filters* are examples of a type of filter that yields sharp transitions by

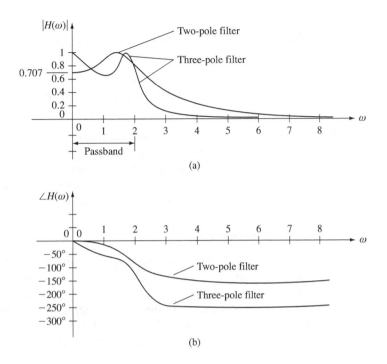

FIGURE 8.37
Frequency curves of two- and three-pole Chebyshev filters with $\omega_c = 2$ radians:
(a) magnitude curves; (b) phase curves.

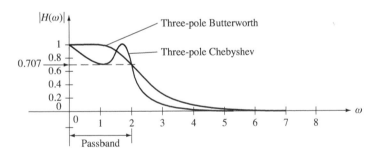

FIGURE 8.38
Magnitude curves of three-pole Butterworth and three-pole Chebyshev filters with
cutoff frequency equal to 2 radians.

permitting ripple in the passband and stopband. This type of filter is not considered here. (See Ludeman [1986].)

We can use the MATLAB command `cheb1ap` from the Signal Processing Toolbox to design type 1 Chebyshev filters. To run this command, we must input the order of the filter and the amount of ripple in dB allowed in the passband. The resulting filter will have a normalized cutoff frequency of 1 rad/sec. For example, to design a two-pole Chebyshev filter that allows a 3-dB ripple in the passband, we would use the following commands:

```
[z,p,k] = cheblap(2,3);
[b,a] = zp2tf(z,p,k);        % convert to polynomials
w = 0:0.01:4;
```

The frequency response curves of the resulting filter are plotted in Figure 8.39. Also plotted in Figure 8.39 are the response curves of the Chebyshev three- and five-pole filters with a maximum 3-dB ripple in the passband. Note that the transition is sharper and the ripple is more pronounced as the number of poles is increased. Also note that in all three cases, the ripple remains within the 3-dB (0.707) limit.

8.6.3 Frequency Transformations

The Butterworth and Chebyshev filters previously discussed are examples of lowpass filters. Starting with any lowpass filter having transfer function $H(s)$, we can modify the cutoff frequency of the filter or construct highpass, bandpass, and bandstop filters, by transforming the frequency variable s. For example, if the cutoff frequency of a lowpass filter is $\omega_c = \omega_1$ and we desire to have the cutoff frequency changed to ω_2, we replace s in $H(s)$ by $s\omega_1/\omega_2$. To convert a lowpass filter with a cutoff frequency of ω_1 to a highpass

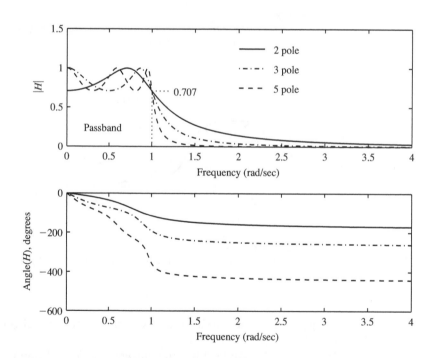

FIGURE 8.39
Frequency response curves of the two-, three-, and five-pole Chebyshev filters.

filter with a 3-dB passband running from $\omega = \omega_2$ to $\omega = \infty$, we replace s in $H(s)$ by $\omega_1\omega_2/s$. To obtain a bandpass filter with a 3-dB passband running from $\omega = \omega_1$ to $\omega = \omega_2$, we replace s in $H(s)$ by

$$\omega_c\frac{s^2 + \omega_1\omega_2}{s(\omega_2 - \omega_1)}$$

Finally, to obtain a bandstop filter with a 3-dB passband running from $\omega = 0$ to $\omega = \omega_1$ and from $\omega = \omega_2$ to $\omega = \infty$, we replace s in $H(s)$ by

$$\omega_c\frac{s(\omega_2 - \omega_1)}{s^2 + \omega_1\omega_2}$$

Example 8.23   Three-Pole Butterworth Filter

Consider the three-pole Butterworth filter with transfer function

$$H(s) = \frac{\omega_c^3}{s^3 + 2\omega_c s^2 + 2\omega_c^2 s + \omega_c^3}$$

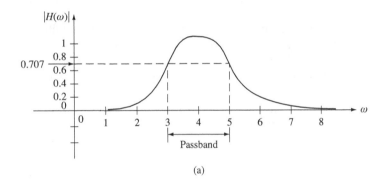

(a)

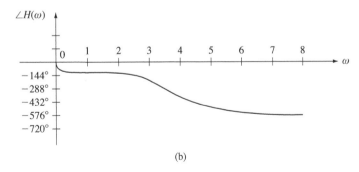

(b)

FIGURE 8.40
Frequency response curves of six-pole bandpass filter: (a) magnitude; (b) phase.

where ω_c is the cutoff frequency. Suppose that the objective is to design a bandpass filter with passband running from $\omega_1 = 3$ to $\omega_2 = 5$. With s replaced by

$$\omega_c \frac{s^2 + 15}{2s}$$

the transfer function of the resulting bandpass filter is

$$H(s) = \frac{\omega_c^3}{\omega_c^3\left(\dfrac{s^2 + 15}{2s}\right)^3 + 2\omega_c^3\left(\dfrac{s^2 + 15}{2s}\right)^2 + 2\omega_c^3\left(\dfrac{s^2 + 15}{2s}\right) + \omega_c^3}$$

$$H(s) = \frac{8s^3}{(s^2 + 15)^3 + 4s(s^2 + 15)^2 + 8s^2(s^2 + 15) + 8s^3}$$

$$= \frac{8s^3}{s^6 + 4s^5 + 53s^4 + 128s^3 + 795s^2 + 900s + 3375}$$

The frequency curves for this filter are displayed in Figure 8.40.

Frequency transformations are very useful with MATLAB, since the standard filter design programs produce a lowpass filter with a normalized cutoff frequency of 1 rad/sec. We can start the design process by first generating a Butterworth or Chebyshev lowpass filter with normalized cutoff frequency. Then the resulting filter can be transformed to a lowpass filter with a different cutoff frequency, or transformed to a highpass, bandpass, or bandstop filter, with the commands lp2lp, lp2hp, lp2bp, and lp2bs. The following examples show how to use MATLAB to design various types of filters:

Example 8.24   Lowpass Filter Design

To design a three-pole Butterworth lowpass filter with a bandwidth of 5 Hz, first design a three-pole filter with cutoff frequency of 1 rad/sec by using the buttap command. Then transform the frequency by using the command lp2lp. The MATLAB commands are as follows:

```
[z,p,k] = buttap(3);  % 3 pole filter
[b,a] = zp2tf(z,p,k);  % convert to polynomials
wb = 5*2*pi;  % new bandwidth in rad/sec
[b,a] = lp2lp(b,a,wb);    % transforms to the new bandwidth
f = 0:15/200:15;   % define the frequency in Hz for plotting
w = 2*pi*f;
H = tf(b,a);
bode(H,w);
```

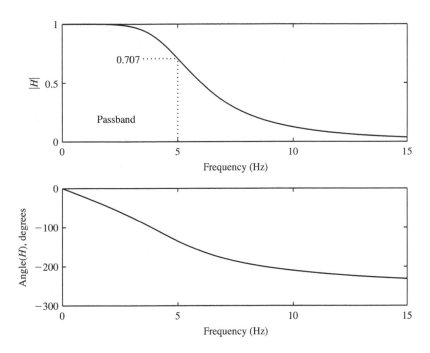

FIGURE 8.41
Frequency response curves for the lowpass filter in Example 8.24.

The coefficients of the numerator and denominator are found to be given by b = [0 0 0 31006] and a = [1 63 1974 31006], respectively. Hence, the 5-Hz bandwidth filter is given by

$$H(s) = \frac{31{,}006}{s^3 + 63s^2 + 1974s + 31{,}006}$$

The resulting frequency response is plotted in Figure 8.41.

Example 8.25   Highpass Filter Design

To design a three-pole highpass filter with cutoff frequency $\omega = 4$ rad/sec, first design a three-pole Chebyshev or Butterworth filter with cutoff frequency of $\omega_c = 1$ rad/sec. Then transform it to a highpass filter. The commands are

```
w0 = 4;   % cutoff frequency
% create a 3 pole Chebyshev Type I filter with 3-dB passband
[z,p,k] = cheblap(3,3);
[b,a] = zp2tf(z,p,t); % convert to polynomials
[b,a] = lp2hp(b,a,w0);   % cutoff frequency is w0
```

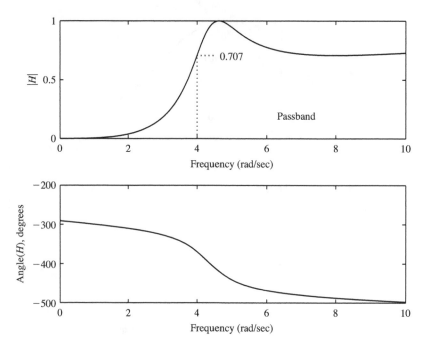

FIGURE 8.42
Frequency response of filter in Example 8.25.

The coefficients of the numerator and denominator, located in the vectors b and a, respectively, are found to be b = [1 0 0 0] and a = [1 14.8 39.1 255.4]. Hence, the resulting highpass filter is given by

$$H(s) = \frac{s^3}{s^3 + 14.8s^2 + 38.1s + 255.4}$$

The frequency response plot is shown in Figure 8.42.

Example 8.26   Design Using MATLAB

In Example 8.23, a three-pole Butterworth lowpass filter was transformed to a bandpass filter with the passband centered at $\omega = 4$ rad/sec and with the bandwidth equal to 2 rad/sec. To perform this conversion via MATLAB, use the following commands:

```
w0 = 4;                  % center of band
wb = 2;                  % bandwidth
[z,p,k] = buttap(3);     % 3 pole Butterworth filter
[b,a] = zp2tf(z,p,k);    % convert to polynomials
[b,a] = lp2bp(b,a,w0,wb); % passband centered at w0, bandwidth=wb
```

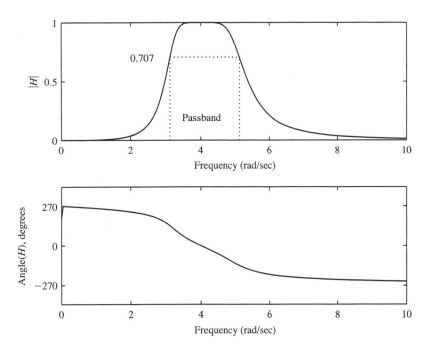

FIGURE 8.43
Response curves for filter in Example 8.26.

The numerator coefficients stored in the vector b match those obtained analytically in Example 8.23. However, the denominator coefficients calculated as a = [1 4 56 136 896 1024 4096] are different from those determined analytically. Although the coefficients have large differences, the corresponding poles are very close, resulting in very little error in the frequency response plot shown in Figure 8.43. The difference is due to roundoff errors in the computations.

Example 8.27   Bandstop Filter

To convert a three-pole Chebyshev lowpass filter to a bandstop filter with a stopband from $\omega = 4$ to $\omega = 6$, use the following MATLAB commands:

```
w0 = 5;    % center of stopband
wb = 2;    % width of stopband
[z,p,k] = cheblap(3,3);    % 3 pole filter with 3-dB ripple
[b,a] = zp2tf(z,p,k);      % convert to polynomials
[b,a] = lp2bs(b,a,w0,wb);  % converts to a bandstop filter
```

The coefficients are calculated to be b = [1 0 75 0 1875 0 15625] and a = [1 7 85 402 2113 4631 15625]. Hence, the bandstop filter is given by

$$H(s) = \frac{s^6 + 75s^4 + 1875s^2 + 15{,}625}{s^6 + 7s^5 + 85s^4 + 402s^3 + 2113s^2 + 4631s + 15{,}625}$$

The resulting frequency response plot is shown in Figure 8.44.

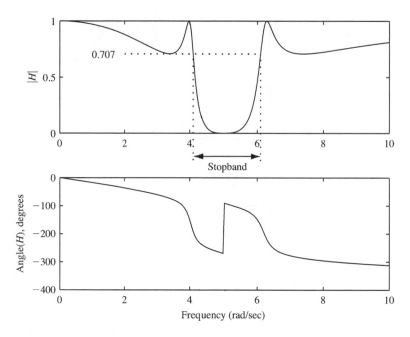

FIGURE 8.44
Frequency response of filter in Example 8.27.

8.7   CHAPTER SUMMARY

Chapter 6 introduced the Laplace transform and the transfer function representation of a system. Chapter 8 shows how a transfer function is used to analyze the behavior of continuous-time systems. Concepts studied include stability, step response, response to sinusoids, frequency response and Bode plots, and filtering. A system is stable if its impulse response decays to zero, which is the case when all of the poles of the transfer function are in the open left-half plane. The system is unstable if its impulse response grows without bound, which results when there is at least one pole in the open right-half plane or repeated poles on the $j\omega$-axis. Otherwise, the system is marginally stable, which results when the system is not unstable and has at least one nonrepeated pole on the $j\omega$-axis. We can determine stability by finding the poles of the transfer function or by using the Routh–Hurwitz stability test. This test is particularly useful when the degree of the denominator of the transfer function is large, or when one or more of the coefficients of the denominator polynomial is unknown. This latter case is common when part of the system is being designed, and the ranges of stability for the coefficients must be determined as part of the design.

The response of a stable system is composed of transient and steady-state components. The steady-state component of the response matches the form of the input signal; so a step input gives rise to a constant steady-state value, and a sinusoidal input results in a sinusoidal steady-state response at the same frequency of the input. The transient component results from the transfer function, and so a general form of the transient of a system can be determined on the basis of the system poles without regard to the form of the input. The step response of a system is a well-accepted means of characterizing the system's behavior and of specifying its performance. There are three main characteristics in a step response: the time constant, the steady-state value, and the possible existence of oscillations or overshoot in the response. The transient response of a system with real poles, such as a first-order system, does not oscillate. Oscillations are a result of complex poles, where the frequency of oscillations is the imaginary part of the pole and the rate of decay or growth is determined from the real part of the pole. The time constant is a measure of the speed of response, that is, it indicates how long it takes for the transient to decay.

PROBLEMS

8.1. For the following linear time-invariant continuous-time systems, determine if the system is stable, marginally stable, or unstable:

(a) $H(s) = \dfrac{s - 4}{s^2 + 7s + 3}$

(b) $H(s) = \dfrac{s + 3}{s^2 + 3}$

(c) $H(s) = \dfrac{2s + 3}{s^2 + 2s - 12}$

(d) $H(s) = \dfrac{3s^3 - 2s + 6}{s^3 + s^2 + s + 1}$

(e) $H(s) = \dfrac{4s + 8}{(s^2 + 4s + 13)(s + 4)}$

8.2. Consider the field-controlled dc motor given by the input/output differential equation

$$L_f I \frac{d^3 y(t)}{dt^3} + (L_f k_d + R_f I)\frac{d^2 y(t)}{dt^2} + R_f k_d \frac{dy(t)}{dt} = kx(t)$$

Assume that all the parameters I, L_f, k_d, R_f, and k are strictly positive (> 0). Determine if the motor is stable, marginally stable, or unstable.

8.3. Consider the model for the ingestion and metabolism of a drug defined in Problem 6.19. Assuming that $k_1 > 0$ and $k_2 > 0$, determine if the system is stable, marginally stable, or unstable. What does your answer imply regarding the behavior of the system? Explain.

8.4. Determine if the mass–spring system in Problem 2.23 is stable, marginally stable, or unstable. Assume that k_1, k_2, and k_3 are strictly positive (> 0).

8.5. Consider the single-eye system studied in Problem 2.35. Assuming that $T_e > 0$, determine if the system is BIBO stable.

8.6. For each of the linear time-invariant continuous-time systems with impulse response $h(t)$ given as follows, determine if the system is BIBO stable.

(a) $h(t) = [2t^3 - 2t^2 + 3t - 2][u(t) - u(t - 10)]$

(b) $h(t) = \dfrac{1}{t}$ for $t \geq 1$, $h(t) = 0$ for all $t < 1$

(c) $h(t) = \sin 2t$ for $t \geq 0$

(d) $h(t) = e^{-t} \sin 2t$ for $t \geq 0$

(e) $h(t) = e^{-t^2}$ for $t \geq 0$

8.7. Using the Routh–Hurwitz test, determine all values of the parameter k for which the following systems are stable:

(a) $H(s) = \dfrac{s^2 + 60s + 800}{s^3 + 30s^2 + (k + 200)s + 40k}$

(b) $H(s) = \dfrac{2s^3 - 3s + 4}{s^4 + s^3 + ks^2 + 2s + 3}$

(c) $H(s) = \dfrac{s^2 + 3s - 2}{s^3 + s^2 + (k + 3)s + 3k - 5}$

(d) $H(s) = \dfrac{s^4 - 3s^2 + 4s + 6}{s^5 + 10s^4 + (9 + k)s^3 + (90 + 2k)s^2 + 12ks + 10k}$

8.8. Suppose that a system has the following transfer function:

$$H(s) = \frac{8}{s + 4}$$

(a) Compute the system response to the inputs (i)–(iv). Identify the steady-state solution and the transient solution.
  - **(i)** $x(t) = u(t)$
  - **(ii)** $x(t) = tu(t)$
  - **(iii)** $x(t) = 2\,(\sin 2t)\,u(t)$
  - **(iv)** $x(t) = 2\,(\sin 10t)\,u(t)$
(b) Use MATLAB to compute the response numerically from $x(t)$ and $H(s)$. Plot the responses, and compare them with the responses obtained analytically in part (a).

Pole Positions and Step Response

8.9. Consider three systems which have the following transfer functions:

  - **(i)** $H(s) = \dfrac{32}{s^2 + 4s + 16}$
  - **(ii)** $H(s) = \dfrac{32}{s^2 + 8s + 16}$
  - **(iii)** $H(s) = \dfrac{32}{s^2 + 10s + 16}$

For each system, do the following:
(a) Determine if the system is critically damped, underdamped, or overdamped.
(b) Calculate the step response of the system.
(c) Use MATLAB to compute the step response numerically. Plot the response, and compare it with the plot of the response obtained analytically in part (b).

8.10. A first-order system has the step response shown in Figure P8.10. Determine the transfer function.

8.11. A second-order system has the step response shown in Figure P8.11. Determine the transfer function.

8.12. Consider the mass–spring–damper system with the input/output differential equation

$$M\frac{d^2 y(t)}{dt^2} + D\frac{dy(t)}{dt} + Ky(t) = x(t)$$

where M is the mass, D is the damping constant, K is the stiffness constant, $x(t)$ is the force applied to the mass, and $y(t)$ is the displacement of the mass relative to the equilibrium position.

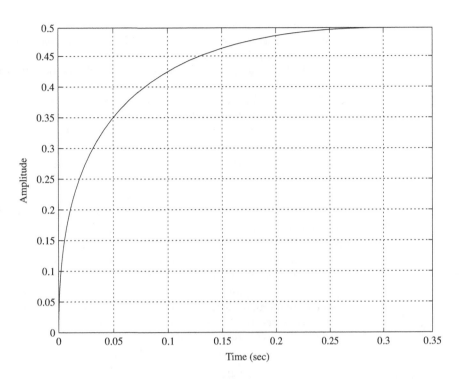

FIGURE P8.10

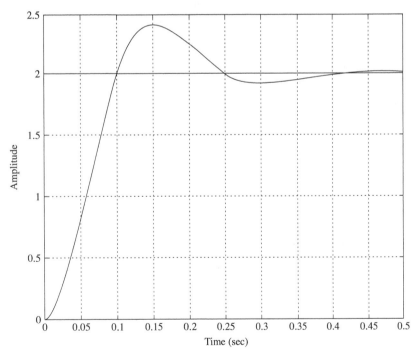

FIGURE P8.11

(a) Determine the pole locations for the cases **(i)** $M = 1$, $D = 50.4$, and $K = 3969$; and **(ii)** $M = 2$, $D = 50.4$, and $K = 3969$. Show the location of the poles on a pole-zero plot. Compute the natural frequency and the time constant for each of the cases. Which has the higher frequency of response? For which case does the transient response decay faster?

(b) Use MATLAB to compute the impulse response of the system for the two cases, and compare your results with the predictions made in part (a).

(c) Repeat parts (a) and (b) for the cases **(i)** $M = 1$, $D = 50.4$, and $K = 15,876$ and **(ii)** $M = 2$, $D = 50.4$, and $K = 15,876$.

Mass–
Spring–
Damper
System

8.13. Again consider the mass–spring–damper system in Problem 8.12. Let $M = 1$, $D = 50.4$, and $K = 3969$.

(a) Compute the response to a unit step in the force.

(b) Compute the steady-state response to an input of $x(t) = 10 \cos(20\pi t)u(t)$.

(c) Compute the steady-state response to an input of $x(t) = 10 \cos(2\pi t)u(t)$.

(d) Use MATLAB to simulate the system with the inputs given in parts (a)–(c). Verify that your answers in parts (a)–(c) are correct by plotting them along with the corresponding results obtained from the simulation.

(e) Use the Mass–Spring–Damper demo available on the textbook Web page to simulate the system with the inputs given in parts (a)–(c), and compare the responses with those plotted in part (d). Change the damping parameter to $D = 127$, and use the applet to simulate the step response. Sketch the response.

8.14. Consider the two systems given by the following transfer functions:

(i) $H(s) = \dfrac{242.5(s + 8)}{(s + 2)[(s + 4)^2 + 81](s + 10)}$

(ii) $H(s) = \dfrac{115.5(s + 8)(s + 2.1)}{(s + 2)[(s + 4)^2 + 81](s + 10)}$

(a) Identify the poles and zeros of the system.

(b) Without computing the actual response, give the general form of the step response.

(c) Determine the steady-state value for the step response.

(d) Determine the dominant pole(s).

(e) Use MATLAB to compute and plot the step response of the system. Compare the plot with the answers expected in parts (b) to (d).

8.15. For each of the circuits in Figure P8.15, compute the steady-state response $y_{ss}(t)$ resulting from the following inputs with zero initial conditions:

(a) $x(t) = u(t)$

(b) $x(t) = (5 \cos 2t)u(t)$

(c) $x(t) = [2 \cos(3t + 45°)]u(t)$

8.16. Consider the mass–spring system in Problem 2.23. Assume that $M_1 = 1$, $M_2 = 10$, and $k_1 = k_2 = k_3 = 0.1$. Compute the steady-state response $y_{ss}(t)$ resulting from the following inputs with zero initial conditions:

(a) $x(t) = u(t)$

(b) $x(t) = (10 \cos t)u(t)$

(c) $x(t) = [\cos(5t - 30°)]u(t)$

8.17. A linear time-invariant continuous-time system has transfer function $H(s) = 2/(s + 1)$. Compute the transient response $y_{tr}(t)$ resulting from the input $x(t) = 3 \cos 2t - 4 \sin t, t \geq 0$, with zero initial conditions.

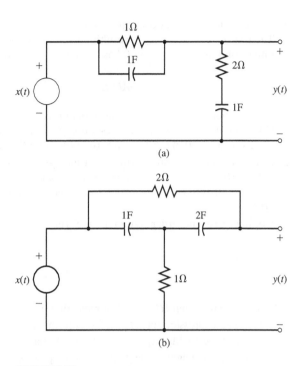

FIGURE P8.15

8.18. A linear time-invariant continuous-time system has transfer function

$$H(s) = \frac{s^2 + 16}{s^2 + 7s + 12}$$

Compute the steady-state and transient responses resulting from the input $x(t) = 2\cos 4t, t \geq 0$, with zero initial conditions.

8.19. A linear time-invariant continuous-time system has transfer function

$$H(s) = \frac{s^2 + 1}{(s + 1)(s^2 + 2s + 17)}$$

Compute both the steady-state response $y_{ss}(t)$ and the transient response $y_{tr}(t)$ when the input $x(t)$ is
(a) $x(t) = u(t)$, with zero initial conditions.
(b) $x(t) = \cos t, t \geq 0$, with zero initial conditions.
(c) $x(t) = \cos 4t, t \geq 0$, with zero initial conditions.

8.20. A linear time-invariant continuous-time system has transfer function

$$H(s) = \frac{s + 2}{(s + 1)^2 + 4}$$

The input $x(t) = C\cos(\omega_0 t + \theta)$ is applied to the system for $t \geq 0$ with zero initial conditions. The resulting steady-state response $y_{ss}(t)$ is

$$y_{ss}(t) = 6\cos(t + 45°), t \geq 0$$

(a) Compute C, ω_0, and θ.
(b) Compute the Laplace transform $Y_{tr}(s)$ of the transient response $y_{tr}(t)$ resulting from this input.

8.21. A linear time-invariant continuous-time system has transfer function $H(s)$ with $H(0) = 3$. The transient response $y_{\text{tr}}(t)$ resulting from the step-function input $x(t) = u(t)$ with zero initial conditions at time $t = 0$ has been determined to be

$$y_{\text{tr}}(t) = -2e^{-t} + 4e^{-3t}, t \geq 0$$

(a) Compute the system's transfer function $H(s)$.

(b) Compute the steady-state response $y_{\text{ss}}(t)$ when the system's input $x(t)$ is equal to $2\cos(3t + 60°), t \geq 0$, with zero initial conditions.

8.22. A linear time-invariant continuous-time system has transfer function $H(s)$. The input $x(t) = 3(\cos t + 2) + \cos(2t - 30°), t \geq 0$, produces the steady-state response $y_{\text{ss}}(t) = 6\cos(t - 45°) + 8\cos(2t - 90°), t \geq 0$, with zero initial conditions. Compute $H(1)$ and $H(2)$.

8.23. Consider a second-order system in the form

$$H(s) = \frac{\omega_n^2}{s^2 + 2\zeta\omega_n s + \omega_n^2}$$

Let $s \to j\omega$ to obtain $H(\omega)$, and suppose that $0 < \zeta < 1$. Without factoring the denominator, find an expression for $|H(\omega)|$. To determine if a peak exists in $|H(\omega)|$, take the derivative of $|H(\omega)|$ with respect to ω. Show that a peak exists for $\omega \neq 0$ only if $\zeta \leq 1/\sqrt{2}$. Determine the height of the peak. What happens to the peak as $\zeta \to 0$?

8.24. Sketch the magnitude and phase plots for the systems that follow. In each case, compute $|H(\omega)|$ and $\angle H(\omega)$ for $\omega = 0$, $\omega = $ 3-dB points, $\omega = \omega_p$ and $\omega \to \infty$. Here, ω_p is the value of ω for which $|H(\omega)|$ is maximum. Verify your calculations by plotting the frequency response, using MATLAB.

(a) $H(s) = \dfrac{10}{s + 5}$

(b) $H(s) = \dfrac{5(s + 1)}{s + 5}$

(c) $H(s) = \dfrac{s + 10}{s + 5}$

(d) $H(s) = \dfrac{4}{(s + 2)^2}$

(e) $H(s) = \dfrac{4s}{(s + 2)^2}$

(f) $H(s) = \dfrac{s^2 + 2}{(s + 2)^2}$

(g) $H(s) = \dfrac{4}{s^2 + \sqrt{2}(2s) + 4}$

8.25. Sketch the magnitude and phase plots for the circuits shown in Figure P8.25. In each case, compute $|H(\omega)|$ and $\angle H(\omega)$ for $\omega = 0$, $\omega = $ 3-dB points, and $\omega \to \infty$.

8.26. Repeat Problem 8.25 for the circuits in Figure P8.15.

8.27. Consider the RLC circuit shown in Figure P8.27. Choose values for R and L such that the damping ratio $\zeta = 1$ and the circuit is a lowpass filter with approximate 3-dB bandwidth equal to 20 rad/sec; that is, $|H(\omega)| \geq (0.707)|H(0)|$ for $0 \leq \omega \leq 20$.

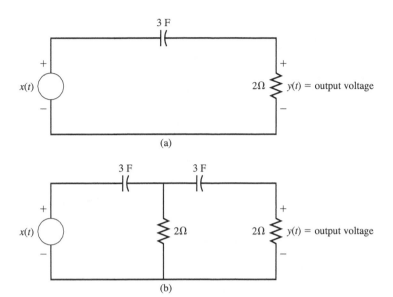

FIGURE P8.25

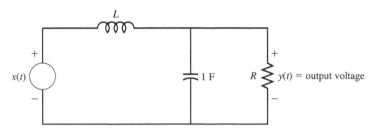

FIGURE P8.27

8.28. A linear time-invariant continuous-time system has transfer function $H(s)$. It is known that $H(0) = 1$ and that $H(s)$ has two poles and no zeros. In addition, the magnitude function $|H(\omega)|$ is shown in Figure P8.28. Determine $H(s)$.

8.29. A linear time-invariant continuous-time system has transfer function $H(s) = K/(s + a)$, where $K > 0$ and $a > 0$ are unknown. The steady-state response to $x(t) = 4 \cos t, t \geq 0$, is $y_{ss}(t) = 20 \cos(t + \phi_1), t \geq 0$. The steady-state response to $x(t) = 5 \cos 4t, t \geq 0$, is $y_{ss}(t) = 10 \cos(4t + \phi_2), t \geq 0$. Here ϕ_1, ϕ_2 are unmeasurable phase shifts. Find K and a.

8.30. Using MATLAB, determine the frequency response curves for the mass–spring system in Problem 2.23. Take $M_1 = 1$, $M_2 = 10$, and $k_1 = k_2 = k_3 = 0.1$.

8.31. Draw the asymptotic Bode plots (both magnitude and phase plots) for the accompanying systems. Compare your plots with the actual Bode plots obtained from MATLAB.

(a) $H(s) = \dfrac{16}{(s + 1)(s + 8)}$

(b) $H(s) = \dfrac{10(s + 4)}{(s + 1)(s + 10)}$

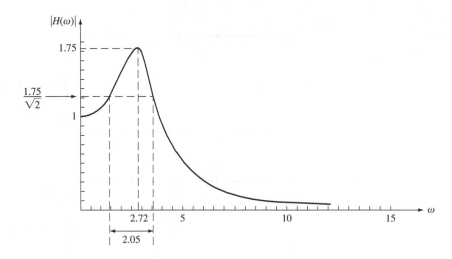

FIGURE P8.28

(c) $H(s) = \dfrac{10}{s(s+6)}$

(d) $H(s) = \dfrac{10}{(s+1)(s^2+4s+16)}$

(e) $H(s) = \dfrac{10}{(s+1)(s^2+s+16)}$

(f) $H(s) = \dfrac{1000(s+1)}{(s+20)^2}$

8.32. A linear time-invariant continuous-time system has a rational transfer function $H(s)$ with two poles and two zeros. The frequency function $H(\omega)$ of the system is given by

$$H(\omega) = \frac{-\omega^2 + j3\omega}{8 + j12\omega - 4\omega^2}$$

Determine $H(s)$.

8.33. Consider the three-pole Butterworth filter given by the transfer function

$$H(s) = \frac{\omega_c^3}{s^3 + 2\omega_c s^2 + 2\omega_c^2 s + \omega_c^3}$$

(a) Derive an expression for the impulse response $h(t)$ in terms of the 3-dB bandwidth ω_c. Plot $h(t)$ when $\omega_c = 1$ rad/sec.

(b) Compare your result in part (a) with the impulse response of an ideal lowpass filter with frequency function $H(\omega) = p_2(\omega)$. Discuss the similarities and differences in the two impulse responses.

8.34. Again consider the three-pole Butterworth filter defined in Problem 8.33.

(a) For the case when $\omega_c = 2\pi$, compute the output response of the filter when the input is $x(t) = u(t) - u(t-1)$ with zero initial conditions.

(b) Repeat part (a) for the case when $\omega_c = 4\pi$.

(c) Using MATLAB, plot the responses found in parts (a) and (b).

(d) Are the results obtained in part (c) expected? Explain.

8.35. For the three-pole Butterworth filter with $\omega_c = 1$, compute the output response $y(t)$ when the input $x(t)$ is

(a) $x(t) = 1, -\infty < t < \infty$

(b) $x(t) = 2 \cos t, -\infty < t < \infty$

(c) $x(t) = \cos(10t + 30°), -\infty < t < \infty$

(d) $x(t) = 2(\cos t)(\sin t), -\infty < t < \infty$

8.36. Again, consider the three-pole Butterworth filter with $\omega_c = 1$. The output response resulting from the input $x(t) = \cos 0.5t, -\infty < t < \infty$, can be expressed in the form $y(t) = B \cos[0.5(t - t_d)], -\infty < t < \infty$, where t_d is the time delay through the filter. The response resulting from the input $x(t) = \cos t, -\infty < t < \infty$, can be expressed in the form $y(t) = C \cos(t - t_d + \phi), -\infty < t < \infty$, where ϕ is the phase distortion resulting from the nonlinear phase characteristic of the filter. Compute t_d and ϕ.

8.37. Repeat Problem 8.34 for the three-pole Chebyshev filter with transfer function

$$H(s) = \frac{0.251\omega_c^3}{s^3 + 0.597\omega_c s^2 + 0.928\omega_c^2 s + 0.251\omega_c^3}$$

8.38. Repeat Problem 8.35 for the three-pole Chebyshev filter with $\omega_c = 1$.

8.39. The objective of this problem is to design both a highpass and a bandpass filter, starting from the two-pole Butterworth filter with transfer function

$$H(s) = \frac{\omega_c^2}{s^2 + \sqrt{2}\omega_c s + \omega_c^2}$$

(a) Design the highpass filter so that the 3-dB bandwidth runs from $\omega = 10$ to $\omega = \infty$.

(b) Design the bandpass filter so that the 3-dB bandwidth runs from $\omega = 10$ to $\omega = 20$.

(c) Using MATLAB, determine the frequency response curves of the filters constructed in parts (a) and (b).

8.40. Repeat Problem 8.39 for the two-pole Chebyshev filter with transfer function

$$H(s) = \frac{0.50\omega_c^2}{s^2 + 0.645\omega_c s + 0.708\omega_c^2}$$

8.41. Design a three-pole Butterworth stopband filter with a stopband from $\omega = 10$ to $\omega = 15$ rad/sec.

(a) Plot the frequency response curves for the resulting filter.

(b) From the magnitude curve plotted in part (a), determine the expected amplitude of the steady-state responses $y_{ss}(t)$ to the following signals: **(i)** $x(t) = \sin 5t$, **(ii)** $x(t) = \sin 12t$, and **(iii)** $x(t) = \sin 5t + \sin 12t$.

(c) Verify your prediction in part (b) by using MATLAB to compute and plot the response of the system to the signals defined in part (b). You may use `lsim` and integrate long enough for the response to reach steady state, or use Simulink. [*Note:* When simulating a continuous-time system to find the response, computers approximate the system as being discrete time. Therefore, when defining the signals $x(t)$ for a time vector `t=0:T:tf`, make sure that the time increment `T` for which $x(t)$ is defined satisfies the Nyquist sampling theorem; that is, $2\pi/T$ is at least twice the highest frequency in $x(t)$. See the comments in Problem 1.2 for further information.]

8.42. Design a three-pole Chebyshev bandpass filter with a passband from $\omega = 10$ to $\omega = 15$ rad/sec. Allow a 3-dB ripple in the passband.

(a) Plot the frequency response curves for the resulting filter.

(b) From the magnitude curve plotted in part (a), determine the expected amplitude of the steady-state responses $y_{ss}(t)$ to the following signals: (i) $x(t) = \sin 5t$, (ii) $x(t) = \sin 12t$, and (iii) $x(t) = \sin 5t + \sin 12t$.

(c) Verify your prediction in part (b) by using MATLAB to compute and plot the response of the system to the signals defined in part (b). (Consider the comment in Problem 8.41 regarding the selection of the time increment when using MATLAB.)

8.43. Design a lowpass Butterworth filter with a bandwidth of 10 rad/sec. Select an appropriate number of poles so that a 25-rad/sec sinusoidal signal is attenuated to a level that is no more than 5% of its input amplitude. Use MATLAB to compute and plot the response of the system to the following signals. (Consider the comments in Problem 8.41 regarding the selection of the time increment when using MATLAB.)

(a) $x(t) = \sin 5t$

(b) $x(t) = \sin 25t$

(c) $x(t) = \sin 5t + \sin 25t$

(d) $x(t) = w(t)$ where $w(t)$ is a random signal whose values are uniformly distributed between 0 and 1. (Use x = rand(201,1) to generate the signal for the time vector t = 0:.05:10.) Plot the random input $x(t)$, and compare it with the system response.

8.44. Design a highpass type 1 Chebyshev filter with a bandwidth of 10 rad/sec. Select an appropriate number of poles so that a 5-rad/sec sinusoidal signal is attenuated to a level that is no more than 10% of its input amplitude and there is at most a 3-dB ripple in the passband. Use MATLAB to compute and plot the response of the system to the signals that follow. (Consider the comment in Problem 8.41 regarding the selection of the time increment when using MATLAB.)

(a) $x(t) = \sin 5t$

(b) $x(t) = \sin 25t$

(c) $x(t) = \sin 5t + \sin 25t$

(d) $x(t) = w(t)$, where $w(t)$ is a random signal whose values are uniformly distributed between 0 and 1. (Use x = rand(201,1) to generate the signal for the time vector t = 0:.05:10.) Plot the random input $x(t)$, and compare it with the system response.

Application to Control

One of the major applications of the transfer function framework is in the study of control. A very common type of control problem is forcing the output of a system to be equal to a desired reference signal, which is referred to as *tracking*. The tracking problem arises in a multitude of applications such as in industrial control and automation, where the objective is to control the position and/or velocity of a physical object. Examples given in this chapter involve velocity control of a vehicle and the control of the angular position of the shaft of a motor. The development begins in Section 9.1 with an introduction to the tracking problem, and then in Section 9.2 conditions are given for solving this problem in terms of a feedback control configuration. Here the focus is on the case when the reference is a constant signal, called a *set point*. In Section 9.3 the study of closed-loop system behavior as a function of a controller gain is addressed in terms of the root locus, and then in Section 9.4 the root locus is applied to the problem of control system design. Section 9.5 summarizes the chapter.

9.1 INTRODUCTION TO CONTROL

Consider a causal linear time-invariant continuous-time system with input $x(t)$ and output $y(t)$. The system is given by its transfer function representation,

$$Y(s) = G_p(s)X(s) \tag{9.1}$$

where $Y(s)$ is the Laplace transform of the output $y(t)$, $X(s)$ is the Laplace transform of the input, and $G_p(s)$ is the transfer function of the system, which is often called the *plant*. Note the change in notation from $H(s)$ to $G_p(s)$ in denoting the transfer function. Throughout this chapter the transfer function of the given system will be denoted by $G_p(s)$, where the subscript "p" stands for "plant."

In many applications, the objective is to force the output $y(t)$ of the system to follow a desired signal $r(t)$, called the *reference signal*. This is called the *tracking problem;* that is, the objective is to find an input $x(t)$ so that the system output $y(t)$ is equal to (tracks) a desired reference signal $r(t)$. In this problem the input $x(t)$ is called a *control input*. In many cases the reference $r(t)$ is a constant r_0, which is called a *set point*. Hence, in *set-point control*, the objective is to find a control input $x(t)$ so that $y(t) = r_0$ for all t in some desired range of values. The simplest form of control is *open loop control*, where the input $x(t)$ depends only on the reference signal $r(t)$ and not on the output $y(t)$. Ideal tracking for a system with zero initial conditions is $Y(s) = R(s)$, and the corresponding open loop control is found from (9.1):

$$X(s) = \frac{R(s)}{G_p(s)} \qquad (9.2)$$

Here, $R(s)$ is the Laplace transform of the reference $r(t)$. For setpoint control, $R(s) = \dfrac{r_0}{s}$. The expression in (9.2) is known as *model inversion*, since the plant is inverted in the control.

There are practical limitations to implementing the control in (9.2). For example, consider set-point control in the case when $G_p(s) = B(s)/A(s)$, where the degree of $A(s)$ is N and the degree of $B(s)$ is M. If $N > M + 1$, then $X(s)$ in (9.2) has degree of numerator larger than the degree of the denominator. Taking the inverse Laplace transform of $X(s)$ would yield an impulse or a derivative of an impulse, neither of which is possible to generate. The other limitation in (9.2) is that the poles of $sX(s)$ must be in the open left-half plane in order for a limit of $x(t)$ to exist. This result comes from the final value theorem introduced in Chapter 6. If $G_p(s)$ has zeros that are not in the open left-half plane, then the ideal control input $x(t)$ cannot be generated.

A nonideal control $x(t)$ can often be found for set-point control such that $y(t)$ is asymptotic to r_0, that is, $y(t) \to r_0$ as $t \to \infty$. Consider the following general form for $X(s)$:

$$X(s) = \frac{r_0}{s}G_c(s) \qquad (9.3)$$

In this form, $G_c(s)$ is some rational function of s. The resulting system output is

$$Y(s) = \frac{r_0}{s}G_p(s)G_c(s) \qquad (9.4)$$

The speed at which $y(t)$ converges to r_0 is determined by the poles of (9.4). A general approach to choosing $X(s)$ in (9.3) is to select zeros of $G_c(s)$ that cancel slow, but stable, poles of $G_p(s)$. The poles of $G_c(s)$ are then chosen to be faster than those of $G_p(s)$. The following example demonstrates the application of open loop control:

Example 9.1   Open-Loop Control

Consider the system consisting of a vehicle moving on a horizontal surface. The output $y(t)$ at time t is the position of the vehicle at time t relative to some reference. The input $x(t)$ is the drive or braking force applied to the vehicle at time t. It follows from Newton's second law of motion (see Section 2.4) that $y(t)$ and $x(t)$ are related by the following input/output differential equation:

$$\frac{d^2y(t)}{dt^2} + \frac{k_f}{M}\frac{dy(t)}{dt} = \frac{1}{M}x(t)$$

Here, M is the mass of the vehicle and k_f is the coefficient representing frictional losses. In terms of the velocity $v(t) = dy(t)/dt$, the differential equation reduces to

$$\frac{dv(t)}{dt} + \frac{k_f}{M}v(t) = \frac{1}{M}x(t) \tag{9.5}$$

The differential equation (9.5) specifies the velocity model of the vehicle. From (9.5) the transfer function of the velocity model is

$$G_p(s) = \frac{1/M}{s + k_f/M} \tag{9.6}$$

Now, with the output of the system defined to be the velocity $v(t)$, the objective in velocity control is to force $v(t)$ to be equal to a desired speed v_0. Hence, in this problem the reference signal $r(t)$ is equal to the constant v_0, and v_0 is the set point. Suppose that the initial velocity $v(0)$ is zero; then the ideal control is found from (9.2):

$$X(s) = \frac{v_0}{sG_p(s)} = \frac{v_0}{s(1/M)/(s + k_f/M)} = \frac{v_0M(s + k_f/M)}{s}$$

Taking the inverse transform results in the control input

$$x(t) = v_0M\delta(t) + v_0k_f, t \geq 0 \tag{9.7}$$

where $\delta(t)$ is the impulse. Obviously, the control (9.7) cannot be implemented, since it contains an impulse. The presence of the impulse in the control input is a result of the requirement that $v(t) = v_0$ for all $t > 0$ starting from $v(0) = 0$. In other words, the impulse is needed to change the velocity instantaneously from zero to the desired set point v_0.

An asymptotic controller in the form of (9.3) is sought, which can be implemented. A simple form of the control is to let $G_c(s) = K$ where K is a (real) constant. The corresponding control input is $x(t) = v_0Ku(t)$, where $u(t)$ is the unit-step function. Substitute $X(s) = v_0K/s$ into the transfer function representation, which gives

$$V(s) = G_p(s)X(s) = \frac{v_0K/M}{s(s + k_f/M)} = \frac{v_0K/k_f}{s} - \frac{v_0K/k_f}{s + k_f/M}$$

If K is set equal to k_f, then the inverse Laplace transform of $V(s)$ yields

$$v(t) = v_0[1 - e^{-(k_f/M)t}], \quad t \geq 0 \tag{9.8}$$

and since $k_f/M > 0$, it is seen that $v(t)$ converges to v_0 as $t \rightarrow \infty$. Therefore, with the input

$$x(t) = (v_0k_f)u(t) \tag{9.9}$$

set point control is achieved in the limit as $t \rightarrow \infty$. Note that the implementation of the control (9.9) requires that the coefficient k_f in the velocity model (9.5) must be known. Also, the ratio k_f/M will be rather small since the mass of the vehicle will be large and the coefficient k_f corresponding to the viscous friction will be relatively small. Hence, the control $x(t) = v_0k_fu(t)$ will result in a large time constant, $\tau = \dfrac{M}{k_f}$, in the system response given by (9.8).

A faster response can be achieved by proper selection of the poles and zeros of $G_c(s)$. Let the Laplace transform $X(s)$ of the control signal $x(t)$ be given by

$$X(s) = \frac{B(s + k_f/M)}{s + C} R(s) \tag{9.10}$$

where B and C are real constants that are to be determined and $R(s)$ is the transform of the reference signal $r(t) = v_0, t \geq 0$. Inserting (9.10) into the transfer function representation $V(s) = G_p(s)X(s)$ and taking the inverse transform of $V(s)$ yield

$$v(t) = \frac{Bv_0}{CM}(1 - e^{-Ct}), \quad t \geq 0$$

If $C > 0$, it is clear that the velocity $v(t)$ will converge to v_0 if $B = CM$. In addition, we can make the rate at which $v(t)$ converges to v_0 as fast as desired by choosing C to be a suitably large positive number. Hence, with $B = CM$ the control signal with transform $X(s)$ given by (9.10) achieves the objective of forcing the velocity $v(t)$ to converge to the set point v_0 with any desired rate of convergence. However, as discussed subsequently, this type of control (i.e., open-loop control) is susceptible to influences on the system output that may result from unknown disturbances applied to the system. For example, the control defined by (9.10) may not work well when gravity acts on the vehicle in going up or down hills.

A major problem with open-loop control is that it is sensitive to modeling errors and to disturbances that may be applied to the system. Consider the control designed in Example 9.1 and given in (9.10) where $B = CM$. The values of plant parameters k_f and M need to be known exactly for the control to work perfectly. There is always some error in plant parameters, and a control law should be designed that is *robust*, that is, not sensitive to these modeling errors. The control law should also be robust to disturbances that may be applied to the system. A block diagram of open-loop control is given in Figure 9.1a, where $d(t)$ is a disturbance. Note that the transform $Y(s)$ of the plant output is given by

$$Y(s) = G_p(s)G_c(s)R(s) + G_p(s)D(s)$$

where $R(s)$ is the transform of the reference signal $r(t)$ and $D(s)$ is the transform of the disturbance $d(t)$. It is clear that the output $y(t)$ of the plant will be "perturbed" by the disturbance input $d(t)$.

To improve robustness, it is necessary that the control signal $x(t)$ depend directly on the plant output $y(t)$. This requires that $y(t)$ be measurable by some type of a sensor, in which case the measured output can be compared with the desired output $r(t)$. This results in the *tracking error* $e(t)$ given by

$$e(t) = r(t) - y(t) \tag{9.11}$$

which can be "fed back" to form the control signal $x(t)$. More precisely, in *feedback control* the error signal $e(t)$ is applied to the controller or compensator with transfer function $G_c(s)$ to yield the control signal $x(t)$ for the given system. A block diagram for the closed-loop control process is given in Figure 9.1b. The overall system shown in Figure 9.1b is called the *closed-loop system*, since it is formed from the given system with transfer function $G_p(s)$ by "closing the loop" around $G_p(s)$ and $G_c(s)$. In the remainder of this chapter, the focus is on closed-loop control.

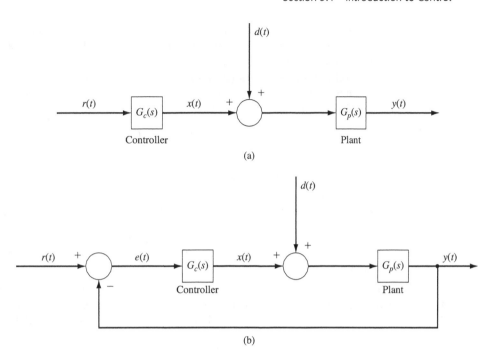

FIGURE 9.1
Block diagram of (a) open-loop control and (b) closed-loop control.

In the feedback control scheme shown in Figure 9.1b, the objective is to design the transfer function $G_c(s)$ of the controller so that the tracking error $e(t)$ converges to zero as $t \to \infty$, which is equivalent to requiring that the output $y(t)$ converges to the reference $r(t)$ as $t \to \infty$. This is sometimes referred to as *asymptotic tracking*, since $y(t) = r(t)$ occurs in the limit as $t \to \infty$. The solution of the tracking problem by the use of the configuration shown in Figure 9.1b is referred to as *output feedback control*, since the system output $y(t)$ is fed back to the input; that is, the control signal $x(t)$ applied to the given system depends on the system output $y(t)$. The dependence of $x(t)$ on $y(t)$ is seen from the transform relationship

$$X(s) = G_c(s)E(s) = G_c(s)[R(s) - Y(s)] \tag{9.12}$$

which follows directly from the block diagram in Figure 9.1b. In (9.12), $R(s)$ is the transform of the reference $r(t)$ and $E(s)$ is the transform of the tracking error $e(t)$.

The simplest type of controller is the one with transfer function $G_c(s) = K_P$, where K_P is a (real) constant. In this case, (9.12) becomes

$$X(s) = K_P E(s) = K_P[R(s) - Y(s)]$$

and taking the inverse transform yields the control signal

$$x(t) = K_P e(t) = K_P[r(t) - y(t)] \tag{9.13}$$

The control given by (9.13) is called *proportional control*, since the control signal $x(t)$ is directly proportional to the error signal $e(t)$. This explains the subscript "P" in K_P, which stands for "proportional."

Combining (9.13) with (9.1) and using the result on the feedback connection in Section 6.6 yield the following closed-loop transfer function for a plant $G_p(s)$ with proportional control:

$$\frac{Y(s)}{R(s)} = \frac{K_P G_P(s)}{1 + K_P G_P(s)} \tag{9.14}$$

Perfect tracking for all input signals would be achieved if $Y(s) = R(s)$, in other words, if the closed-loop transfer function is equal to 1. A proportional controller results in a nonzero tracking error; however, the closed-loop transfer function gets closer to unity as the value of K_P increases, thus improving the tracking accuracy of the controller.

Example 9.2   Proportional Control

Again, consider a vehicle on a level surface with the velocity model given by (9.5), and suppose that the goal is to force the velocity $v(t)$ to track a desired speed v_0, so that $r(t) = v_0 u(t)$. In this case the velocity $v(t)$ can be measured by a speedometer, and thus the tracking error

$$e(t) = r(t) - v(t) = v_0 - v(t)$$

can be computed. With proportional feedback control, the control signal $x(t)$ applied to the car is given by

$$x(t) = K_P[v_0 - v(t)] \tag{9.15}$$

With the control (9.15), the closed-loop transfer function from (9.14) is

$$\frac{V(s)}{R(s)} = \frac{K_P/M}{s + k_f/M + K_P/M}$$

With $R(s) = v_0/s$, $V(s)$ is found to be

$$V(s) = \frac{-K_P v_0/(k_f + K_P)}{s + k_f/M + K_P/M} + \frac{K_P v_0/(k_f + K_P)}{s} \tag{9.16}$$

Inverse transforming (9.16) yields the response

$$v(t) = -\frac{K_P v_0}{k_f + K_P} e^{-[(k_f + K_P)/M]t} + \frac{K_P v_0}{k_f + K_P}, \quad t \geq 0 \tag{9.17}$$

From (9.17) it is seen that, if $(k_f + K_P)/M > 0$, then $v(t)$ converges to $K_P v_0/(k_f + K_P)$. Since there is no finite value of K_P for which $K_P/(k_f + K_P) = 1$, the proportional controller will result in a *steady-state tracking error* equal to

$$v_0 - \frac{K_P v_0}{k_f + K_P} = \left(1 - \frac{K_P}{k_f + K_P}\right)v_0 = \frac{k_f}{k_f + K_P}v_0 \tag{9.18}$$

However, we can make the tracking error given by (9.18) as small as desired by taking K_P to be suitably large, compared with k_f. As will be seen from results given in the next section, it is possible to obtain a zero steady-state error by modifying the proportional controller.

From (9.17) we see that we can make the rate at which $v(t)$ converges to the steady-state value as fast as desired by again taking K_P to be suitably large. To see this, suppose that $k_f = 10$, $M = 1000$, and $v_0 = 60$. Then the transform $V(s)$ given by (9.16) becomes

$$V(s) = \frac{0.06K_P}{[s + 0.01(1 + 0.1K_P)]s}$$

The resulting velocity $v(t)$ can be computed in MATLAB by the step command, the Symbolic Math Toolbox, or Simulink, as demonstrated in Examples 6.34–6.36. The following commands are used with step:

```
num = 0.06*Kp;
den = [1 0.01+0.001*Kp];
H = tf(num,den);
step(H)
```

Running the MATLAB software with $K_P = 100$, 200, and 500 results in the velocity responses shown in Figure 9.2. Note that the fastest response with the smallest steady-state error is achieved when $K_P = 500$, the largest value of K_P. For $K_P = 500$, the steady-state error is

$$\frac{k_f}{k_f + K_P}v_0 = \frac{10}{510}60 = 1.176$$

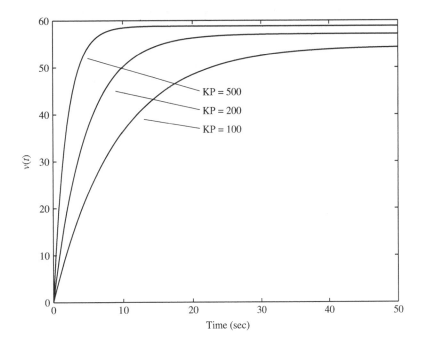

FIGURE 9.2
Velocity responses in Example 9.2 for $K_P = 100, 200,$ and 500.

Now, suppose that a step disturbance force $d(t) = 50u(t)$ is applied to the vehicle at time $t = 0$, where the disturbance may be a result of the vehicle going down a long incline. If the open-loop control scheme shown in Figure 9.1a is used, the transform $V(s)$ of the velocity will be given by

$$V(s) = G_p(s)G_c(s)R(s) + G_p(s)\frac{50}{s}$$

Thus, the step disturbance will result in a perturbation in the velocity $v(t)$ of the car equal to the inverse transform of

$$G_p(s)\frac{50}{s} = \frac{(0.001)50}{(s + 0.01)s} = \frac{5}{s} - \frac{5}{s + 0.01}$$

which is equal to

$$5(1 - e^{-0.01t}), t \geq 0$$

Hence, the disturbance will result in a sizable error in achieving the desired set point of $v_0 = 60$. In contrast, if a step disturbance input $d(t)$ is applied to the vehicle with the feedback control $x(t)$ given by (9.15), the transform $V(s)$ of the velocity is given by

$$V(s) = G_p(s)X(s) + G_p(s)\frac{50}{s} = K_P G_p(s)\left[\frac{v_0}{s} - V(s)\right] + G_p(s)\frac{50}{s} \qquad (9.19)$$

Solving (9.19) for $V(s)$ gives

$$V(s) = \frac{K_P G_p(s)}{1 + K_P G_p(s)}\frac{v_0}{s} + \frac{G_p(s)}{1 + K_P G_p(s)}\frac{50}{s}$$

and thus the perturbation of the velocity resulting from a step disturbance is equal to the inverse transform of

$$\frac{G_p(s)}{1 + K_P G_p(s)}\frac{50}{s} = \frac{50/M}{(s + k_f/M + K_P/M)s}$$

For $k_f = 10$, $M = 1000$, and $K_P = 500$, the perturbation is

$$0.098(1 - e^{-0.51t}), t \geq 0$$

Obviously, this term is much smaller than in the case of open-loop control, and thus in this example, closed-loop control is much more "robust" to step a disturbance than is open-loop control.

9.1.1 MATLAB Simulation of the Closed-Loop System

One method of simulating a closed-loop system is demonstrated in Example 9.2, which uses the `step` command available with the Control System Toolbox. The demonstrated method requires computing the closed-loop transfer function analytically. As mentioned in Example 9.2, if the closed-loop transfer function is known, the simulation methods demonstrated in Examples 6.34–6.36 can be used. Alternatively, we can use MATLAB without explicitly determining the closed-loop transfer function.

If the Control System Toolbox is available, use the following commands:

```
GcGp = tf(conv(Bp,Bc),conv(Ap,Ac));
Gcl = feedback(GcGp,1);
step(Gcl);
```

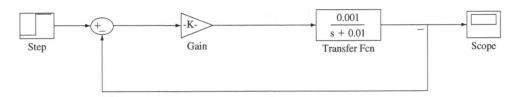

FIGURE 9.3
Simulink model of closed-loop system with proportional feedback in Example 9.2

Here, the vectors Bp and Bc hold the numerator coefficients of $G_p(s)$ and $G_c(s)$, respectively; and Ap and Ac hold the denominator coefficients of $G_p(s)$ and $G_c(s)$, respectively.

If Simulink is available, build the block diagram of the feedback connection shown in Figure 9.3. Use the description of Simulink given in Example 6.36 to build the model. The summing junction and the gain block are found from the "Math Operations" menu in the Simulink library. Note that the default for the summing junction is to have two positive inputs, but negative feedback requires that the feedback term be negated. Double click on the summing junction to set the sign of the second term to be negative. The feedback shown in Figure 9.3 is for proportional control. To simulate a more general control given by $G_c(s)$, replace the gain block with a transfer function block. Click "Simulation," then "Start" to simulate the system. Double click on the scope to see the results.

9.2   TRACKING CONTROL

Given a system with transfer function $G_p(s)$, a controller with transfer function $G_c(s)$, and a reference signal $r(t)$, again consider the feedback control configuration shown in Figure 9.1b. Throughout this section it is assumed that both the system and the controller are finite dimensional, and thus $G_p(s)$ and $G_c(s)$ are rational functions of s given by

$$G_p(s) = \frac{B_p(s)}{A_p(s)} \quad \text{and} \quad G_c(s) = \frac{B_c(s)}{A_c(s)} \qquad (9.20)$$

where $B_p(s)$, $A_p(s)$, $B_c(s)$, and $A_c(s)$ are polynomials in s with the degree of $A_p(s)$ equal to N and the degree of $A_c(s)$ is equal to q. Hence, the given system has N poles, and the controller has q poles.

As discussed in Section 9.1, in tracking control the objective is to design the controller transfer function $G_c(s)$ so that the tracking error $e(t) = r(t) - y(t)$ converges to zero as $t \to \infty$. The solution to this problem involves the closed-loop poles; that is, the poles of the closed-loop system. The closed-loop poles are determined as follows:

First, from the block diagram in Figure 9.1b, when $d(t) = 0$, it is clear that the transform $Y(s)$ of the output $y(t)$ is given by

$$Y(s) = G_p(s)G_c(s)[R(s) - Y(s)] \qquad (9.21)$$

where $R(s)$ is the transform of the reference input $r(t)$. Solving (9.21) for $Y(s)$ yields

$$Y(s) = \frac{G_p(s)G_c(s)}{1 + G_p(s)G_c(s)}R(s) \tag{9.22}$$

It is worth noting that the closed-loop transfer function representation (9.22) follows directly from the results in Section 6.6.

With $G_{cl}(s)$ defined to be the closed-loop transfer function, from (9.22),

$$G_{cl}(s) = \frac{G_p(s)G_c(s)}{1 + G_p(s)G_c(s)} \tag{9.23}$$

Then, inserting (9.20) into (9.23) results in the following expression for the closed-loop transfer function:

$$G_{cl}(s) = \frac{[B_p(s)/A_p(s)][B_c(s)/A_c(s)]}{1 + [B_p(s)/A_p(s)][B_c(s)/A_c(s)]}$$

$$G_{cl}(s) = \frac{B_p(s)B_c(s)}{A_p(s)A_c(s) + B_p(s)B_c(s)} \tag{9.24}$$

From (9.24), it is seen that p is a pole of the closed-loop system if and only if

$$A_p(p)A_c(p) + B_p(p)B_c(p) = 0$$

Therefore, the closed-loop poles are the roots of the polynomial equation

$$A_p(s)A_c(s) + B_p(s)B_c(s) = 0 \tag{9.25}$$

Note that the degree of the polynomial in (9.25) is equal to $N + q$, where N is the degree of $A_p(s)$ and q is the degree of $A_c(s)$. Hence, the number of closed-loop poles is equal to $N + q$, which is the sum of the number of poles of the given system and the number of poles of the controller.

Example 9.3   Calculation of Closed-Loop Transfer Function

Suppose that

$$G_p(s) = \frac{s + 2}{s^2 + 4s + 10}, \qquad G_c(s) = \frac{s + 1}{s(s + 10)}$$

Then, both $G_p(s)$ and $G_c(s)$ have two poles, which implies that the closed-loop system has four poles. Now, we see that

$$B_p(s) = s + 2, \quad A_p(s) = s^2 + 4s + 10$$
$$B_c(s) = s + 1, \quad A_c(s) = s(s + 10)$$

and inserting this into (9.24) gives the closed-loop transfer function

$$G_{cl}(s) = \frac{\left(\dfrac{s+1}{s(s+10)}\right)\left(\dfrac{s+2}{s^2+4s+10}\right)}{1 + \left(\dfrac{s+1}{s(s+10)}\right)\left(\dfrac{s+2}{s^2+4s+10}\right)}$$

$$G_{cl}(s) = \frac{s^2 + 3s + 2}{s^4 + 14s^3 + 51s^2 + 103s + 2}$$

The closed-loop poles are the roots of the equation

$$s^4 + 14s^3 + 51s^2 + 103s + 2 = 0$$

Using the MATLAB command `roots` reveals that the closed-loop poles are -0.0196, -9.896, $-2.042 \pm j2.477$.

Again, consider the tracking error $e(t) = r(t) - y(t)$. With $E(s)$ equal to the Laplace transform of the tracking error $e(t)$, from the block diagram in Figure 9.1b it is seen that

$$E(s) = R(s) - Y(s) \tag{9.26}$$

Inserting the expression (9.22) for $Y(s)$ into (9.26) yields

$$E(s) = R(s) - \frac{G_p(s)G_c(s)}{1 + G_p(s)G_c(s)} R(s)$$

$$= \frac{1}{1 + G_p(s)G_c(s)} R(s) \tag{9.27}$$

Then, inserting (9.20) into (9.27) gives

$$E(s) = \frac{1}{1 + [B_p(s)/A_p(s)][B_c(s)/A_c(s)]} R(s)$$

$$= \frac{A_p(s)A_c(s)}{A_p(s)A_c(s) + B_p(s)B_c(s)} R(s) \tag{9.28}$$

From the analysis given in Section 8.3, the tracking error $e(t)$ converges to zero as $t \rightarrow \infty$ if and only if all the poles of $E(s)$ are located in the open left-half plane. (This result also follows directly from the final value theorem.) From (9.28) it is seen that the poles of $E(s)$ include the closed-loop poles, that is, the values of s for which

$$A_p(s)A_c(s) + B_p(s)B_c(s) = 0$$

As a result, a necessary condition for $e(t) \rightarrow 0$ is that the closed-loop system must be stable, and thus all the closed-loop poles must be located in the open left-half plane. It is important to stress that, although stability of the closed-loop system is necessary for tracking, it is not sufficient for tracking. Additional conditions that guarantee tracking depend on the reference signal $r(t)$. This is investigated in detail next for the case of a step input.

9.2.1   Tracking a Step Reference

Suppose that the reference input $r(t)$ is equal to $r_0 u(t)$, where r_0 is a real constant and $u(t)$ is the unit-step function. As discussed in Section 9.1, this case corresponds to set-point control, where the constant r_0 is the set point. Note that when $r(t) = r_0 u(t)$, the resulting output response $y(t)$ with zero initial conditions is equal to r_0 times the step response of the closed-loop system. When $R(s) = r_0/s$, the expression (9.28) for the transform $E(s)$ of the tracking error becomes

$$E(s) = \frac{A_p(s)A_c(s)}{A_p(s)A_c(s) + B_p(s)B_c(s)} \frac{r_0}{s} \tag{9.29}$$

In this case, the poles of $E(s)$ are equal to the poles of the closed-loop system plus a pole at $s = 0$. Therefore, if the closed-loop system is stable, the conditions for applying the final value theorem to $E(s)$ are satisfied. Thus, the limiting value of $e(t)$ as $t \to \infty$ can be computed by the final-value theorem, which gives

$$\lim_{t \to \infty} e(t) = \lim_{s \to 0} sE(s) = \frac{A_p(0)A_c(0)r_0}{A_p(0)A_c(0) + B_p(0)B_c(0)} \tag{9.30}$$

With the steady-state error e_{ss} defined by

$$e_{ss} = \lim_{t \to \infty} e(t)$$

from (9.30) it is seen that

$$e_{ss} = \frac{A_p(0)A_c(0)r_0}{A_p(0)A_c(0) + B_p(0)B_c(0)} \tag{9.31}$$

The steady-state error e_{ss} can be written in the form

$$e_{ss} = \frac{1}{1 + [B_p(0)/A_p(0)][B_c(0)/A_c(0)]} r_0 \tag{9.32}$$

and since

$$G_p(0) = \frac{B_p(0)}{A_p(0)} \quad \text{and} \quad G_c(0) = \frac{B_c(0)}{A_c(0)}$$

(9.32) can be expressed in the form

$$e_{ss} = \frac{1}{1 + G_p(0)G_c(0)} r_0 \tag{9.33}$$

From (9.33) it is clear that the steady-state error is zero if and only if

$$G_p(0)G_c(0) = \infty \tag{9.34}$$

which is the case if $G_p(s)G_c(s)$ has a pole at $s = 0$. The open-loop system defined by the transfer function $G_p(s)G_c(s)$ is said to be a *type 1 system* if $G_p(s)G_c(s)$ has a single pole at $s = 0$. Hence, the system given by (9.22) will track the step input $r_0 u(t)$ if

$G_p(s)G_c(s)$ is a type 1 system. In addition, it follows from the results in Section 8.3 that the rate at which the error $e(t)$ approaches zero depends on the location of the closed-loop poles in the open left-half plane. In particular, the farther over in the left-half plane the closed-loop poles are, the faster the rate of convergence of $e(t)$ to zero will be. Also note that, since the system output $y(t)$ is equal to $r_0 - e(t)$, the rate at which $y(t)$ converges to the set point r_0 is the same as the rate at which $e(t)$ converges to zero.

The open-loop system $G_p(s)G_c(s)$ is said to be a *type 0 system* if $G_p(s)G_c(s)$ does not have any poles at $s = 0$. This is equivalent to requiring that $G_p(0)G_c(0) \neq \infty$. Thus, from the prior analysis, it is clear that when the reference signal $r(t)$ is a step function and $G_p(s)G_c(s)$ is a type 0 system, the closed-loop system (9.22) will have a nonzero steady-state tracking error e_{ss} given by (9.33).

Suppose that the goal of the controller is to have zero tracking error for a step reference, but the original plant $G_p(s)$ does not have a pole at $s = 0$. To achieve the goal, the controller must have a pole at $s = 0$, so that the product $G_p(s)G_c(s)$ is type 1. A common type of controller used to achieve this goal is a *proportional plus integral* (PI) controller, which is given by

$$G_c(s) = K_P + \frac{K_I}{s} = \frac{K_P s + K_I}{s} \tag{9.35}$$

where K_P and K_I are real constants. In this case, the transform $X(s)$ of the control input $x(t)$ applied to the plant is given by

$$X(s) = G_c(s)E(s) = K_p E(s) + \frac{K_I}{s}E(s) \tag{9.36}$$

Inverse transforming (9.36) gives

$$x(t) = K_p e(t) + K_I \int_0^t e(\tau)\, d\tau \tag{9.37}$$

The first term on the right-hand side of (9.37) corresponds to proportional control (as discussed in Section 9.1), while the second term corresponds to integral control, since this term is given in terms of the integral of the error $e(t)$. Thus, the subscript "I" in K_I stands for "integral." With the controller transfer function (9.35), the transform $E(s)$ of the error given by equation (9.28) becomes

$$E(s)\frac{A_p(s)s}{A_p(s)s + B_p(s)(K_p s + K_I)}R(s)$$

Obviously, the coefficients of the denominator polynomial of $E(s)$ depend upon K_P and K_I. Thus, the poles of $E(s)$, which are the poles of the closed-loop system, can be modified through the selection of K_P and K_I. Hence, the objective of the control designer is to select values for K_P and K_I that result in closed loop poles that have an acceptable rate of convergence to zero of the error $e(t)$, or, equivalently, the rate of convergence of the output $y(t)$ to the reference r_0.

These results are illustrated by the following examples:

Example 9.4   Proportional plus Integral Control

Again consider velocity control of a vehicle moving on a level surface, as studied in Examples 9.1 and 9.2. The goal is to have the velocity $v(t)$ of the vehicle converge to a desired velocity v_0 as $t \to \infty$. Recall that the transfer function $G_p(s)$ of the velocity model of the vehicle is given by $G_p(s) = (1/M)/(s + k_f/M)$. In Example 9.2 proportional control was considered where $G_c(s) = K_P$. In this case

$$G_p(s)G_c(s) = \frac{K_P/M}{s + k_f/M}$$

Clearly, $G_p(s)G_c(s)$ does not have a pole at $s = 0$, and the open-loop system $G_p(s)G_c(s)$ is type 0. Hence, as first observed in Example 9.2, there is a steady-state error e_{ss} in tracking the step input $v_0 u(t)$. From (9.33), e_{ss} is given by

$$e_{ss} = \frac{1}{1 + K_P/k_f} v_0 = \frac{k_f v_0}{k_f + K_P}$$

This checks with the result obtained in Example 9.2. [See (9.18).]

In order to meet the objective of zero steady-state error to a step reference, a PI controller of the form given in (9.35) is used. The transform of the error $E(s)$ is found from (9.27) to be

$$E(s) = \left[\frac{s(s + k_f/M)}{s(s + k_f/M) + (K_P s + K_I)(1/M)} \right] \frac{v_0}{s}$$

$$= \frac{(s + k_f/M)v_0}{s^2 + (k_f/M + K_P/M)s + K_I/M}$$

In this example, the design parameters K_P and K_I can be chosen to place the poles of $E(s)$ (equivalently, the closed-loop poles) arbitrarily in the open left-half plane. Hence, we can make the rate of convergence to zero of the error $e(t)$, or equivalently, the rate of convergence of $v(t)$ to v_0, as fast as desired by selecting values for K_P and K_I. To investigate this, suppose that $k_f = 10$, $M = 1000$, and $v_0 = 60$. Then, the transform $V(s)$ of the velocity response is

$$V(s) = G_{cl}(s)\frac{v_0}{s} = \frac{G_p(s)G_c(s)}{1 + G_p(s)G_c(s)}\frac{60}{s} = \frac{0.06(K_P s + K_I)}{[s^2 + 0.01(1 + 0.1K_P)s + 0.001K_I]s}$$

When $K_I = 0$, so that there is no integral control, the response $v(t)$ was computed in Example 9.2 for three different values of K_P. (See Figure 9.2.) As K_I is taken to be suitably large, the integral control action will eliminate the steady-steady state error seen in Figure 9.2. For instance, for $K_P = 500$ and $K_I = 1, 5$, and 10, the responses are plotted in Figure 9.4. Note that when $K_I = 1$, the integral action is not sufficiently strong to bring the velocity up to 60 during the 50-second time interval in the plot, although it is true that $v(t)$ is converging to 60 in the limit as $t \to \infty$. Note that for $K_I = 5$ the velocity reaches 60 in about 12 seconds, and increasing K_I to 10 results in $v(t)$ getting up to 60, but now there is some overshoot in the response. Clearly, $K_I = 5$ yields the best response, and there is a reason for this; namely, when $K_I = 5$, the controller transfer function $G_c(s) = (500s + 5)/s$ has a zero at $s = -5/500 = -0.01$, and this zero cancels the pole of $G_p(s)$ at $s = -0.01$. This results in a first-order closed-loop system with transfer function

$$G_{cl}(s) = \frac{\left(\dfrac{500s + 5}{s}\right)\left(\dfrac{0.001}{s + 0.01}\right)}{1 + \left(\dfrac{500s + 5}{s}\right)\left(\dfrac{0.001}{s + 0.01}\right)} = \frac{0.5}{s + 0.5}$$

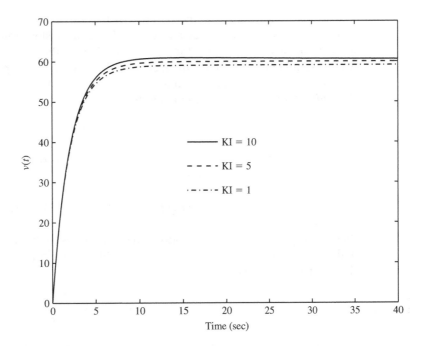

FIGURE 9.4
Velocity responses in Example 9.4.

Thus, when $K_P = 500$ and $K_I = 5$, the prior expression for $V(s)$ reduces to a first-order rational function; in other words, there is a pole–zero cancellation in the expression for $V(s)$. For these values of K_P and K_I,

$$V(s) = G_{cl}(s)\frac{v_0}{s} = \left(\frac{0.5}{s + 0.5}\right)\frac{60}{s} = 60\left(\frac{1}{s} - \frac{1}{s + 0.5}\right)$$

Then, taking the inverse Laplace transform yields

$$v(t) = 60(1 - e^{-0.5t}), t \geq 0$$

Hence, the transient $y_{tr}(t) = -60 \exp(-0.5t)$ is a simple exponential that decays to zero with time constant $1/0.5 = 2$ seconds. Therefore, the velocity $v(t)$ converges to 60 with the rate of convergence corresponding to a time constant of 2 seconds. The key point here is that, by choosing $G_c(s)$ so that it cancels the pole in $G_p(s)$, we cause the closed-loop system to become a first-order system, which is much easier to deal with than a second-order system. It is common in practice to design the controller transfer function $G_c(s)$ so that it cancels one or more stable poles of the plant transfer function $G_p(s)$. Another example of this is given in Section 9.4.

In Example 9.4, the design parameters K_I and K_P of the PI controller could be chosen to place the closed-loop poles arbitrarily in the open left-half plane. This was true with the vehicle-control example, because the plant was first order. When the plant is higher than first order, the closed-loop poles cannot be made arbitrarily fast (that is, placed farther left in the s-plane) by the use of a PI controller or a proportional controller, $G_c(s) = K_P$. In fact, a PI controller tends to slow down the response, that is, to result in closed-loop poles that are to the right of the closed-loop poles achievable when a simple proportional controller is used. If the goal is to speed up the closed-loop response, even over that achievable with a proportional controller, a *proportional plus derivative controller* is often used; that is,

$$G_c(s) = K_P + K_D s \tag{9.38}$$

where K_P and K_D are constants. In this case, the transform $X(s)$ of the control signal applied to the plant is given by

$$X(s) = G_c(s)E(s) = K_P E(s) + K_D s E(s) \tag{9.39}$$

Since multiplication by s in the s-domain corresponds to differentiation in the time domain, inverse transforming both sides of (9.39) results in the following expression for the control signal $x(t)$:

$$x(t) = K_P e(t) + K_D \frac{de(t)}{dt} \tag{9.40}$$

The first term on the right-hand side of (9.40) corresponds to proportional control, while the second term corresponds to derivative control, since this term is given in terms of the derivative of the tracking error $e(t)$. (Thus, the subscript "D" in K_D stands for "derivative.")

Example 9.5   Proportional plus Derivative Control

A problem that arises in many applications is controlling the position of an object. An example is controlling the angular position of a valve in some chemical process, or controlling the angular position of a circular plate used in some manufacturing operation, such as drilling or component insertion. In such applications, a fundamental problem is controlling the angular position of the shaft of a motor used to drive a specific mechanical structure (such as a valve or plate). Often, the motor used is a field-controlled dc (direct current) motor, which is illustrated in Figure 9.5. The load indicated in the figure is the structure (valve, plate, etc.) to which the motor shaft is connected. The input to the motor is the voltage $v_f(t)$ applied to the field circuit, and the output is the angle $\theta(t)$ of the motor shaft. As shown in Section 2.4, the torque $T(t)$ developed by the motor is related to the angle $\theta(t)$ by the differential equation

$$I\frac{d^2\theta(t)}{dt^2} + k_d\frac{d\theta(t)}{dt} = T(t) \tag{9.41}$$

where I is the moment of inertia of the motor and load and k_d is the viscous friction coefficient of the motor and load. In the usual approximation of motor dynamics, it is assumed that the torque $T(t)$ is given approximately by

$$T(t) = k_m v_f(t) \tag{9.42}$$

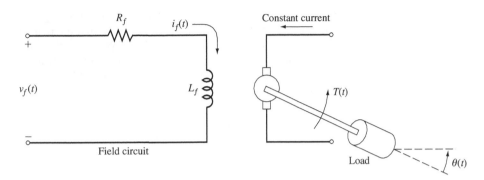

FIGURE 9.5
Field-controlled dc motor with load.

where k_m is the motor constant, which is strictly positive ($k_m > 0$). Inserting (9.42) into (9.41) gives

$$I\frac{d^2\theta(t)}{dt^2} + k_d\frac{d\theta(t)}{dt} = k_m v_f(t) \tag{9.43}$$

which is the input/output differential equation for the dc motor. Taking the Laplace transform of both sides of (9.43) results in the following transfer function representation for the dc motor:

$$\Theta(s) = \frac{k_m/I}{(s + k_d/I)s}V_f(s) \tag{9.44}$$

In this equation, $\Theta(s)$ is the transform of $\theta(t)$ and $V_f(s)$ is the transform of $v_f(t)$. In this case the system (the dc motor with load) has a pole at $s = 0$, and thus for any controller transfer function $G_c(s) = B_c(s)/A_c(s)$, the open-loop system $G_p(s)G_c(s)$ will be type 1. Hence, it is not necessary that $G_c(s)$ have a pole at $s = 0$ in order to track a step input $r(t) = \theta_0 u(t)$, where θ_0 is the desired angular position of the motor shaft. Choosing the simplest possible $G_c(s)$ results in the proportional controller given by $G_c(s) = K_P$, where K_P is a real constant. In this case, from (9.28) the transform $E(s)$ of the tracking error is

$$E(s) = \left[\frac{(s + k_d/I)s}{(s + k_d/I)s + k_m K_P/I}\right]\frac{\theta_0}{s}$$

$$= \frac{(s + k_d/I)\theta_0}{s^2 + (k_d/I)s + k_m K_P/I} \tag{9.45}$$

It follows from the Routh–Hurwitz stability test that the two poles of $E(s)$ are in the open left-half plane if and only if $k_d/I > 0$ and $k_m K_P/I > 0$. Since $k_d > 0$, $I > 0$, and $k_m > 0$, this condition is equivalent to $K_P > 0$. Therefore, for any value of $K_P > 0$, the tracking error $e(t) = \theta_0 - \theta(t)$ converges to zero, which implies that $\theta(t) \to \theta_0$.

Although the error $e(t)$ converges to zero for any $K_P > 0$, we cannot obtain an arbitrarily fast rate of convergence to zero by choosing K_P. In other words, we cannot place the poles of $E(s)$ arbitrarily far over in the left-half plane by choosing K_P. This follows directly from the expression (9.45) for $E(s)$, from which it is seen that the real parts of both of the two poles of $E(s)$ cannot be more negative than $-k_d/2I$. We can verify this by applying the quadratic formula to the polynomial $s^2 + (k_d/I)s + k_m K_P/I$ in the denominator of $E(s)$. A suitably fast rate of convergence of $e(t)$ to zero can be achieved by the use of a PD controller of the form (9.38).

With the PD controller given by (9.38), the transform $E(s)$ of the tracking error becomes

$$E(s) = \left[\frac{(s + k_d/I)s}{(s + k_d/I)s + (k_m/I)(K_P + K_D s)}\right]\frac{\theta_0}{s}$$

$$= \frac{(s + k_d/I)\theta_0}{s^2 + (k_d/I + k_m K_D/I)s + k_m K_P/I} \tag{9.46}$$

From (9.46) it is clear that the coefficients of the denominator polynomial of $E(s)$ can be chosen arbitrarily by the selection of K_P and K_D, and thus the poles of $E(s)$ and the poles of the closed-loop system can be placed anywhere in the open left-half plane. Therefore, the rate of convergence of $\theta(t)$ to θ_0 can be made as fast as desired by the choice of appropriate values of K_P and K_D. The form of the transient part of the response to a step input depends on the location of closed-loop poles. To investigate this, the output response $\theta(t)$ will be computed in the case when $I = 1$, $k_d = 0.1$, and $k_m = 10$. With these values for the system parameters, the closed-loop transfer function is

$$G_{\text{cl}}(s) = \frac{G_p(s)G_c(s)}{1 + G_p(s)G_c(s)}$$

$$= \frac{[10/s(s + 0.1)](K_P + K_D s)}{1 + [10/s(s + 0.1)](K_P + K_D s)}$$

$$= \frac{10(K_P + K_D s)}{s^2 + (0.1 + 10K_D)s + 10K_P}$$

Setting

$$s^2 + (0.1 + 10K_D)s + 10K_P = s^2 + 2\zeta\omega_n s + \omega_n^2 \tag{9.47}$$

results in the following form for $G_{\text{cl}}(s)$:

$$G_{\text{cl}}(s) = \frac{10(K_P + K_D s)}{s^2 + 2\zeta\omega_n s + \omega_n^2}$$

Except for the zero at $s = -K_P/K_D$, $G_{\text{cl}}(s)$ has the same form as the second-order transfer function studied in Section 8.3. If the effect of the zero is ignored, the analysis in Section 8.3 of the step response of the second-order case can be applied here. In particular, from the results in Section 8.3, it was shown that when the damping ratio ζ is between 0 and 1, the transient in the step response decays to zero at a rate corresponding to the exponential factor $\exp(-\zeta\omega_n t)$. [See (8.24).] Selecting $\zeta\omega_n = 1$ and using (9.47) gives

$$\zeta\omega_n = 1 = \frac{0.1 + 10K_D}{2}$$

Solving for K_D yields $K_D = 0.19$. Now, to avoid a large overshoot in the step response, the damping ratio ζ should not be smaller than $1/\sqrt{2}$. With $\zeta = 1/\sqrt{2}$ and $\zeta\omega_n = 1$, then $\omega_n = \sqrt{2}$, and using (9.47) gives

$$10K_P = \omega_n^2 = 2$$

and thus $K_P = 0.2$. The closed-loop transfer function $G_{cl}(s)$ is then

$$G_{cl}(s) = \frac{1.9s + 2}{s^2 + 2s + 2}$$

With $\theta_0 = 1$, the step response was then computed by the MATLAB command step(Gcl) with

```
num = [1.9 2]; den = [1 2 2];
Gcl = tf(num,den);
```

A plot of the response is given in Figure 9.6. Note that the overshoot is fairly pronounced. To reduce this, the damping ratio ζ should be increased. For example, setting $\zeta = 0.9$, but keeping $\zeta\omega_n = 1$ results in the values $K_P = 0.123$ and $K_D = 0.19$. The resulting step response is also shown in Figure 9.6. Note that the overshoot is smaller, but now the response is more "sluggish." To achieve a faster response, $\zeta\omega_n$ could be made larger than 1. The reader is invited to try this.

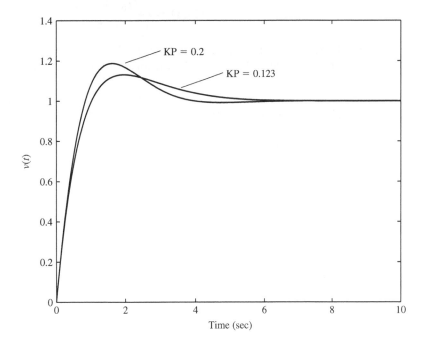

FIGURE 9.6
Step response with $K_D = 0.19$, and $K_P = 0.2$ and 0.123.

Consider the control signal $x(t)$ given in (9.40). For a step reference, $e(t) = r_0 - y(t)$. Therefore, (9.40) can be expressed in the form

$$x(t) = K_P e(t) - K_D \frac{dy(t)}{dt} \tag{9.48}$$

Hence, to implement the control (9.48), it is necessary to measure the derivative $dy(t)/dt$ of the output $y(t)$. Unfortunately, this is often not possible to do in practice as a result of the presence of high-frequency noise in $y(t)$. For example, suppose that $y(t)$ contains a very small noise component equal to $10^{-4} \sin 10^6 t$. When this term is differentiated, the result is $100 \sin 10^6 t$, which is not small in magnitude and can "swamp out" the signal terms. In practice, PD controllers are often implemented with an additional high-frequency filter to mitigate the effects of high-frequency noise.

9.3   ROOT LOCUS

Again, consider the feedback control system with the transfer function representation

$$Y(s) = G_{cl}(s)R(s) \tag{9.49}$$

where the closed-loop transfer function $G_{cl}(s)$ is given by

$$G_{cl}(s) = \frac{G_p(s)G_c(s)}{1 + G_p(s)G_c(s)} \tag{9.50}$$

The closed-loop system is shown in Figure 9.7.

It is still assumed that the plant transfer function $G_p(s)$ has N poles and the controller transfer function $G_c(s)$ has q poles. Then, if there are no pole–zero cancellations, the product $G_p(s)G_c(s)$ has $N + q$ poles, which are equal to the poles of the plant plus the poles of the controller. In addition, the zeros of $G_p(s)G_c(s)$ are equal to the zeros of the plant plus the zeros of the controller. With the zeros of $G_p(s)G_c(s)$ denoted by $z_1, z_2, \ldots, z_r$ and the poles denoted by $p_1, p_2, \ldots, p_{N+q}$, $G_p(s)G_c(s)$ can be expressed in factored form:

$$G_p(s)G_c(s) = K \frac{(s - z_1)(s - z_2) \cdots (s - z_r)}{(s - p_1)(s - p_2) \cdots (s - p_{N+q})} \tag{9.51}$$

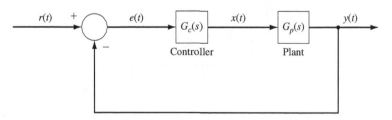

FIGURE 9.7
Feedback control system.

In (9.51), K is a constant that contains the leading coefficients of the numerator and denominator polynomials of $G_p(s)G_c(s)$. Inserting (9.51) into (9.50) yields the following expression for the closed-loop transfer function:

$$G_{cl}(s) = \frac{K\dfrac{(s - z_1)(s - z_2)\cdots(s - z_r)}{(s - p_1)(s - p_2)\cdots(s - p_{N+q})}}{1 + K\dfrac{(s - z_1)(s - z_2)\cdots(s - z_r)}{(s - p_1)(s - p_2)\cdots(s - p_{N+q})}}$$

$$G_{cl}(s) = \frac{K(s - z_1)(s - z_2)\cdots(s - z_r)}{(s - p_1)(s - p_2)\cdots(s - p_{N+q}) + K(s - z_1)(s - z_2)\cdots(s - z_r)} \qquad (9.52)$$

From (9.52) it is seen that the closed-loop poles are the $N + q$ roots of the equation

$$(s - p_1)(s - p_2)\cdots(s - p_{N+q}) + K(s - z_1)(s - z_2)\cdots(s - z_r) = 0 \qquad (9.53)$$

Obviously, the values of the $N + q$ closed-loop poles depend on the value of the constant K. In particular, note that when $K = 0$, the closed-loop poles are the same as the poles of $G_p(s)G_c(s)$.

Since the characteristics of the tracking error $e(t) = r(t) - y(t)$ depend directly on the values (or locations) of the closed-loop poles, in feedback control system design it is of major interest to know what possible closed-loop pole locations we can obtain by varying K. For example, K may correspond to a parameter (e.g., gain) in the controller that can be chosen by the designer, in which case the question arises as to whether or not there is a value of K that results in "good" pole locations. To answer this, it is first necessary to determine all closed-loop pole locations as K is varied over some range of values. This leads to the "180° root locus" (or the "$K > 0$ root locus"), which is the plot in the complex plane of the $N + q$ closed-loop poles as K is varied from 0 to ∞. Since only the $K > 0$ case is considered here, the 180° root locus or the $K > 0$ root locus will be referred to as the root locus. In the root locus construction, the constant K is called the *root-locus gain*.

Since there are $N + q$ closed-loop poles, the root locus has $N + q$ branches, where each branch corresponds to the movement in the complex plane of a closed-loop pole as K is varied from 0 to ∞. Since the closed-loop poles are the poles of $G_p(s)G_c(s)$ when $K = 0$, the root locus begins (when $K = 0$) at the poles of $G_p(s)G_c(s)$. As K is increased from zero, the branches of the root locus depart from the poles of $G_p(s)G_c(s)$, one branch per pole. As K approaches ∞, r of the branches move to the r zeros of $G_p(s)G_c(s)$, one branch per zero, and the other $N + q - r$ branches approach ∞.

A real or complex number p is on the root locus if and only if p is a root of (9.53) for some value of $K > 0$. That is, p is on the root locus if and only if for some $K > 0$,

$$(p - p_1)(p - p_2)\cdots(p - p_{N+q}) + K(p - z_1)(p - z_2)\cdots(p - z_r) = 0 \qquad (9.54)$$

Dividing both sides of (9.54) by $(p - p_1)(p - p_2)\cdots(p - p_{N+q})$ gives

$$1 + K\frac{(p - z_1)(p - z_2)\cdots(p - z_r)}{(p - p_1)(p - p_2)\cdots(p - p_{N+q})} = 0 \qquad (9.55)$$

Dividing both sides of (9.55) by K and rearranging terms yield

$$\frac{(p - z_1)(p - z_2) \cdots (p - z_r)}{(p - p_1)(p - p_2) \cdots (p - p_{N+q})} = -\frac{1}{K} \tag{9.56}$$

Thus, p is on the root locus if and only if (9.56) is satisfied for some $K > 0$.

Now, if $P(s)$ is defined by

$$P(s) = \frac{(s - z_1)(s - z_2) \cdots (s - z_r)}{(s - p_1)(s - p_2) \cdots (s - p_{N+q})} \tag{9.57}$$

then $KP(s) = G_p(s)G_c(s)$, and in terms of P, (9.56) becomes

$$P(p) = -\frac{1}{K} \tag{9.58}$$

Thus, p is on the root locus if and only if (9.58) is satisfied for some $K > 0$. Since $P(p)$ is a complex number in general, (9.58) is equivalent to the following two conditions:

$$|P(p)| = \frac{1}{K} \tag{9.59}$$

$$\angle P(p) = \pm 180° \tag{9.60}$$

The condition (9.59) is called the *magnitude criterion*, and the condition (9.60) is called the *angle criterion*. Any real or complex number p that satisfies the angle criterion (9.60) is on the root locus; that is, if (9.60) is satisfied, then (9.59) is also satisfied if

$$K = \frac{1}{|P(p)|} \tag{9.61}$$

In other words, for the value of K given by (9.61), p is on the root locus. This result shows that the root locus consists of all those real or complex numbers p such that the angle criterion (9.60) is satisfied. The use of the angle criterion is illustrated in the following example:

Example 9.6   Root Locus for First-Order System

Inter-
active
Root
Locus

Consider the closed-loop system with plant equal to the vehicle with velocity model given by $G_p(s) = 0.001/(s + 0.01)$, and with proportional controller given by $G_c(s) = K_P$. Then,

$$G_p(s)G_c(s) = \frac{0.001K_P}{s + 0.01} = K\frac{1}{s + 0.01} \tag{9.62}$$

where $K = 0.001K_P$. In this case, $G_p(s)G_c(s)$ has no zeros and one pole at $s = -0.01$, and thus $N + q = 1$ and $r = 0$. The root locus therefore has one branch that begins (when $K = 0$) at $s = -0.01$ and goes to ∞ as $K \to \infty$. From (9.62), $P(s)$ is

$$P(s) = \frac{1}{s + 0.01}$$

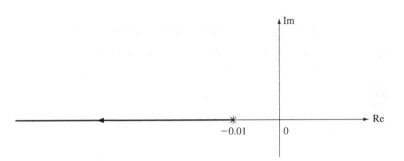

FIGURE 9.8
Root locus in Example 9.6.

and thus

$$\angle P(p) = -\angle(p + 0.01)$$

Then $\angle P(p) = \pm 180°$ if and only if p is a negative real number with $p < -0.01$. Hence, the root locus consists of all negative real numbers p such that $p < -0.01$. The root locus is plotted in Figure 9.8. The arrow in Figure 9.8 shows the direction of movement of the closed-loop pole as $K \to \infty$. Note that in this case the closed-loop pole goes to ∞ by "moving out" on the negative real axis of the complex plane.

Given a negative real number $p < -0.01$, from (9.61) the value of K for which p is on the root locus is

$$K = \frac{1}{|1/(p + 0.01)|} = |p + 0.01|$$

For instance, for the closed-loop pole to be $p = -0.2$, the value of the root locus gain K is

$$K = |-0.2 + 0.01| = 0.19$$

Then, since $K = 0.001 K_P$ [see (9.62)], the gain K_P of the proportional controller must be

$$K_P = \frac{0.19}{0.001} = 190$$

This is the value of K_P that puts the closed-loop pole at $s = -0.2$. The reader is invited to check this result by computing the closed-loop transfer function with $G_c(s) = 190$.

9.3.1   Root-Locus Construction

Again consider the feedback control system in Figure 9.7, with $G_p(s)G_c(s)$ expressed in the factored form (9.51) and with $P(s)$ defined by (9.57) so that $KP(s) = G_p(s)G_c(s)$. Note that the zeros (respectively, poles) of $P(s)$ are the same as the zeros (respectively, poles) of $G_p(s)G_c(s)$. The closed-loop poles are the roots of equation (9.53), where the z_i are the zeros of $P(s)$ and the p_i are the poles of $P(s)$. A sketch of the root locus for $K > 0$ can be generated by numerical computation of the poles of the closed-loop transfer function for specified values of K.

Example 9.7   Root Locus for Second-Order System

Inter-
active
Root
Locus

Consider the dc motor with transfer function $G_p(s) = 10/(s + 0.1)s$ and with proportional controller $G_c(s) = K_P$. Then,

$$G_p(s)G_c(s) = \frac{10K_P}{(s + 0.1)s} = K\frac{1}{(s + 0.1)s}$$

and thus,

$$K = 10K_P \quad \text{and} \quad P(s) = \frac{1}{(s + 0.1)s}$$

Since $P(s)$ has two poles at $p_1 = 0$ and $p_2 = -0.1$, there are two closed-loop poles, and the root locus has two branches, which start at $s = 0$ and $s = -0.1$. The closed-loop transfer function is computed to be

$$G_{cl}(s) = \frac{KP(s)}{1 + KP(s)}$$

$$= \frac{K/s(s + 0.1)}{1 + K/s(s + 0.1)}$$

$$= K\frac{1}{s^2 + 0.1s + K}$$

The closed-loop poles are given by the roots of

$$s^2 + 0.1s + K = 0 \tag{9.63}$$

We can obtain the root locus numerically by substituting specific values for K into (9.63) and finding the roots of the resulting equation. In this case the quadratic formula can be used to solve for the roots:

$$s = -0.05 \pm 0.5\sqrt{0.01 - 4K}$$

For $K = 0$, the closed-loop poles are at $p_1 = 0$ and $p_2 = -0.1$, as expected. A computer program can be written that computes the roots of the equation in (9.63), starting at $K = 0$ and incrementing K by small amounts until a specified upper limit. In this particular example, it is obvious that for $0 < K < 0.0025$, the closed-loop poles are real and negative; for $K = 0.0025$, the closed-loop poles are real and equal. Finally, for $K > 0.0025$, the poles are complex and located at p, $\bar{p} = -0.05 \pm j0.5\sqrt{4K - 0.01}$. A plot of the resulting root locus is given in Figure 9.9.

The MATLAB M-file `rlocus` in the Control System Toolbox computes and plots the root locus, given the numerator and denominator of $P(s)$. The commands to generate this plot are given by

```
num = 1; den  = [1 0.1 0];
P = tf(num,den);
rlocus(P)
```

where num contains the coefficients of the numerator of $P(s)$ and den contains the coefficients of the denominator of $P(s)$. A version of this command called `rootlocus` that works with the Student Version of MATLAB is available on the book website. The corresponding commands are

```
num = 1; den  = [1 0.1 0];
rootlocus(num,den)
```

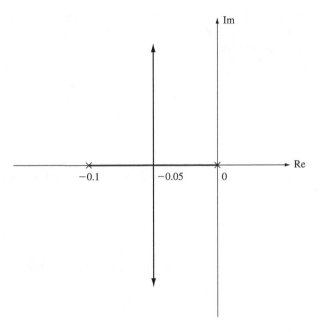

FIGURE 9.9
Root locus in Example 9.7.

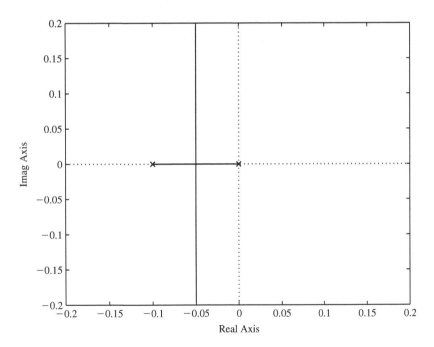

FIGURE 9.10
MATLAB plot of root locus in Example 9.7.

The resulting plot is given in Figure 9.10. Both rootlocus commands automatically generate the values of K that will yield a good plot. To customize the plot, we can compute the root locus for gains specified in a vector K:

```
num = 1; den   = [1 0.1 0];
P = tf(num,den);
K = 0:.0005:.04;
p = rlocus(P,K);
plot(p)
```

Example 9.8   Root Locus for Second-Order System with Zero

Inter-
active
Root
Locus

Now, suppose that

$$G_p(s) = \frac{2}{(s - 1)(s^2 + 2s + 5)} \quad \text{and} \quad G_c(s) = A(s + 3)$$

where A is a real constant (a gain in the controller). Then,

$$G_p(s)G_c(s) = \frac{2A(s + 3)}{(s - 1)(s^2 + 2s + 5)} = K\frac{s + 3}{(s - 1)(s^2 + 2s + 5)}$$

and thus,

$$K = 2A \quad \text{and} \quad P(s) = \frac{s + 3}{(s - 1)(s^2 + 2s + 5)} = \frac{s + 3}{s^3 + s^2 + 3s - 5}$$

In this case, $P(s)$ has three poles at $p_1 = -1 + j2$, $p_2 = -1 - j2$, and $p_3 = 1$, and $P(s)$ has one zero at $z_1 = -3$. Therefore, the root locus has three branches beginning (when $K = 0$ or $A = 0$) at $-1 \pm j2$ and 1. A precise sketch of the root locus can be produced by the MATLAB command `rlocus`. In this example, the following commands compute the root locus and then generate a plot:

```
num = [1 3]; den   = [1 1 3 -5];
P = tf(num,den);
rlocus(P);
```

The resulting root locus is shown in Figure 9.11. The branches starting at the poles $-1 \pm j2$ go to infinity as $K \to \infty$, while the branch starting at $p_3 = 1$ goes to the zero at $z_1 = -3$ as $K \to \infty$.

   The root locus shown in Figure 9.11 can be used to determine the range of values of K (or A) such that the closed-loop system is stable, that is, the range of values of K for which all three closed-loop poles are in the open left-half plane. First, note that since one of the branches starts at 1 when $K = 0$ and moves to the origin, the closed-loop system is not stable for $0 < K \le c$, where c is the value of K for which there is a closed-loop pole at $s = 0$. The constant c can be determined by the magnitude criterion (9.61), which gives

$$c = \frac{1}{|P(0)|} = \left|\frac{-5}{3}\right| = \frac{5}{3}$$

From the root locus in Figure 9.11, it is also clear that the closed-loop system is not stable for $K > b$, where b is the value of K for which the two complex poles are equal to $\pm j\omega_c$, where $\pm j\omega_c$

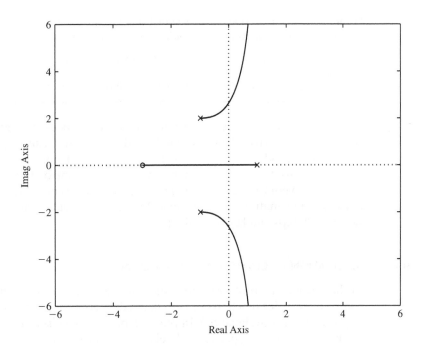

FIGURE 9.11
MATLAB plot of root locus.

are the points on the imaginary axis where the two branches cross over into the right-half plane. From Figure 9.11, an approximate value of ω_c is 2.6. Then,

$$b = \frac{1}{|P(j2.6)|} = \left| \frac{-j17.576 - 6.76 + j7.8 - 5}{j2.6 + 3} \right| = \sqrt{\frac{(11.76)^2 + (9.776)^2}{(2.6)^2 + 9}} = 3.85$$

Thus, an approximate range for stability is $\frac{5}{3} < K < 3.85$; or since $K = 2A$, the range on A is $\frac{5}{6} < A < 1.925$. We can determine the exact range on K or A for stability by computing the exact value of K for which the two complex poles are on the imaginary axis. We can carry this out by using the Routh–Hurwitz test as follows. First, the closed-loop transfer function is

$$G_{cl}(s) = \frac{KP(s)}{1 + KP(s)}$$

$$= \frac{K(s + 3)}{s^3 + s^2 + (K + 3)s + (3K - 5)}$$

and thus the Routh array is

s^3	1	$K + 3$
s^2	1	$3K - 5$
s^1	$\dfrac{(K + 3) - (3K - 5)}{1}$	
s^0	$3K - 5$	

From the results in Section 8.2, we see that there are two poles on the $j\omega$-axis when the term in the first column of the Routh array indexed by s^1 is zero. Thus,

$$K + 3 - (3K - 5) = 0$$

which gives $K = 4$. Therefore, the range for stability is $\frac{5}{3} < K < 4$, or in terms of $A, \frac{5}{6} < A < 2$.

An alternative means of sketching the root locus uses graphical construction rules that are derived from the angle and magnitude criteria (9.59) and (9.60). The graphical method provides insight into the effect of pole and zero locations on the shape of the root locus. Such insight is valuable for control design; however, only an introduction to control design that applies the root locus method is covered in this book. For more information on the graphical construction rules and their use in control design, see Phillips and Harbor [2000].

9.4   APPLICATION TO CONTROL SYSTEM DESIGN

In this section the root-locus construction is applied to the problem of designing the controller transfer function $G_c(s)$ so that a desired performance is achieved in tracking a specific reference signal $r(t)$. In practice, performance is usually specified in terms of accuracy, speed of response, and relative stability. One measure of accuracy is the steady-state error. If the reference $r(t)$ is a step function (the case of set-point control), the results in Section 9.2 show that the steady-state error will be zero if the open-loop system $G_p(s)G_c(s)$ is type 1. As noted before, if the plant transfer function $G_p(s)$ does not have a pole at $s = 0$, the controller transfer function $G_c(s)$ must have a pole at zero in order to have a type 1 system. Hence, in the case of set-point control, the best possible steady-state performance (i.e., zero steady-state error) is easily obtained by inclusion of (if necessary) a pole at zero in $G_c(s)$.

Measures of speed of response are defined from the step response of a stable system. As discussed in Section 8.3, one measure of speed of response is the time constant τ, which can be computed as $-1/\mathrm{Re}(p)$ where p is the dominant pole. Another measure is the *settling time*, which is the time that it takes for the response to reach and stay within a 5% band of the steady-state value. A common approximation for settling time is 3τ. Both the settling time and the time constant become smaller as the dominant pole is moved farther to the left in the s-plane, that is, as the real part of the pole becomes larger in magnitude and more negative.

Relative stability refers to the robustness of the control, essentially, how much modeling error is tolerated before the closed-loop system becomes unstable. The system becomes unstable when modeling errors cause a shift in the location of the closed poles from the open left-half plane onto the $j\omega$-axis or into the right-half plane. As complex second-order poles of a system move closer to the $j\omega$-axis, the system response becomes more oscillatory. This is seen in Example 8.11 where it is noted that the smaller the value of ζ is, the more pronounced the oscillation will be. A specific measure of relative stability is the *percent overshoot*, (P.O.), which is defined by

$$\mathrm{P.O.} = \frac{M_p - y_{ss}}{y_{ss}} \times 100\%$$

where M_p is the peak value of the response $y(t)$ and y_{ss} is the steady-state value. We can compute the peak value analytically by taking the derivative of the expression for $y(t)$ and setting it equal to zero to find the time t_p at which the peak occurs in $y(t)$. Then, $M_p = y(t_p)$. For a second-order system with transfer function given in (8.23), that is, no zeros and poles equal to $-\zeta\omega_n \pm j\omega_d$, the step response is given in (8.24). Following the foregoing procedure to find $t_p = \pi/\omega_d$ results in the following approximation for P.O.:

$$\text{P.O.} = \exp\left(-\frac{\pi\zeta}{\sqrt{1-\zeta^2}}\right) \times 100 \tag{9.64}$$

Thus, a specific requirement on P.O., such as P.O. $< 10\%$, can be translated to a specification on ζ. The approximation in (9.64) is often generalized to second-order systems with zeros and higher-order systems with dominant second-order poles. Generally, ζ is chosen such that $\zeta \geq 0.4$, which yields P.O. $\leq 25.5\%$.

To summarize, speed of response and relative stability can be related to closed-loop pole positions. Moreover, specifications on the time constant or settling time and on the P.O. can be used to determine the region of the complex plane in which the dominant closed-loop poles must lie. The root locus can then be plotted to determine if there is any value of K so that a closed-loop pole lies in the acceptable region of the complex plane. Generally, this procedure is first performed with a proportional controller, $G_c(s) = K_P$, since it is a simple controller to design and implement. If a dominant branch of the root locus does not lie in the desired region of the complex plane, a more complex controller $G_c(s)$ is used that reshapes the root locus.

Example 9.9   Proportional Controller Design

A controller is to be designed for the dc motor considered in Example 9.7, where

$$G_p(s) = \frac{10}{(s+0.1)s}$$

The specifications for the closed-loop system are that the time constant τ be less than or equal to 25 seconds and the damping ratio ζ be greater than or equal to 0.4. Since $\tau \approx -1/\text{Re}(p)$ and $\tau \leq 25$, $\text{Re}(p) \leq -0.04$. Also, recall that the damping ratio is defined by $\zeta = \cos\theta$, where θ is the angle of the pole position measured with respect to the negative real axis. These specifications can be transferred to the complex plane as shown in Figure 9.12. Any closed-loop pole that lies in the shaded region is acceptable. Next, examine a proportional controller, $G_c(s) = K_P$. Let $K = 10K_p$ and $P(s) = 1/(s+0.1)s$. The root locus for the dc motor with a proportional controller is given in Figure 9.9 and is redrawn in Figure 9.13, along with the specifications. The pole designated as p_L in Figure 9.13 marks the point at which the root locus enters the desired region. The pole designated as p_H marks the point at which the root locus exits the desired region. The values of gain K, which give closed-loop poles at p_L and p_H, then specify the range of values of K for which the specifications are satisfied. To find the gain K_L that yields a closed-loop pole at p_L, use the magnitude criterion given in (9.61) for $p = p_L$, where $p_L = -0.04$, is obtained from the graph:

$$K_L = |(p+0.1)p|_{p=-0.04} = 0.0024$$

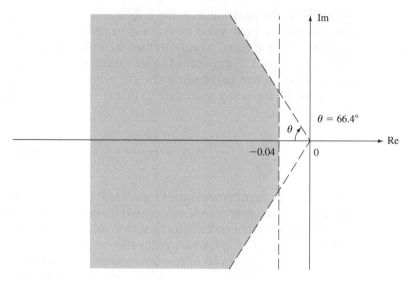

FIGURE 9.12
Complex plane showing region of acceptable pole positions.

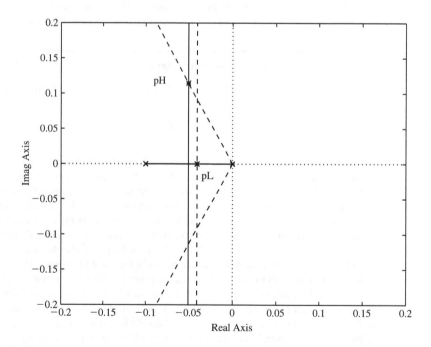

FIGURE 9.13
Root locus for Example 9.9 and region of acceptable pole positions.

Repeat the procedure to find the gain K_H that yields a closed-loop pole at p_H. Using the magnitude criterion on $p_H = -0.05 + j0.114$ yields $K_H = 0.0156$. Since the root-locus plot is continuous with respect to K, the range of K that satisfies the specifications is $0.0024 \le K \le 0.0156$ or $0.00024 \le K_P \le 0.00156$.

The step response of the closed-loop system for $K = 0.0024$ is obtained by the following MATLAB commands:

```
K = 0.0024;
num = K; den = [1 .1 0];
P = tf(num,den);
Gcl = feedback(P,1);
step(Gcl)
```

The command `feedback` computes the closed-loop transfer function from $P(s)$. Alternately, Simulink can be used to simulate the response, as demonstrated in Section 9.2. The resulting step responses for $K = K_L$ and $K = K_H$ are shown in Figure 9.14. Note that the transient response for $K = K_L$ decays more slowly than might be expected for a system with a time constant of $\tau = 25$ sec. This is because the poles for $K = K_L$ are close to each other, one pole at $s = -0.04$ and the other at $s = -0.06$.

In the approximation $\tau \approx -1/\mathrm{Re}(p)$, it is assumed that p is the dominant pole and that the rest of the system poles are much farther left, so that their effect is negligible. The resulting transient response resembles a pure exponential with a decay rate of $\mathrm{Re}(p)$. When the poles are close, as in this example, the resulting transient response is not nearly exponential, so that the

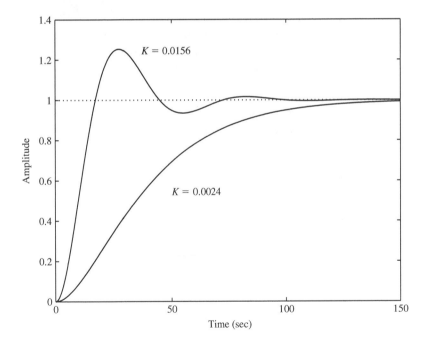

FIGURE 9.14
Step responses for closed-loop system with proportional control for $K = 0.0024$ and for $K = 0.0156$.

approximation $\tau \approx 1/0.04$ is not very accurate. Approximations like the one for the time constant are often employed to obtain an initial control design that gives a reasonable time response, and iteration is then used to fine-tune the results. The reader is invited to investigate the step responses for other values of K between K_L and K_H.

Suppose that the specifications were modified so that $\tau \leq 10$ seconds and $\zeta \geq 0.4$. This requires that the real part of the dominant pole must lie to the left of -0.1. Examination of the root locus with a proportional controller shows that no gain exists so that a dominant branch of the root locus lies to the left of -0.1. Hence, a proportional controller cannot satisfy the specifications. In this case, a more complex controller should be chosen, as discussed next.

It is often the case that a proportional controller cannot satisfy the specifications. Since the root locus is defined by the poles and zeros of the transfer function $G_p(s)G_c(s)$, adding zeros and/or poles to $G_c(s)$ will change the shape of the root locus. An important part of control design is to determine where to put the controller zeros and poles to yield a desirable response. A further consideration for the control design is satisfying the specifications on steady-state errors. The discussion here is limited to PD, PI, and PID controllers. The effect of each of these controllers on a root locus is examined as follows.

Consider the PD controller introduced in Example 9.5 and given by

$$G_c(s) = K_P + K_D s$$

$$= K_D\left(s + \frac{K_P}{K_D}\right)$$

This type of controller contributes a zero to the rational function $P(s)$. The addition of this zero tends to pull the root locus to the left, when compared with the root locus with a proportional gain. Thus, the PD controller is generally used to speed up the transient response of a system over that obtainable with a proportional controller.

There are two design parameters for a PD controller, K_D and K_P. The ratio $-K_P/K_D$ defines the zero location of the controller. One rule of thumb is to choose the zero location to be in the left-hand plane and to the left of the rightmost pole. In some cases, the zero may be chosen to cancel a stable real pole, typically, the second pole from the right. Once the zero location has been determined, the regular root-locus design method illustrated in Example 9.9 can be used to select K_D.

Example 9.10   PD Controller Design

Consider the dc motor of Examples 9.5, 9.7, and 9.9. Now choose $G_c(s)$ as a PD controller and let $z = -K_P/K_D$. Then,

$$G_p(s)G_c(s) = \frac{10K_D(s-z)}{(s+0.1)s} = K\frac{s-z}{(s+0.1)s}$$

and thus,

$$K = 10K_D \quad \text{and} \quad P(s) = \frac{s-z}{(s+0.1)s}$$

Consider four different values for z: $= -0.05, -0.1, -0.2,$ and -1. The root locus for $z = -0.05$ is obtained by the MATLAB commands

```
num = [1 0.05]; den = [1 .1 0];
P = tf(num,den);
rlocus(P)
```

Use `rootlocus` in the Student Version of MATLAB where `rootlocus` is found from the book website. The root loci for the four different cases, $z = -0.05$, $z = -0.1$, $z = -0.2$, and $z = -1$, are given in Figure 9.15. For $z = -0.05$, there is a branch from $s = 0$ to $s = -0.05$. As K is increased, the closed-loop pole on this branch gets closer to the zero at -0.05. Recall from Section 8.3 that, if a zero is very near a pole, the residue is small and the pole is not dominant. Hence, as K is increased, the residue of the pole on this branch gets smaller, making it less dominant. The other branch goes to $-\infty$; therefore, the response can be made suitably fast by choosing a large value of K. For $z = -0.1$, the zero cancels a plant pole, making the closed-loop system behave as a first-order response. The resulting branch starts at $s = 0$ and goes to $-\infty$ along the negative real axis as $K \to \infty$; hence, the transient response can be made as fast as desired by an increase in the value of K. For

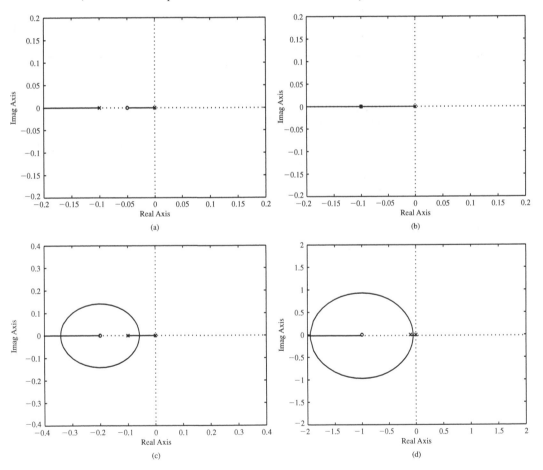

FIGURE 9.15
Root loci for PD controllers in Example 9.10: (a) $z = -0.05$; (b) $z = -0.1$; (c) $z = -0.2$; (d) $z = -1$.

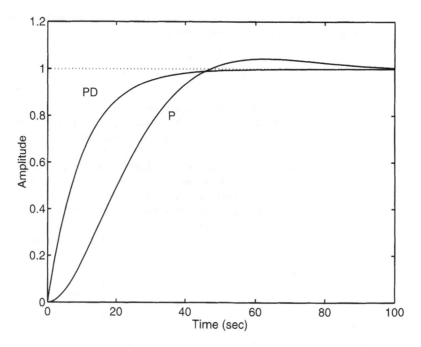

FIGURE 9.16
Closed-loop step response for PD and P controllers.

$z = -0.2$, the root locus has two branches starting at $s = 0$ and $s = -0.1$, which come together along the real axis to meet at $s = -0.06$, and then the branches split apart, forming a circular arc, which breaks into the real axis at $s = -0.34$; one branch then moves toward the zero at -0.2, and the other branch goes to $-\infty$ along the real axis. The circular arc means that this choice of zero location allows for an underdamped response. The dominant poles are farthest to the left when the closed-loop poles are both equal to -0.34. For $z = -1$, the circular arc has a large radius and crosses the real axis at a point that is farther left than for that obtained for $z = -0.2$. In this case the fastest response is obtained for the value of K that yields closed-loop poles at $s = -1.95$.

Now, let $z = -0.1$, which cancels a plant pole. Suppose that the specification requires that $\tau \leq 10$ seconds, so that the dominant pole must lie to the left of -0.1. Choose a desired closed-loop pole to be $p = -0.1$. Then use the magnitude criterion (9.61) to solve for the corresponding K, $K = 0.1$. The resulting control is $G_c(s) = 0.01(s + 0.1)$. Figure 9.16 shows the closed-loop step response of the dc motor with this PD controller. For comparison's sake, the closed-loop dc motor with a proportional controller, $G_c(s) = 0.0005$, is also shown. The gain for the proportional control was chosen to give closed-loop poles at $-0.05 \pm j0.05$, which are as far left as possible with proportional control. The damping ratio of $\zeta = 0.707$ is large enough to give a reasonably small oscillation of the transient.

Consider the PI controller introduced in Example 9.5 and given by

$$G_c(s) = \frac{K_P s + K_I}{s} = K_P \frac{s + K_I/K_P}{s}$$

This controller increases the system type and is generally used to reduce the steady-state error, that is, increase the steady-state accuracy of the tracking between $y(t)$ and $r(t)$.

The addition of a pole at the origin and a zero at $-K_I/K_P$ affects the shape of the root locus and therefore may affect the transient response. In general, the PI controller results in a root locus that is to the right of a root locus drawn for a proportional controller. Hence, the transient response is generally slower than is possible with a proportional controller. The ratio of K_I/K_P is usually chosen so that the resulting zero is closer to the origin than any plant pole. The smaller the ratio is, and hence the closer the zero is to the origin, the smaller the effect of the PI compensator on the root locus will be. Therefore, if a proportional controller can be found that gives desirable transient response, but unacceptable steady-state error, a PI controller can be used to obtain nearly the same closed-loop pole location, but much smaller steady-state errors.

Example 9.11   PI Controller Design

Consider a system with a transfer function

$$G_p(s) = \frac{1}{(s+1)(s+4)}$$

A controller is to be designed so that the output $y(t)$ tracks a reference input $r(t)$ with a small error. The root locus with a proportional controller is given in Figure 9.17a. To reduce the steady-state error, design a PI control of the form

$$G_c(s) = K_P + \frac{K_I}{s} = K_P \frac{s-z}{s}$$

where $z = -K_I/K_P$. In this case,

$$K = K_P \quad \text{and} \quad P(s) = \frac{s-z}{s(s+1)(s+4)}$$

Consider three choices for z: -0.01, -1, and -3. The corresponding root loci are given in Figure 9.17b–d. Note that the root-locus branches for the proportional controller are farther left than the root loci for any of the PI controllers. Also notice that the closer the zero z is to the origin, the closer the PI root locus plot is to that of the P root locus.

A proportional plus integral plus derivative (PID) controller combines the benefits of a PI and a PD controller; that is, it increases the system type so that it decreases the steady-state error, plus it improves the transient response by moving the root locus to the left. The general form of this controller is

$$G_c(s) = K_P + K_D s + \frac{K_I}{s}$$

$$G_c(s) = \frac{K_D s^2 + K_P s + K_I}{s}$$

This controller has one pole at the origin and two zeros. With the zeros denoted as z_1 and z_2, the controller has the general form

$$G_c(s) = K_D \frac{(s-z_1)(s-z_2)}{s}$$

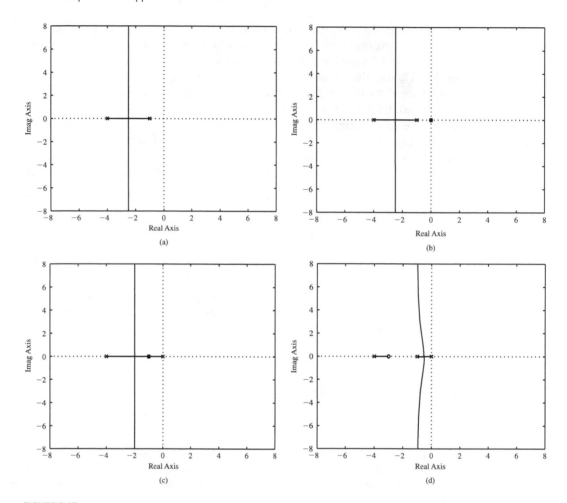

FIGURE 9.17
Root loci for Example 9.11: (a) P control; (b) PI control with $z = -0.01$; (c) PI control with $z = -1$; (d) PI control with $z = -3$.

Generally, one of the zeros is chosen to be near the origin (like that of a PI controller), and the other is chosen farther to the left (like a PD controller).

Example 9.12   PID Controller Design

Again consider the system given Example 9.11. A PI controller with $z_1 = -1$ was designed in Example 9.11. To this control, add a PD controller with a zero at $z_2 = -8$. Note that the zero of the PD controller can be chosen arbitrarily so that the root locus can be moved arbitrarily far to the left. The resulting PID controller has the form

$$G_c(s) = K_D \frac{(s + 1)(s + 8)}{s}$$

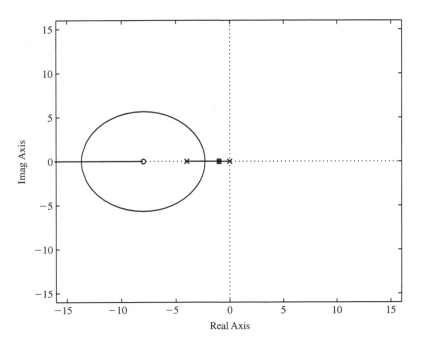

FIGURE 9.18
Root locus of PID controller.

Then,

$$K = K_D \quad \text{and} \quad P(s) = \frac{s + 8}{(s + 4)s}$$

The root locus, drawn in Figure 9.18, is farther to the left than any of the root loci drawn for a PI controller, as shown in Figure 9.17.

9.5   CHAPTER SUMMARY

This chapter is an introduction to continuous-time control theory, which uses the transfer function representation of a system. Control is used to meet specifications on the performance of a dynamical system. A common goal in control applications is tracking, where the system output is forced to track a desired signal. In the case of a constant signal, this is known as setpoint control. Open-loop control is a simple approach where an appropriate input signal is determined that results in the desired system response. Feedback control, where the control signal depends on the measured output signal, is much more robust to modeling errors and external disturbances than is open loop-control. Furthermore, feedback control can be used to stabilize an unstable plant.

The four common types of control are proportional control, PI control, PD control, and PID control. Proportional control is the simplest to implement, since the control consists only of a gain. However, proportional control is the most limited in terms of the performance achievable. PI control is used when accuracy of the response is needed, but PI control tends to slow down the response. PD control is used to speed up the system response, while PID control is used to both speed up the response and to improve accuracy.

A root locus, which is the plot of the closed-loop poles as a function of the system gain, is a common tool used for control design. The specifications on a system, such as time constant and percent overshoot, can be translated into desired closed-loop pole positions. The root locus is used to determine the range of gain values that meet these specifications. If no gain exists that meets the specifications for the type of control selected, the shape of the root locus must be changed by the use of a different control. For example, the root locus of a system with a PD control tends to have root locus branches that are to the left of the root locus branches for the system with proportional control.

PROBLEMS

9.1. Consider the following system transfer function:

$$G_p(s) = \frac{1}{s + 0.1}$$

(a) An open-loop control is shown in Figure P9.1a. Design the control, $G_c(s)$, so that the combined plant and controller $G_c(s)G_p(s)$ has a pole at $p = -2$, and the output $y(t)$ tracks a constant reference signal $r(t) = r_0 u(t)$ with zero steady-state error, where $e_{ss} = r_0 - y_{ss}$.

(b) Now, suppose that the plant pole at $p = -0.1$ was modeled incorrectly and that the actual pole is $p = -0.2$. Apply the control designed in part (a) and the input $r(t) = r_0 u(t)$ to the actual plant, and compute the resulting steady-state error.

(c) A feedback controller $G_c(s) = 2(s + 0.1)/s$ is used in place of open-loop control, as shown in Figure P9.1b. Verify that the closed-loop pole of the nominal system is at $p = -2$. (The nominal system has the plant pole at $p = -0.1$.) Let the input to the closed-loop system be $r(t) = r_0 u(t)$. Verify that the steady-state error $e_{ss} = r_0 - y_{ss}$ is zero.

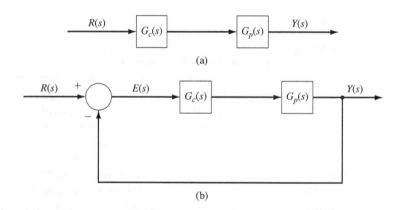

(a)

(b)

FIGURE P9.1

(d) Compute the steady-state error of the actual closed-loop system (with plant pole at $p = -0.2$) when $r(t) = r_0 u(t)$. Compare this error with that of the actual open-loop system computed in part (b).

(e) Simulate the responses of the systems in parts (a) to (d) when $r_0 = 1$. Explain the differences (and similarities) in responses.

9.2. Examine the effect of a disturbance on the performance of open- and closed-loop control systems by performing the following analysis:

(a) Consider an open-loop control system with a disturbance $D(s)$, as shown in Figure P9.2a. Define an error $E(s) = R(s) - Y(s)$, where $R(s)$ is a reference signal. Derive an expression for $E(s)$ in terms of $D(s)$, $X(s)$, and $R(s)$. Suppose that $D(s)$ is known. Can its effect be removed from $E(s)$ by proper choice of $X(s)$ and/or $G_c(s)$? Now, suppose that $D(s)$ represents an unknown disturbance. Can its effect be removed (or reduced) from $E(s)$ by proper choice of $X(s)$ and/or $G_c(s)$? Justify your answers.

(b) Now consider the feedback system shown in Figure P9.2b. Derive an expression for $E(s)$ in terms of $D(s)$ and $R(s)$. Suppose that $D(s)$ represents an unknown disturbance and that $G_c(s) = K$. Can the effect of $D(s)$ be removed (or reduced) from $E(s)$ by proper choice of K? Justify your answer.

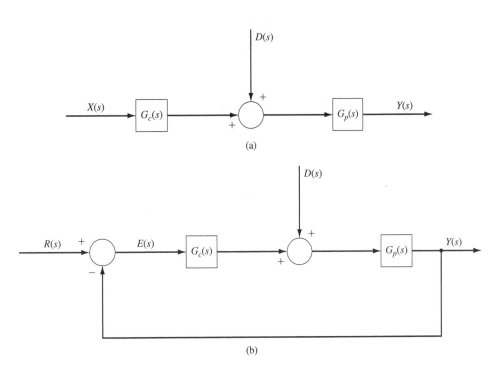

FIGURE P9.2

9.3. A rocket is drawn in Figure P9.3a, where $\theta(t)$ represents the angle between the rocket's orientation and its velocity, $\phi(t)$ represents the angle of the thrust engines, and $w(t)$ represents wind gusts, which act as a disturbance to the rocket. The goal of the control design is to have the angle $\theta(t)$ track a reference angle $\theta_r(t)$. The angle of the thrust engines can be directly

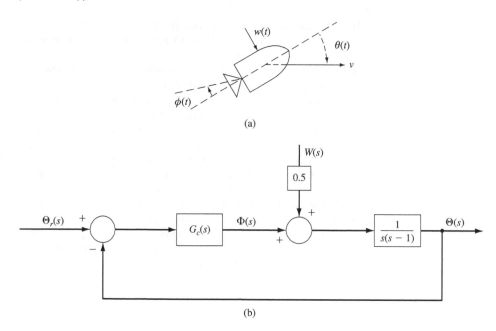

FIGURE P9.3

controlled by motors that position the engines; therefore, the plant output is $\theta(t)$ and the controlled input is $\phi(t)$. The system can be modeled by the following equation:

$$\Theta(s) = \frac{1}{s(s-1)}\Phi(s) + \frac{0.5}{s(s-1)}W(s)$$

(a) Consider an open-loop control $\Theta(s) = G_c(s)X(s)$, where $G_c(s)$, the controller transfer function, and $x(t)$, the command signal, can be chosen as desired. Is such a controller practical for having $\theta(t)$ track $\theta_r(t)$? Justify your answer.

(b) Now consider a feedback controller as shown in Figure P9.3b, where

$$G_c(s) = K(s+2)$$

Find an expression for the output $\Theta(s)$ of the closed-loop system in terms of $W(s)$ and $\Theta_r(s)$. Consider the part of the response due to $W(s)$; the lower this value is, the better the disturbance rejection will be. How does the magnitude of this response depend on the magnitude of K?

(c) Suppose that $\Theta_r(s) = 0$ and $w(t)$ is a random signal uniformly distributed between 0 and 1. Define a vector w in MATLAB as $w = \text{rand}(201,1)$, and define the time vector as $t = 0:0.05:10$. Use w as the input to the closed-loop system, and simulate the response for the time interval $0 \le t \le 10$. Perform the simulation for $K = 5, 10,$ and 20, and plot the responses. Explain how the magnitude of the response is affected by the magnitude of K. Does this result match your prediction in part (b)?

9.4. Consider the feedback control system shown in Figure P9.4. Assume that the initial conditions are zero.

(a) Derive an expression for $E(s)$ in terms of $D(s)$ and $R(s)$, where $E(s)$ is the Laplace transform of the error signal $e(t) = r(t) - y(t)$.

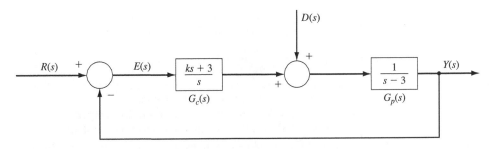

FIGURE P9.4

    (b) Suppose that $r(t) = u(t)$ and $d(t) = 0$ for all t. Determine all (real) values of K so that $e(t) \rightarrow 0$ as $t \rightarrow \infty$.

    (c) Suppose that $r(t) = u(t)$ and $d(t) = u(t)$. Determine all (real) values of K so that $e(t) \rightarrow 0$ as $t \rightarrow \infty$.

    (d) Suppose that $r(t) = u(t)$ and $d(t) = (\sin t)u(t)$. Determine all (real) values of K so that $e(t) \rightarrow 0$ as $t \rightarrow \infty$.

    (e) Again suppose that $r(t) = u(t)$ and $d(t) = (\sin t)u(t)$. With the controller transfer function given by

$$G_c(s) = \frac{7s^3 + K_1 s + K_2}{s(s^2 + 1)}$$

    determine all (real) values of K_1 and K_2 so that $e(t) \rightarrow 0$ as $t \rightarrow \infty$.

9.5. Consider a feedback connection as shown in Figure P9.1(b). The impulse response of the system with transfer function $G_p(s)$ is $h(t) = (\sin t)u(t)$.

    (a) Determine the transfer function $G_c(s)$ so that the impulse response of the feedback connection is equal to $(\sin t)e^{-t}u(t)$.

    (b) For $G_c(s)$ equal to your answer in part (a), compute the step response of the feedback connection.

9.6. Each of the following systems is to be controlled by feedback:

    (i) $G_p(s) = \dfrac{s + 5}{s + 1}$

    (ii) $G_p(s) = \dfrac{1}{s(s + 4)}$

    For each system, do the following:

    (a) Use the angle condition to determine which part of the real axis is on the root locus when $G_c(s) = K$. For the system in (ii), verify by using the angle condition that $s = -2 + j\omega$ is on the root locus for all real ω.

    (b) Calculate the closed-loop poles for specific values of $K > 0$, then use this information to plot the root locus.

    (c) Verify the answers in parts (a) and (b) by using MATLAB to plot the root locus.

9.7. Use MATLAB to plot the root locus for each of the following systems:

    (a) $G_p(s) = \dfrac{1}{(s + 1)(s + 10)}; \quad G_c(s) = K$

    (b) $G_p(s) = \dfrac{1}{(s + 1)(s + 4)(s + 10)}; \quad G_c(s) = K$

(c) $G_p(s) = \dfrac{(s + 4)^2 + 4}{[(s + 2)^2 + 16](s + 8)}$; $G_c(s) = K$

(d) $G_p(s) = \dfrac{s + 4}{(s + 6)^2 + 64}$; $G_c(s) = K$

9.8. For each of the systems given in Problem 9.7, determine the following:
(a) The range of K that gives a stable response.
(b) The value of K (if any) that gives a critically damped response.
(c) The value(s) of K that gives the smallest time constant.

9.9. For each of the closed-loop systems defined in Problems 9.7, do the following:
(a) Compute the steady-state error e_{ss} to a unit step input when $K = 100$.
(b) Verify your answer in part (a) by simulating the responses of the closed-loop systems to a step input.

9.10. Use MATLAB to plot the root locus for each of the following systems:

(a) $G_p(s) = \dfrac{1}{(s + 1)(s + 10)}$; $G_c(s) = \dfrac{K(s + 1.5)}{s}$

(b) $G(s) = \dfrac{1}{(s + 1)(s + 10)}$; $G_c(s) = K(s + 15)$

(c) $G_p(s) = \dfrac{1}{s(s - 2)}$; $G_c(s) = K(s + 4)$

(d) $G_p(s) = \dfrac{1}{(s + 2)^2 + 9}$; $G_c(s) = \dfrac{K(s + 4)}{s + 10}$

(e) $G_p(s) = \dfrac{1}{(s + 1)(s + 3)}$; $G_c(s) = \dfrac{K(s + 6)}{s + 10}$

9.11. Repeat Problem 9.9 for the closed-loop systems defined in Problem 9.10.

9.12. The transfer function of a dc motor is

$$G_p(s) = \dfrac{\Theta(s)}{V_a(s)} = \dfrac{60}{s(s + 50)}$$

where $\theta(t)$ is the angle of the motor shaft and $v_a(t)$ is the input voltage to the armature. A closed-loop system is used to try to make the angle of the motor shaft track a desired motor angle $\theta_r(t)$. Unity feedback is used, as shown in Figure P9.12, where $G_c(s) = K_P$ is the gain of an amplifier. Let $K = K_P(60)$.
(a) Plot the root locus for the system.
(b) Calculate the closed-loop transfer function for the following values of K: $K = 500$, 625, 5000, and 10,000. For each value of K, identify the corresponding closed-loop poles on the root locus plotted in part (a).
(c) Plot the step response for each value of K in part (b). For which value of K does the closed-loop response have the smallest time constant? The smallest overshoot?

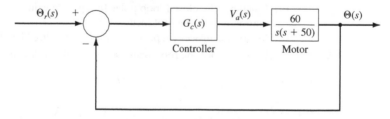

FIGURE P9.12

9.13. A system has the transfer function

$$G_p(s) = \frac{1}{(s + 1)(s + 7)}$$

(a) Sketch the root locus for a closed-loop system with a proportional controller, $G_c(s) = K_P$.

(b) Compute the closed-loop poles for $K_P = 5, 9, 73, 409$, and mark these pole positions on the root locus. Describe what type of closed-loop behavior you will expect for each of these selections of K_P. Calculate the steady-state error to a unit step function for each of these values of K_P.

(c) Verify the results of part (b) by using MATLAB to compute and plot the closed-loop step response for each value of K_P.

9.14. A system has the transfer function

$$G_p(s) = \frac{s + 4}{s(s + 2)(s + 8)}$$

(a) Sketch the root locus, using MATLAB for a proportional controller, $G_c(s) = K_P$.

(b) Find a value of K_P that yields a closed-loop damping ratio of $\zeta = 0.707$ for the dominant poles. Give the corresponding closed-loop pole.

(c) Use MATLAB to compute and plot the closed-loop step response for the value of K_P found in part (b).

9.15. A third-order system has the transfer function

$$G_p(s) = \frac{1}{(s + 1)(s + 3)(s + 5)}$$

The performance specifications are that the dominant second-order poles have a damping ratio of $0.4 \leq \zeta \leq 0.707$ and $\zeta\omega_n > 1$.

(a) Plot the root locus for $G_c(s) = K_P$.

(b) From the root locus, find the value(s) of K_P that satisfy the criteria.

9.16. A system has the transfer function

$$G_p(s) = \frac{1}{s^2}$$

(a) Sketch the root locus for a closed-loop system with a proportional controller, $G_c(s) = K_P$. Describe what type of closed-loop response you will expect.

(b) Sketch the root locus for a PD controller of the form $G_c(s) = K_D s + K_P = K_D(s + 2)$. Describe what type of closed-loop response you will expect as K_D is varied.

(c) Give the steady-state error of the closed-loop system with a PD controller for a step input when $K_D = 10$.

(d) Verify your result in part (c) by simulating the system.

9.17. A system has the transfer function

$$G_p(s) = \frac{s + 4}{(s + 1)(s + 2)}$$

(a) Sketch the root locus for a closed-loop system with a proportional controller, $G_c(s) = K_P$. Determine the value of K_P that will give closed-loop poles with a time constant of $\tau = 0.5$ seconds.

(b) Compute the steady-state error of the step response for the value of K_P chosen in part (a).

(c) Design a PI controller so that the closed-loop system has a time constant of approximately 0.5 sec. For simplicity of design, select the zero of the controller to cancel the pole of the system. What is the expected steady-state error to a step input?

(d) Simulate the closed-loop system with two different controllers designed in parts (a) and (c) to verify the results of parts (b) and (c).

9.18. The system shown in Figure P9.18 is a temperature control system where the output temperature $T(t)$ should track a desired set-point temperature $r(t)$. The open-loop system has the transfer function

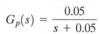

$$G_p(s) = \frac{0.05}{s + 0.05}$$

(a) Sketch the root locus for a closed-loop system with a proportional controller, $G_c(s) = K_P$. Suppose that the desired temperature is 70°F. Let $r(t) = 70u(t)$, and compute the gain required to yield a steady-state error of 2°. What is the resulting time constant of the closed-loop system?

(b) Design a PI controller so that the closed-loop system has the same time constant as that computed in part (a). For simplicity of design, select the zero of the controller to cancel the pole of the system.

(c) To verify the results, simulate the response of the closed-loop system to $r(t) = 70u(t)$ for the two different controllers designed in parts (a) and (b).

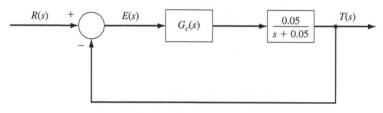

FIGURE P9.18

9.19. A system is given by the transfer function

$$G_p = \frac{10}{s(s + 1)}$$

Suppose that the desired closed-loop poles are at $-3 \pm j3$.

(a) Design a PD controller to obtain the desired poles. Use the angle criterion (9.60) evaluated at the desired closed-loop pole (i.e., $p = -3 + j3$) to determine the zero position.

(b) Simulate the step response of the closed-loop system.

9.20. A dc motor has the transfer function

$$\frac{\Omega(s)}{V_i(s)} = G_p(s) = \frac{2}{(s + 2)(s + 10)}$$

where $\Omega(s)$ represents the motor speed and $V_i(s)$ represents the input voltage.

(a) Design a proportional controller to have a closed-loop damping ratio of $\zeta = 0.707$. For this value of K_P, determine the steady-state error for a unit step input.

(b) Design a PID controller so that the dominant closed-loop poles are at $-10 \pm 10j$. For simplicity, select one of the zeros of the controller to cancel the pole at -2. Then, use the angle criterion (9.60) with $p = -10 + j10$ to determine the other zero position. What is the expected steady-state error to a step input?

(c) To verify your results, simulate the step response of the closed-loop system with the two different controllers designed in (a) and (b).

9.21. Consider the rocket described in Problem 9.3. A feedback loop measures the angle $\theta(t)$ and determines the corrections to the thrust engines.

    (a) Design a PD controller to have the closed loop poles at $-0.5 \pm 0.5j$. (*Hint:* See the comment regarding the use of the angle criterion in Problem 9.19.)

    (b) Simulate the response of the resulting closed-loop system to a unit impulse input.

9.22. An inverted pendulum shown in Figure P9.22 has the transfer function

$$G_p(s) = \frac{\Theta(s)}{T(s)} = \frac{2}{s^2 - 2}$$

Inter-
active
Root
Locus

where $\Theta(s)$ represents the angle of the rod and $T(s)$ represents the torque applied by a motor at the base.

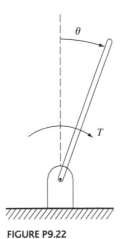

FIGURE P9.22

    (a) Sketch the root locus for a proportional controller, $G_c(s) = K_P$. What type of closed-loop response would you expect for different values of K_P?

    (b) Design a controller of the form

$$G_c(s) = K_L \frac{s - z_c}{s - p_c}$$

Choose $z_c = -3$, and solve for p_c from the angle criterion so that the dominant closed-loop poles are at $-3 \pm 3j$. (*Note:* The resulting controller is called a *lead controller.*) Draw the resulting root locus for this system, and calculate the gain K_L that results in the desired closed-loop poles.

    (c) Simulate the impulse response of the closed-loop system with the controller designed in part (b). (The impulse is equivalent to someone bumping the pendulum.)

9.23. A system has the transfer function

$$G_p(s) = \frac{1}{s(s + 2)}$$

(a) Sketch the root locus for a proportional controller, $G_c(s) = K_P$.

(b) Design a controller of the form

$$G_c(s) = K_L \frac{s - z_c}{s - p_c}$$

Inter-
active
Root
Locus

Select the zero of the controller to cancel the pole at -2. Solve for p_c from the angle criterion so that the dominant closed loop poles are at $-2 \pm 3j$. Draw the resulting root locus for this system, and calculate the gain K_L that results in the desired closed-loop poles.

(c) Design another controller by using the method described in part (b); except choose the zero to be at $z_c = -3$. Draw the resulting root locus for this system, and calculate the gain K_L that results in the desired closed-loop poles.

(d) Compare the two controllers designed in parts (b) and (c) by simulating the step response of the two resulting closed-loop systems. Since both systems have the same dominant poles at $-2 \pm 3j$, speculate on the reason for the difference in the actual response.

9.24. Design a feedback controller that sets the position of a table tennis ball suspended in a plastic tube, as illustrated in Figure P9.24. Here, M is the mass of the ball, g the gravity constant, $y(t)$ the position of the ball at time t, and $x(t)$ the wind force on the ball due to the fan. The position $y(t)$ of the ball is continuously measured in real time by an ultrasonic sensor. The system is modeled by the differential equation

$$M\ddot{y}(t) = x(t) - Mg$$

The objective is to design the feedback controller so that $y(t) \rightarrow y_0$ as $t \rightarrow \infty$, where y_0 is the desired position (set point).

(a) Can the control objective be met by the use of a proportional controller given by $G_c(s) = K_P$? Justify your answer.

(b) Can the control objective be met by the use of a PI controller given by $G_c(s) = K_P + K_I/s$? Justify your answer.

(c) Design a PID controller that achieves the desired objective when $M = 1$ and $g = 9.8$.

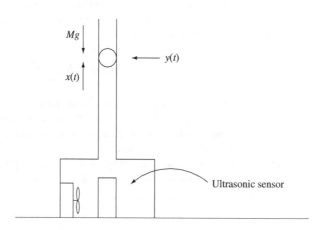

FIGURE P9.24

9.25. A proportional controller can be implemented by the use of a simple amplifier. However, PD, PI, and PID controllers require a compensating network. Often, this is achieved in analog with the use of operational amplifier (op amp) circuits. Consider the ideal op amp in Figure P9.25a. This op amp is an infinite impedance circuit element, so that $v_a = 0$ and $i_a = 0$. These relationships also hold when the op amp is embedded in a circuit, as shown in Figure P9.25b.

(a) Suppose that $R_1 = 1000\ \Omega$, $R_2 = 2000\ \Omega$, $C_1 = C_2 = 0$ in Figure P9.25b. Compute the transfer function between the input v_1 and the output v_2. (This circuit is known as an inverting circuit.)

(b) Suppose that $R_1 = 10\ k\Omega$, $R_2 = 20\ k\Omega$, $C_1 = 10\ \mu F$, and $C_2 = 0$ in Figure 9.25b. The resulting circuit is a PD controller. Compute the transfer function of the circuit.

(c) Suppose that $R_1 = 10\ k\Omega$, $R_2 = \infty$ (removed from circuit), $C_1 = 200\ \mu F$, and $C_2 = 10\ \mu F$ in Figure 9.25b. The resulting circuit is a PI controller. Compute the transfer function of the circuit.

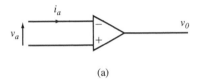

(a)

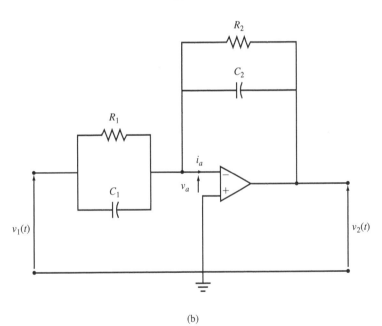

(b)

FIGURE P9.25

Design of Digital Filters and Controllers

In this chapter the continuous- and discrete-time techniques developed in Chapters 5 to 9 are utilized to design digital filters and digital controllers. The development begins in Section 10.1 with the study of the discretization of continuous-time signals and systems. Here a Fourier analysis is given for analog signal discretization, which is then utilized to generate a frequency-domain condition for the discretization of continuous-time systems. In Section 10.2 the design of digital filters is pursued in terms of the discretization of analog prototype filters. As with analog filters discussed in Chapter 8, digital filters can be lowpass, highpass, bandpass, and bandstop, or can take on an arbitrary frequency response function characteristic. As noted in Section 7.5, digital filters can have an infinite-duration impulse (i.e., unit-pulse) response, in which case they are referred to as IIR filters, or the impulse response may decay to zero in a finite number of steps, in which case the filter is an FIR filter. In Section 10.2 the design of IIR filters is developed in terms of analog prototypes that are then mapped into digital filters by use of the bilinear transformation. The application of MATLAB to carry out this design process is considered in Section 10.3. Then in Section 10.4 the design of FIR filters is developed by the truncating or windowing of the impulse response of an IIR filter.

The mapping concept discussed in Section 10.1 for transforming analog filters to digital filters can also be used to map continuous-time controllers to discrete-time (digital) controllers. This is discussed in Section 10.5 along with a brief development of the response matching technique. Here part of the emphasis is on step response matching, which is commonly used in digital control. The mapping of an analog controller into a digital controller is illustrated in the dc motor application considered in Chapter 9. Section 10.5 includes the description of a lab project involving digital control of a dc motor based on a LEGO MINDSTORMS kit. A summary of the chapter is presented in Section 10.6.

10.1 DISCRETIZATION

Let $x(t)$ be a continuous-time signal that is to be sampled, and let $X(\omega)$ denote the Fourier transform of $x(t)$. As discussed in Section 3.4, the plots of the magnitude $|X(\omega)|$ and the angle $\angle X(\omega)$ versus ω display the amplitude and phase spectra of the signal $x(t)$. Now with the sampling interval equal to T, as discussed in Section 5.4, the sampled signal $x_s(t)$ can be represented by the multiplication of the signal $x(t)$ by the impulse train $p(t)$; that is, $x_s(t) = x(t)p(t)$, where

$$p(t) = \sum_{n=-\infty}^{\infty} \delta(t - nT)$$

From Chapter 10 of *Fundamentals of Signals and Systems Using the Web and MATLAB*, Third Edition. Edward W. Kamen, Bonnie S. Heck. Copyright © 2007 by Pearson Education, Inc. Publishing as Prentice Hall. All rights reserved.

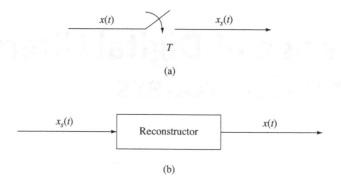

FIGURE 10.1
Signal sampler (a) and signal reconstructor (b).

The sampling operation is illustrated in Figure 10.1a. Recall that the sampling frequency ω_s is equal to $2\pi/T$.

In Section 5.4, it was shown that the Fourier transform $X_s(\omega)$ of the sampled signal $x_s(t)$ is given by

$$X_s(\omega) = \sum_{n=-\infty}^{\infty} \frac{1}{T} X(\omega - n\omega_s) \tag{10.1}$$

Note that $X_s(\omega)$ consists of scaled replicas of $X(\omega)$ shifted in frequency by multiples of the sampling frequency ω_s. Also note that $X_s(\omega)$ is a periodic function of ω with period equal to ω_s.

Physically, sampling is accomplished through the use of an analog-to-digital (A/D) converter, which first samples the signal to obtain $x_s(t)$ and then converts it into a string of pulses with amplitude 0 or 1. The process of sampling an analog signal $x(t)$ and then converting the sampled signal $x_s(t)$ into a binary-amplitude signal is also called *pulse-code modulation*. The binary-amplitude signal is constructed by the quantizing and then encoding of the sampled signal $x_s(t)$. A detailed description of the process is left to a more advanced treatment of sampling.

The continuous-time signal $x(t)$ can be regenerated from the sampled signal $x_s(t) = x(t)p(t)$ by the use of a *signal reconstructor*, illustrated in Figure 10.1b. As shown in the figure, the output of the signal reconstructor is exactly equal to $x(t)$, in which case the signal reconstructor is said to be "ideal." Note that, if an (ideal) signal reconstructor is put in cascade with a sampler, the result is the analog signal $x(t)$. More precisely, for the cascade connection shown in Figure 10.2, if the input is $x(t)$, the output is $x(t)$. This shows that signal reconstruction is the inverse of signal sampling, and conversely.

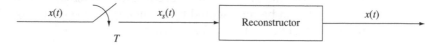

FIGURE 10.2
Cascade of a signal sampler and reconstructor.

It was shown in Section 5.4 that, if the signal $x(t)$ is bandlimited with bandwidth B, that is, $|X(\omega)| = 0$ for $\omega > B$, and if the sampling frequency ω_s is greater than or equal to $2B$, then $x(t)$ can be exactly recovered from the sampled signal $x_s(t)$ by the application of $x_s(t)$ to an ideal lowpass filter with bandwidth B. In most practical situations, signals are not perfectly bandlimited, and as a result aliasing may occur. However, as discussed in Section 5.4 the distortion due to aliasing can be reduced significantly if the signal is lowpass filtered before it is sampled. In particular, if the frequency content of the signal is not very small for frequencies larger than half of the sampling frequency ω_s, the effect of aliasing can be reduced by first lowpass filtering with an analog filter, often called an *anti-aliasing filter,* having a bandwidth less than or equal to $\omega_s/2$.

10.1.1   Hold Operation

Instead of lowpass filtering, there are other methods for reconstructing a continuous-time signal $x(t)$ from the sampled signal $x_s(t)$. One of these is the hold operation illustrated in Figure 10.3. The output $\widetilde{x}(t)$ of the hold device is given by

$$\widetilde{x}(t) = x(nT), \quad nT \leq t < nT + T \tag{10.2}$$

From (10.2) it is seen that the hold operation "holds" the value of the sampled signal at time nT until it receives the next value of the sampled signal at time $nT + T$. The output $\widetilde{x}(t)$ of the hold device is a piecewise-constant analog signal; that is, $\widetilde{x}(t)$ is constant over each T-second interval $nT \leq t < nT + T$. Since the amplitude of $\widetilde{x}(t)$ is constant over each T-second interval, the device is sometimes called a *zero-order hold.*

It turns out that the hold device corresponds to a type of lowpass filter. To see this, the frequency response function of the hold device will be computed. First, the Fourier transform $X_s(\omega)$ of the sampled signal $x_s(t) = x(t)p(t)$ will be expressed in a form different from that given in (10.1): Using the definition of the Fourier transform gives

$$X_s(\omega) = \int_{-\infty}^{\infty} x_s(t)e^{-j\omega t}\, dt$$

$$= \int_{-\infty}^{\infty} \sum_{n=-\infty}^{\infty} x(t)\delta(t - nT)e^{-j\omega t}\, dt$$

$$= \sum_{n=-\infty}^{\infty} \int_{-\infty}^{\infty} x(t)\delta(t - nT)e^{-j\omega t}\, dt$$

$$= \sum_{n=-\infty}^{\infty} x(nT)e^{-j\omega nT} \tag{10.3}$$

FIGURE 10.3
Hold operation.

Now the Fourier transform $\widetilde{X}(\omega)$ of the output $\widetilde{x}(t)$ of the hold device is given by

$$\widetilde{X}(\omega) = \int_{-\infty}^{\infty} \widetilde{x}(t)e^{-j\omega t}\, dt$$

$$= \sum_{n=-\infty}^{\infty} \int_{nT}^{nT+T} x(nT)e^{-j\omega t}\, dt$$

$$= \sum_{n=-\infty}^{\infty}\left[\int_{nT}^{nT+T} e^{-j\omega t}\, dt\right]x(nT)$$

$$= \sum_{n=-\infty}^{\infty}\left[-\frac{1}{j\omega}e^{-j\omega t}\right]_{t=nT}^{t=nT+T} x(nT)$$

$$= \frac{1 - e^{-j\omega T}}{j\omega}\sum_{n=-\infty}^{\infty} e^{-j\omega nT}x(nT) \qquad (10.4)$$

Then using (10.3) in (10.4) yields

$$\widetilde{X}(\omega) = \frac{1 - e^{-j\omega T}}{j\omega}X_s(\omega) \qquad (10.5)$$

From (10.5) it is seen that the frequency function $H_{\text{hd}}(\omega)$ of the hold device is given by

$$H_{\text{hd}}(\omega) = \frac{1 - e^{-j\omega T}}{j\omega}$$

A plot of the magnitude function $|H_{\text{hd}}(\omega)|$ is given in Figure 10.4 for the case when $T = 0.2$. As seen from the plot, the frequency response function of the hold device does correspond to that of a lowpass filter. The hold device is often used in digital control, which is considered briefly in Section 10.5.

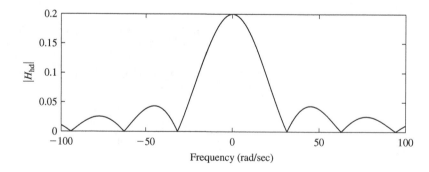

FIGURE 10.4
Magnitude of the frequency response function of the hold device with $T = 0.2$.

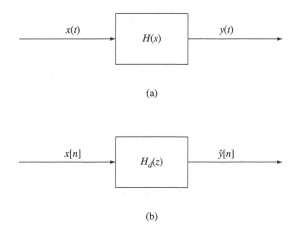

FIGURE 10.5
Given continuous-time system (a) and discretization (b).

10.1.2 System Discretization

Now consider a linear time-invariant continuous-time system with transfer function $H(s)$. As illustrated in Figure 10.5a, $y(t)$ is the output response of the system resulting from input $x(t)$ with zero initial conditions in the system prior to the application of the input. Now given $T > 0$, let $x[n]$ and $y[n]$ denote the discrete-time signals formed from the values of $x(t)$ and $y(t)$ at the times $t = nT$; that is,

$$x[n] = x(t)|_{t=nT} \quad \text{and} \quad y[n] = y(t)|_{t=nT}$$

A discretization of the given continuous-time system with transfer function $H(s)$ is a linear time-invariant discrete-time system with input $x[n]$, output $\hat{y}[n]$, and transfer function $H_d(z)$, where the subscript "d" stands for "discrete."

The discrete-time system is illustrated in Figure 10.5b. To be a discretization, the behavior of the discrete-time system must correspond to that of the continuous-time system at the sample times $t = nT$. One way to specify this correspondence is to require that the output $\hat{y}[n]$ of the discrete-time system satisfy the condition

$$\hat{y}[n] = y[n] = y(nT), \quad n = 0, \pm 1, \pm 2, \ldots \tag{10.6}$$

In other words, the output of the discretization should be exactly equal to (i.e., should match) the output values of the continuous-time system at the sample times $t = nT$. As shown next, this matching condition can be expressed in the frequency domain.

Let $\hat{Y}(\Omega)$ denote the discrete-time Fourier transform (DTFT) of $\hat{y}[n]$ given by

$$\hat{Y}(\Omega) = \sum_{n=-\infty}^{\infty} \hat{y}[n]e^{-j\Omega n}$$

Then, for (10.6) to be satisfied, it must be true that

$$\hat{Y}(\Omega) = \sum_{n=-\infty}^{\infty} y(nT)e^{-j\Omega n} \tag{10.7}$$

Now let $Y_s(\omega)$ denote the Fourier transform of the sampled output signal $y_s(t) = y(t)p(t)$. Then replacing x by y in (10.3) yields

$$Y_s(\omega) = \sum_{n=-\infty}^{\infty} y(nT)e^{-j\omega nT} \tag{10.8}$$

and setting $\omega = \Omega/T$ in (10.8) gives

$$Y_s(\omega)|_{\omega=\Omega/T} = \sum_{n=-\infty}^{\infty} y(nT)e^{-j\Omega n} \tag{10.9}$$

But the right-hand sides of (10.7) and (10.9) are identical, and thus for (10.6) to be satisfied it must be true that

$$\hat{Y}(\Omega) = Y_s(\omega)|_{\omega=\Omega/T} = Y_s\left(\frac{\Omega}{T}\right) \tag{10.10}$$

Since $\hat{Y}(\Omega)$ is periodic in Ω with period 2π, it is only necessary to consider (10.10) for $-\pi \leq \Omega \leq \pi$; that is,

$$\hat{Y}(\Omega) = Y_s(\omega)|_{\omega=\Omega/T} = Y_s\left(\frac{\Omega}{T}\right), \quad -\pi \leq \Omega \leq \pi \tag{10.11}$$

From (10.11) it is seen that the DTFT $\hat{Y}(\Omega)$ of the discrete-time system output $\hat{y}[n]$ must equal the Fourier transform $Y_s(\omega)$ of the sampled output $y_s(t)$ with $Y_s(\omega)$ evaluated at $\omega = \Omega/T$. This is the key "matching condition" in the frequency domain.

To find a discretization $H_d(\Omega)$ of a continuous-time system $H(\omega)$ that satisfies the matching condition, first note that in the ω domain

$$Y(\omega) = H(\omega)X(\omega) \tag{10.12}$$

Throughout this section it is assumed that the continuous-time system is stable so that $H(\omega)$ is equal to the transfer function $H(s)$ with $s = j\omega$. In order to avoid distortion due to aliasing, we also assume that

$$|H(\omega)| \leq c \quad \text{for } \omega > \omega_s/2 \tag{10.13}$$

for some finite value c, and

$$|X(\omega)| = 0 \quad \text{for } \omega > \omega_s/2 \tag{10.14}$$

where again $\omega_s = 2\pi/T$ is the sampling frequency.

With the condition (10.14), it follows that

$$Y_s(\omega) = H(\omega)X_s(\omega) \text{ for } -\omega_s/2 \leq \omega \leq \omega_s/2 \tag{10.15}$$

where $Y_s(\omega)$ is the Fourier transform of the sampled output $y_s(t)$ and $X_s(\omega)$ is the Fourier transform of the sampled input $x_s(t)$.

Combining the frequency matching condition (10.11) and (10.15) reveals that the DTFT $\hat{Y}(\Omega)$ of the discrete-time system output $\hat{y}[n]$ is given by

$$\hat{Y}(\Omega) = [H(\omega)X_s(\omega)]_{\omega = \Omega/T}, \quad -\pi \leq \Omega \leq \pi$$

$$= H\left(\frac{\Omega}{T}\right)X_s\left(\frac{\Omega}{T}\right), \quad -\pi \leq \Omega \leq \pi \tag{10.16}$$

But,

$$X_s\left(\frac{\Omega}{T}\right) = X_d(\Omega) \tag{10.17}$$

where $X_d(\Omega)$ is the DTFT of $x[n]$. We find the proof of (10.17) by setting $\omega = \Omega/T$ in (10.3).

Then inserting (10.17) into (10.16) yields

$$\hat{Y}(\Omega) = H\left(\frac{\Omega}{T}\right)X_d(\Omega), \quad -\pi \leq \Omega \leq \pi \tag{10.18}$$

But, from the results of Chapter 5 [see (5.60)], we see that

$$\hat{Y}(\Omega) = H_d(\Omega)X_d(\Omega), \quad -\pi \leq \Omega \leq \pi \tag{10.19}$$

where

$$H_d(\Omega) = H_d(z)|_{z=e^{j\Omega}}$$

is the frequency response function of the discrete-time system. Finally, comparing (10.18) and (10.19) reveals that

$$H_d(\Omega) = H\left(\frac{\Omega}{T}\right), \quad -\pi \leq \Omega \leq \pi \tag{10.20}$$

This is the fundamental design requirement specified in the frequency domain. Note that (10.20) defines a transformation that maps the continuous-time filter with frequency function $H(\omega)$ into the discrete-time system with frequency function $H_d(\Omega)$. More precisely, as we see from (10.20), we construct $H_d(\Omega)$ simply by setting $\omega = \Omega/T$ in $H(\omega)$.

The transformation given by (10.20) can be expressed in terms of the Laplace transform variable s and the z-transform variable z as follows:

$$H_d(z) = H(s)|_{s=(1/T)\ln z} \tag{10.21}$$

Here, $\ln z$ is the natural logarithm of the complex variable z. To verify that (10.21) does imply (10.20), simply set $z = e^{j\Omega}$ in both sides of (10.21). The relationship (10.21) shows how the continuous-time system with transfer function $H(s)$ can be transformed into the discrete-time system with transfer function $H_d(z)$. Unfortunately, due to the nature of the function $(1/T)\ln z$, it is not possible to use (10.21) to derive an expression for $H_d(z)$ that is rational in z (i.e., a ratio of polynomials in z). However, the log function $(1/T)\ln z$ can be approximated by

$$\frac{1}{T}\ln z \approx \frac{2}{T}\frac{z-1}{z+1} \tag{10.22}$$

This leads to the following transformation from the z-domain to the s-domain:

$$s = \frac{2}{T} \frac{z - 1}{z + 1} \tag{10.23}$$

Solving (10.23) for s in terms of z yields the inverse transformation given by

$$z = \frac{1 + (T/2)s}{1 - (T/2)s} \tag{10.24}$$

The transformation defined by (10.24) is called the *bilinear transformation* from the complex plane into itself. The term *bilinear* means that the relationship in (10.24) is a bilinear function of z and s. For a detailed development of bilinear transformations in complex function theory, see Churchill and Brown [2003]. For a derivation of the bilinear transformation by use of the trapezoidal integration approximation, see Problem 10.3.

The bilinear transformation (10.24) has the property that it maps the open left-half plane into the open unit disk. In other words, if Re $s < 0$, then z given by (10.24) is located in the open unit disk of the complex plane; that is, $|z| < 1$. In addition, the transformation maps the $j\omega$-axis of the complex plane onto the unit circle of the complex plane.

Now, using the approximation (10.22) in (10.21) results in

$$H_d(z) = H\left(\frac{2}{T} \frac{z - 1}{z + 1}\right) \tag{10.25}$$

By (10.25) the transfer function of the digital filter is (approximately) equal to the transfer function of the given continuous-time system, with s replaced by $(2/T)[(z - 1)/(z + 1)]$.

Note that, since the bilinear transformation maps the open left-half plane into the open unit disk, the poles of $H_d(z)$ given by (10.25) are in the open unit disk if and only if the given continuous-time system is stable (which is assumed in the foregoing). Thus, the resulting discrete-time system is stable, which shows that the bilinear transformation preserves stability. This is a very desirable property for any discretization process.

The bilinear transformation is commonly used in the design of digital filters, as shown in the next section, and in the design of digital controllers, as shown in Section 10.5. It is worth noting that in practical applications such as digital filtering and digital control, the assumption given in (10.14) is rarely met. Instead, it is assumed that $|X(\omega)| \approx 0$ for $\omega > \omega_s/2$ so that the effect of aliasing is negligible.

10.2   DESIGN OF IIR FILTERS

If a digital filter is to be used to filter continuous-time signals, the design specifications are usually given in terms of the continuous-time frequency spectrum, for example, bandwidth and passband ripple. Design methods are well established to meet these specifications, with prototype lowpass, highpass, bandpass, or bandstop analog filters. Two examples of these are the Butterworth and Chebyshev filters discussed in Section 8.6.

It is a reasonable strategy, therefore, first to design an analog prototype filter with frequency response function $H(\omega)$, and then to select a digital filter that best approximates the behavior of the desired analog filter. To obtain a good approximation, the frequency response function $H_d(\Omega)$ of the digital filter should be designed so that the condition given in (10.20) is satisfied, that is, $H_d(\Omega) = H(\omega)$ in the range $-\pi \leq \Omega \leq \pi$, where $\omega = \Omega/T$. This approximation, however, does not result in a digital filter $H_d(z)$ that is rational in z. The bilinear transformation defined in (10.23) is used instead, which results in an approximation given by (10.25) that satisfies the condition in (10.20) only approximately. The resulting digital filter will have an infinite impulse response (IIR), and thus this design approach yields an IIR filter. There are other methods for designing IIR filters; however, only the analog-to-digital transformation approach is considered here. See Oppenheim and Schafer [1989] for a discussion of other design methods, especially numerical methods.

Example 10.1   Two-Pole Butterworth Filter

Consider the two-pole Butterworth filter with transfer function

$$H(s) = \frac{\omega_c^2}{s^2 + \sqrt{2}\omega_c s + \omega_c^2}$$

Constructing the discretization $H_d(z)$ defined by (10.25) yields

$$H_d(z) = H\left(\frac{2}{T}\frac{z-1}{z+1}\right)$$

$$= \frac{\omega_c^2}{\left(\dfrac{2}{T}\dfrac{z-1}{z+1}\right)^2 + \sqrt{2}\omega_c\left(\dfrac{2}{T}\dfrac{z-1}{z+1}\right) + \omega_c^2}$$

$$= \frac{(T^2/4)(z+1)^2\omega_c^2}{(z-1)^2 + \left(T/\sqrt{2}\right)\omega_c(z+1)(z-1) + (T^2/4)(z+1)^2\omega_c^2}$$

$$= \frac{(T^2\omega_c^2/4)(z^2 + 2z + 1)}{\left(1 + \dfrac{\omega_c T}{\sqrt{2}} + \dfrac{T^2\omega_c^2}{4}\right)z^2 + \left(\dfrac{T^2\omega_c^2}{2} - 2\right)z + \left(1 - \dfrac{\omega_c T}{\sqrt{2}} + \dfrac{T^2\omega_c^2}{4}\right)}$$

For $\omega_c = 2$ and $T = 0.2$, the transfer function is given by

$$H_d(z) = \frac{0.0302(z^2 + 2z + 1)}{z^2 - 1.4514z + 0.5724}$$

The magnitude function $|H_d(\Omega)|$ plotted in Figure 10.6 is obtained by the following MATLAB commands:

```
OMEGA = -pi:2*pi/300:pi;
z = exp(j*OMEGA);
H = 0.0302*(z.^2 + 2*z + 1)./(z.^2 - 1.4514*z + 0.5724);
plot(OMEGA,abs(H))
```

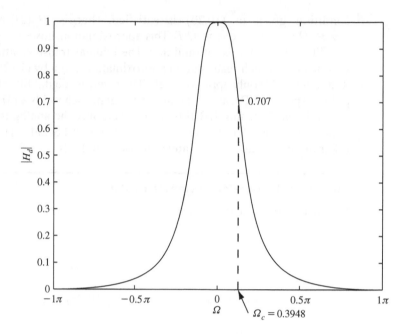

FIGURE 10.6
Magnitude function of discretization.

In this case, the cutoff frequency of the discretization is

$$\Omega_c = 2 \tan^{-1} \frac{(2)(0.2)}{2} = 0.3948$$

which is very close to the desired value $\omega_c T = 0.4$.

For the digital filter with transfer function $H_d(z)$ given by (10.25), in general, the matching condition (10.20) is not satisfied, since the bilinear transformation is an approximation to $s = (1/T)\ln z$. As a result, if $H(s)$ is the transfer function of an analog lowpass filter with cutoff frequency ω_c, in general, the corresponding cutoff frequency of the discretization $H_d(z)$ will not be equal to $\omega_c T$. The cutoff frequency of the discretization is said to be *warped* from the desired value $\omega_c T$, which results in an error in the digital filter realization of the given analog filter. The amount of warping can be computed as follows.

Setting $z = e^{j\Omega}$ in (10.25) gives

$$H_d(\Omega) = H_d(z)|_{z=e^{j\Omega}} = H\left(\frac{2}{T} \frac{e^{j\Omega} - 1}{e^{j\Omega} + 1} \right) \tag{10.26}$$

Now, since the inverse of the bilinear transformation maps the unit circle onto the $j\omega$-axis, the point $(2/T)[(e^{j\Omega} - 1)/(e^{j\Omega} + 1)]$ must be equal to some point on the $j\omega$-axis; that is,

$$j\omega = \frac{2}{T} \frac{e^{j\Omega} - 1}{e^{j\Omega} + 1} \tag{10.27}$$

for some value of ω. Hence,

$$\omega = \frac{2}{T} \frac{(1/j)(e^{j\Omega} - 1)}{e^{j\Omega} + 1}$$

$$= \frac{2}{T} \frac{(1/j2)(e^{j(\Omega/2)} - e^{-j(\Omega/2)})}{(1/2)(e^{j(\Omega/2)} + e^{-j(\Omega/2)})}$$

$$= \frac{2}{T} \tan \frac{\Omega}{2} \tag{10.28}$$

The inverse relationship is

$$\Omega = 2 \tan^{-1} \frac{\omega T}{2} \tag{10.29}$$

Combining (10.26) and (10.27) yields

$$H_d(\Omega) = H(\omega)$$

where Ω is given by (10.29).

Therefore, if ω_c is the cutoff frequency of the given analog filter [with transfer function $H(s)$], the corresponding cutoff frequency Ω_c of the discretization $H_d(z)$ is given by

$$\Omega_c = 2 \tan^{-1} \frac{\omega_c T}{2}$$

The amount of warping from the desired value $\Omega_c = \omega_c T$ depends on the magnitude of $(\omega_c T)/2$. If $(\omega_c T)/2$ is small so that $\tan^{-1}[(\omega_c T)/2)] \approx (\omega_c T)/2$, then

$$\Omega_c \approx 2\left(\frac{\omega_c T}{2}\right) = \omega_c T$$

Thus, in this case the warping is small.

We can eliminate the effect of warping by *prewarping* the analog filter prior to applying the bilinear transformation. In this process the cutoff frequency of the analog filter is designed so that the corresponding cutoff frequency Ω_c of the digital filter is equal to $\omega_c T$, where ω_c is the desired value of the analog filter cutoff frequency. The relationship $\Omega_c = \omega_c T$ follows directly from the desired matching condition (10.20). Hence, from (10.29) it is seen that the prewarped analog cutoff frequency, denoted by ω_p, should be selected as

$$\omega_p = \frac{2}{T} \tan \frac{\Omega_c}{2} \tag{10.30}$$

Thus, the analog filter should be designed to have the analog cutoff frequency ω_p (instead of ω_c) so that the distortion introduced by the bilinear transformation will be canceled by the prewarping. The procedure is illustrated by the following example:

Example 10.2   Prewarping

We can redesign the two-pole lowpass filter designed in Example 10.1 so that the cutoff frequencies match, by prewarping the analog frequency as discussed previously. For a desired analog cutoff frequency of $\omega_c = 2$ and $T = 0.2$, the desired digital cutoff frequency is $\Omega_c = 0.4$. The prewarped analog frequency is calculated from (10.30) to be $\omega_p = 2.027$. Substituting ω_p for ω_c in the transfer function of the digital filter constructed in Example 10.1 yields the redesigned filter with transfer function

$$H_d(z) = \frac{0.0309(z^2 + 2z + 1)}{z^2 - 1.444z + 0.5682} \tag{10.31}$$

10.2.1   Application to Noise Removal

Consider the problem of filtering out (or reducing in magnitude) the noise $e[n]$ contained in a signal $x[n] = s[n] + e[n]$, where $s[n]$ is the smooth part of $x[n]$. Some simple designs for reducing the noise that may be present in a signal $x[n]$ were considered in Sections 5.6 and 7.5. Since the digital cutoff frequency Ω_c of the IIR digital filter given by (10.31) is equal to 0.4 radians per unit time, this filter should be very effective in reducing any high-frequency noise that may be present in a signal. To see if this is the case, in the following example we apply the filter to closing price data for the stock index fund QQQQ.

Example 10.3   Application of IIR Filter Design to Stock Price Data

The filter with transfer function (10.31) can be applied to the closing price $c[n]$ of the stock index fund QQQQ for the 50-day period from March 1, 2004, to May 10, 2004, by the following MATLAB commands:

```
c = csvread('QQQQdata2.csv',1,4,[1 4 50 4]);
y(1)=c(1);y(2)=c(2);
for i=3:50;
 y(i)=1.444*y(i-1)-0.5682*y(i-2)+0.0309*[c(i)+2*c(i-1)+c(i-2)];
end;
```

The result is displayed in Figure 10.7, where the closing price $c[n]$ is plotted by the use of o's and the filter output $y[n]$ is plotted by the use of *'s. Note that $y[n]$ is very smooth, but it is delayed in comparison with $c[n]$ by about three to four days. However, we can reduce the time delay through the filter by two days by first multiplying the transfer function of the filter by z^2. From the properties of the z-transform (see Table 7.2), this corresponds to taking a two-step left shift of the impulse response of the filter. This results in the transfer function

$$\frac{0.0309(z^2 + 2z + 1)z^2}{z^2 - 1.444z + 0.5682}$$

Expanding by long division gives

$$\frac{0.0309(z^2 + 2z + 1)z^2}{z^2 - 1.444z + 0.5682} = 0.0309z^2 + 0.1064z + \frac{0.2021z^2 - 0.0605z}{z^2 - 1.444z + 0.5682} \tag{10.32}$$

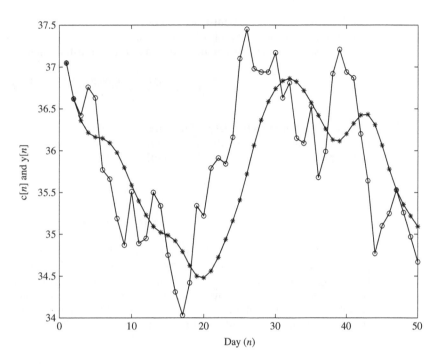

FIGURE 10.7
Filter input and output in Example 10.3.

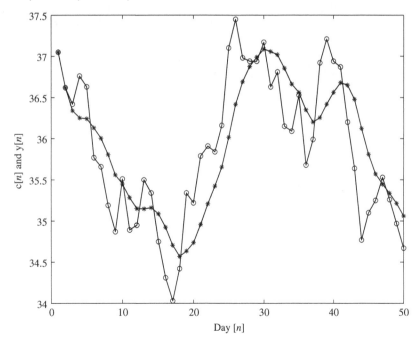

FIGURE 10.8
Modified filter input and output.

The first two terms on the right-hand side of (10.32) correspond to values of the filter's impulse response at times $n = -2$ and $n = -1$. Since the filter we are attempting to design must be causal, these two values must be deleted, and thus the transfer function of the filter becomes

$$\frac{0.2021z^2 - 0.0605z}{z^2 - 1.444z + 0.5682} \tag{10.33}$$

Finally, the transfer function $H(z)$ of the filter must equal 1 when $z = 1$, since there should be no gain or attenuation through the filter when the input is equal to a constant. Setting $z = 1$ in (10.33) gives $H(1) = 0.1416/0.1242 = 1.1401$. Hence, the transfer function of the desired filter is

$$\frac{\left(\dfrac{1}{1.1401}\right)\left(0.2021z^2 - 0.0605z\right)}{z^2 - 1.444z + 0.5682} \tag{10.34}$$

The output $y[n]$ of this filter, along with the input $c[n]$, is displayed in Figure 10.8. Again, $c[n]$ is plotted by the use of o's, and $y[n]$ is plotted by the use of *'s. From Figure 10.8 it is seen that the time delay through the modified filter is approximately one to two days, as opposed to three to four days in the filter with transfer function (10.31). However, comparing Figures 10.7 and 10.8 reveals that the output of the modified filter is not quite as smooth as the output of the filter with transfer function (10.31), and so a price is paid for reducing the time delay by two days.

Application to trading stocks.   In Section 7.5 the MACD approach to trading was considered by the use of the IIR EWMA filters with different values of the parameter b. Recall that the "MACD signal" is the difference between the output of the "faster filter" and the output of the "slower filter." Instead of using the IIR EWMA filters, we can form the difference signal by subtracting the output of the filter with transfer function (10.31) from the output of the filter with transfer function (10.34). It is possible that this difference signal will yield a higher profit for some periods than that obtained by the use of the MACD signal formed from the outputs of the EWMA filters. There are also other examples of digital filters that could be used to form a difference signal for trading. The interested reader is invited to investigate this, using historical data for QQQQ or for some other stock.

10.3   DESIGN OF IIR FILTERS USING MATLAB

In this section it is shown that the Signal Processing Toolbox in MATLAB can be used to design a digital filter by the Butterworth and Chebyshev analog prototypes discussed in Section 8.6. It should be noted that there are other analog prototype filters that are available in MATLAB, but these are not considered in this book. To design an IIR filter via MATLAB, the first step is to use MATLAB to design an analog filter that meets the desired criteria, and then the analog filter is mapped to a discrete-time (digital) filter by the bilinear transformation. Recall from Chapter 8 that the design of an analog filter by the use of MATLAB begins with the design of an N-pole lowpass filter with a bandwidth normalized to 1 rad/sec. If the analog filter is a Butterworth, the command used is `buttap`, while the command for a Chebyshev filter is `cheb1ap`. Then the filter is transformed via frequency transformations into a lowpass filter with a different bandwidth or into a highpass, bandpass, or bandstop filter with the desired frequency requirements. In MATLAB, the resulting filter analog transfer function is stored with the numerator and denominator coefficients in vectors. This transfer function can be mapped to a digital-filter transfer function by the command `bilinear`.

Example 10.4   MATLAB Design of Butterworth Filter

Digital Filtering of Continuous Time Signals

The two-pole lowpass Butterworth filter with $\omega_c = 2$ and $T = 0.2$ designed in Example 10.1 can be found by the following commands:

```
[z,p,k] = buttap(2); % creates a 2-pole filter
[num,den] = zp2tf(z,p,k);
wc = 2;  % desired cutoff frequency
[num,den] = lp2lp(num,den,wc);
T = 0.2;
[numd,dend] = bilinear(num,den,1/T)
```

The program designs a two-pole lowpass Butterworth filter with cutoff frequency of 1 rad/sec and then transforms it to a lowpass filter with $\omega_c = 2$. Recall from Section 8.6 that the frequency transformation is performed by the command `lp2lp`. Then the bilinear transformation is used to map the filter to the z-domain. The resulting vectors containing the coefficients of the digital filter are given by

```
numd = [0.0302 0.0605 0.0302]
dend = [1 -1.4514 0.5724]
```

This result corresponds exactly to the filter generated in Example 10.1.

The Signal Processing Toolbox includes the M-files `butter` and `cheby1` that already contain all the steps needed for design on the basis of analog prototypes. These commands first design the appropriate analog filter, then transform it to discrete time by the bilinear transformation. The available filter types are lowpass, highpass, and bandstop. The M-files require that the number of poles be determined and the digital cutoff frequencies be specified. Recall that the continuous-time frequency ω is related to the discrete-time frequency Ω by $\Omega = \omega T$. Hence, the digital cutoff frequency is $\Omega_c = \omega_c T$, where ω_c is the desired analog cutoff frequency. The M-files also require that the cutoff frequency be normalized by π.

Example 10.5   Alternative Design

Consider the lowpass filter design in Example 10.1. This design can be accomplished by the following commands:

```
N = 2;       % number of poles
T = 0.2;     % sampling time
wc = 2;      % analog cutoff frequency
Wc = wc*T/pi;   % normalized digital cutoff frequency
[numd,dend] = butter(N,Wc)
```

The resulting filter is defined by the numerator and denominator coefficients

$$\text{numd} = [0.0309 \; 0.0619 \; 0.0309];$$
$$\text{dend} = [1 \; -1.444 \; 0.5682];$$

This matches the results found in Example 10.2, which uses the prewarping method.

Digital
Filtering
of Con-
tinuous
Time
Signals

Example 10.6   Chebyshev Type 1 Highpass Filter

Now the objective is to design a Chebyshev type 1 highpass filter with an analog cutoff frequency of $\omega_c = 2$ rad/sec and sampling interval $T = 0.2$, and a passband ripple of 3 dB. The MATLAB commands are

```
N = 2;      % number of poles
Rp = 3;     % passband ripple
T = .2;     % sampling period
wc = 2;     % analog cutoff frequency
Wc = wc*T/pi;  % normalized digital cutoff frequency
[numd,dend] = cheby1(N,Rp,Wc,'high')
```

The filter is given by

$$\text{numd} = [0.5697 \; -1.1394 \; 0.5697]$$
$$\text{dend} = [1 \; -1.516 \; 0.7028]$$

The frequency response of the digital filter is given in Figure 10.9.

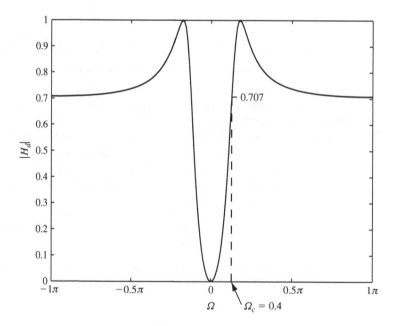

FIGURE 10.9
Frequency response function of digital filter in Example 10.6.

Example 10.7   Filtering Specific Frequencies

For the signal

$$x(t) = 1 + \cos t + \cos 5t$$

the objective is to remove the $\cos 5t$ component by the use of a two-pole digital lowpass Butterworth filter. Since the highest-frequency component of $x(t)$ is 5 rad/sec, to avoid aliasing, the sampling frequency should be at least 10 rad/sec. Thus, a sampling period of $T = 0.2$ is sufficiently small to avoid aliasing (equivalently, $\omega_s = 2\pi/T = 10\pi$ rad/sec). In addition, a filter cutoff frequency ω_c of 2 rad/sec should result in attenuation of the component $\cos 5t$ with little attenuation of $1 + \cos t$. Hence, the filter designed in Examples 10.4 and 10.5 should be adequate for the filtering task.

The following commands create the sampled version of $x(t)$ and then filter it by the command filter:

```
numd = [0.0309 0.0619 0.0309]; % define digital filter
dend = [1 -1.444 0.5682];
n = 0:80;
T = 0.2;
x = 1 + cos(T*n) + cos(T*5*n);
y = filter(numd,dend,x);
% plot x(t) with more resolution
t = 0:0.06:15;
xa = 1 + cos(t) + cos(5*t);
subplot(211),plot(t,xa); % analog input, x(t)
subplot(212),plot(n*T,y); % analog output, y(t)
pause
subplot(211),stem(n*T,x); % sampled input, x[n]
subplot(212),stem(n*T,y); % output of filter, y[n]
```

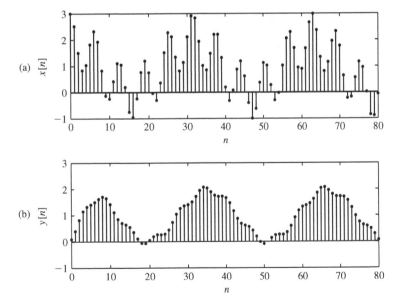

FIGURE 10.10
Plot of (a) discrete-time signal $x[n]$ and (b) output $y[n]$ of the digital filter.

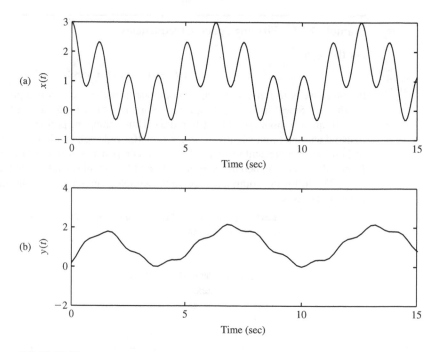

FIGURE 10.11
Plot of (a) analog signal $x(t)$ and (b) analog output $y(t)$.

The discrete-time signal $x[n]$ and the digital filter output $y[n]$ are plotted in Figure 10.10. In the preceding program, the analog output $y(t)$ is generated by the command `plot`, which approximates the output of an ideal reconstructor. The plots of the analog input $x(t)$ and analog output $y(t)$ are shown in Figure 10.11. Note that there is an initial transient in $y(t)$ due to the effect of the initial conditions, and as t increases, the analog output $y(t)$ quickly settles to steady-state behavior, since the poles of the digital filter are inside the unit circle and thus the filter is stable. From the plot of $y(t)$ in Figure 10.11b, it is clear that the frequency component $\cos 5t$ has been significantly attenuated; however, there is still some small component of $\cos 5t$ present in $y(t)$. We could have achieved a better result by using a larger-order filter, since additional poles can yield a sharper transition between the passband and the stopband. (See Chapter 8.) To achieve better rejection of the $\cos 5t$ term, the reader is invited to rewrite the previous program with a five-pole filter instead of the two-pole filter.

Example 10.8   Removing Signal Components

Again consider the signal $x(t)$ defined in Example 10.7:

$$x(t) = 1 + \cos t + \cos 5t$$

In this example, we desire to remove the component $1 + \cos t$ by using the digital highpass filter designed in Example 10.6. The following commands implement the digital filter and plot the results:

```
numd = [0.5697 -1.1394 0.5697]; % define the digital filter
dend = [1 -1.516 0.7028];
n = 0:75;
T = 0.2;
x = 1 + cos(T*n) + cos(T*5*n);
y = filter(numd,dend,x);
% plot x(t) with more resolution
t = 0:0.1:15;
xa = 1 + cos(t) + cos(5*t);
subplot(211),plot(t,xa); % analog input, x(t)
subplot(212),plot(n*T,y); % analog output, y(t)
```

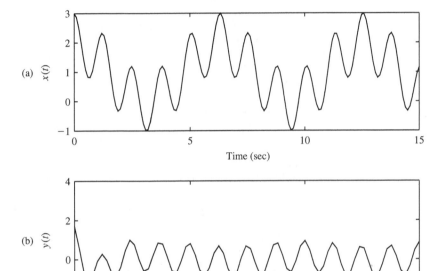

FIGURE 10.12
Plot of (a) analog signal $x(t)$ and (b) analog output $y(t)$ in Example 10.8.

The plots of $x(t)$ and $y(t)$ are shown in Figure 10.12. In the plot of $y(t)$, notice that the dc component of $x(t)$ has been filtered out and the cos t component has been reduced by about 85%, while the cos $5t$ component is left intact. Had the filter been designed with an analog cutoff frequency of a larger value, say, $\omega_c = 3$, the component cos t would be further reduced along with some attenuation of cos $5t$. The reader is invited to try this as an exercise.

Digital
Filtering
of Con-
tinuous
Time
Signals

Example 10.9   Filtering Random Signals

Consider the random continuous-time signal $x(t)$ shown in Figure 10.13a. It is assumed that the signal is bandlimited to 5π rad/sec, so that a sampling time of $T = 0.2$ is acceptable. The sampled signal is first sent through the digital lowpass filter designed in Example 10.5. The resulting analog

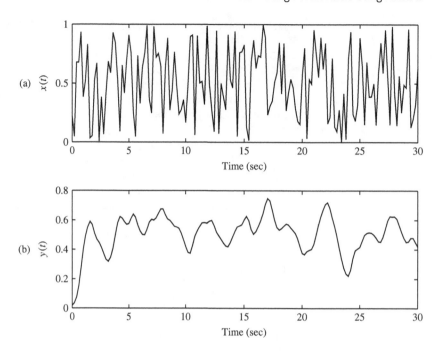

FIGURE 10.13
Plots of (a) the signal $x(t)$ and (b) the lowpass-filtered analog output $y(t)$ in Example 10.9.

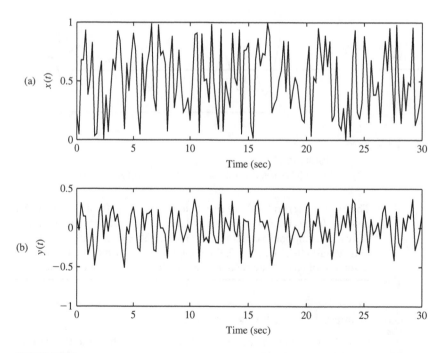

FIGURE 10.14
Plots of (a) the signal $x(t)$ and (b) the highpass-filtered analog output $y(t)$ in Example 10.9.

output $y(t)$ is shown in Figure 10.13b. Notice that the filtered signal $y(t)$ is smoother than the input signal $x(t)$, which is a result of the removal of the higher-frequency components in $x(t)$.

The sampled signal is then sent through the digital highpass filter designed in Example 10.6, and the resulting analog output is shown in Figure 10.14b. Notice that the dc component has been removed, while the peak-to-peak amplitude of the signal $x(t)$ remains approximately equal to 1.

To generate the filtered output $y(t)$, replace the definition of $x(t)$ in the M-files for Examples 10.7 and 10.8 with the command $x = \text{rand}(1, \text{length}(n));$. This creates a vector with random numbers that are uniformly distributed between 0 and 1.

10.4   DESIGN OF FIR FILTERS

As noted previously, in contrast to an IIR filter, an FIR filter is a digital filter where the impulse response (i.e., the unit-pulse response) $h[n]$ is zero for all $n \geq N$. We can design an FIR filter by truncating the impulse response of an IIR filter. In particular, let $H(\Omega)$ represent a desired IIR filter with impulse response $h[n]$. A corresponding FIR filter is given by

$$h_d[n] = \begin{cases} h[n], & 0 \leq n \leq N - 1 \\ 0, & \text{otherwise} \end{cases}$$

where N is the length of the filter. The transfer function of the FIR filter is given by

$$H_d(z) = \sum_{n=0}^{N-1} h_d[n]z^{-n}$$

The corresponding frequency response can be calculated directly from the definition of the DTFT:

$$H_d(\Omega) = \sum_{n=0}^{N-1} h_d[n]e^{-jn\Omega} \tag{10.35}$$

Ideally, $H_d(\Omega)$ should be a close approximation to the DTFT of $h[n]$ (the desired IIR filter); that is, $H_d(\Omega) \approx H(\Omega)$. However, the truncation of $h[n]$ introduces some errors in the frequency response so that $H_d(\Omega)$ may be significantly different from $H(\Omega)$.

Analytically, the truncation of the infinite impulse response can be expressed as a multiplication by a signal $w[n]$ called a window:

$$h_d[n] = \omega[n]h[n] \tag{10.36}$$

In this case,

$$\omega[n] = \begin{cases} 1 & \text{if } 0 \leq n \leq N - 1 \\ 0, & \text{otherwise} \end{cases} \tag{10.37}$$

Recall from Chapter 4 that multiplication in the time domain corresponds to a convolution in the frequency domain. Hence, taking the DTFT of both sides of (10.36) yields

$$H_d(\Omega) = \frac{1}{2\pi} \int_{-\pi}^{\pi} H(\Omega - \lambda)W(\lambda)d\lambda \tag{10.38}$$

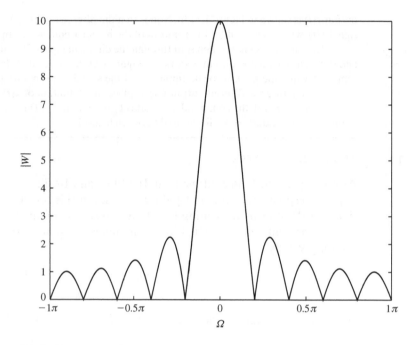

FIGURE 10.15
Plot of the magnitude $|W(\Omega)|$ of the DTFT of $w[n]$.

where the DTFT $W(\Omega)$ of $w[n]$ is given by

$$W(\Omega) = \frac{\sin(\Omega N/2)}{\sin(\Omega/2)} e^{-j\Omega(N-1)/2}$$

The plot of $|W(\Omega)|$ is shown in Figure 10.15 for $N = 10$. Notice that there is a main lobe and sidelobes with regularly spaced zero crossings at $\Omega = 2\pi m/N$ for $m = 0$, $\pm 1, \pm 2, \ldots$.

To achieve a perfect match between $H_d(\Omega)$ and $H(\Omega)$, we see from (10.38) that $W(\Omega)$ would have to be equal to the impulse $2\pi\delta(\Omega)$. This corresponds to $\omega[n] = 1$ for all n, and thus for such a $w[n]$, there is no signal truncation and the filter given by (10.36) would be IIR.

However, it is possible to have $W(\Omega)$ be a close approximation to the impulse function; in particular, the narrower the main lobe and the smaller the size of the sidelobes of $|W(\Omega)|$ are, the closer $W(\Omega)$ approximates an impulse function. As N is increased, the width of the main lobe of $|W(\Omega)|$ becomes narrower. Since N is the length of the filter, then the larger the value of N is, the closer $W(\Omega)$ approximates an impulse, and hence the closer $H_d(\Omega)$ approximates $H(\Omega)$. From a practical standpoint, however, a filter with a very large length may be too complex to implement.

One procedure for designing an FIR filter is summarized in the following steps: First, select an ideal real-valued filter frequency response function $H_i(\Omega)$ with the desired frequency characteristics. From the results of Chapter 5, taking the inverse DTFT of a real-valued frequency function $H_i(\Omega)$ will produce a time function $h_i[n]$ that is symmetrical about $n = 0$; that is,

$$h_i[n] = h_i[-n], n = 0, 1, 2, \dots \tag{10.39}$$

Hence, $h_i[n]$ must have nonzero values for $n < 0$, and therefore the ideal filter is noncausal. To produce a causal filter, the impulse response $h_i[n]$ must be delayed a sufficient number of time units so that the important characteristics of the delayed $h_i[n]$ occur for $n \geq 0$. Let the delay be equal to an integer m so that the delayed impulse response is equal to $h_i[n - m]$. Then truncate $h_i[n - m]$ for $n < 0$ and for $n > N - 1$, where $N = 2m + 1$, to yield the FIR filter with the following impulse response:

$$h_d[n] = \begin{cases} h_i[n - m] & \text{for } 0 \leq n \leq N - 1 \\ 0, & \text{otherwise} \end{cases} \tag{10.40}$$

Note that the symmetry in $h_i[n]$ about $n = 0$ as given by (10.39) results in $h_d[n]$ being symmetrical about $n = m$, so that $h_d[n] = h_d[2m - n], n = 0, 1, \dots, N - 1$. Examples of the symmetry in $h_d[n]$ are shown in Figure 10.16.

Due to the symmetry in $h_d[n]$, it turns out that the frequency response function $H_d(\Omega)$ can be expressed in the form

$$H_d(\Omega) = A(\Omega)e^{-jm\Omega} \tag{10.41}$$

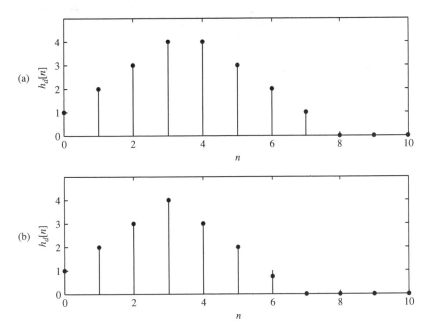

FIGURE 10.16
FIR filters with symmetry in the case when (a) $m = 7/2, N = 8$; and (b) $m = 3, N = 7$.

where $A(\Omega)$ is a real-valued function of Ω. If $A(\Omega) \geq 0$ for $0 \leq \Omega \leq \pi$, then from (10.41) it is seen that the FIR filter has a linear phase function given by $\angle A(\Omega) = -m\Omega$ for $0 \leq \Omega \leq \pi$. Hence, in this case the filter delays any input by m time units. If $A(\Omega) < 0$ for $0 < \Omega_1 < \Omega < \Omega_2 < \pi$, then $\angle A(\Omega) = \pi - m\Omega$ for $\Omega_1 < \Omega < \Omega_2$. In this case, the filter changes the sign of any sinusoidal input whose frequency Ω is in the interval $\Omega_1 < \Omega < \Omega_2$.

In the foregoing development, it was assumed that the delay m is an integer that results in an odd-length filter (i.e., $N = 2m + 1$ is odd). An even-length FIR filter can also be designed by the selection of m to be half of an integer. The filter would still be defined as in (10.40), but it would no longer be a delayed and truncated version of the ideal filter.

10.4.1   Alternative FIR Design Procedure

A more standard, but equivalent FIR filter design procedure is to reorder the steps in the previous procedure as follows. Since shifting in the time domain by m samples is equivalent to multiplying by $e^{-j\Omega m}$ in the frequency domain, we can perform the shift in time prior to taking the inverse DTFT. The difference in the approaches relates to a preference between working in the time domain or in the frequency domain. This design procedure can be summarized in the following steps: Select an ideal filter with real-valued frequency function $H_i(\Omega)$, and then multiply by $e^{-j\Omega m}$, where m is either an integer or an integer divided by 2. The inverse DTFT of the product $e^{-j\Omega m} H_i(\Omega)$ is then computed, and the resulting sequence is truncated for $n < 0$ and $n > N - 1$, where $N = 2m + 1$. If m is selected to be an integer, the resulting filter matches the form in (10.40) and has odd length (i.e., $N = 2m + 1$). If m is selected to be half of an integer, an even-length FIR filter is obtained that would no longer be a delayed and truncated version of the ideal filter. Thus, while the first design procedure previously described is very intuitive, the alternative design procedure is more general, resulting in either odd-length or even-length filters.

Example 10.10   FIR Lowpass Filter

Consider the ideal lowpass filter with frequency response function $H_i(\Omega)$ shown in Figure 10.17. Note that the cutoff frequency is Ω_c. To make the filter causal, introduce a phase shift of $e^{-j\Omega m}$ in $H_i(\Omega)$. The frequency response function $H(\Omega)$ of the resulting filter is then given by $H(\Omega) = H_i(\Omega)e^{-j\Omega m}$. From the definition of $H_i(\Omega)$, $H(\Omega)$ can be written in the form

$$H(\Omega) = \begin{cases} e^{-j\Omega m} & \text{if } |\Omega| \leq \Omega_c \\ 0 & \text{if } |\Omega| > \Omega_c \end{cases} \qquad (10.42)$$

We can compute the impulse response $h[n]$ of this filter by taking the inverse DTFT of (10.42), using the results in Chapter 5. This yields

$$h[n] = \frac{\sin(\Omega_c(n - m))}{\pi(n - m)} = \frac{\Omega_c}{\pi} \text{sinc}\left[\frac{\Omega_c(n - m)}{\pi}\right]$$

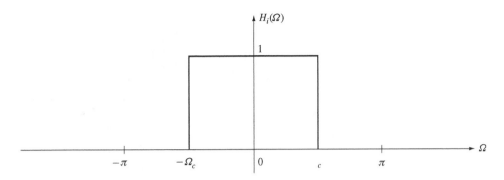

FIGURE 10.17
Frequency response function $H_i(\Omega)$ of the ideal lowpass filter in Example 10.10.

We obtain the FIR filter by truncating the response $h[n]$ for $n < 0$ and for $n > N - 1 = 2m$, which gives

$$h_d[n] = \begin{cases} \dfrac{\Omega_c}{\pi}\, \mathrm{sinc}\left[\dfrac{\Omega_c(n - m)}{\pi}\right] & \text{for } 0 \leq n \leq N - 1 \\ 0, & \text{otherwise} \end{cases} \tag{10.43}$$

For $\Omega_c = 0.4$, the impulse response of the ideal filter with a zero phase shift (i.e., in the case $m = 0$) is shown in Figure 10.18. Note the nonzero values of the impulse response for $n < 0$, which results from the noncausal nature of this filter. Shown in Figure 10.19 are the impulse responses of the resulting FIR filter defined by (10.43) for the cases when $m = 10$ and $m = \frac{21}{2}$. Notice that the FIR filter lengths are $N = 21$ and $N = 22$, respectively.

The frequency response function $H_d(\Omega)$ of the FIR filter with impulse response $h_d[n]$ is found by direct computation from the definition given by (10.35). This yields the magnitude function $|H_d(\Omega)|$ shown in Figure 10.20a for the case when $N = 21$. Comparing Figures 10.17 and 10.20a reveals that, in contrast to the frequency function $H(\Omega)$ of the ideal IIR filter, the frequency function $H_d(\Omega)$ of the FIR filter does not have a sharp transition between the passband

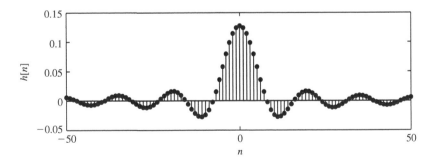

FIGURE 10.18
Impulse response of the ideal filter in Example 10.10.

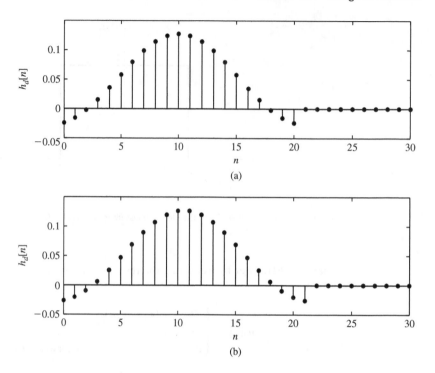

FIGURE 10.19
Impulse responses of FIR filter for (a) $N = 21$ and (b) $N = 22$.

and the stopband. Also, there is some ripple in the magnitude plot of $H_d(\Omega)$, which is a result of the truncation process. For $N = 41$, the magnitude of $H_d(\Omega)$ is plotted in Figure 10.20b. Note that the ripple has approximately the same magnitude as in the case when $N = 21$, but now the transition is much sharper.

The MATLAB commands to compute the frequency response function of the FIR filter for $N = 21$ are

```
Wc = .4; % digital cutoff frequency
N = 21; % filter length
m = (N-1)/2; % phase shift
n = 0:2*m + 10; % define points for plot
h = Wc/pi*sinc(Wc*(n-m)/pi); % delayed ideal filter
w = [ones(1,N) zeros(1,length(n)-N)]; % window
hd = h.*w;
W = -pi:2*pi/300:pi; % plot the frequency response
Hd = freqz(hd,1,W);
plot(W,abs(Hd));
```

To conclude this example, consider the analog signal $x(t)$ shown in Figure 10.21a. Recall that this signal was used in Examples 10.7 and 10.8. The lowpass FIR filter designed earlier for $N = 21$ can be used to filter the signal to remove the cos 5t component. The lowpass filter previously constructed was designed with $\Omega_c = 0.4$, which is the same cutoff frequency as for the IIR

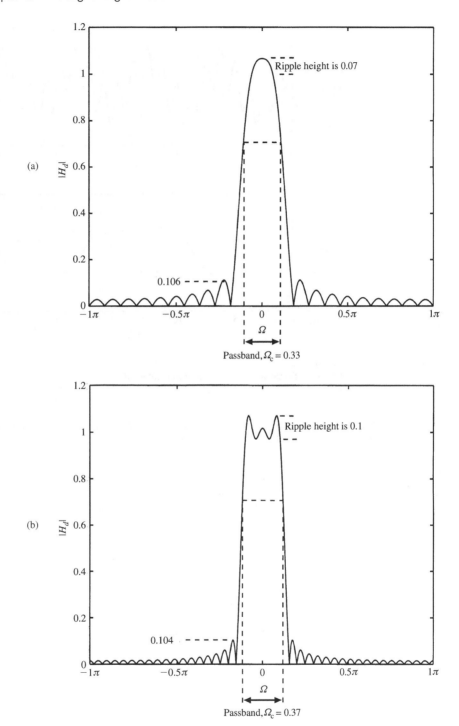

FIGURE 10.20
FIR filter magnitude function $|H_d(\Omega)|$ for the case (a) $N = 21$ and (b) $N = 41$.

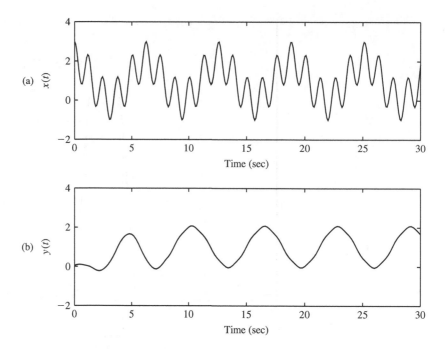

FIGURE 10.21
Plots of (a) the signal $x(t)$ and (b) the analog output $y(t)$.

filter designed in Example 10.7. The following MATLAB commands show how the signal $x(t)$ is filtered by the FIR filter designed earlier:

```
n = 0:150;
T = .2;
x = 1 + cos(n*T) + cos(5*T*n); % sampled input, x[n]
y = filter(hd,1,x); % sampled output, y[n]
t = 0:.1:30; % plot x(t) with more resolution
x = 1 + cos(t) + cos(5*t);
subplot(211),plot(t,x) % input, x(t)
subplot(212),plot(n*T,y) % output, y(t)
```

The resulting analog output $y(t)$ is shown in Figure 10.21b. Notice that the cos $5t$ component of $x(t)$ is filtered and the dc value and low-frequency component are passed without attenuation. Had a larger value of N been used in the prior FIR filter design, the analog response $y(t)$ would be even smoother than that seen in Figure 10.21b, since more of the high-frequency component would be filtered. The reader is invited to repeat the problem of filtering $x(t)$ by using an FIR filter having length $N = 41$.

10.4.2 Windows

As previously discussed, in FIR filter design the infinite-length impulse response $h[n]$ is multiplied by a window $w[n]$ to yield the truncated impulse response given by

$h_d[n] = w[n]h[n]$. The particular window $w[n]$ defined by (10.37) is referred to as the *rectangular window*, since it produces an abrupt truncation of $h[n]$. It turns out that the ripple in $H_d(\Omega)$ resulting from the use of the rectangular window (e.g., see Figure 10.20) can be reduced by the use of a window that tapers off gradually. There are several types of windows that have a gradual transition, each producing a different effect on the resulting FIR filter. Two examples are the Hanning and Hamming windows, which are defined next, along with the rectangular window.

Rectangular:
$$w[n], \qquad\qquad 0 \le n \le N - 1$$

Hanning:
$$w[n] = \frac{1}{2}\left(1 - \cos\frac{2\pi n}{N - 1}\right), \qquad 0 \le n \le N - 1$$

Hamming:
$$w[n] = 0.54 - 0.46\cos\frac{2\pi n}{N - 1}, \qquad 0 \le n \le N - 1$$

All three of these windows are plotted in Figure 10.22 for $N = 21$. (Stem plotting has been suppressed so that the comparisons between the functions are more apparent.)

The log of the magnitude of the DTFT of the window function $w[n]$ is plotted in decibels for the rectangular, Hanning, and Hamming windows in Figure 10.23 for $N = 21$. As discussed in the first part of this section, the frequency response function $H_d(\Omega)$ of an FIR filter with impulse response $h_d[n] = h[n]w[n]$ is a better approximation to the desired frequency response $H(\Omega)$ when the main lobe of $|W(\Omega)|$ is narrow

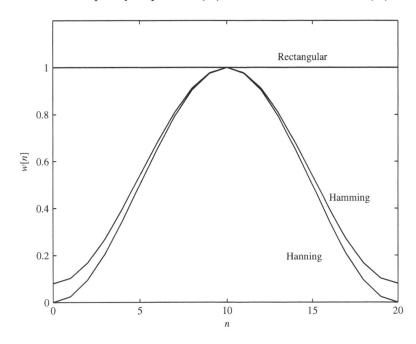

FIGURE 10.22
The rectangular, Hanning, and Hamming window functions for $N = 21$.

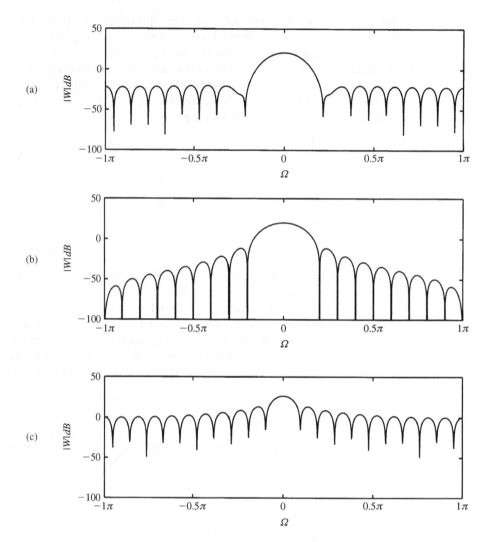

FIGURE 10.23
Log magnitude of the DTFT of the (a) rectangular, (b) Hanning, and (c) Hamming window.

and the sidelobes are small in value. The type of window function $w[n]$ that is used is based on these criteria. In particular, the nonrectangular windows have much smaller sidelobes than the rectangular window, and as a result, there is much less ripple in the frequency response function of the FIR filter.

However, for nonrectangular windows, the main lobes are wider, which means that the transition region between the passband and stopband of the FIR filter is more gradual. A more sophisticated window called a *Kaiser window* is generally used for design of practical filters, since it allows the designer the freedom to trade off the sharpness of the pass-to-stopband transitions with the magnitude of the ripples. See Oppenheim and Schafer [1989] for a more detailed discussion of windows.

Example 10.11   Lowpass Filtering by the Use of Hanning and Hamming Windows

Consider the lowpass filter designed in Example 10.10. Instead of using a rectangular window to truncate the infinite impulse response, we will use Hanning and Hamming windows. The MAT-LAB commands given in Example 10.10 can be rerun, where the definition of w is replaced by the following statement for the Hanning window:

```
w = [0 hanning(N-2)' zeros (1,length(n)-N+1)];
```

and the following statement for the Hamming window:

```
w = [hamming(N)' zeros(1,length(n)-N)];
```

With $N = 41$ ($m = 20$), the impulse response for the FIR filter designed by use of the rectangular window is shown in Figure 10.24a, while Figure 10.24b shows the corresponding frequency function. Figures 10.25 and 10.26 show the impulse response and the frequency response of the FIR filter designed by use of the Hanning and Hamming windows, respectively. Note that the ripple in the frequency response designed by use of the rectangular window is very noticeable, while the ripple in the other filter frequency responses is negligible. Also note that the transition region between the passband and stopband is more gradual for the nonrectangular windows.

The Signal Processing Toolbox contains the command `fir1`, which automatically performs the commands required in Examples 10.10 and 10.11. See the tutorial that is available on the website.

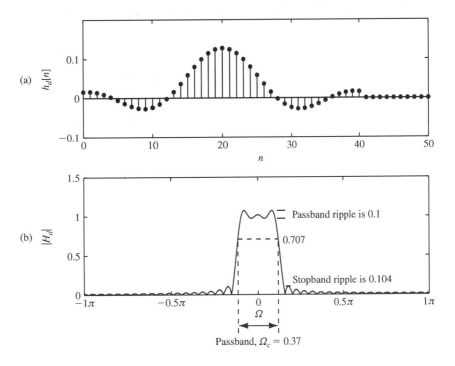

FIGURE 10.24
Plot of (a) impulse response and (b) filter frequency response for the rectangular window.

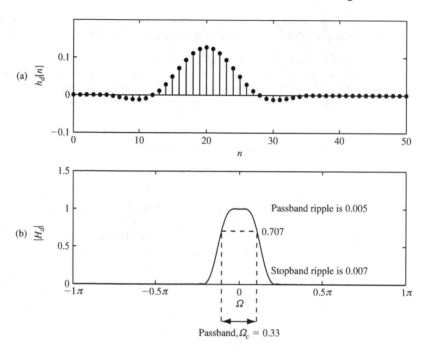

FIGURE 10.25
Plot of (a) impulse response and (b) filter frequency response for the Hanning window.

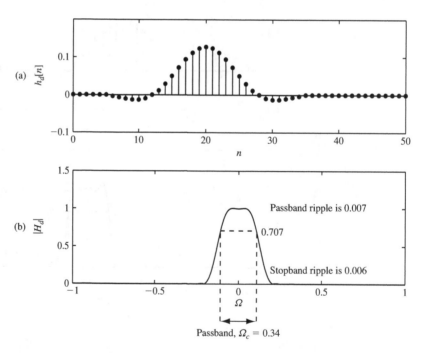

FIGURE 10.26
Plot of (a) impulse response and (b) filter frequency response for the Hamming window.

An alternative way of designing FIR filters is to use numerical techniques to derive the filter coefficients to match arbitrary frequency characteristics. For example, MATLAB includes a command that utilizes the Parks–McClellan algorithm, but this topic is beyond the scope of this book. See Oppenheim and Schafer [1989] for a discussion of various algorithmic techniques.

10.5   DESIGN OF DIGITAL CONTROLLERS

Digital control of a continuous-time system has become very standard in recent years as computer processors have become smaller, cheaper, and more powerful. Very complicated control structures can be implemented easily by a digital signal processor, whereas an equivalent analog controller may require very complex hardware. Applications where digital control has been used are engine controllers in many automobiles, flight controls on aircraft, equipment control in manufacturing systems, robotics, climate control in buildings, and process controllers in chemical plants.

Digital control started becoming commonplace in the 1970s and early 1980s as computers were becoming cheaper and more compact. The theory for continuous-time control design was already mature at that time, so the first method for digital controller design was developed on the basis of discretizing a standard continuous-time controller and then implementing the discretization by the use of a sampler-and-hold circuit. A continuous-time system (or plant) with digital controller is illustrated in Figure 10.27b, while Figure 10.27a shows the standard analog controller configuration that was studied in Chapter 9.

The method of digital controller design where a continuous-time controller is discretized (i.e., mapped to a discrete-time controller) is often referred to as *analog emulation*. More recently, direct design methods have become more established that entail mapping the continuous-time plant into the discrete-time domain and then designing the controller by the discrete-time counterparts to the root-locus method discussed in Chapter 9, as well as frequency-domain design techniques based on Bode plots. The discussion in this section is limited to the analog emulation method of design. For details on direct design methods, see Franklin et al. [1997].

The analog emulation method for designing digital controllers is very similar to the design of digital filters by the use of analog prototypes. In fact, in both methods a continuous-time system (either a filter or a controller) is first designed and then is mapped to a discrete-time transfer function. The bilinear transformation that was developed in Section 10.1 can be used to map an analog controller with transfer function $G_c(s)$ to a digital controller with transfer function $G_d(z)$. The implementation of the digital controller is achieved by the use of a computer or digital signal processor, whose output is converted to a continuous-time signal by the use of a D/A converter. This is the same process that is used for the implementation of a digital filter.

In addition to the bilinear transformation developed in Section 10.1, in applications involving digital control there are a number of other techniques that are often used to transform a continuous-time transfer function to a discrete-time function.

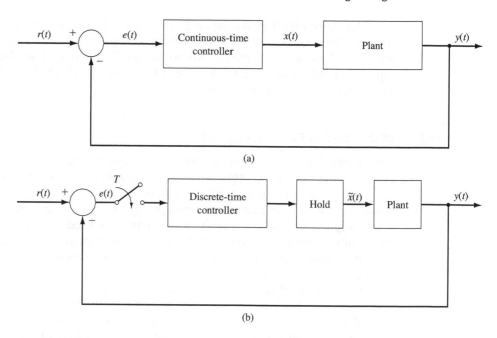

r(t) + e(t) Continuous-time controller x(t) Plant y(t) −

(a)

r(t) + e(t) T Discrete-time controller Hold $\tilde{x}(t)$ Plant y(t) −

(b)

FIGURE 10.27
Block diagram of control system with (a) analog controller and (b) digital controller.

One such method, termed *response matching,* involves matching the output of a continuous-time system to the output of a discrete-time system when the input is a specific function $x(t)$. In particular, consider a continuous-time system with transfer function $G(s)$, and let $y(t)$ be the output resulting from a specific input $x(t)$ with zero initial conditions in the system. In response matching, the objective is to construct a discrete-time system with transfer function $G_d(z)$ such that when the input $x[n]$ to the discrete-time system is $x[n] = x(nT) = x(t)|_{t=nT}$, the output $y[n]$ of the discrete-time system is $y[n] = y(nT) = y(t)|_{t=nT}$, where T is the sampling interval. In other words, for the specific input under consideration, the output $y(t)$ of the continuous-time system matches the output $y[n]$ of the discrete-time system at the sample times $t = nT$. Clearly, the transfer function $G_d(z)$ of the desired discrete-time system is given by

$$G_d(z) = \frac{Y(z)}{X(z)}$$

where $X(z)$ and $Y(z)$ are the z-transforms of the discretized input $x(nT)$ and output $y(nT)$, respectively.

Of particular interest in digital control is step-response matching, where the input $x(t)$ is a step function. In this case, we compute the output $y(t)$ by taking the inverse Laplace transform of $Y(s) = G(s)/s$, where $G(s)$ is the transfer function of the given continuous-time system. To determine the corresponding discrete-time system, $y(t)$ is discretized to obtain $y[n] = y(nT)$, and then the z-transform $Y(z)$ of $y[n]$ is computed. The transfer function $G_d(z)$ of the discrete-time system is given by

$$G_d(z) = Y(z)\frac{z-1}{z} \tag{10.44}$$

The process is illustrated by the following example:

Example 10.12   Step-Response Matching

Consider the continuous-time system with transfer function

$$G(s) = 0.2\frac{s+0.1}{s+2}$$

The transform of the step response of this system is

$$Y(s) = 0.2\frac{s+0.1}{s(s+2)}$$

$$= \frac{0.01}{s} + \frac{0.19}{s+2} \tag{10.45}$$

and taking the inverse Laplace transform of (10.45) gives the following step response:

$$y(t) = 0.01 + 0.19e^{-2t}, t \geq 0$$

The discretized version of $y(t)$ is

$$y[n] = 0.01 + 0.19e^{-2nT}, n \geq 0 \tag{10.46}$$

and taking the z-transform of (10.46) gives

$$Y(z) = 0.01\frac{z}{z-1} + 0.19\frac{z}{z-e^{-2T}}$$

$$= \frac{0.2z^2 - (0.01e^{-2T} + 0.19)z}{(z-1)(z-e^{-2T})}$$

Hence, using (10.44) yields the following transfer function for the corresponding discrete-time system:

$$G_d(z) = \frac{0.2z - (0.01e^{-2T} + 0.19)}{z - e^{-2T}}$$

The Signal Processing Toolbox in MATLAB can be used to perform step-response matching. For instance, for the system in Example 10.12, the coefficients of the transfer function $G(s)$ are stored in num and den, a value is defined for the sampling time T, and the command c2dm is used, as follows:

```
num = .2*[1 .1];
den = [1 2];
[numd,dend] = c2dm(num,den,T,'zoh');
```

The reader is invited to run this and check the results against those obtained in Example 10.12. It is worth noting that in the command c2dm, if the option `'zoh'` is replaced by `'tustin'`, the resulting computer computation uses the bilinear transformation to obtain the discrete-time system.

The main differences between the design of digital filters and the design of digital controllers are that in digital control, the determination of the sampling period must take into account the effect of the feedback, and a delay introduced by a digital controller may affect stability in the feedback loop. These considerations are discussed in more detail next.

Given a continuous-time plant with transfer function $G_p(s)$, suppose that an analog controller with transfer function $G_c(s)$ has been designed by a method such as the root-locus technique discussed in Section 9.3. To map $G_c(s)$ to a digital equivalent $G_d(z)$ by using the bilinear transformation or the response matching technique, we must determine the appropriate sampling period T. Generally, the smaller the sampling time is, the better the matching between the desired continuous-time controller and the digital controller that is implemented will be. Improving the efficiency of the computer program that performs the discrete-time calculations can reduce the sampling time; however, a reduction in sampling time generally requires the use of faster (and more expensive) A/D converters and digital signal processors. Thus, it is important that we compute the maximum sampling time that yields a good approximation when using the digital controller in place of the analog controller. An appropriate sampling frequency can be determined via the following analysis:

In the block diagram shown in Figure 10.27b, note that the signal to be sampled is the error $e(t) = r(t) - y(t)$, where $r(t)$ is the reference and $y(t)$ is the measured output signal. The reference signal has a frequency content that is usually known; however, the frequency content of $y(t)$ depends on the controller given by the transfer function $G_d(z)$. In general, $y(t)$ is not strictly bandlimited, but higher frequencies are attenuated sufficiently so that sampling will not cause substantial aliasing errors. To determine an appropriate sampling frequency, the frequency content of $y(t)$ for the system with the analog controller can first be found by application of the ω-domain relationship:

$$Y(\omega) = G_{cl}(\omega)R(\omega) \tag{10.47}$$

Here, the closed-loop frequency response function $G_{cl}(\omega)$ is given by

$$G_{cl}(\omega) = \frac{G_c(\omega)G_p(\omega)}{1 + G_c(\omega)G_p(\omega)}$$

Thus, from (10.47) it is seen that the frequency content of $y(t)$ depends on the frequency response characteristics of the closed-loop system and on the frequency content of the reference input $r(t)$. Now for sufficiently small sampling times, the digital implementation of the controller closely approximates the designed analog controller so that the frequency content of the measured output $y(t)$ (when the digital controller is utilized) is closely approximated by (10.47). Hence, the sampling frequency should be chosen to be higher than twice the largest significant frequency component of $Y(\omega)$

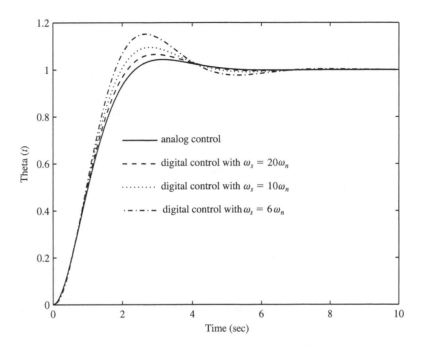

FIGURE 10.28
Response to a step reference signal in Example 10.13.

given by (10.47), and it should be sufficiently large so that (10.47) gives a good approximation of the frequency content of the output when the digital controller is used. One rule of thumb is to choose the sampling frequency to be greater than 10 to 20 times the bandwidth of the analog closed-loop system.

Once a specific sampling interval T is chosen, the continuous-time controller $G_c(s)$ can be mapped to a discrete-time equivalent by the step-response matching method or the bilinear transformation. As mentioned previously, other methods of mapping are also available. Most will yield nearly the same answer if the sampling period is small with respect to the system's natural frequencies. The discretization process is illustrated next.

Example 10.13   Digital Control of dc Motor

A digital controller implementation will be given for an analog controller for the dc motor defined in Example 9.5. Here the plant (the dc motor) is given by

$$G_p(s) = \frac{10}{(s + 0.1)s}$$

and the analog controller has transfer function

$$G_c(s) = 0.2\frac{s + 0.1}{s + 2}$$

The corresponding closed-loop transfer function is given by

$$G_{cl}(s) = \frac{2}{s^2 + 2s + 2}$$

The natural frequency of the closed-loop system is $\omega_n = \sqrt{2} \approx 1.41$ rad/sec. Since this is a second-order system with no zeros, the natural frequency is approximately the same as the bandwidth. Thus, to get good performance from the digital controller, the sampling frequency is chosen to be $\omega_s \approx 20\omega_n$ or $T = 0.22$ second.

From the results in Example 10.12, we see that the digital controller found by using step-response matching for $T = 0.22$ is

$$G_d(z) = 0.2\frac{z - 0.982}{z - 0.644} \tag{10.48}$$

To use the bilinear transformation to find the digital controller, substitute $s = 2(z - 1)/T(z + 1)$ in $G_c(s)$ for s:

$$G_d(z) = G_c(s)|_{s=2(z-1)/T(z+1)} = \frac{0.166(z - 0.978)}{z - 0.639} \tag{10.49}$$

Due to the small sampling time, the digital controllers in (10.48) and (10.49) found from the two different mapping methods are very similar.

To simulate the response of the plant with digital control, the M-file `hybrid` can be used. This M-file is available from the website. The commands to derive $G_d(z)$ from $G_c(s)$ by use of the bilinear transformation and to simulate the response to a step input are as follows:

```
T = 0.22;
Nc = .2*[1 .1]; % analog controller
Dc = [1 2];
[Nd,Dd] = bilinear(Nc,Dc,1/T); % digital controller
t = 0:.5*T:10;
u = ones(1,length(t)); % step input
Np = 10; % plant
Dp = [1 .1 0];
[theta,uc] = hybrid(Np,Dp,Nd,Dd,T,t,u);
```

The responses to a step reference are shown in Figure 10.28 for three different digital controllers that are implemented with different sampling frequencies. Also shown in Figure 10.28 is the step response found by the use of the analog controller. Notice that there is very little degradation in the response, due to the digital implementation when $\omega_s = 20\omega_n$, but the degradation is more apparent with a smaller sampling frequency such as $\omega_s = 10\omega_n$ and $\omega_s = 6\omega_n$.

A simulation of the plant and digital controller can also be performed by use of the Simulink model shown in Figure 10.29. We obtain the plant transfer function by clicking and dragging the Transfer Function icon found in the Continuous menu item in the library browser. Similarly, the transfer function for the digital controller is found in the Discrete menu, the scope is found in the Sinks menu, the step function is found in the Sources menu, and the summing junction is found in Math Operations. The parameters for a particular object are set by double-clicking on the object to open a Function Block Parameter Window. The sampling time is chosen to be 0.22 and must be entered as a parameter for several of the blocks.

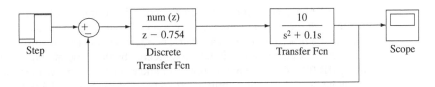

FIGURE 10.29
Simulink model of the digital controller and continuous-time plant in Example 10.13.

To implement a digital controller in a physical system by using a standard programming language like C, we must convert the control transfer function to a difference equation. The difference equation can be implemented recursively by the use of a loop. The implementation procedure outlined in the following example applies to digital filters as well as digital controllers:

Example 10.14   Digital Control or Digital Filter Implementation

To implement a digital controller defined by a transfer function $G_d(z)$, first take the inverse z-transform of the transfer function for the controller to obtain a difference equation with input $e[n]$ and output $u[n]$. Here, $u[n]$ represents a command to the motor, not a step signal. Consider the lead controller for a dc motor given in Example 10.13. Multiply $G_d(z)$ by z^{-1}/z^{-1}, and set it equal to $U(z)/E(z)$ to obtain

$$U(z) = \frac{0.166(1 - 0.978z^{-1})}{1 - 0.639z^{-1}} E(z)$$

$$(1 - 0.639z^{-1})U(z) = 0.166(1 - 0.978z^{-1})E(z)$$

$$U(z) - 0.639z^{-1}U(z) = 0.166E(z) - 0.162z^{-1}E(z)$$

Take the inverse z-transform of the last expression to obtain the difference equation for the controller:

$$u[n] = 0.166e[n] - 0.162e[n - 1] + 0.639u[n - 1]$$

The following pseudocode can be used to write a program to implement the control:

```
ulast = 0;
elast = 0;
loop{
    y = input(sensor);
    e = r - y;
    u = 0.166*e - 0.162*elast + 0.639*ulast;
    output(u);
    ulast = u;
    elast = e;
}
```

579

In this implementation, the sampling time is the amount of computation time that it takes to run through the loop one time. A more sophisticated program would put the controller computations into a procedure that is called at a periodic rate in order to ensure constant sampling times.

An IIR digital filter can be implemented by the same methodology. Consider the second-order IIR filter designed in Example 10.1 and given by a transfer function denoted as $H_d(z)$. Let the input be $x[n]$ and the output be $y[n]$. Due to the second-order poles and zeros, the difference equation will depend on the last two values of the input and output, that is, $x[n-1]$, $x[n-2]$, $y[n-1]$, and $y[n-2]$, which will both have to be saved for the recursion.

10.5.1   Digital Control Project

An experiment that utilizes a LEGO® MINDSTORMS® kit is described in this subsection. The experiment can be used for a project if the students or the instructor have the LEGO kits, or this section can be used as a case study for implementing controls.

LEGO MINDSTORMS kits provide a cheap, portable experimental platform for simple projects in digital control and for digital filtering. Several experiments for digital control are described in the IEEE Control Systems Magazine, Special Issue on Innovations in Undergraduate Education, Vol. 24, No. 5, October, 2004. One experiment in particular is simple enough for students who are taking an introductory systems and control course (see Heck et al. [2004]). The experiment is a feedback control of a dc motor designed to control the angle of an output shaft. The cost and portability of this experiment make it possible for students to check out an assembly to use at home, and so, just a few experimental assemblies can be used for an entire class.

The experimental assembly described in Heck et al. [2004] is shown in Figure 10.30 and consists of a dc motor, a rotation sensor, LEGO building pieces, and the LEGO RCX brick. The LEGO brick consists of a microcontroller and input and output ports. This platform provides all the essential components for *embedded controls*—that is, control systems implemented by the use of *embedded computers*. Embedded computers,

FIGURE 10.30
LEGO-based experiment for digital control of a motor.

unlike desktop computers, are processors that are part of a larger system where the processor is not the focal point of the system. In embedded controls, a computer program is written to perform the digital control computations. This program is generally written and compiled on a desktop or laptop machine and then downloaded to a target machine, the LEGO brick in this application. As in many embedded computing applications, the capabilities of the LEGO brick are limited. The RCX brick, for example, runs at a small clock rate (8MHz), is an 8-bit processor, and does not have a floating point processor. A newer version of the LEGO brick, the NXT brick, has more capabilities.

The goal of this project is to control the angle of an output shaft that is attached to the motor shaft through gears. The angle of the motor shaft is measured by a rotation sensor. The transfer function for the motor assembly is given by

$$G_p(s) = \frac{Y(s)}{V(s)} = \frac{A}{Ts^2 + s} \frac{1}{N}$$

where y is the angle of the output shaft, v is the voltage to the motor, T is the time constant of the motor-load system, A is a constant, and N is the gear ratio. For the system shown in Figure 10.29, $N = 5$.

The simplest control is a proportional control, which is implemented digitally from the following pseudocode:

```
loop{
    y = input(rotation sensor);
    e = r − y;
    u = K*u;
    output(u);
}
```

This program can be implemented in C. The program, as well as detailed instructions for building the apparatus, is available from the website for this book.

Students are instructed to build the apparatus to be used as a project and to program the LEGO brick with the proportional controller, using various values of K, such as $K = \frac{1}{5}, \frac{1}{4}, \frac{1}{3}$, and $\frac{1}{2}$. With a reference r equal to specific value, such as 32, run the program. While it is running, try to turn the output shaft manually away from its set point. The controller resists this movement and acts to return the shaft to its set point position. This is the main difference between open-loop control and closed-loop control: The closed-loop controller is robust to disturbances acting on the system. Use the datalog function to save the sensor data for a step response, and upload the data to a computer for plotting.

The next part of the project is to design and implement a lead controller of the form

$$G_c(s) = K\frac{s + b}{s + a}$$

where $b < a$. Control design requires first identifying the parameters A and T of the transfer function of the motor assembly. Often, we can accomplish system identification by inputting a step function into the system and determining the time constant and damping ratio of the response. Dominant second-order poles can be determined from a step response by the method given in Section 8.3. In this case, however, the open-loop transfer function has a pole at the origin, and so a step response would include a ramp, which would make system identification difficult. An alternative is to apply proportional control to the system in order to stabilize it, and then perform system identification of the closed-loop system. A value of $K = \frac{1}{3}$ works well for this system. Find the poles and the dc value of the closed-loop system from the step response. Either analytically from the closed-loop transfer function or using the root locus, find the parameters T and A of the plant transfer function.

Once the open-loop system is identified, then design the lead controller to have a faster time constant than the proportional controller. A rule of thumb is to choose the zero at $-b$ to cancel the second plant pole from the right, and then use the root locus method to select an appropriate value for a. Placing $-a$ farther left in the s-plane pulls the root locus to the left, making the closed loop response faster. A practical limitation is the sampling time. In the proportional controller implementation described by the pseudocode, the sampling time is the actual time that it takes for the loop to be computed. Consider that a lead controller is more complicated and will take more time to compute, and so the sampling time will be slightly slower. Therefore, choosing a arbitrarily large would result in a very fast closed-loop response, which would require a sampling time that is smaller than is possible with this hardware. The speed of response of the closed-loop system in this project is limited because of the sampling time limitations.

Once a lead controller $G_c(s)$ is chosen that gives acceptable closed-loop response, the controller can be discretized by either the bilinear transformation or the step-response matching method. These methods require the actual sampling time. A lower limit on the sampling time is the processor execution time for the loop that performs the computations shown in Example 10.14. Determine the sampling time with a dummy set of coefficients in the controller. With the actual sampling time recorded, discretize the controller to determine the actual coefficients. Program the RCX brick with the lead controller, and record the step response. Again, try to turn the shaft away from the set point, and notice the resistance due to the feedback action.

10.6   CHAPTER SUMMARY

This chapter demonstrates the use of the fundamental concepts described in Chapters 5–9 to design digital filters and digital controllers, both of which are commonly used in engineering applications. In these practical applications, we interface a computer with a physical system by sampling an analog signal and then processing the sampled signal numerically, usually with a digital signal processor or a microcontroller.

The chapter begins with the discretization of a signal by sampling. Reconstruction of the sampled signal into an analog signal can be done through an ideal lowpass filter if aliasing did not occur in the sampling process. This condition is ensured if the original analog signal was bandlimited and if the sampling frequency was at least twice as high as the highest frequency of the analog signal. A practical method to reconstruct an analog signal from a digital signal is a hold operation, which is employed in digital-to-analog converters. In the frequency domain, the hold operation corresponds to a nonideal lowpass filter.

We accomplish system discretization by finding a transformation between the continuous-time domain and the discrete-time domain. An exact transformation that satisfies the matching condition is given by $\omega = \Omega/T$, or correspondingly, $s = (1/T)\ln z$.

Application of this transformation to a transfer function $H(s)$, however, yields a transfer function $H_d(z)$ that is not rational in z. A commonly used approximation to the exact transformation is the bilinear transformation, where we find the transfer function of the discrete-time system by replacing s with $(2/T)[(z-1)/(z+1)]$ in the continuous-time system. The bilinear transformation is used commonly in both digital filter design and digital control design to obtain a digital system that emulates a desired analog filter or analog controller. An alternative discretization method that is used in digital control design is step-response matching. While the exact matching condition ensures that the sampled response of the analog system matches the response of the digital system for all inputs, step-response matching ensures only that the outputs match when the inputs are step functions. This relaxation of the matching condition results in a discretization that is rational in z.

Digital filters are classified as having an infinite impulse response or a finite impulse response. An IIR filter, which is characterized by having nonzero poles in the transfer function $H(z)$, is obtained when an analog prototype filter is discretized. Common digital filters based on analog prototypes include Butterworth filters and Chebyshev filters. The infinite impulse response results from a dependence on previous values of the output y to compute the current value of the output. In contrast, FIR filters do not depend on previous values of the output. The resulting transfer function has only poles at the origin. We can obtain an FIR filter by truncating and shifting the impulse response of an IIR filter that has desirable frequency domain characteristics. Equivalently, we can obtain an FIR filter by multiplying a filter $H_i(\Omega)$ that has the desired characteristics by $e^{-j\Omega m}$ where m is an integer. We obtain the FIR filter by truncating the inverse DTFT of $e^{-j\Omega m}H_i(\Omega)$. Both of these design approaches require a truncation, which results in a ripple in the frequency response of the FIR filter. We may decrease this ripple by performing a more gradual truncation, which we achieve by using a window in place of a sharp truncation. Two common windows are the Hanning and Hamming windows.

We can obtain digital controllers by discretizing a continuous-time controller. A common consideration in digital control is the selection of the sampling frequency. To avoid aliasing, the sampled signal should have negligible frequency content beyond half of the sampling frequency. Since the output signal is normally not bandlimited, a rule of thumb exists to ensure that the level of aliasing is low. In particular, as the desired closed-loop frequency response most typically resembles a lowpass filter, you should choose the sampling frequency to be 10 to 20 times the bandwidth of the closed-loop continuous-time system. In the absence of numerical precision considerations, the performance of the digital controller approaches that of the continuous-time controller as the sampling frequency increases.

PROBLEMS

10.1. Determine an appropriate sampling frequency that avoids aliasing for the following signals:

   (a) $x(t) = 3\,\mathrm{sinc}^2(t/2\pi),\ -\infty < t < \infty$

   (b) $x(t) = 4\,\mathrm{sinc}(t/\pi)\cos 2t,\ -\infty < t < \infty$

   (c) $x(t) = e^{-5t}u(t)$

10.2. Digitize the following systems by using the bilinear transformation. Assume that $T = 0.2$ second.

    (i) $H(s) = 2/(s + 2)$

    (ii) $H(s) = 4(s + 1)/(s^2 + 4s + 4)$

    (iii) $H(s) = 2s/(s^2 + 1.4s + 1)$

(a) For each continuous-time system, simulate the step response, using `step`.

(b) For each discrete-time system derived in part (a), simulate the step response, using `dstep`. Compare these responses, $y[n] = y(nT)$, with the corresponding responses $y(t)$ obtained in part (a) by plotting the results.

10.3. The bilinear transformation was introduced in (10.23) as a means of approximating the exact mapping $s = (1/T)\ln(z)$. An alternative derivation for the bilinear transformation involves the trapezoidal approximation of an integral, as follows:

$$\int_{nT}^{(n+1)T} f(t)\, dt \approx \frac{T}{2}[f([n + 1]T) + f(nT)]$$

Here, the right-hand side of the expression represents the area of the trapezoid that best fits under the curve $f(t)$ from $t = nT$ to $t = (n + 1)T$. (See Figure P10.3.) Now consider a first-order continuous-time system:

$$H(s) = \frac{Y(s)}{X(s)} = \frac{a}{s + a}$$

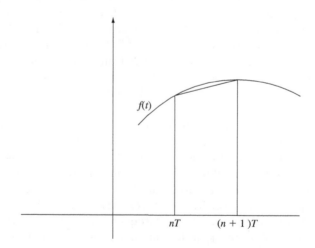

FIGURE P10.3

To derive the bilinear transformation, perform the following steps: (1) Find the differential equation that relates the input $x(t)$ to the output $y(t)$; (2) integrate both sides of the differential equation from $t = nT$ to $t = (n + 1)T$, using the trapezoidal approximation where appropriate; (3) obtain a corresponding difference equation by letting $y[n] = y(nT)$ and $x[n] = x(nT)$; (4) compute the digital transfer function $H_d(z)$ of the difference equation; and (5) find a relationship between s and z so that $H(s) = H_d(z)$.

10.4. Consider the sampled data system shown in Figure P10.4. Compute exact values for $y(1)$, $y(2)$, and $y(3)$ when the following conditions are met:

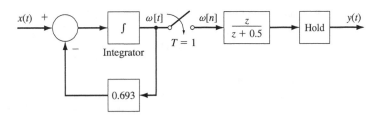

FIGURE P10.4

(a) $\omega(0) = -1$, $y(0) = 1$, and $x(t) = \delta(t) =$ unit impulse
(b) $\omega(0) = y(0) = 0$ and $x(t) = e^{-2t}u(t)$
(c) $\omega(0) = 2$, $y(0) = -1$, and $x(t) = u(t)$

10.5. Consider the one-pole lowpass filter given by the transfer function

$$H(s) = \frac{B}{s + B}$$

(a) Design a discrete-time system that realizes this filter, using the bilinear transformation for $\omega_s = B$, $\omega_s = 2B$, and $\omega_s = 5B$.
(b) Sketch the frequency response of the continuous-time system for $B = 10$, and mark the three different sampling frequencies. Which should give a more accurate discrete-time realization?
(c) Use MATLAB to compute the step responses of the three discrete-time systems obtained in part (a) with $B = 10$. Plot the responses versus $t = nT$, where T is the sampling period. On the same graph, plot the step response of the original continuous-time system, and compare the accuracy of the three discretizations.

10.6. Suppose that the sampled sinusoid $x(nT) = \cos \omega_0 nT$, $n = 0, \pm 1, \pm 2, \ldots$, is applied (with $\omega_s = 2B$) to the discretization constructed in Problem 10.5. Determine the range of values of ω_0 for which the peak magnitude of the resulting output response is greater than or equal to 0.707. In other words, determine the "effective" 3-dB bandwidth of the discretization. Compare your results with those obtained in part (b) of Problem 10.5.

10.7. A two-pole bandpass filter is given by the transfer function

$$H(s) = \frac{100}{s^2 + 2s + 101}$$

(a) Determine the 3-dB points of the filter.
(b) Digitize the filter, using the bilinear transformation for an arbitrary sampling interval T.
(c) Plot the frequency response function of the digital filter $H_d(z)$ obtained in part (b). Take $T = 0.02$ second.

10.8. Consider the two-pole highpass filter given by the transfer function

$$H(s) = \frac{s^2}{s^2 + \sqrt{2}s + 1}$$

(a) Determine the 3-dB point of the filter.
(b) Discretize the filter, using the bilinear transformation.
(c) With the sampling interval $T = 0.1$, plot the frequency response of the digital filter obtained in part (b).

10.9. Consider the two-pole Chebyshev lowpass filter given by

$$H(s) = \frac{0.5\omega_c^2}{s^2 + 0.645\omega_c s + 0.708\omega_c^2}$$

where ω_c is the 3-dB point.

(a) Discretize the filter, using the bilinear transformation.

(b) Determine the output response $y(nT)$ of the discretized filter to the inputs $x(nT)$ that follow. Take $\omega_c = 6\pi$ and $T = 0.01$. Plot the resulting analog signal $y(t)$ generated from an ideal reconstructor [i.e., plot $y(nT)$, using the `plot` command].

(i) $x(nT) = p_1(nT)$, where $p_1(t) = u(t + 1/2) - u(t - 1/2)$

(ii) $x(nT) = p_1(nT) + 0.5w(nT)$, where $w(nT)$ is a noise signal whose values are random numbers between 0 and 1. (Use `rand` in MATLAB to generate the signal.)

(iii) $x(nT) = p_1(nT) + w(nT)$

(iv) $x(nT) = (1 - 2|nT|)p_1(nT)$

(v) $x(nT) = (1 - 2|nT|)p_1(nT) + 0.5w(nT)$

10.10. Design a three-pole lowpass IIR filter to have an analog cutoff frequency of $\omega_c = 15$. The sampling interval is $T = 0.1$. Perform the design twice, once without prewarping the frequency and once with prewarping the frequency. Plot the frequency curves of the two filters, and compare the actual digital cutoff frequencies. Base your design on a Butterworth analog prototype filter.

10.11. Consider the closing prices of QQQQ for the 149-business-day period from July 1, 2004, up to February 1, 2005.

(a) Using the filters with transfer functions (10.31) and (10.34) and the trading strategy given in Section 7.5, determine the days when there is a buy and the days when there is a sell.

(b) Using your result in part (a), determine the net gain or loss per share.

10.12. Use analytical methods to design a three-pole lowpass IIR digital filter that has an analog cutoff frequency of $\omega_c = 10$ rad/sec, and assume a sampling period of $T = 0.1$ second. Base your design on a Chebyshev analog prototype filter with a passband ripple of 3 dB.

(a) Specify the desired digital cutoff frequency, Ω_c. Also, give the largest frequency component of an input $x(t)$ that would be allowed, to avoid aliasing.

(b) Verify your analytical design by using MATLAB to design the filter numerically. Plot the frequency response function for the resulting digital filter. Measure the actual digital cutoff frequency.

(c) Use MATLAB to simulate the response to the following sampled signal: $x(nT) = 1 + \sin \pi nT + \sin 6\pi nT$. Plot the analog input and outputs, $x(t)$ and $y(t)$, and the digital filter inputs and outputs, $x[n]$ and $y[n]$.

10.13. Use analytical methods to design a three-pole highpass IIR digital filter that has an analog cutoff frequency of $\omega_c = 10$ rad/sec, and assume a sampling period of $T = 0.1$ second. Base your design on a Butterworth analog prototype filter.

(a) Specify the desired digital cutoff frequency, Ω_c. Also, give the largest frequency component of $x(t)$ that would be allowed, to avoid aliasing.

(b) Verify your analytical design by using MATLAB to design the filter numerically. Plot the frequency response function for the resulting digital filter. Measure the actual digital cutoff frequency.

(c) Use MATLAB to simulate the response to the following sampled signal: $x(nT) = 1 + \sin \pi nT + \sin 6\pi nT$. Plot the analog input and outputs, $x(t)$ and $y(t)$, and the digital filter inputs and outputs, $x[n]$ and $y[n]$.

10.14. Use MATLAB to design a three-pole bandpass IIR digital filter that has an analog pass-band of $\omega_c = 1$ to $\omega_c = 5$ rad/sec; assume a sampling period of $T = 0.1$ second. Base your design on a Butterworth analog prototype filter. Specify the filter in terms of $H_d(z)$.

   (a) Plot the frequency response function for the resulting digital filter.

   (b) From the frequency response curve plotted in part (a), estimate the amplitude of the response to the following input signals: $x(t) = 1$, $x(t) = \sin \pi t$, and $x(t) = \sin 6\pi t$.

   (c) Simulate the response of the filter to the sampled signal $x(nT) = 1 + \sin \pi nT + \sin 6\pi nT$. Compare the response with the expected amplitudes derived in part (b).

   (d) Simulate the response of the filter to the sampled random signal $x(nT)$ generated from the MATLAB command $x = \text{rand}(1, 200)$. Plot the input and corresponding response in continuous time.

10.15. Design a lowpass FIR filter of length $N = 30$ that has a cutoff frequency of $\Omega_c = \pi/3$. Use the rectangular window.

   (a) Plot the impulse response of the filter.

   (b) Plot the magnitude of the frequency response curve, and determine the amount of ripple in the passband.

   (c) Compute and plot the response of the filter to an input of $x[n] = 1 + 2 \cos(\pi n/6) + 2 \cos(2\pi n/3)$. Discuss the filtering effect on the various components of $x[n]$.

10.16. Design a highpass FIR filter of length $N = 15$ that has a cutoff frequency of $\Omega_c = \pi/2$. Use the rectangular window.

   (a) Plot the impulse response of the filter.

   (b) Plot the magnitude of the frequency response curve, and determine the amount of ripple in the passband.

   (c) Repeat the design for length $N = 31$, and compare the two filters in terms of ripple and transition region.

   (d) Compute and plot the response of each filter to an input of $x[n] = 2 + 2 \cos(\pi n/3) + 2 \cos(2\pi n/3)$. Compare the filters in terms of their effect on the various components of $x[n]$.

10.17. Design a lowpass FIR filter of length $N = 10$ that has a cutoff frequency of $\Omega_c = \pi/3$. Perform your design by using **(i)** a rectangular window, **(ii)** a Hamming window, and **(iii)** a Hanning window.

   (a) Plot the impulse responses of each filter.

   (b) Plot the magnitude of the frequency response for each filter, and compare the filters in terms of the stopband and passband ripples.

   (c) Compute and plot the response of each filter to an input of $x[n] = 2 + 2 \cos(\pi n/4) + 2 \cos(\pi n/2)$. Compare the filters in terms of their effect on the various components of $x[n]$.

10.18. Consider the vehicle system described in Example 9.1 and given by the input/output differential equation

$$\frac{d^2y(t)}{dt^2} + \frac{k_f}{M}\frac{dy(t)}{dt} = \frac{1}{M}x(t)$$

where $y(t)$ is the position of the car at time t. Suppose that $M = 1$ and $k_f = 0.1$.

   (a) Discretize the system, using the bilinear transformation for $T = 5$ seconds.

   (b) Discretize the system, using step-response matching for $T = 5$ seconds.

   (c) For each digitization obtained in parts (a) and (b), plot the analog output $y(nT)$ of the discrete-time step response. Compare these results with the step response for the original continuous-time system. Plot all results for $t = 0$ to $t = 50$ seconds.

   (d) Repeat parts (a) to (c) for $T = 1$ second.

10.19. Each continuous-time system given next represents a desired closed-loop transfer function that will be achieved by the use of a digital feedback controller. Determine a range of appropriate sampling frequencies ω_s for each system.

 (a) $G_{cl}(s) = 10/(s + 10)$

 (b) $G_{cl}(s) = 4/(s^2 + 2.83s + 4)$

 (c) $G_{cl}(s) = 9(s + 1)/(s^2 + 5s + 9)$

10.20. Consider the following continuous-time system:

$$G_p(s) = \frac{2}{s + 2}$$

Digital control is to be applied to this system so that the resulting closed-loop system has a pole at $s = -4$.

 (a) Design a continuous-time feedback controller $G_c(s)$ that achieves the desired closed-loop pole.

 (b) Obtain a digital controller from part (a), where $T = 0.25$ second. Using this control, simulate the step response of the closed-loop system, using either Simulink or the command `hybrid`.

 (c) Repeat part (b) with $T = 0.1$ second. Compare the resulting response with that obtained in part (b), and determine which is closer to the desired closed-loop response.

10.21. Consider the following continuous-time system:

$$G_p(s) = \frac{1}{(s + 1)(s + 2)}$$

The following lead controller is designed that gives closed-loop poles at $s = -4 \pm 8j$:

$$G_c(s) = \frac{73(s + 2)}{s + 5}$$

 (a) Digitize the controller, using step-response matching for $T = 0.1$ second. Using this control, simulate the step response of the closed-loop system, using either Simulink or the command `hybrid`.

 (b) Repeat part (a) with $T = 0.05$ second. Compare the resulting response with that obtained in part (a), and determine which is closer to the desired closed-loop response.

The models that have been considered so far in this book are mathematical representations of the input/output behavior of the system under study. In this chapter a new type of model is defined, which is specified in terms of a collection of variables that describe the internal behavior of the system. These variables are called the state variables of the system. The model defined in terms of the state variables is called the state or state-variable representation. The objective of this chapter is to define the state model and to study the basic properties of this model for both continuous-time and discrete-time systems. An in-depth development of the state approach to systems can be found in a number of textbooks. For example, the reader may want to refer to Kailath [1980], Brogan [1991], or Rugh [1996].

The state model is given in terms of a matrix equation, and thus the reader should be familiar with matrix algebra. A brief review is given in Appendix B. As a result of the matrix form of the state model, it is easily implemented on a computer. A number of commercially available software packages, such as MATLAB, can be used to carry out the matrix operations arising in the state model. In particular, MATLAB is built around operations involving matrices and vectors and thus is well suited for the study of the state model. MATLAB's definition of operations in terms of matrices and vectors results in state model computations that are as easy as standard calculator operations.

The development of the state model begins in Section 11.1 with the notion of state and the definition of the state equations for a continuous-time system. The construction of state models from input/output differential equations is considered in Section 11.2. The solution of the state equations is studied in Section 11.3. Then in Section 11.4 the discrete-time version of the state model is presented. In Section 11.5 the notion of equivalent state representations is studied, and in Section 11.6 the discretization of continuous-time state models is pursued. A summary of the chapter is given in Section 11.7.

11.1 STATE MODEL

Consider a single-input single-output causal continuous-time system with input $v(t)$ and output $y(t)$. Throughout this chapter the input will be denoted by $v(t)$, rather than $x(t)$, since the symbol "$x(t)$" will be used to denote the system state as defined subsequently.

Given a value t_1 of the time variable t, in general, it is not possible to compute the output response $y(t)$ for $t \geq t_1$ from only knowledge of the input $v(t)$ for $t \geq t_1$. The reason for this is that the application of the input $v(t)$ for $t < t_1$ may put energy into the system that affects the output response for $t \geq t_1$. For example, a voltage or current applied to an RLC circuit for $t < t_1$ may result in voltages on the capacitors and/or currents in the inductors at time t_1. These voltages and currents at time t_1 can then affect the output of the RLC circuit for $t \geq t_1$.

Given a system with input $v(t)$ and output $y(t)$, for any time point t_1 the state $x(t_1)$ of the system at time $t = t_1$ is defined to be that portion of the past history $(t \le t_1)$ of the system required to determine the output response $y(t)$ for all $t \ge t_1$, given the input $v(t)$ for $t \ge t_1$. A nonzero state $x(t_1)$ at time t_1 indicates the presence of energy in the system at time t_1. In particular, the system is in the zero state at time t_1 if and only if there is no energy in the system at time t_1. If the system is in the zero state at time t_1, the response $y(t)$ for $t \ge t_1$ can be computed from knowledge of the input $v(t)$ for $t \ge t_1$. If the state at time t_1 is not zero, knowledge of the state is necessary to compute the output response for $t \ge t_1$.

If the given system is finite dimensional, the state $x(t)$ of the system at time t is an N-element column vector given by

$$x(t) = \begin{bmatrix} x_1(t) \\ x_2(t) \\ \vdots \\ x_N(t) \end{bmatrix}$$

The components $x_1(t), x_2(t), \ldots, x_N(t)$ are called the *state variables* of the system, and the number N of state variables is called the *dimension* of the state model (or system). For example, suppose that the given system is an *RLC* circuit. From circuit theory, any energy in the circuit at time t is completely characterized by the voltages across the capacitors at time t and the currents in the inductors at time t. Thus, the state of the circuit at time t can be defined to be a vector whose components are the voltages across the capacitors at time t and the currents in the inductors at time t. If the number of capacitors in the circuit is equal to N_C and the number of inductors in the circuit is equal to N_L, the total number of state variables is equal to $N_C + N_L$.

Now suppose that the system is an integrator with the input/output relationship

$$y(t) = \int_{t_0}^{t} v(\lambda)\, d\lambda, \quad t > t_0 \tag{11.1}$$

where $y(t_0) = 0$. It will be shown that the state $x(t)$ of the integrator can be chosen to be the output $y(t)$ of the integrator at time t. To see this, let t_1 be an arbitrary value of time with $t_1 > t_0$. Then rewriting (11.1) yields

$$y(t) = \int_{t_0}^{t_1} v(\lambda)\, d\lambda + \int_{t_1}^{t} v(\lambda)\, d\lambda, \, t \ge t_1 \tag{11.2}$$

From (11.2) it is seen that the first term on the right-hand side of (11.2) is equal to $y(t_1)$. Therefore,

$$y(t) = y(t_1) + \int_{t_1}^{t} v(\lambda)\, d\lambda, \, t \ge t_1 \tag{11.3}$$

The relationship (11.3) shows that $y(t_1)$ characterizes the energy in the system at time t_1. More precisely, from (11.3) it is seen that $y(t)$ can be computed for all $t \ge t_1$ from knowledge of $v(t)$ for $t \geq t_1$ and knowledge of $y(t_1)$. Thus, the state at time t_1 can be taken to be $y(t_1)$.

Now consider an interconnection of integrators, adders, subtracters, and scalar multipliers. Since adders, subtracters, and scalar multipliers are memoryless devices, the energy in the interconnection is completely characterized by the values of the outputs of the integrators. Thus, the state at time t can be defined to be a vector whose components are the outputs of the integrators at time t.

11.1.1 State Equations

Consider a single-input single-output finite-dimensional continuous-time system with state $x(t)$ given by

$$x(t) = \begin{bmatrix} x_1(t) \\ x_2(t) \\ \vdots \\ x_N(t) \end{bmatrix}$$

The state $x(t)$ is a vector-valued function of time t. In other words, for any particular value of t, $x(t)$ is an N-element column vector. The vector-valued function $x(t)$ is called the *state trajectory of the system.*

If the system with N-dimensional state vector $x(t)$ is linear and time invariant, the system can be modeled by the state equations given by

$$\dot{x}(t) = Ax(t) + bv(t) \tag{11.4}$$

$$y(t) = cx(t) + dv(t) \tag{11.5}$$

where A is an $N \times N$ matrix, b is an N-element column vector, c is an N-element row vector, d is a real-valued constant, and $\dot{x}(t)$ is the derivative of the state vector with the derivative taken component by component; that is,

$$\dot{x}(t) = \begin{bmatrix} \dot{x}_1(t) \\ \dot{x}_2(t) \\ \vdots \\ \dot{x}_N(t) \end{bmatrix}$$

Since $x(t)$ and $\dot{x}(t)$ are N-element column vectors, (11.4) is a vector differential equation. In particular, (11.4) is a first-order vector differential equation. Equation (11.5) is called the *output equation* of the system. The term $dv(t)$ in (11.5) is a "direct feed" between the input $v(t)$ and output $y(t)$. If $d = 0$, there is no direct connection between $v(t)$ and $y(t)$.

Equations (11.4) and (11.5) constitute the state model of the system. This representation is a time-domain model of the system, since the equations are in terms of functions of time. Note that the state model is specified in two parts; (11.4) describes the state response resulting from the application of an input $v(t)$ with initial state $x(t_0) = x_0$, while (11.5) gives the output response as a function of the state and input. The two parts of the state model correspond to a cascade decomposition of the system

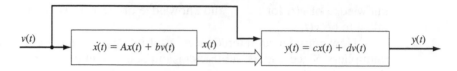

FIGURE 11.1
Cascade structure corresponding to state model.

as illustrated in Figure 11.1. The double line for $x(t)$ in Figure 11.1 indicates that $x(t)$ is a vector signal.

From Figure 11.1 it is seen that the system state $x(t)$ is an "internal" vector variable of the system; that is, the state variables [the components of $x(t)$] are signals within the system. Since the state model is specified in terms of the internal vector variable $x(t)$, the representation is an internal model of the system. The form of this model is quite different from that of the external or input/output models studied in Chapter 2.

With a_{ij} equal to the ij entry of the matrix A, and b_i equal to the ith component of the column vector b, (11.4) can be written in the expanded form

$$\dot{x}_1(t) = a_{11}x_1(t) + a_{12}x_2(t) + \cdots + a_{1N}x_N(t) + b_1v(t)$$
$$\dot{x}_2(t) = a_{21}x_1(t) + a_{22}x_2(t) + \cdots + a_{2N}x_N(t) + b_2v(t)$$
$$\vdots \qquad\qquad \vdots$$
$$\dot{x}_N(t) = a_{N1}x_1(t) + a_{N2}x_2(t) + \cdots + a_{NN}x_N(t) + b_Nv(t)$$

With $c = [c_1 \ c_2 \ \cdots \ c_N]$, the expanded form of (11.5) is

$$y(t) = c_1x_1(t) + c_2x_2(t) + \cdots + c_Nx_N(t) + dv(t)$$

From the expanded form of the state equations, it is seen that the derivative $\dot{x}_i(t)$ of the ith state variable and the output $y(t)$ are equal to linear combinations of all the state variables and the input.

11.2   CONSTRUCTION OF STATE MODELS

In the first part of this section it is shown how to construct a state model from the input/output differential equation of the system. The development begins with the first-order case.

Consider the single-input single-output continuous-time system given by the first-order input/output differential equation

$$\dot{y}(t) + ay(t) = bv(t) \tag{11.6}$$

where a and b are scalar constants. Defining the state $x(t)$ of the system to be equal to $y(t)$ results in the state model

$$\dot{x}(t) = -ax(t) + bv(t)$$
$$y(t) = x(t)$$

Thus, it is easy to construct a state model from a first-order input/output differential equation. In terms of the notation of (11.4) and (11.5), the coefficients A, b, c, and d of this state model are

$$A = -a, \quad b = b, \quad c = 1, \quad d = 0$$

Now suppose that the given system has the second-order input/output differential equation

$$\ddot{y}(t) + a_1\dot{y}(t) + a_0 y(t) = b_0 v(t) \tag{11.7}$$

Defining the state variables by

$$x_1(t) = y(t), \quad x_2(t) = \dot{y}(t)$$

yields the state equations

$$\dot{x}_1(t) = x_2(t)$$
$$\dot{x}_2(t) = -a_0 x_1(t) - a_1 x_2(t) + b_0 v(t)$$
$$y(t) = x_1(t)$$

Writing these equations in matrix form yields the following state model:

$$\begin{bmatrix} \dot{x}_1(t) \\ \dot{x}_2(t) \end{bmatrix} = \begin{bmatrix} 0 & 1 \\ -a_0 & -a_1 \end{bmatrix} \begin{bmatrix} x_1(t) \\ x_2(t) \end{bmatrix} + \begin{bmatrix} 0 \\ b_0 \end{bmatrix} v(t)$$

$$y(t) = \begin{bmatrix} 1 & 0 \end{bmatrix} \begin{bmatrix} x_1(t) \\ x_2(t) \end{bmatrix}$$

The definition of state variables in terms of the output and derivatives of the output extends to any system given by the Nth-order input/output differential equation

$$y^{(N)}(t) + \sum_{i=0}^{N-1} a_i y^{(i)}(t) = b_0 v(t) \tag{11.8}$$

where $y^{(i)} = d^i y/dt^i$. To express (11.8) in state-equation form, first define the state variables by

$$x_i(t) = y^{(i-1)}(t), i = 1, 2, \ldots, N$$

Then, from (11.8), the resulting state equations are

$$\dot{x}_1(t) = x_2(t)$$
$$\dot{x}_2(t) = x_3(t)$$
$$\vdots$$
$$\dot{x}_{N-1}(t) = x_N(t)$$
$$\dot{x}_N(t) = -\sum_{i=0}^{N-1} a_i x_{i+1}(t) + b_0 v(t)$$
$$y(t) = x_1(t)$$

Writing these equations in matrix form yields the N-dimensional state model $\dot{x}(t) = Ax(t) + bv(t), y(t) = cx(t)$, where

$$A = \begin{bmatrix} 0 & 1 & 0 & \cdots & 0 \\ 0 & 0 & 1 & & 0 \\ \vdots & \vdots & & \ddots & \vdots \\ 0 & 0 & 0 & & 1 \\ -a_0 & -a_1 & -a_2 & \cdots & -a_{N-1} \end{bmatrix}, \quad b = \begin{bmatrix} 0 \\ 0 \\ \vdots \\ 0 \\ b_0 \end{bmatrix}, \quad c = [1 \ 0 \ 0 \ \cdots \ 0]$$

From the preceding constructions, it is tempting to conclude that the state variables of a system can always be defined to be equal to the output $y(t)$ and derivatives of $y(t)$. Unfortunately, this is not the case. For instance, suppose that the system is given by the linear second-order input/output differential equation

$$\ddot{y}(t) + a_1\dot{y}(t) + a_0y(t) = b_1\dot{v}(t) + b_0v(t) \tag{11.9}$$

where $b_1 \neq 0$. Note that (11.9) is not a special case of (11.8), since $\ddot{y}(t)$ depends on $\dot{v}(t)$.

If $x_1(t) = y(t)$ and $x_2(t) = \dot{y}(t)$, it is not possible to eliminate the term $b_1\dot{v}(t)$ in (11.9). Thus, there is no state model with respect to this definition of state variables. But the system does have the following state model:

$$\begin{bmatrix} \dot{x}_1(t) \\ \dot{x}_2(t) \end{bmatrix} = \begin{bmatrix} 0 & 1 \\ -a_0 & -a_1 \end{bmatrix}\begin{bmatrix} x_1(t) \\ x_2(t) \end{bmatrix} + \begin{bmatrix} 0 \\ 1 \end{bmatrix}v(t), \quad y(t) = [b_0 \ \ b_1]\begin{bmatrix} x_1(t) \\ x_2(t) \end{bmatrix} \tag{11.10}$$

To verify that (11.10) is a state model, it must be shown that the input/output differential equation corresponding to (11.10) is the same as (11.9). Expanding (11.10) gives

$$\dot{x}_1(t) = x_2(t) \tag{11.11}$$

$$\dot{x}_2(t) = -a_0x_1(t) - a_1x_2(t) + v(t) \tag{11.12}$$

$$y(t) = b_0x_1(t) + b_1x_2(t) \tag{11.13}$$

Differentiating both sides of (11.13) and using (11.11) and (11.12) yields

$$\dot{y}(t) = b_0x_2(t) + b_1[-a_0x_1(t) - a_1x_2(t) + v(t)]$$
$$= -a_1y(t) + (a_1b_0 - a_0b_1)x_1(t) + b_0x_2(t) + b_1v(t) \tag{11.14}$$

Differentiating both sides of (11.14) and again using (11.11) and (11.12) gives

$$\ddot{y}(t) = -a_1\dot{y}(t) + (a_1b_0 - a_0b_1)x_2(t)$$
$$+ b_0[-a_0x_1(t) - a_1x_2(t) + v(t)] + b_1\dot{v}(t)$$
$$= -a_1\dot{y}(t) - a_0y(t) + b_0v(t) + b_1\dot{v}(t)$$

This is the same as the input/output differential equation (11.9) of the given system, and thus (11.10) is a state model.

The state variables $x_1(t)$ and $x_2(t)$ in the state model (11.10) can be expressed in terms of $v(t), y(t)$, and $\dot{y}(t)$. The derivation of these expressions is left to the interested reader. (See Problem 11.3.)

Now consider the linear time-invariant system given by the Nth-order input/output differential equation

$$y^{(N)}(t) + \sum_{i=0}^{N-1} a_i y^{(i)}(t) = \sum_{i=0}^{N-1} b_i v^{(i)}(t) \tag{11.15}$$

Note that this input/output differential equation includes derivatives of the input $v(t)$, whereas there are no derivatives in the input/output differential equation (11.8). The system given by (11.15) has the N-dimensional state model $\dot{x}(t) = Ax(t) + bv(t), y(t) = cx(t)$, where

$$A = \begin{bmatrix} 0 & 1 & 0 & \cdots & 0 \\ 0 & 0 & 1 & & 0 \\ \vdots & \vdots & & \ddots & \vdots \\ 0 & 0 & 0 & & 1 \\ -a_0 & -a_1 & -a_2 & \cdots & -a_{N-1} \end{bmatrix}, \quad b = \begin{bmatrix} 0 \\ 0 \\ \vdots \\ 0 \\ 1 \end{bmatrix}, \quad c = [b_0 \ b_1 \ \cdots \ b_{N-1}]$$

The verification that this is a state model is omitted.

11.2.1 Integrator Realizations

Any linear time-invariant system given by the N-dimensional state model

$$\dot{x}(t) = Ax(t) + bv(t)$$
$$y(t) = c(t) + dv(t)$$

can be realized by an interconnection of N integrators and combinations of adders, subtracters, and scalar multipliers. The steps of the realization process are as follows:

Step 1. For each state variable $x_i(t)$, construct an integrator and define the output of the integrator to be $x_i(t)$. The input to the ith integrator will then be equal to $\dot{x}_i(t)$. Note that, if there are N state variables, the integrator realization will contain N integrators.

Step 2. Put an adder/subtracter in front of each integrator. Feed into the adders/subtracters scalar multiples of the state variables and input according to the vector equation $\dot{x}(t) = Ax(t) + bv(t)$.

Step 3. Put scalar multiples of the state variables and input into an adder/subtracter to realize the output $y(t)$ in accordance with the equation $y(t) = cx(t) + dv(t)$.

Example 11.1   Integrator Realization

Consider a two-dimensional state model with arbitrary coefficients; that is,

$$\begin{bmatrix} \dot{x}_1(t) \\ \dot{x}_2(t) \end{bmatrix} = \begin{bmatrix} a_{11} & a_{12} \\ a_{21} & a_{22} \end{bmatrix} \begin{bmatrix} x_1(t) \\ x_2(t) \end{bmatrix} + \begin{bmatrix} b_1 \\ b_2 \end{bmatrix} v(t)$$

$$y(t) = \begin{bmatrix} c_1 & c_2 \end{bmatrix} \begin{bmatrix} x_1(t) \\ x_2(t) \end{bmatrix}$$

Following the previous steps results in the realization shown in Figure 11.2.

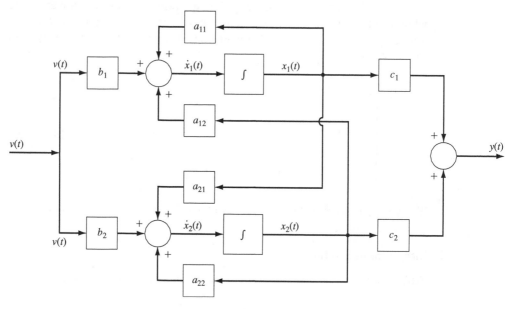

FIGURE 11.2
Realization in Example 11.1.

There is a converse to the result that any linear time-invariant system given by a state model has an integrator realization. Namely, any system specified by an interconnection consisting of N integrators and combinations of adders, subtracters, and scalar multipliers has a state model of dimension N. We can compute a state model directly from the interconnection by employing the following steps:

Step 1. Define the output of each integrator in the interconnection to be a state variable. Then, if the output of the ith integrator is $x_i(t)$, the input to this integrator is $\dot{x}_i(t)$.

Step 2. By looking at the interconnection, express each $\dot{x}_i(t)$ in terms of a sum of scalar multiples of the state variables and input. Writing these relationships in matrix form yields the vector equation $\dot{x}(t) = Ax(t) + bv(t)$.

Step 3. Again looking at the interconnection, express the output $y(t)$ in terms of scalar multiples of the state variables and input. Writing this in vector form yields the output equation $y(t) = cx(t) + dv(t)$.

Example 11.2   State Equations from an Integrator Realization

Consider the system shown in Figure 11.3. With the output of the first integrator denoted by $x_1(t)$ and the output of the second integrator denoted by $x_2(t)$, from Figure 11.3, we find that

$$\dot{x}_1(t) = -x_1(t) - 3x_2(t) + v(t)$$
$$\dot{x}_2(t) = x_1(t) + 2v(t)$$

Also from Figure 11.3, we see that

$$y(t) = x_1(t) + x_2(t) + 2v(t)$$

Thus, the coefficient matrices of the state model are

$$A = \begin{bmatrix} -1 & -3 \\ 1 & 0 \end{bmatrix}, \quad b = \begin{bmatrix} 1 \\ 2 \end{bmatrix}, \quad c = \begin{bmatrix} 1 & 1 \end{bmatrix}, \quad d = 2$$

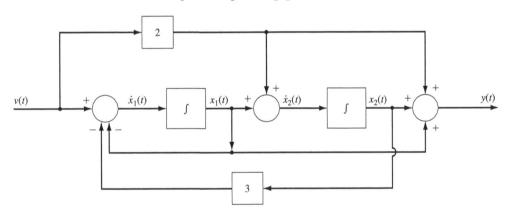

FIGURE 11.3
System in Example 11.2.

From the foregoing results it is seen that there is a one-to-one correspondence between integrator realizations and state models.

11.2.2   Multi-Input Multi-Output Systems

The state model generalizes very easily to multi-input multi-output systems. In particular, the state model of a p-input r-output linear time-invariant finite-dimensional continuous-time system is given by

$$\dot{x}(t) = Ax(t) + Bv(t)$$
$$y(t) = Cx(t) + Dv(t)$$

where now B is an $N \times p$ matrix of real numbers, C is an $r \times N$ matrix of real numbers, and D is an $r \times p$ matrix of real numbers. The matrix A is still $N \times N$, as in the single-input single-output case.

If a p-input r-output system is specified by a collection of coupled input/output differential equations, we can construct a state model by generalizing the procedure given previously in the single-input single-output case. The process is illustrated by the following example:

Example 11.3 A Coupled Two-Car System

Consider two cars moving along a level surface as shown in Figure 11.4. It is assumed that the mass of both cars is equal to M and that the coefficient corresponding to viscous friction is the same for both cars and is equal to k_f. As illustrated, $d_1(t)$ is the position of the first car at time t, $d_2(t)$ is the position of the second car at time t, $f_1(t)$ is the drive or braking force applied to the first car, and $f_2(t)$ is the drive or braking force applied to the second car. Thus, the motion of the two cars is given by the differential equations

$$\ddot{d}_1(t) + \frac{k_f}{M}\dot{d}_1(t) = \frac{1}{M}f_1(t)$$

$$\ddot{d}_2(t) + \frac{k_f}{M}\dot{d}_2(t) = \frac{1}{M}f_2(t)$$

The first car also has a radar, which gives a measurement of the distance

$$w(t) = d_2(t) - d_1(t)$$

between the two cars at time t. The purpose of the distance measurement is to allow for automatic control of the speed of the car with the radar so that the car maintains a safe distance behind the car in front of it.

The inputs of the two-car system are defined to be $f_1(t)$ and $f_2(t)$, and thus the system is a two-input system. The outputs are defined to be the velocity $\dot{d}_1(t)$ of the first car and the separation $w(t)$ between the two cars, both of which can be measured at the car with the radar. Thus the output (viewed from the car with the radar) is a vector given by

$$y(t) = \begin{bmatrix} \dot{d}_1(t) \\ w(t) \end{bmatrix}$$

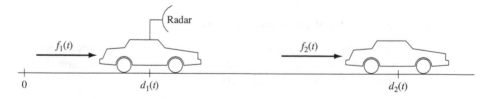

FIGURE 11.4
Coupled two-car system.

With this definition of the output, the system is a two-output system. Now, if the state variables are defined by

$$x_1(t) = \dot{d}_1(t)$$

$$x_2(t) = \dot{d}_2(t)$$

$$x_3(t) = w(t)$$

then the state model of the system is

$$\begin{bmatrix} \dot{x}_1(t) \\ \dot{x}_2(t) \\ \dot{x}_3(t) \end{bmatrix} = \begin{bmatrix} \dfrac{-k_f}{M} & 0 & 0 \\ 0 & \dfrac{-k_f}{M} & 0 \\ -1 & 1 & 0 \end{bmatrix} \begin{bmatrix} x_1(t) \\ x_2(t) \\ x_3(t) \end{bmatrix} + \begin{bmatrix} \dfrac{1}{M} & 0 \\ 0 & \dfrac{1}{M} \\ 0 & 0 \end{bmatrix} \begin{bmatrix} f_1(t) \\ f_2(t) \end{bmatrix} \tag{11.16}$$

$$y(t) = \begin{bmatrix} 1 & 0 & 0 \\ 0 & 0 & 1 \end{bmatrix} \begin{bmatrix} \dot{d}_1(t) \\ \dot{d}_2(t) \\ w(t) \end{bmatrix} \tag{11.17}$$

If a p-input r-output system is given by an interconnection of integrators, adders, subtracters, and scalar multipliers, a state model can be constructed directly from the interconnection. The process is very similar to the steps given in the single-input single-output case.

Example 11.4   Two-Input Two-Output System

Consider the two-input two-output system shown in Figure 11.5. From the figure, we find that

$$\dot{x}_1(t) = -3y_1(t) + v_1(t)$$

$$\dot{x}_2(t) = v_2(t)$$

$$y_1(t) = x_1(t) + x_2(t)$$

$$y_2(t) = x_2(t)$$

Inserting the expression for $y_1(t)$ into the expression for $\dot{x}_1(t)$ gives

$$\dot{x}_1(t) = -3[x_1(t) + x_2(t)] + v_1(t)$$

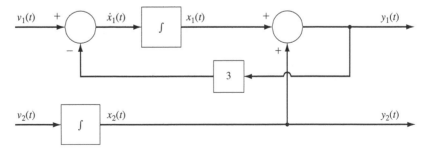

FIGURE 11.5
System in Example 11.4.

Putting these equations in matrix form results in the state model

$$\begin{bmatrix} \dot{x}_1(t) \\ \dot{x}_2(t) \end{bmatrix} = \begin{bmatrix} -3 & -3 \\ 0 & 0 \end{bmatrix} \begin{bmatrix} x_1(t) \\ x_2(t) \end{bmatrix} + \begin{bmatrix} 1 & 0 \\ 0 & 1 \end{bmatrix} \begin{bmatrix} v_1(t) \\ v_2(t) \end{bmatrix}$$

$$\begin{bmatrix} y_1(t) \\ y_2(t) \end{bmatrix} = \begin{bmatrix} 1 & 1 \\ 0 & 1 \end{bmatrix} \begin{bmatrix} x_1(t) \\ x_2(t) \end{bmatrix}$$

11.3   SOLUTION OF STATE EQUATIONS

Consider the p-input r-output linear time-invariant continuous-time system given by the state model

$$\dot{x}(t) = Ax(t) + Bv(t) \tag{11.18}$$

$$y(t) = Cx(t) + Dv(t) \tag{11.19}$$

Recall that the matrix A is $N \times N$, B is $N \times p$, C is $r \times N$, and D is $r \times p$. Given an initial state $x(0)$ at initial time $t = 0$ and an input $v(t)$, $t \geq 0$, in this section an analytical expression is derived for the solution $x(t)$ of (11.18). Then, from this, (11.19) will be used to derive an expression for the output response $y(t)$. The numerical solution of (11.18) and (11.19) by MATLAB will be demonstrated in examples that follow. The numerical solution procedure used by MATLAB is discussed in Section 11.6.

We begin the development by considering the free (unforced) vector differential equation

$$\dot{x}(t) = Ax(t), \quad t > 0 \tag{11.20}$$

with initial state $x(0)$. To solve (11.20), it is necessary to define the *matrix exponential* e^{At}, which is a generalization of the scalar exponential e^{at}. For each real value of t, e^{At} is defined by the matrix power series

$$e^{At} = I + At + \frac{A^2 t^2}{2!} + \frac{A^3 t^3}{3!} + \frac{A^4 t^4}{4!} + \cdots \tag{11.21}$$

where I is the $N \times N$ identity matrix. The matrix exponential e^{At} is an $N \times N$ matrix of time functions. Later it will be shown how the elements of e^{At} can be computed by the Laplace transform.

There are a couple of properties of e^{At} that are needed: First, for any real numbers t and λ,

$$e^{A(t+\lambda)} = e^{At} e^{A\lambda} \tag{11.22}$$

We can check the relationship (11.22) by setting $t = t + \lambda$ in (11.21).

Taking $\lambda = -t$ in (11.22) gives

$$e^{At} e^{-At} = e^{A(t-t)} = I \tag{11.23}$$

The relationship (11.23) shows that the matrix e^{At} always has an inverse, which is equal to the matrix e^{-At}.

The derivative $d/dt(e^{At})$ of the matrix exponential e^{At} is defined to be the matrix that we form by taking the derivative of the components of e^{At}. We can compute the derivative $d/dt(e^{At})$ by taking the derivative of the terms composing the matrix power series in (11.21). The result is

$$\frac{d}{dt}e^{At} = A + A^2 t + \frac{A^3 t^2}{2!} + \frac{A^4 t^3}{3!} + \cdots$$

$$= A\left(I + At + \frac{A^2 t^2}{2!} + \frac{A^3 t^3}{3!} + \cdots \right)$$

$$\frac{d}{dt}e^{At} = Ae^{At} = e^{At}A \tag{11.24}$$

Again consider the problem of solving (11.20). It is claimed that the solution is

$$x(t) = e^{At}x(0), \qquad t \geq 0 \tag{11.25}$$

To verify that the expression (11.25) for $x(t)$ is the solution, take the derivative of both sides of (11.25). This gives

$$\frac{d}{dt}x(t) = \frac{d}{dt}[e^{At}x(0)]$$

$$= \left[\frac{d}{dt}e^{At} \right]x(0)$$

Then using (11.24) gives

$$\frac{d}{dt}x(t) = Ae^{At}x(0) = Ax(t)$$

Thus, the expression (11.25) for $x(t)$ does satisfy the vector differential equation (11.20).

From (11.25) we see that we can compute the state $x(t)$ at time t resulting from state $x(0)$ at time $t = 0$ with no input applied for $t \geq 0$ by multiplying $x(0)$ by the matrix e^{At}. As a result of this property, the matrix e^{At} is called the *state-transition matrix* of the system.

11.3.1 Solution to Forced Equation

An expression for the solution to the forced equation (11.18) will now be derived. We can compute the solution by using a matrix version of the integrating factor method used in solving a first-order scalar differential equation.

Multiplying both sides of (11.18) on the left by e^{-At} and rearranging terms yields

$$e^{-At}[\dot{x}(t) - Ax(t)] = e^{-At}Bv(t) \tag{11.26}$$

From (11.24), we find that the left-hand side of (11.26) is equal to the derivative of $e^{-At}x(t)$. Thus,

$$\frac{d}{dt}[e^{-At}x(t)] = e^{-At}Bv(t) \tag{11.27}$$

Integrating both sides of (11.27) gives

$$e^{-At}x(t) = x(0) + \int_0^t e^{-A\lambda}Bv(\lambda)\,d\lambda$$

Finally, multiplying both sides on the left by e^{At} gives

$$x(t) = e^{At}x(0) + \int_0^t e^{A(t-\lambda)}Bv(\lambda)\,d\lambda, \quad t \geq 0 \tag{11.28}$$

Equation (11.28) is the complete solution of (11.18) resulting from initial state $x(0)$ and input $v(t)$ applied for $t \geq 0$.

11.3.2 Output Response

Inserting (11.28) into (11.19) results in the following expression for the complete output response $y(t)$ resulting from initial state $x(0)$ and input $v(t)$:

$$y(t) = Ce^{At}x(0) + \int_0^t Ce^{A(t-\lambda)}Bv(\lambda)\,d\lambda + Dv(t), t \geq 0 \tag{11.29}$$

From the definition of the unit impulse $\delta(t)$, (11.29) can be rewritten in the form

$$y(t) = Ce^{At}x(0) + \int_0^t (Ce^{A(t-\lambda)}Bv(\lambda) + D\delta(t-\lambda)v(\lambda))\,d\lambda, t \geq 0 \tag{11.30}$$

Then defining

$$y_{zi}(t) = Ce^{At}x(0) \tag{11.31}$$

and

$$y_{zs}(t) = \int_0^t (Ce^{A(t-\lambda)}Bv(\lambda) + D\delta(t-\lambda)v(\lambda))\,d\lambda \tag{11.32}$$

from (11.30) we have that

$$y(t) = y_{zi}(t) + y_{zs}(t)$$

The term $y_{zi}(t)$ is called the *zero-input response*, since it is the complete output response when the input $v(t)$ is zero. The term $y_{zs}(t)$ is called the *zero-state response*, since it is the complete output response when the initial state $x(0)$ is zero.

The zero-state response $y_{zs}(t)$ is the same as the response to input $v(t)$ with no initial conditions in the system at time $t = 0$. In the single-input single-output case, from the results in Chapter 2, we obtain

$$y_{zs}(t) = h(t) * v(t) = \int_0^t h(t - \lambda)v(\lambda)\, d\lambda, \quad t \geq 0 \tag{11.33}$$

where $h(t)$ is the impulse response of the system. Equating the right-hand sides of (11.32) and (11.33) yields

$$\int_0^t (Ce^{A(t-\lambda)}Bv(\lambda) + D\delta(t - \lambda)v(\lambda))\, d\lambda = \int_0^t h(t - \lambda)v(\lambda)\, d\lambda \tag{11.34}$$

For (11.34) to hold for all inputs $v(t)$, it must be true that

$$h(t - \lambda) = Ce^{A(t-\lambda)}B + D\delta(t - \lambda),\, t \geq \lambda$$

or

$$h(t) = Ce^{At}B + D\delta(t),\, t \geq 0 \tag{11.35}$$

From the relationship (11.35), it is possible to compute the impulse response directly from the coefficient matrices of the state model of the system.

11.3.3 Solution via the Laplace Transform

Again consider a p-input r-output N-dimensional system given by the state model (11.18) and (11.19). To compute the state and output responses resulting from initial state $x(0)$ and input $v(t)$ applied for $t \geq 0$, the expressions (11.28) and (11.29) can be used. To avoid having to evaluate the integrals in (11.28) and (11.29), we can use the Laplace transform to compute the state and output responses of the system. The transform approach is a matrix version of the procedure that was given for solving a scalar first-order differential equation. The steps are as follows.

Taking the Laplace transform of the state equation (11.18) gives

$$sX(s) - x(0) = AX(s) + BV(s) \tag{11.36}$$

where $X(s)$ is the Laplace transform of the state vector $x(t)$ with the transform taken component by component; that is,

$$X(s) = \begin{bmatrix} X_1(s) \\ X_2(s) \\ \vdots \\ X_N(s) \end{bmatrix}$$

where $X_i(s)$ is the Laplace transform of $x_i(t)$. The term $V(s)$ in (11.36) is the Laplace transform of the input $v(t)$, where again the transform is taken component by component.

Now (11.36) can be rewritten in the form

$$(sI - A)X(s) = x(0) + BV(s) \tag{11.37}$$

where I is the $N \times N$ identity matrix. Note that in factoring out s from $sX(s)$, we must multiply s by I. The reason for this is that A cannot be subtracted from the scalar s, since A is an $N \times N$ matrix. However, A can be subtracted from the diagonal matrix sI. By definition of the identity matrix, the product $(sI - A)X(s)$ is equal to $sX(s) - AX(s)$.

It turns out that the matrix $sI - A$ in (11.37) always has an inverse $(sI - A)^{-1}$. (It will be seen later why this is true.) Thus, both sides of (11.37) can be multiplied on the left by $(sI - A)^{-1}$. This gives

$$X(s) = (sI - A)^{-1}x(0) + (sI - A)^{-1}BV(s) \tag{11.38}$$

The right-hand side of (11.38) is the Laplace transform of the state response resulting from initial state $x(0)$ and input $v(t)$ applied for $t \geq 0$. We can then compute the state response $x(t)$ by taking the inverse Laplace transform of the right-hand side of (11.38).

Comparing (11.38) with (11.28) reveals that $(sI - A)^{-1}$ is the Laplace transform of the state-transition matrix e^{At}. Since e^{At} is a well-defined function of t, this shows that the inverse $(sI - A)^{-1}$ must exist. Also note that

$$e^{At} = \text{inverse Laplace transform of } (sI - A)^{-1} \tag{11.39}$$

The relationship (11.39) is very useful for computing e^{At}. This will be illustrated in a subsequent example.

Taking the Laplace transform of the output equation (11.19) yields

$$Y(s) = CX(s) + DV(s) \tag{11.40}$$

Inserting (11.38) into (11.40) gives

$$Y(s) = C(sI - A)^{-1}x(0) + [C(sI - A)^{-1}B + D]V(s) \tag{11.41}$$

The right-hand side of (11.41) is the Laplace transform of the complete output response resulting from initial state $x(0)$ and input $v(t)$.

If $x(0) = 0$ (no initial energy in the system), (11.41) reduces to

$$Y(s) = Y_{zs}(s) = H(s)V(s) \tag{11.42}$$

where $H(s)$ is the transfer function matrix of the system given by

$$H(s) = C(sI - A)^{-1}B + D \tag{11.43}$$

So, the transfer function matrix can be computed directly from the coefficient matrices of the state model of the system.

Example 11.5   Application to the Two-Car System

Again consider the two-car system in Example 11.3 with the state model of the system given by (11.16) and (11.17). Recall that $x_1(t)$ is the velocity of the first car, $x_2(t)$ is the velocity of the second car, $x_3(t)$ is the distance between the cars, and $y_1(t) = x_1(t)$, $y_2(t) = x_3(t)$. The state-transition matrix e^{At} of the system will be computed first. Since the A matrix for this system is 3×3, the inverse of the matrix

$$sI - A = \begin{bmatrix} s + \dfrac{k_f}{M} & 0 & 0 \\ 0 & s + \dfrac{k_f}{M} & 0 \\ 1 & -1 & s \end{bmatrix}$$

can be computed by the method given in Appendix B. The determinant of $sI - A$ is given by

$$\det(sI - A) = \left(s + \frac{k_f}{M} \right) \begin{vmatrix} s + \dfrac{k_f}{M} & 0 \\ -1 & s \end{vmatrix} = \left(s + \frac{k_f}{M} \right)^2 s$$

and the cofactor matrix of $sI - A$ is given by

$$\text{cof}(sI - A) = \begin{bmatrix} s\left(s + \dfrac{k_f}{M} \right) & 0 & -\left(s + \dfrac{k_f}{M} \right) \\ 0 & s\left(s + \dfrac{k}{M} \right) & s + \dfrac{k_f}{M} \\ 0 & 0 & \left(s + \dfrac{k_f}{M} \right)^2 \end{bmatrix}$$

Hence,

$$(sI - A)^{-1} = \frac{1}{\det(sI - A)} \text{cof}(A)' = \begin{bmatrix} \dfrac{1}{s + \dfrac{k_f}{M}} & 0 & 0 \\ 0 & \dfrac{1}{s + \dfrac{k}{M}} & 0 \\ -\dfrac{1}{s\left(s + \dfrac{k}{M} \right)} & \dfrac{1}{s\left(s + \dfrac{k_f}{M} \right)} & \dfrac{1}{s} \end{bmatrix}$$

By (11.39), e^{At} is equal to the inverse Laplace transform of $(sI - A)^{-1}$. Using partial fraction expansions and table lookup gives

$$e^{At} = \begin{bmatrix} e^{-(k_f/M)t} & 0 & 0 \\ 0 & e^{-(k_f/M)t} & 0 \\ -\dfrac{M}{k_f} + \dfrac{M}{k_f} e^{-(k_f/M)t} & \dfrac{M}{k_f} - \dfrac{M}{k_f} e^{-(k_f/M)t} & 1 \end{bmatrix}$$

Now the state response $x(t)$ resulting from any initial state $x(0)$ with zero input is given by

$$x(t) = e^{At} x(0), \quad t \geq 0$$

For example, if the initial velocity of the first car is $x_1(0) = 60$, the initial velocity of the second car is $x_2(0) = 60$, and the initial separation between the cars is $x_3(0) = 100$, then the initial state is

$$x(0) = \begin{bmatrix} 60 \\ 60 \\ 100 \end{bmatrix} \tag{11.44}$$

and

$$x(t) = \begin{bmatrix} e^{-(k_f/M)t} & 0 & 0 \\ 0 & e^{-(k_f/M)t} & 0 \\ -\dfrac{M}{k_f} + \dfrac{M}{k_f}e^{-(k_f/M)t} & \dfrac{M}{k_f} - \dfrac{M}{k_f}e^{-(k_f/M)t} & 1 \end{bmatrix} \begin{bmatrix} 60 \\ 60 \\ 100 \end{bmatrix}$$

which gives

$$x(t) = \begin{bmatrix} 60e^{-(k_f/M)t} \\ 60e^{-(k_f/M)t} \\ 100 \end{bmatrix}, \quad t \geq 0 \tag{11.45}$$

This result shows that both cars slow down at the same exponential rate and that the separation between the cars remains equal to the initial value of 100 for all $t \geq 0$.

Now suppose that $M = 1000$, $k_f = 10$, and the initial state $x(0)$ is given by (11.44), but now the second car receives a force of $f_2(t) = -100e^{-0.1t}$ for $t \geq 0$, which is a result of the driver suddenly stepping on the brake and then gradually taking his foot off the brake. The key question is whether or not the first car hits the second car due to the braking of the second car. To determine this, the state response $x(t)$ resulting from only the input

$$v(t) = \begin{bmatrix} 0 \\ -100e^{-0.1t} \end{bmatrix}, t \geq 0 \tag{11.46}$$

will be computed first. The Laplace transform of $v(t)$ is

$$V(s) = \begin{bmatrix} 0 \\ \dfrac{-100}{s + 0.1} \end{bmatrix}$$

Then using (11.38) with $x(0) = 0$ yields

$$X(s) = \begin{bmatrix} s + 0.01 & 0 & 0 \\ 0 & s + 0.01 & 0 \\ 1 & -1 & s \end{bmatrix}^{-1} \begin{bmatrix} 0.001 & 0 \\ 0 & 0.001 \\ 0 & 0 \end{bmatrix} \begin{bmatrix} 0 \\ \dfrac{-100}{s + 0.1} \end{bmatrix}$$

$$= \begin{bmatrix} \dfrac{1}{s + 0.01} & 0 & 0 \\ 0 & \dfrac{1}{s + 0.01} & 0 \\ -\dfrac{1}{s(s + 0.01)} & \dfrac{1}{s(s + 0.01)} & \dfrac{1}{s} \end{bmatrix} \begin{bmatrix} 0 \\ \dfrac{-0.1}{s + 0.1} \\ 0 \end{bmatrix}$$

$$= \begin{bmatrix} 0 \\ \dfrac{-0.1}{(s + 0.01)(s + 0.1)} \\ \dfrac{-0.1}{s(s + 0.01)(s + 0.1)} \end{bmatrix}$$

Expanding the components of $X(s)$ gives

$$X(s) = \begin{bmatrix} 0 \\ \dfrac{-1.11}{s + 0.01} + \dfrac{1.11}{s + 0.1} \\ \dfrac{-100}{s} + \dfrac{111.11}{s + 0.01} - \dfrac{11.11}{s + 0.1} \end{bmatrix}$$

Taking the inverse Laplace transform of $X(s)$ yields the following state response resulting from the input $v(t)$:

$$x(t) = \begin{bmatrix} 0 \\ 1.11(e^{-0.1t} - e^{-0.01t}) \\ -100 + 111.11e^{-0.01t} - 11.11e^{-0.1t} \end{bmatrix}, t \geq 0 \qquad (11.47)$$

The complete state response resulting from the input $v(t)$ given by (11.46) and the initial state $x(0)$ given by (11.44) is equal to the sum of the responses given by (11.45) and (11.47). Adding these responses gives

$$x(t) = \begin{bmatrix} 60e^{-0.01t} \\ 60e^{-0.01t} + 1.11(e^{-0.1t} - e^{-0.01t}) \\ 111.11e^{-0.01t} - 11.11e^{-0.1t} \end{bmatrix}, t \geq 0 \qquad (11.48)$$

From (11.48) it is seen that the distance between the cars is given by

$$x_3(t) = 111.11e^{-0.01t} - 11.11e^{-0.1t}, t \geq 0 \qquad (11.49)$$

Since $111.11e^{-0.01t} > 11.11e^{-0.1t}$ for all $t \geq 0$, and assuming that the cars have point mass, it follows from (11.49) that the first car does not hit the second car, but the two cars do collide in the limit as $t \to \infty$.

MATLAB can be used to solve for the state and output responses of the system resulting from any initial state $x(0)$ and any input that can be expressed in closed form. For example, when $x(0)$ is given by (11.44) and $v(t)$ is given by (11.46), the complete state response can be computed by the following commands:

```
A = [-0.01 0 0;0 -0.01 0;-1 1 0]; B = [0.001 0;0 0.001;0 0];
C = [1 0 0;0 0 1]; D = zeros(2,2);
sys = ss(A,B,C,D);
t = 0:1:200; % simulate for 0 < t < 200 sec
x0 = [60 60 100]'; % define the I.C.
v(:,1)=zeros(201,1);
v(:,2)=-100*exp(-0.1*t)';
[y,t,x] = lsim(sys,v,t,x0);
plot(t,x(:,1),t,x(:,2),t,x(:,3))
```

Running the program produces the plots of $x_1(t)$, $x_2(t)$, and $x_3(t)$ shown in Figure 11.6.

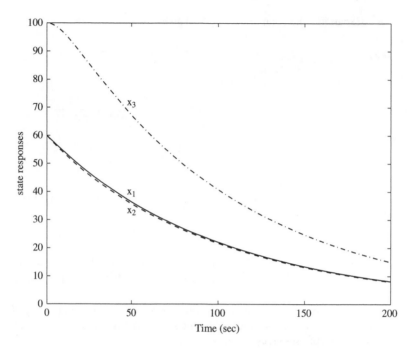

FIGURE 11.6
Plots of the state responses in Example 11.5.

When $x(0) = 0$, we may compute the output response $y(t)$ by taking the inverse Laplace transform of the transfer function representation $Y(s) = H(s)V(s)$. In this example, the transfer function is

$$H(s) = C(sI - A)^{-1}B$$

$$= \begin{bmatrix} 1 & 0 & 0 \\ 0 & 0 & 1 \end{bmatrix} \begin{bmatrix} s + 0.01 & 0 & 0 \\ 0 & s + 0.01 & 0 \\ 1 & -1 & s \end{bmatrix}^{-1} \begin{bmatrix} 0.001 & 0 \\ 0 & 0.001 \\ 0 & 0 \end{bmatrix}$$

$$= \begin{bmatrix} 1 & 0 & 0 \\ 0 & 0 & 1 \end{bmatrix} \begin{bmatrix} \dfrac{1}{s + 0.01} & 0 & 0 \\ 0 & \dfrac{1}{s + 0.01} & 0 \\ -\dfrac{1}{s(s + 0.01)} & \dfrac{1}{s(s + 0.01)} & \dfrac{1}{s} \end{bmatrix} \begin{bmatrix} 0.001 & 0 \\ 0 & 0.001 \\ 0 & 0 \end{bmatrix}$$

$$= \begin{bmatrix} \dfrac{1}{s + 0.01} & 0 & 0 \\ -\dfrac{1}{s(s + 0.01)} & \dfrac{1}{s(s + 0.01)} & \dfrac{1}{s} \end{bmatrix} \begin{bmatrix} 0.001 & 0 \\ 0 & 0.001 \\ 0 & 0 \end{bmatrix}$$

$$H(s) = \begin{bmatrix} \dfrac{0.001}{s + 0.01} & 0 \\ -\dfrac{0.001}{s(s + 0.01)} & \dfrac{0.001}{s(s + 0.01)} \end{bmatrix}$$

MATLAB can be used to calculate the transfer function matrix $H(s)$. To compute $H(s)$ type

```
tf(sys)
```

Using this command for the preceding system results in

```
Transfer function from input 1 to output...
            0.001
#1:   ----------
        s + 0.01


            -0.001
#2:   --------------
        s^2 + 0.01 s


Transfer function from input 2 to output...
#1:   0

            0.001
#2:   ------------
        s^2 + 0.01 s
```

These results match those found analytically.

11.4   DISCRETE-TIME SYSTEMS

A p-input r-output finite-dimensional linear time-invariant discrete-time system can be modeled by the state equations

$$x[n+1] = Ax[n] + Bv[n] \tag{11.50}$$

$$y[n] = Cx[n] + Dv[n] \tag{11.51}$$

The state vector $x[n]$ is the N-element column vector

$$x[n] = \begin{bmatrix} x_1[n] \\ x_2[n] \\ \vdots \\ x_N[n] \end{bmatrix}$$

As in the continuous-time case, the state $x[n]$ at time n represents the past history (before time n) of the system.

The input $v[n]$ and output $y[n]$ are the column vectors

$$v[n] = \begin{bmatrix} v_1[n] \\ v_2[n] \\ \vdots \\ v_p[n] \end{bmatrix}, \quad y[n] = \begin{bmatrix} y_1[n] \\ y_2[n] \\ \vdots \\ y_r[n] \end{bmatrix}$$

The matrices A, B, C, and D in (11.50) and (11.51) are $N \times N$, $N \times p$, $r \times N$, and $r \times p$, respectively. Equation (11.50) is a first-order vector difference equation. Equation (11.51) is the output equation of the system.

11.4.1   Construction of State Models

Consider a single-input single-output linear time-invariant discrete-time system with the input/output difference equation

$$y[n + N] + \sum_{i=0}^{N-1} a_i y[n + i] = b_0 v[n] \tag{11.52}$$

Defining the state variables

$$x_{i+1}[n] = y[n + i], \quad i = 0, 1, 2, \ldots, N - 1 \tag{11.53}$$

results in a state model of the form (11.50) and (11.51) with

$$A = \begin{bmatrix} 0 & 1 & 0 & \cdots & 0 \\ 0 & 0 & 1 & & 0 \\ \vdots & \vdots & & \ddots & \vdots \\ 0 & 0 & 0 & \cdots & 1 \\ -a_0 & -a_1 & -a_2 & \cdots & -a_{N-1} \end{bmatrix}, \quad B = \begin{bmatrix} 0 \\ 0 \\ \vdots \\ 0 \\ b_0 \end{bmatrix}$$

$$C = [1 \quad 0 \quad 0 \quad \cdots \quad 0], \quad D = 0$$

If the right-hand side of (11.52) is modified so that it is in the more general form

$$\sum_{i=0}^{N-1} b_i v[n + i]$$

the foregoing state model is still a state model, except that B and C must be modified so that

$$B = \begin{bmatrix} 0 \\ 0 \\ \vdots \\ 0 \\ 1 \end{bmatrix}$$

$$C = [b_0 \quad b_1 \quad \cdots \quad b_{N-1}]$$

In this case, the state variables $x_i[n]$ are functions of $y[n]$, $v[n]$, and left shifts of $y[n]$ and $v[n]$; that is, $x_i[n]$ is no longer given by (11.53). This state model is the discrete-time counterpart of the state model that was generated from a linear constant-coefficient input/output differential equation in the continuous-time case. (See Section 11.2.)

11.4.2   Realizations Using Unit-Delay Elements

Given a discrete-time system with the N-dimensional state model (11.50) and (11.51), a realization of the system can be constructed that consists of an interconnection of N unit delay elements and combinations of adders, subtracters, and scalar multipliers. Conversely, if a discrete-time system is specified by an interconnection of unit delays, adders, subtracters, and scalar multipliers, a state model of the form (11.50) and (11.51) can be generated directly from the interconnection. The procedure for going from interconnection diagrams to state models (and conversely) is analogous to that given in the continuous-time case.

Example 11.6   Interconnection of Unit Delay Elements

Consider the three-input two-output three-dimensional discrete-time system given by the interconnection in Figure 11.7. From the diagram, we find that

$$x_1[n + 1] = -x_2[n] + v_1[n] + v_3[n]$$
$$x_2[n + 1] = x_1[n] + v_2[n]$$
$$x_3[n + 1] = x_2[n] + v_3[n]$$
$$y_1[n] = x_2[n]$$
$$y_2[n] = x_1[n] + x_3[n] + v_2[n]$$

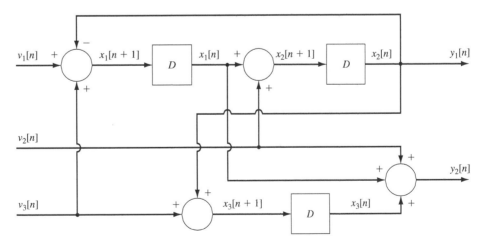

FIGURE 11.7
System in Example 11.6.

Writing these equations in matrix form results in the state model

$$\begin{bmatrix} x_1[n+1] \\ x_2[n+1] \\ x_3[n+1] \end{bmatrix} = \begin{bmatrix} 0 & -1 & 0 \\ 1 & 0 & 0 \\ 0 & 1 & 0 \end{bmatrix} \begin{bmatrix} x_1[n] \\ x_2[n] \\ x_3[n] \end{bmatrix} + \begin{bmatrix} 1 & 0 & 1 \\ 0 & 1 & 0 \\ 0 & 0 & 1 \end{bmatrix} \begin{bmatrix} v_1[n] \\ v_2[n] \\ v_3[n] \end{bmatrix}$$

$$\begin{bmatrix} y_1[n] \\ y_2[n] \end{bmatrix} = \begin{bmatrix} 0 & 1 & 0 \\ 1 & 0 & 1 \end{bmatrix} \begin{bmatrix} x_1[n] \\ x_2[n] \\ x_3[n] \end{bmatrix} + \begin{bmatrix} 0 & 0 & 0 \\ 0 & 1 & 0 \end{bmatrix} \begin{bmatrix} v_1[n] \\ v_2[n] \\ v_3[n] \end{bmatrix}$$

11.4.3 Solution of State Equations

Again consider the p-input r-output discrete-time system with the state model

$$x[n+1] = Ax[n] + Bv[n] \tag{11.54}$$

$$y[n] = Cx[n] + Dv[n] \tag{11.55}$$

We can solve the vector difference equation (11.54) by using a matrix version of recursion. The process is a straightforward generalization of the recursive procedure for solving a first-order scalar difference equation. (See Chapter 2.) The steps are as follows.

It is assumed that the initial state of the system is the state $x[0]$ at initial time $n = 0$. Then setting $n = 0$ in (11.54) gives

$$x[1] = Ax[0] + Bv[0] \tag{11.56}$$

Setting $n = 1$ (11.54) and using (11.56) yields

$$\begin{aligned} x[2] &= Ax[1] + Bv[1] \\ &= A[Ax[0] + Bv[0]] + Bv[1] \\ &= A^2 x[0] + ABv[0] + Bv[1] \end{aligned}$$

If this process is continued, for any integer value of $n \geq 1$,

$$x[n] = A^n x[0] + \sum_{i=0}^{n-1} A^{n-i-1} Bv[i], \quad n \geq 1 \tag{11.57}$$

The right-hand side of (11.57) is the state response resulting from initial state $x[0]$ and input $v[n]$ applied for $n \geq 0$. Note that, if $v[n] = 0$ for $n \geq 0$, then

$$x[n] = A^n x[0], n \geq 0 \tag{11.58}$$

From (11.58) it is seen that the state transition from initial state $x[0]$ to state $x[n]$ at time n (with no input applied) is equal to A^n times $x[0]$. Therefore, in the discrete-time case the state-transition matrix is the matrix A^n.

Example 11.7   Combination Bank Loan and Savings Account

Some banks will allow customers to take out a loan, with the loan payments automatically deducted from the customer's savings account at the same bank, resulting in a "combination bank loan/savings account." This is done for convenience (since the loan payments are automatically deducted) and to generate income from the interest on the amount in the savings account. To set up a state model for this process, let $x_1[n]$ denote the loan balance after the nth month, and let $x_2[n]$ denote the amount in the savings account after the nth month. The initial amount of the loan is $x_1[0]$, and the initial amount in the savings account is $x_2[0]$. Then the state equations for the loan/savings account are given by

$$x_1[n] = \left(1 + \frac{I_1}{12}\right)x_1[n-1] - p[n], \; n = 1, 2, \ldots \tag{11.59}$$

$$x_2[n] = \left(1 + \frac{I_2}{12}\right)x_2[n-1] - p[n] + d[n], \; n = 1, 2, \ldots \tag{11.60}$$

where I_1 (respectively, I_2) is the yearly interest rate in decimal form for the loan (respectively, savings account), $p[n]$ is the loan payment at the end of the nth month, and $d[n]$ is the amount deposited in the savings account at the end of the nth month. Note from (11.60) that it is assumed that the interest in the savings account is compounded monthly. Taking the input to the system to be

$$v[n] = \begin{bmatrix} p[n+1] \\ d[n+1] \end{bmatrix}$$

and replacing n by $n+1$ in (11.59) and (11.60) results in the following state equation in matrix form:

$$\begin{bmatrix} x_1[n+1] \\ x_2[n+1] \end{bmatrix} = \begin{bmatrix} 1 + \dfrac{I_1}{12} & 0 \\ 0 & 1 + \dfrac{I_2}{12} \end{bmatrix} \begin{bmatrix} x_1[n] \\ x_2[n] \end{bmatrix} + \begin{bmatrix} -1 & 0 \\ -1 & 1 \end{bmatrix} v[n], \; n = 0, 1, 2, \ldots \tag{11.61}$$

The state equation (11.61) can then be solved recursively as discussed previously. For example, with $I_1 = 0.06$ (6%) and $I_2 = 0.03$ (3%), the loan is a mortgage with $x_1[0] = 300{,}000.00$; the initial amount in the savings account is $x_2[0] = 20{,}000.00$; and $p[n] = d[n] = 2{,}000.00$ for $n = 1, 2, \ldots$; then the values of the state $x[n]$ at $n = 1$ and $n = 2$ are

$$\begin{bmatrix} x_1[1] \\ x_2[1] \end{bmatrix} = \begin{bmatrix} 1.005 & 0 \\ 0 & 1.0025 \end{bmatrix} \begin{bmatrix} 300000 \\ 20000 \end{bmatrix} + \begin{bmatrix} -1 & 0 \\ -1 & 1 \end{bmatrix} \begin{bmatrix} 2000 \\ 2000 \end{bmatrix}$$

$$\begin{bmatrix} x_1[1] \\ x_2[1] \end{bmatrix} = \begin{bmatrix} 301500 \\ 20050 \end{bmatrix} + \begin{bmatrix} -2000 \\ 0 \end{bmatrix}$$

$$\begin{bmatrix} x_1[1] \\ x_2[1] \end{bmatrix} = \begin{bmatrix} 299500 \\ 20050 \end{bmatrix}$$

$$\begin{bmatrix} x_1[2] \\ x_2[2] \end{bmatrix} = \begin{bmatrix} 1.005 & 0 \\ 0 & 1.0025 \end{bmatrix} \begin{bmatrix} 299500 \\ 20050 \end{bmatrix} + \begin{bmatrix} -2000 \\ 0 \end{bmatrix}$$

$$\begin{bmatrix} x_1[2] \\ x_2[2] \end{bmatrix} = \begin{bmatrix} 298997.50 \\ 20100.125 \end{bmatrix}$$

Note that, as expected, the loan balance is decreasing and the amount in the savings account is increasing. In Problem 11.26, the reader is asked to determine the number of months it takes to pay off the mortgage and the amount in the savings account at the time of the payoff.

11.4.4 Computation of Output Response

Again consider the state model given by (11.54) and (11.55). Inserting the state response (11.57) into the output equation (11.55) gives

$$y[n] = CA^n x[0] + \sum_{i=0}^{n-1} CA^{n-i-1} Bv[i] + Dv[n], \quad n \geq 1 \tag{11.62}$$

The right-hand side of (11.62) is the complete output response resulting from initial state $x[0]$ and input $v[n]$. The term

$$y_{zi}[n] = CA^n x[0], n \geq 0$$

is the zero-input response, and the term

$$y_{zs}[n] = \sum_{i=0}^{n-1} CA^{n-i-1} Bv[i] + Dv[n], \quad n \geq 1$$

$$= \sum_{i=0}^{n} \{CA^{n-i-1} Bu[n - i - 1] + \delta[n - i]D\}v[i], \quad n \geq 1 \tag{11.63}$$

is the zero-state response (where $\delta[n] =$ unit pulse located at $n = 0$).

In the single-input single-output case,

$$y_{zs}[n] = \sum_{i=0}^{n} h[n - i]v[i] \tag{11.64}$$

where $h[n]$ is the unit-pulse response of the system. Comparing (11.63) and (11.64) reveals that

$$h[n - i] = CA^{n-i-1} Bu[n - i - 1] + \delta[n - i]D$$
$$h[n] = CA^{n-1} Bu[n - 1] + \delta[n]D$$

or

$$h[n] = \begin{cases} D, & n = 0 \\ CA^{n-1}B, & n \geq 1 \end{cases}$$

11.4.5 Solution via the z-Transform

Taking the z-transform of the vector difference equation (11.54) gives

$$zX(z) - zx[0] = AX(z) + BV(z) \tag{11.65}$$

where $X(z)$ and $V(z)$ are the z-transforms of $x[n]$ and $v[n]$, respectively, with transforms taken component by component. Solving (11.65) for $X(z)$ gives

$$X(z) = (zI - A)^{-1}zx[0] + (zI - A)^{-1}BV(z) \tag{11.66}$$

The right-hand side of (11.66) is the z-transform of the state response $x[n]$ resulting from initial state $x[0]$ and input $v[n]$.

Comparing (11.66) and (11.57) shows that $(zI - A)^{-1}z$ is the z-transform of the state-transition matrix A^n. Thus,

$$A^n = \text{inverse } z\text{-transform of } (zI - A)^{-1}z \tag{11.67}$$

Taking the z-transform of (11.55) and using (11.66) yields

$$Y(z) = C(zI - A)^{-1}zx[0] + [C(zI - A)^{-1}B + D]V(z) \tag{11.68}$$

The right-hand side of (11.68) is the z-transform of the complete output response resulting from initial state $x[0]$ and input $v[n]$.

If $x[0] = 0$,

$$Y(z) = Y_{zs}(z) = [C(zI - A)^{-1}B + D]V(z) \tag{11.69}$$

Since

$$Y_{zs}(z) = H(z)V(z)$$

where $H(z)$ is the transfer function matrix, then by (11.69),

$$H(z) = C(zI - A)^{-1}B + D \tag{11.70}$$

Example 11.8   Computation of Transfer Function Matrix

Again consider the system in Example 11.6. The state-transition matrix A^n will be computed first.

$$(zI - A)^{-1} = \begin{bmatrix} z & 1 & 0 \\ -1 & z & 0 \\ 0 & -1 & z \end{bmatrix}^{-1} = \frac{1}{(z^2 + 1)z}\begin{bmatrix} z^2 & -z & 0 \\ z & z^2 & 0 \\ 1 & z & z^2 + 1 \end{bmatrix}$$

Expanding the components of $(zI - A)^{-1}z$ and using table lookup gives

$$A^n = \begin{bmatrix} +\cos\dfrac{\pi}{2}n & -\sin\dfrac{\pi}{2}n & 0 \\[2mm] \sin\dfrac{\pi}{2}n & +\cos\dfrac{\pi}{2}n & 0 \\[2mm] -\left(\cos\dfrac{\pi}{2}n\right)u[n-1] & \sin\dfrac{\pi}{2}n & \delta[n] \end{bmatrix}$$

The state response $x[n]$ resulting from any initial state $x[0]$ (with no input applied) can be computed by (11.58). For example, if

$$x[0] = \begin{bmatrix} 1 \\ 1 \\ 0 \end{bmatrix},$$

then

$$x[n] = A^n x[0] = \begin{bmatrix} \cos\frac{\pi}{2}n - \sin\frac{\pi}{2}n \\ \sin\frac{\pi}{2}n + \cos\frac{\pi}{2}n \\ -\left(\cos\frac{\pi}{2}n\right)u[n-1] + \sin\frac{\pi}{2}n \end{bmatrix}, \quad n \geq 0$$

MATLAB contains the command dlsim, which can be used to compute the state response $x[n]$. For example, to compute $x[n]$ for the prior system with $x_1[0] = x_2[0] = 1$ and $x_3[0] = 0$, the commands are

```
A = [0 -1 0;1 0 0;0 1 0]; B = [1 0 1;0 1 0;0 0 1];
C = [0 1 0;1 0 1]; D = [0 0 0;0 1 0];
x0 = [1 1 0]'; % define the I.C.
n = 0:1:10;
v = zeros(length(n),3);
[y,x] = dlsim(A,B,C,D,v,x0);
```

Running the program results in the values of $x[n]$ for $n = 1$ to 5, given in the following table:

	$x[0]$	$x[1]$	$x[2]$	$x[3]$	$x[4]$	$x[5]$
x_1	1	−1	−1	1	1	−1
x_2	1	1	−1	−1	1	1
x_3	0	1	1	−1	−1	1

The reader is invited to verify that these values are an exact match with the values obtained from the analytical solution given previously.

To conclude the example, the transfer function matrix of the system will be computed. From (11.70), we have that

$$H(z) = C(zI - A)^{-1}B + D$$

$$= \frac{1}{(z^2 + 1)z} \begin{bmatrix} 0 & 1 & 0 \\ 1 & 0 & 1 \end{bmatrix} \begin{bmatrix} z^2 & -z & 0 \\ z & z^2 & 0 \\ 1 & z & z^2 + 1 \end{bmatrix} \begin{bmatrix} 1 & 0 & 1 \\ 0 & 1 & 0 \\ 0 & 0 & 1 \end{bmatrix} + \begin{bmatrix} 0 & 0 & 0 \\ 0 & 1 & 0 \end{bmatrix}$$

$$= \frac{1}{(z^2 + 1)z} \begin{bmatrix} z & z^2 & z \\ z^2 + 1 & (z^2 + 1)z & 2(z^2 + 1) \end{bmatrix}$$

To use MATLAB to calculate the transfer function matrix, type

```
sys = ss(A,B,C,D);
tf(sys)
```

This yields

```
Transfer function from input 1 to output...
          1
#1:   -------
      s^2 + 1

      s^2 + 1
#2:   -------
      s^3 + s


Transfer function from input 2 to output...
          s
#1:   -------
      s^2 + 1

      s^3 + s
#2:   -------
      s^3 + s


Transfer function from input 3 to output...
          1
#1:   -------
      s^2 + 1

      2 s^2 + 2
#2:   ---------
       s^3 + s
```

The reader will note that these coefficients match those previously found analytically.

11.5   EQUIVALENT STATE REPRESENTATIONS

Unlike the transfer function model, the state model of a system is not unique. The relationship between state models of the same system is considered in this section. The analysis that follows is developed in terms of continuous-time systems. The theory in the discrete-time case is very similar, and so attention will be restricted to the continuous-time case.

Consider a p-input r-output N-dimensional linear time-invariant continuous-time system given by the state model

$$\dot{x}(t) = Ax(t) + Bv(t) \tag{11.71}$$

$$y(t) = Cx(t) + Dv(t) \tag{11.72}$$

Let P denote a fixed $N \times N$ matrix with entries that are real numbers. It is required that P be invertible, and so the determinant $|P|$ of P is nonzero. (See Appendix B.) The inverse of P is denoted by P^{-1}.

In terms of the matrix P, a new state vector $\overline{x}(t)$ can be defined for the given system, where

$$\overline{x}(t) = Px(t) \tag{11.73}$$

The relationship (11.73) is called a *coordinate transformation*, since it is a mapping from the original state coordinates to the new state coordinates.

Multiplying both sides of (11.73) on the left by the inverse P^{-1} of P gives

$$P^{-1}\overline{x}(t) = P^{-1}Px(t)$$

By definition of the inverse, $P^{-1}P = I$, where I is the $N \times N$ identity matrix. Hence,

$$P^{-1}\overline{x}(t) = Ix(t) = x(t)$$

or

$$x(t) = P^{-1}\overline{x}(t) \tag{11.74}$$

Via (11.74), it is possible to go from the new state vector $\overline{x}(t)$ back to the original state vector $x(t)$. Note that, if P were not invertible, it would not be possible to go back to $x(t)$ from $\overline{x}(t)$. This is the reason P must be invertible.

In terms of the new state vector $\overline{x}(t)$, it is possible to generate a new state-equation model of the given system. The steps are as follows.

Taking the derivative of both sides of (11.73) and using (11.71) yields

$$\dot{\overline{x}}(t) = P\dot{x}(t) = P[Ax(t) + Bv(t)]$$
$$= PAx(t) + PBv(t) \tag{11.75}$$

Inserting the expression (11.74) for $x(t)$ into (11.75) and (11.72) gives

$$\dot{\overline{x}}(t) = PA(P^{-1})\overline{x}(t) + PBv(t) \tag{11.76}$$
$$y(t) = C(P^{-1})\overline{x}(t) + Dv(t) \tag{11.77}$$

Defining the matrices

$$\overline{A} = PA(P^{-1}), \quad \overline{B} = PB, \quad \overline{C} = C(P^{-1}), \quad \overline{D} = D, \tag{11.78}$$

we find that (11.76) and (11.77) can be written in the form

$$\dot{\overline{x}}(t) = \overline{A}\,\overline{x}(t) + \overline{B}v(t) \tag{11.79}$$
$$y(t) = \overline{C}\,\overline{x}(t) + \overline{D}v(t) \tag{11.80}$$

Equations (11.79) and (11.80) are the state equations of the given system in terms of the new state vector $\overline{x}(t)$. Thus, it is possible to generate a new N-dimensional state model from the original N-dimensional state model. Since the preceding construction can be carried out for any invertible $N \times N$ matrix P, and there are an infinite number of such matrices, it is possible to generate an infinite number of new state models from a given state model.

Let the original state model be denoted by the quadruple (A, B, C, D) and the new state model be denoted by the quadruple $(\overline{A}, \overline{B}, \overline{C}, \overline{D})$. The state models (A, B, C, D) and $(\overline{A}, \overline{B}, \overline{C}, \overline{D})$ are said to be related by the coordinate transformation P, since the state vector $\overline{x}(t)$ of the latter is related to the state vector $x(t)$ of the former by the relationship $\overline{x}(t) = Px(t)$. Any two such state models are said to be *equivalent*. The only difference between two equivalent state models is in the labeling of states. More precisely, the states of $(\overline{A}, \overline{B}, \overline{C}, \overline{D})$ are linear combinations [given by $\overline{x}(t) = Px(t)$] of the states of (A, B, C, D).

It should be stressed that the notion of equivalent state models applies only to state models having the same dimension. State models with different dimensions cannot be related by a coordinate transformation.

Any two equivalent state models have exactly the same input/output relationship. In particular, the transfer function matrices corresponding to any two equivalent models are the same. To prove this, let $H(s)$ and $\overline{H}(s)$ denote the transfer function matrices associated with (A, B, C, D) and $(\overline{A}, \overline{B}, \overline{C}, \overline{D})$, respectively; that is,

$$H(s) = C(sI - A)^{-1}B + D \tag{11.81}$$

$$\overline{H}(s) = \overline{C}(sI - \overline{A})^{-1}\overline{B} + \overline{D} \tag{11.82}$$

Inserting the expressions (11.78) for $\overline{A}, \overline{B}, \overline{C}, \overline{D}$ into (11.82) yields

$$\overline{H}(s) = C(P^{-1})[sI - PA(P^{-1})]^{-1}PB + D$$

Now,

$$sI - PA(P^{-1}) = P(sI - A)(P^{-1})$$

In addition, for any $N \times N$ invertible matrices X, Y, W,

$$(WXY)^{-1} = (Y^{-1})(X^{-1})(W^{-1})$$

Thus,

$$\overline{H}(s) = C(P^{-1})P(sI - A)^{-1}(P^{-1})PB + D$$
$$= C(sI - A)^{-1}B + D$$
$$= H(s)$$

So, the transfer function matrices are the same.

Example 11.9   Equivalent State Models

Consider the system given by the second-order differential equation

$$\ddot{y}(t) + 2\dot{y}(t) + 3y(t) = \dot{v}(t) + 2v(t)$$

This equation is in the form of (11.9), and thus a state model for this system is obtained by (11.10). This results in the state model

$$\begin{bmatrix} \dot{x}_1(t) \\ \dot{x}_2(t) \end{bmatrix} = \begin{bmatrix} 0 & 1 \\ -3 & -2 \end{bmatrix} \begin{bmatrix} x_1(t) \\ x_2(t) \end{bmatrix} + \begin{bmatrix} 0 \\ 1 \end{bmatrix} v(t), \quad y(t) = \begin{bmatrix} 2 & 1 \end{bmatrix} \begin{bmatrix} x_1(t) \\ x_2(t) \end{bmatrix}$$

The state variables $x_1(t)$ and $x_2(t)$ in the foregoing state model have no physical meaning, and thus another set of state variables might be just as suitable. For example, let $\bar{x}(t) = Px(t)$, where

$$P = \begin{bmatrix} 1 & 1 \\ 0 & 1 \end{bmatrix}$$

As it is easy to show that $|P|$ is not zero, this is a valid transformation matrix. In terms of $\bar{x}(t)$, the new state model is given by (11.79) and (11.80), where the quadruple $(\bar{A}, \bar{B}, \bar{C}, \bar{D})$ can be found from (11.78). The new coefficient matrices can be determined from the following MAT-LAB commands:

```
A = [0 1;-3 -2]; B = [0 1]'; C = [2 1]; D = 0;
P = [1 1;0 1];
Abar = P*A*inv(P);
Bbar = P*B;
Cbar = C*inv(P);
Dbar = D;
```

MATLAB contains these commands in the M-file ss2ss with the following usage:

```
sys1 = ss(A,B,C,D);
sys2 = ss2ss(sys1,P);
```

Running MATLAB results in the matrices

$$\bar{A} = \begin{bmatrix} -3 & 2 \\ -3 & 1 \end{bmatrix}; \quad \bar{B} = \begin{bmatrix} 1 \\ 1 \end{bmatrix}; \quad \bar{C} = \begin{bmatrix} 2 & -1 \end{bmatrix}; \quad \bar{D} = 0$$

The transfer function matrices for the previous state models can be calculated by the command

```
tf(sys1)
```

Inserting the values for A, B, C, and D results in the transfer function

$$C(sI - A)^{-1}B + D = \frac{s + 2}{s^2 + 2s + 3}$$

Using the same commands on the second state model shows that the transfer functions are indeed identical.

By considering a coordinate transformation, it is sometimes possible to go from one state model (A, B, C, D) to another state model $(\bar{A}, \bar{B}, \bar{C}, \bar{D})$ for which one or more of the coefficients matrices $\bar{A}, \bar{B}, \bar{C}$ has a special form. Such models are called canonical models, or canonical forms. Examples are the diagonal form, control-canonical form, and the observer-canonical form. Due to the special structure of these canonical forms, they can result in a significant simplification in the solution to certain classes of problems. For example, the control-canonical form is very useful in the study of state feedback. For an in-depth development of the various canonical forms, the reader is referred to Kailath [1980], Brogan [1991], or Rugh [1996].

11.5.1   Example of the Diagonal Form

Consider the *RLC* series circuit shown in Figure 11.8. To determine a state model of the circuit, we can define the state variables to be the current $i(t)$ in the inductor and the voltage $v_c(t)$ across the capacitor; that is,

$$x_1(t) = i(t)$$

$$x_2(t) = v_c(t)$$

Summing the voltages around the loop gives

$$Ri(t) + L\frac{di(t)}{dt} + v_c(t) = v(t)$$

Hence,

$$\dot{x}_1(t) = -\frac{R}{L}x_1(t) - \frac{1}{L}x_2(t) + \frac{1}{L}v(t)$$

Also,

$$\dot{x}_2(t) = \frac{1}{C}x_1(t)$$

$$y(t) = x_2(t)$$

So, the circuit has the state model $\dot{x}(t) = Ax(t) + Bv(t)$, $y(t) = Cx(t)$, where

$$A = \begin{bmatrix} -\dfrac{R}{L} & -\dfrac{1}{L} \\ \dfrac{1}{C} & 0 \end{bmatrix}, \qquad B = \begin{bmatrix} \dfrac{1}{L} \\ 0 \end{bmatrix}, \qquad C = \begin{bmatrix} 0 & 1 \end{bmatrix}$$

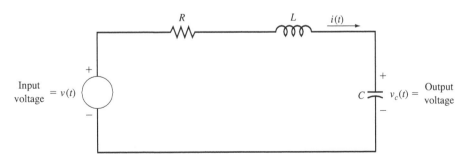

FIGURE 11.8
Series *RLC* circuit.

Now the question is whether or not there is a coordinate transformation $\bar{x}(t) = Px(t)$, such that $\bar{A} = PA(P^{-1})$ is in the diagonal form

$$\bar{A} = \begin{bmatrix} \bar{a}_1 & 0 \\ 0 & \bar{a}_2 \end{bmatrix}$$

Part of the interest in the diagonal form is the simplification that results from this form. For instance, if $\bar{A}$ is in this diagonal form, the state transition matrix has the simple form

$$e^{\bar{A}t} = \begin{bmatrix} e^{\bar{a}_1 t} & 0 \\ 0 & e^{\bar{a}_2 t} \end{bmatrix}$$

Not every matrix A can be put into a diagonal form $\bar{A}$ by a coordinate transformation $\bar{x}(t) = Px(t)$. There are systematic methods for studying the existence and computation of diagonal forms. In the following development a direct procedure is given for determining the existence of a diagonal form for the series RLC circuit:

First, with $\det(sI - A)$ equal to the determinant of $sI - A$ (see Appendix B), inserting the previous A into $\det(sI - A)$ gives

$$\det(sI - A) = \det \begin{bmatrix} s + \dfrac{R}{L} & \dfrac{1}{L} \\[2mm] -\dfrac{1}{C} & s \end{bmatrix} = s^2 + \frac{R}{L}s + \frac{1}{LC} \qquad (11.83)$$

With $\bar{A}$ equal to the preceding diagonal form, we see that

$$\det(sI - \bar{A}) = \det \begin{bmatrix} s - \bar{a}_1 & 0 \\ 0 & s - \bar{a}_2 \end{bmatrix} = (s - \bar{a}_1)(s - \bar{a}_2) \qquad (11.84)$$

From results in matrix algebra, it can be shown that

$$\det(sI - A) = \det(sI - \bar{A})$$

Then, by equating (11.83) and (11.84), we see that $\bar{a}_1$ and $\bar{a}_2$ must be the zeros of $s^2 + (R/L)s + (1/LC)$. Thus,

$$\bar{a}_1^2 + \frac{R}{L}\bar{a}_1 + \frac{1}{LC} = 0$$

$$\bar{a}_2^2 + \frac{R}{L}\bar{a}_2 + \frac{1}{LC} = 0$$

Now, since $\bar{A} = PA(P^{-1})$, then $\bar{A}P = PA$. Setting $\bar{A}P = PA$ gives

$$\begin{bmatrix} \bar{a}_1 & 0 \\ 0 & \bar{a}_2 \end{bmatrix}\begin{bmatrix} p_1 & p_2 \\ p_3 & p_4 \end{bmatrix} = \begin{bmatrix} p_1 & p_2 \\ p_3 & p_4 \end{bmatrix}\begin{bmatrix} -\dfrac{R}{L} & -\dfrac{1}{L} \\[2mm] \dfrac{1}{C} & 0 \end{bmatrix}$$

Equating the entries of $\overline{A}P$ and PA results in the following equations:

$$\overline{a}_1 p_1 = -\frac{R p_1}{L} + \frac{p_2}{C} \tag{11.85}$$

$$\overline{a}_1 p_2 = -\frac{p_1}{L} \tag{11.86}$$

$$\overline{a}_2 p_3 = -\frac{R p_3}{L} + \frac{p_4}{C} \tag{11.87}$$

$$\overline{a}_2 p_4 = -\frac{p_3}{L} \tag{11.88}$$

Equations (11.85) and (11.86) reduce to the single constraint

$$p_2 = -\frac{p_1}{L\overline{a}_1}$$

Equations (11.87) and (11.88) reduce to the single constraint

$$p_4 = -\frac{p_3}{L\overline{a}_2}$$

Therefore,

$$P = \begin{bmatrix} p_1 & -\dfrac{p_1}{L\overline{a}_1} \\[2ex] p_3 & -\dfrac{p_3}{L\overline{a}_2} \end{bmatrix}$$

Finally, since P must be invertible, $\det P \neq 0$, and thus

$$-\frac{p_1 p_3}{L\overline{a}_2} + \frac{p_1 p_3}{L\overline{a}_1} \neq 0 \tag{11.89}$$

Equation (11.89) is satisfied if and only if $\overline{a}_1 \neq \overline{a}_2$. Hence, the diagonal form exists if and only if the zeros of $s^2 + (R/L)s + (1/LC)$ are distinct. Note that the foregoing transformation matrix P is not unique.

11.6   DISCRETIZATION OF STATE MODEL

Again consider a p-input r-output N-dimensional continuous-time system given by the state model

$$\dot{x}(t) = Ax(t) + Bv(t) \tag{11.90}$$

$$y(t) = Cx(t) + Dv(t) \tag{11.91}$$

In this section it will be shown that the state representation can be discretized in time, which results in a discrete-time simulation of the given continuous-time system.

From the results in Section 11.3, we note that the state response $x(t)$ resulting from initial state $x(0)$ and input $v(t)$ is given by

$$x(t) = e^{At}x(0) + \int_0^t e^{A(t-\lambda)}Bv(\lambda) \, d\lambda, \quad t > 0$$

Now suppose that the initial time is changed from $t = 0$ to $t = \tau$, where τ is any real number. Solving the vector differential equation (11.90) with the initial time $t = \tau$ gives

$$x(t) = e^{A(t-\tau)}x(\tau) + \int_\tau^t e^{A(t-\lambda)}Bv(\lambda) \, d\lambda, \quad t > \tau \tag{11.92}$$

Let T be a fixed positive number. Then setting $\tau = nT$ and $t = nT + T$ in (11.92), where n is the discrete-time index, yields

$$x(nT + T) = e^{AT}x(nT) + \int_{nT}^{nT+T} e^{A(nT+T-\lambda)}Bv(\lambda) \, d\lambda \tag{11.93}$$

Equation (11.93) looks like a state equation for a discrete-time system, except that the second term on the right-hand side is not in the form of a matrix times $v(nT)$. This term can be expressed in such a form if the input $v(t)$ is constant over the T-second intervals $nT \leq t \leq nT + T$; that is,

$$v(t) = v(nT), \qquad nT \leq t < nT + T \tag{11.94}$$

If $v(t)$ satisfies (11.94), then (11.93) can be written in the form

$$x(nT + T) = e^{AT}x(nT) + \left\{ \int_{nT}^{nT+T} e^{A(nT+T-\lambda)}B \, d\lambda \right\}v(nT) \tag{11.95}$$

Let B_d denote the $N \times p$ matrix defined by

$$B_d = \int_{nT}^{nT+T} e^{A(nT+T-\lambda)}B \, d\lambda \tag{11.96}$$

Carrying out the change of variables $\bar{\lambda} = nT + T - \lambda$ in the integral in (11.96) gives

$$B_d = \int_T^0 e^{A\bar{\lambda}}B(-d\bar{\lambda}) = \int_0^T e^{A\lambda}B \, d\lambda$$

From this expression it is seen that B_d is independent of the time index n.

Now let A_d denote the $N \times N$ matrix defined by

$$A_d = e^{AT}$$

Then, in terms of A_d and B_d, the difference equation (11.95) can be written in the form

$$x(nT + T) = A_d x(nT) + B_d v(nT) \tag{11.97}$$

Setting $t = nT$ in both sides of (11.91) results in the discretized output equation

$$y(nT) = Cx(nT) + Dv(nT) \tag{11.98}$$

Equations (11.97) and (11.98) are the state equations of a linear time-invariant N-dimensional discrete-time system. This discrete-time system is a discretization in time of the given continuous-time system. If the input $v(t)$ is constant over the T-second intervals $nT \leq t < nT + T$, the values of the state response $x(t)$ and output response $y(t)$ for $t = nT$ can be computed exactly by solution of the state equations (11.97) and (11.98). Since (11.97) can be solved recursively, the discretization process yields a numerical method for solving the state equation of a linear time-invariant continuous-time system.

If $v(t)$ is not constant over the T-second intervals $nT \leq t < nT + T$, the solution of (11.97) and (11.98) will yield approximate values for $x(nT)$ and $y(nT)$. In general, the accuracy of the approximate values will improve as T is made smaller. So, even if $v(t)$ is not piecewise constant, the representation (11.97) and (11.98) serves as a discrete-time simulation of the given continuous-time system.

Since the step function $u(t)$ is constant for all $t > 0$, in the single-input single-output case the step response of the discretization (11.97) and (11.98) will match the values of the step response of the given continuous-time system. Hence, the preceding discretization process is simply a state version of step-response matching. (See Section 10.5.)

The foregoing discretization in time can be used to discretize any system given by an Nth-order linear constant-coefficient input/output differential equation. In particular, a state model of the system can be constructed by the use of the realization given in Section 11.2, and then the coefficient matrices of this state model can be discretized, which yields a discretization in time of the given continuous-time system.

Example 11.10   Car on a Level Surface

Consider a car on a level surface with state model

$$\begin{bmatrix} \dot{x}_1(t) \\ \dot{x}_2(t) \end{bmatrix} = \begin{bmatrix} 0 & 1 \\ 0 & \dfrac{-k_f}{M} \end{bmatrix} \begin{bmatrix} x_1(t) \\ x_2(t) \end{bmatrix} + \begin{bmatrix} 0 \\ \dfrac{1}{M} \end{bmatrix} f(t)$$

$$y(t) = \begin{bmatrix} 1 & 0 \end{bmatrix} \begin{bmatrix} x_1(t) \\ x_2(t) \end{bmatrix}$$

where $x_1(t)$ is the position, $x_2(t)$ is the velocity of the car, and $f(t)$ is the drive or braking force applied to the car. To compute the discretized matrices A_d and B_d for this system, it is first necessary to calculate the state transition matrix e^{At}. For the A in this example, note that

$$(sI - A)^{-1} = \begin{bmatrix} s & -1 \\ 0 & s + \dfrac{k_f}{M} \end{bmatrix}^{-1} = \frac{1}{s(s + k_f/M)} \begin{bmatrix} s + \dfrac{k_f}{M} & 1 \\ 0 & s \end{bmatrix}$$

Thus,

$$
e^{At} = \begin{bmatrix} 1 & \dfrac{M}{k_f}\left[1 - \exp\left(-\dfrac{k_f}{M}t\right)\right] \\[2ex] 0 & \exp\left(-\dfrac{k_f}{M}t\right) \end{bmatrix}
$$

and

$$
e^{AT} = \begin{bmatrix} 1 & \dfrac{M}{k_f}\left[1 - \exp\left(-\dfrac{k_f T}{M}\right)\right] \\[2ex] 0 & \exp\left(-\dfrac{k_f T}{M}\right) \end{bmatrix}
$$

$$
B_d = \int_0^T e^{A\lambda} B \, d\lambda = \begin{bmatrix} \displaystyle\int_0^T \dfrac{1}{k_f}\left[1 - \exp\left(-\dfrac{k_f}{M}\lambda\right)\right] d\lambda \\[3ex] \displaystyle\int_0^T \dfrac{1}{M}\exp\left(-\dfrac{k_f}{M}\lambda\right) d\lambda \end{bmatrix}
$$

$$
= \begin{bmatrix} \dfrac{T}{k_f} - \dfrac{M}{k_f^2}\left[1 - \exp\left(-\dfrac{k_f T}{M}\right)\right] \\[3ex] \dfrac{1}{k_f}\left[1 - \exp\left(-\dfrac{k_f T}{M}\right)\right] \end{bmatrix}
$$

When $M = 1$, $k_f = 0.1$, and $T = 0.1$, the discretized matrices are

$$
A_d = \begin{bmatrix} 1 & 0.09950166 \\ 0 & 0.99004983 \end{bmatrix}, \quad B_d = \begin{bmatrix} 0.00498344 \\ 0.09950166 \end{bmatrix}
$$

The state model of the resulting discrete-time simulation is

$$
\begin{bmatrix} x_1(0.1n + 0.1) \\ x_2(0.1n + 0.1) \end{bmatrix} = \begin{bmatrix} 1 & 0.09950166 \\ 0 & 0.99004983 \end{bmatrix}\begin{bmatrix} x_1(0.1n) \\ x_2(0.1n) \end{bmatrix} + \begin{bmatrix} 0.00498344 \\ 0.09950166 \end{bmatrix}f(0.1n)
$$

where

$$
x_1(0.1n) = y(0.1n) \quad \text{and} \quad x_2(0.1n) = \dot{y}(0.1n)
$$

We can find an approximation for A_d and for B_d numerically by first writing e^{AT} in series form, using (11.21):

$$
e^{AT} = I + AT + \frac{A^2 T^2}{2!} + \frac{A^3 T^3}{3!} + \cdots
$$

We can find the matrix A_d by truncating the preceding series after a few terms, and the matrix B_d by substituting the truncated series into equation (11.96) and evaluating, which yields

$$
B_d = \sum_{k=0}^{N} \frac{A^k T^{k+1}}{(k+1)!} B
$$

MATLAB performs a similar computation of A_d and B_d via the command c2d. For instance, in Example 11.10 the MATLAB commands are

```
kf = .1; m=1;
A = [0 1;0 -kf/m]; B = [0 1/m]'; C = [1 0];
T = 0.1;
[Ad,Bd] = c2d(A,B,T)
```

This program yields

$$A_d = \begin{bmatrix} 1.0 & 0.09950166 \\ 0 & 0.99004983 \end{bmatrix} \quad \text{and} \quad B_d = \begin{bmatrix} 0.004983375 \\ 0.099501663 \end{bmatrix}$$

which is consistent with the result obtained in Example 11.10.

11.7   CHAPTER SUMMARY

This chapter introduces the state representation of linear time-invariant continuous-time and discrete-time systems, which is specified in terms of an N-length vector $x(t)$ of internal variables, called the state variables. The state $x(t_1)$ at time t_1 completely characterizes the past history of the system (up to time t_1), so that the output response for $t \geq t_1$ can be computed from knowledge of $x(t_1)$ and the input $v(t)$ for $t \geq t_1$. Some of the state variables may not be directly measurable or directly determinable from the system outputs, which is the reason the state variables are referred to as the "internal variables" (i.e., they are signals "inside" of the system). The state equations specify the relationships between the state and the system inputs and outputs in the time-domain. Since the state model is given in terms of state variables, it is quite different from the other time-domain models (the input/output differential equation model and the convolution model) that were studied in Chapter 2. As seen from the results in this chapter, in the single input and single output case, it is easy to go from an input/output differential equation model to a state model, and vice versa.

In the systems literature, the state equation representation has been shown to be very useful in many types of applications, especially in the case of systems with multiple inputs and outputs, due in part to the first-order differential form for the state dynamics. However, the state equations are, in general, matrix equations, and thus matrix algebra is required in order for us to work with the state model. As noted in the chapter, this is easily handled by the MATLAB software, as the software is "geared to" the matrix-equation format.

The state model was developed primarily during the 1950s, and then around 1960, it was shown that this formulation provides a very effective framework for solution of the "linear quadratic optimal control problem" and signal "estimation problems." The latter resulted in the Kalman filter, which has been successfully applied to a multitude of practical applications, such as the determination of accurate estimates of positions and velocities in aerospace systems. (See Kamen and Su [1999] for details on state estimation.) In the past couple of decades, it has been discovered that transfer-function-type techniques offer advantages over the state-variable approach in the solution of optimal control problems (such as H-infinity control), but the state approach is still used in control applications and is most certainly used in estimation problems.

PROBLEMS

11.1. For the circuit in Figure P11.1, find the state model with the state variables defined to be $x_1(t) = i_L(t)$, $x_2(t) = v_C(t)$, and with the output defined as $y(t) = i_L(t) + v_C(t)$.

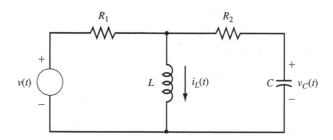

FIGURE P11.1

11.2. For the circuits in Figure P11.2, find the state model with the state variables as defined in the circuit diagrams.

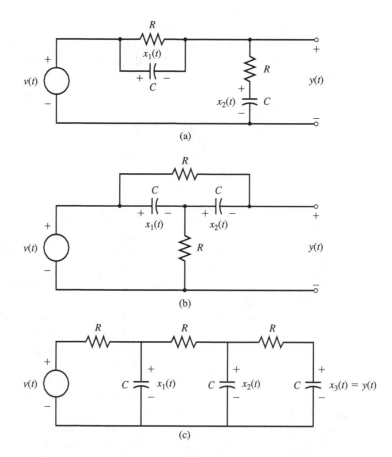

FIGURE P11.2

11.3. A system is described by the input/output differential equation

$$\ddot{y} + a_1\dot{y} + a_0 y = b_1\dot{v} + b_0 v$$

The following state model for this system was given in Section 11.2:

$$\begin{bmatrix} \dot{x}_1 \\ \dot{x}_2 \end{bmatrix} = \begin{bmatrix} 0 & 1 \\ -a_0 & -a_1 \end{bmatrix}\begin{bmatrix} x_1 \\ x_2 \end{bmatrix} + \begin{bmatrix} 0 \\ 1 \end{bmatrix}v$$

$$y = \begin{bmatrix} b_0 & b_1 \end{bmatrix}\begin{bmatrix} x_1 \\ x_2 \end{bmatrix}$$

Find expressions for x_1 and x_2 in terms of v, y, and $\dot{y}$.

11.4. When the input $v(t) = \cos t, t \geq 0$, is applied to a linear time-invariant continuous-time system, the resulting output response (with no initial energy in the system) is

$$y(t) = 2 - e^{-5t} + 3\cos t, \quad t \geq 0$$

Find a state model of the system with the smallest possible number of state variables. Verify the model by simulating the response to $v(t) = \cos t$.

11.5. A linear time-invariant continuous-time system has the transfer function

$$H(s) = \frac{s^2 - 2s + 2}{s^2 + 3s + 1}$$

Find a state model of the system with the smallest possible number of state variables.

11.6. Consider the two-input two-output linear time-invariant continuous-time system shown in Figure P11.6.
(a) Find the state model of the system with the state variables defined to be $x_1(t) = w(t)$, $x_2(t) = y_1(t) - w(t) - v_2(t)$, and $x_3(t) = y_2(t)$.
(b) Find the state model of the system with the state variables defined to be $x_1(t) = y_1(t) - v_2(t)$, $x_2(t) = y_2(t)$, and $x_3(t) = y_1(t) - w(t) - v_2(t)$.

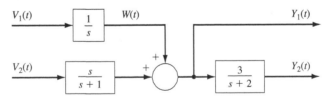

FIGURE P11.6

11.7. For the system shown in Figure P11.7, find a state model with the number of state variables equal to 2.

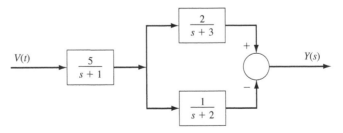

FIGURE P11.7

11.8. A linear time-invariant continuous-time system has the transfer function

$$H(s) = \frac{bs}{s^2 + a_1 s + a_0}$$

Find the state model of the system with the state variables defined to be

$$x_1(t) = y(t) \quad \text{and} \quad x_2(t) = \int_{-\infty}^{t} y(\lambda)\, d\lambda$$

11.9. For the two-input two-output system shown in Figure P11.9, find a state model with the smallest possible number of state variables.

11.10. A linear time-invariant continuous-time system has the state model

$$\begin{bmatrix} \dot{x}_1(t) \\ \dot{x}_2(t) \end{bmatrix} = \begin{bmatrix} 0 & 1 \\ -1 & 2 \end{bmatrix}\begin{bmatrix} x_1(t) \\ x_2(t) \end{bmatrix} + \begin{bmatrix} 0 \\ 1 \end{bmatrix} v(t)$$

$$y(t) = \begin{bmatrix} 1 & 2 \end{bmatrix}\begin{bmatrix} x_1(t) \\ x_2(t) \end{bmatrix}$$

Derive an expression for $x_1(t)$ and $x_2(t)$ in terms of $v(t)$, $y(t)$, and (if necessary) the derivatives of $v(t)$ and $y(t)$.

11.11. Consider the system consisting of two masses and three springs shown in Figure P11.11. The masses are on wheels that are assumed to be frictionless. The input $v(t)$ to the system is the force $v(t)$ applied to the first mass. The position of the first mass is $w(t)$, and the position of the second mass is the output $y(t)$, where both $w(t)$ and $y(t)$ are defined with respect to some equilibrium position. With the state variables defined by $x_1(t) = w(t)$, $x_2(t) = \dot{w}(t)$, $x_3(t) = y(t)$, and $x_4(t) = \dot{y}(t)$, find the state model of the system.

11.12. Consider the "dueling pendulums" shown in Figure P11.12. Each pendulum has mass M and length L. The angular position of the pendulum on the left is the output $y(t)$, and the angular position of the one on the right is $\theta(t)$. Mg is the force on each pendulum due to gravity, where g is the gravity constant. The distance between the two pendulums is $d(t)$, which is assumed to be nonnegative for all t. As shown, a spring is attached between the two pendulums. The force on each mass due to the spring depends on the amount the

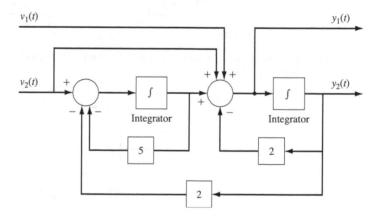

FIGURE P11.9

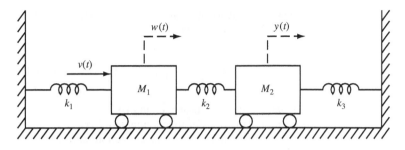

FIGURE P11.11

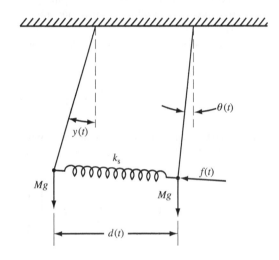

FIGURE P11.12

spring is stretched. The input to the two-pendulum system is a force $f(t)$ applied to the mass on the right, with the force applied tangential to the motion of the mass. *Assuming that the angles $\theta(t)$ and $y(t)$ are small for all t*, derive the state model for the system with the state variables defined by $x_1(t) = \theta(t)$, $x_2(t) = \dot{\theta}(t)$, $x_3(t) = y(t)$, $x_4(t) = \dot{y}(t)$, and $x_5(t) = d(t) - d_0$, where d_0 is the distance between the pendulums when they are in the vertical resting position.

11.13. A two-input two-output linear time-invariant continuous-time system has the transfer function matrix

$$H(s) = \begin{bmatrix} \dfrac{1}{s+1} & 0 \\ \dfrac{1}{s+1} & \dfrac{1}{s+2} \end{bmatrix}$$

Find the state model of the system with the state variables defined to be $x_1(t) = y_1(t)$, $x_2(t) = y_2(t)$, where $y_1(t)$ is the first system output and $y_2(t)$ is the second system output.

11.14. A linear time-invariant continuous-time system is given by the state model $\dot{x}(t) = Ax(t) + Bv(t)$, $y(t) = Cx(t)$, where

$$A = \begin{bmatrix} 0 & 2 \\ 0 & -1 \end{bmatrix}, \quad B = \begin{bmatrix} 1 & 0 \\ 1 & -1 \end{bmatrix}, \quad C = \begin{bmatrix} 1 & 3 \end{bmatrix}$$

(a) Compute the state transition matrix e^{At}.
(b) Compute the transfer function matrix $H(s)$.
(c) Compute the impulse response function matrix $H(t)$.
(d) Compute the state response $x(t)$ for $t > 0$ resulting from initial state $x(0) = [-1 \quad 1]'$ (prime denotes transpose) and input $v(t) = [u(t) \, u(t)]'$.
(e) If possible, find a nonimpulsive input $v(t)$ with $v(t) = 0$ for $t < 0$, such that the state response $x(t)$ resulting from initial state $x(0) = [1 \, -1]'$ and input $v(t)$ is given by $x(t) = [u(t) \, -u(t)]'$.

11.15. Consider a single car moving on a level surface, given by the input/output differential equation

$$\frac{d^2 y(t)}{dt^2} + \frac{k_f}{M}\frac{dy(t)}{dt} = \frac{1}{M}v(t)$$

where $y(t)$ is the position of the car at time t. With state variables $x_1(t) = y(t)$, $x_2(t) = \dot{y}(t)$, the car has the state model $\dot{x}(t) = Ax(t) + bv(t)$, $y(t) = cx(t)$, where

$$A = \begin{bmatrix} 0 & 1 \\ 0 & \dfrac{-k_f}{M} \end{bmatrix}, \quad b = \begin{bmatrix} 0 \\ \dfrac{1}{M} \end{bmatrix}, \quad c = \begin{bmatrix} 1 & 0 \end{bmatrix}$$

In the following independent parts, take $M = 1$, and $k_f = 0.1$:
(a) Using the state model, derive an expression for the state response $x(t)$ resulting from initial conditions $y(0) = y_0$, $\dot{y}(0) = v_0$, with $v(t) = 0$, for all t.
(b) With $v(t) = 0$ for $0 \leq t \leq 10$, it is known that $y(10) = 0$, $\dot{y}(10) = 55$. Compute $y(0)$ and $\dot{y}(0)$.
(c) The force $v(t) = 1$ is applied to the car for $0 \leq t \leq 10$. The state $x(5)$ at time $t = 5$ is known to be $x(5) = [50 \, 20]'$. Compute the initial state $x(0)$ at time $t = 0$.
(d) Now suppose that $v(t) = 0$ for $10 \leq t \leq 20$, and $y(10) = 5$, $y(20) = 50$. Compute the state $x(10)$ at time $t = 10$.
(e) Verify the answers for parts (b)–(d) by simulating the response of the state model.

11.16. Consider the two-car system in Example 11.5, with $k_f = 10$ and $M = 1000$. With $x(0) = [60 \, 60 \, 100]'$, compute the forces $f_1(t)$ and $f_2(t)$ that must be applied to the cars so that $x(t) = [60 \, 60 \, 100]'$ for all $t \geq 0$.

11.17. For the two-car system in Example 11.5, with $k_f = 10$ and $M = 1000$, use MATLAB to determine the velocities of the two cars and the separation between the cars for $t = 0, 1, 2, \ldots, 9$, when $x(0) = [60 \, 60 \, 100]'$ and $v(t) = [600 \, -2000]'$ for $t = 0, 1, 2, \ldots, 9$.
(a) Give the MATLAB code for determining the responses for $t = 0, 1, 2, \ldots, 9$.
(b) Generate a MATLAB plot of the velocities and the separation for $t = 0, 1, 2, \ldots, 9$.
(c) From your result in part (b), what do you conclude? Explain.

11.18. A two-input two-output linear time-invariant continuous-time system is given by the state model $\dot{x}(t) = Ax(t) + Bv(t)$, $y(t) = Cx(t)$, where

$$A = \begin{bmatrix} 1 & -1 \\ 0 & 2 \end{bmatrix}, \quad B = \begin{bmatrix} 1 & 1 \\ 0 & 0 \end{bmatrix}, \quad C = \begin{bmatrix} 1 & -1 \\ 1 & -1 \end{bmatrix}$$

(a) The output response $y(t)$ resulting from some initial state $x(0)$ with $v(t) = 0$ for all $t \geq 0$ is given by

$$y(t) = \begin{bmatrix} 2e^{2t} \\ 2e^{2t} \end{bmatrix}, \quad t \geq 0$$

Compute $x(0)$.

(b) The output response $y(t)$ resulting from some initial state $x(0)$ and input $v(t) = [u(t)\ u(t)]'$ is given by

$$y(t) = \begin{bmatrix} 4e^{t} - 2e^{2t} - 2 \\ 4e^{t} - 2e^{2t} - 2 \end{bmatrix}, \quad t \geq 0$$

Compute $x(0)$.

11.19. A linear time-invariant continuous-time system has state model $\dot{x}(t) = Ax(t) + Bv(t)$, $y(t) = Cx(t)$, where

$$A = \begin{bmatrix} -8 & -4 \\ 12 & 6 \end{bmatrix}, \quad B = \begin{bmatrix} 1 & 1 \\ 2 & 2 \end{bmatrix}, \quad C = [1 \quad -2]$$

The following parts are independent:

(a) Suppose that $y(2) = 3$ and $\dot{y}(2) = 5$. Compute $x(2)$.

(b) Suppose that $v(t) = 0$ for $0 \leq t \leq 1$ and that $x(1) = [1\ 1]'$. Compute $x(0)$.

(c) Suppose that $x(0) = [1\ 1]'$. If possible, find an input $v(t)$ such that the output response resulting from $x(0)$ and $v(t)$ is zero; that is, $y(t) = 0, t > 0$.

11.20. As first noted in Problem 6.19, the ingestion and metabolism of a drug in a human are modeled by the equations

$$\frac{dw(t)}{dt} = -k_1 w(t) + v(t)$$

$$\frac{dy(t)}{dt} = k_1 w(t) - k_2 y(t)$$

where the input $v(t)$ is the ingestion rate of the drug, the output $y(t)$ is the mass of the drug in the bloodstream, and $w(t)$ is the mass of the drug in the gastrointestinal tract. In the following parts, assume that $k_1 \neq k_2$:

(a) With the state variables defined to be $x_1(t) = w(t)$ and $x_2(t) = y(t)$, find the state model of the system.

(b) With A equal to the system matrix found in part (a), compute the state transition matrix e^{At}.

(c) Compute the inverse of e^{At}.

(d) Using your answer in part (c), compute the state $x(t)$ for all $t > 0$, when $v(t) = 0$ for $t \geq 0$ and the initial state is $x(0) = [M_1\ M_2]'$.

(e) Using the state model found in part (a), compute the state response $x(t)$ for all $t > 0$, when $v(t) = e^{at}, t \geq 0$, and $x(0) = [M_1\ M_2]'$. Assume that $a \neq k_1 \neq k_2$.

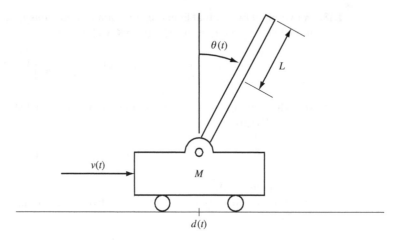

FIGURE P11.21

11.21. Consider an inverted pendulum on a motor-driven cart, as illustrated in Figure P11.21. Here, $\theta(t)$ is the angle of the pendulum from the vertical position, $d(t)$ is the position of the cart at time t, $v(t)$ is the drive or braking force applied to the cart, and M is the mass of the cart. The mass of the pendulum is m. From the laws of mechanics (see Section 2.2), the process is described by the differential equations

$$(J + mL^2)\ddot{\theta}(t) - mgL \sin \theta(t) + mL\ddot{d}(t) \cos \theta(t) = 0$$

$$(M + m)\ddot{d}(t) + mL\ddot{\theta}(t) = v(t)$$

where J is the moment of inertia of the inverted pendulum about the center of mass, g is the gravity constant, and L is one-half the length of the pendulum. We assume that the angle $\theta(t)$ is small, and therefore $\cos \theta(t) \approx 1$ and $\sin \theta(t) \approx \theta(t)$. In the following parts, take $J = 1, L = 1, g = 9.8, M = 1$, and $m = 0.1$:

(a) With the state variables defined to be $x_1 = \theta(t)$, $x_2 = \dot{\theta}(t)$, $x_3 = d(t)$, and $x_4 = \dot{d}(t)$, and the output defined to be $\theta(t)$, find the state model of the inverted pendulum.

(b) With A equal to the system matrix found in part (a), compute the state transition matrix e^{At}.

(c) Compute the inverse $(e^{At})^{-1}$ of the state transition matrix.

(d) Using your answer in part (c), compute the state $x(5)$ at time $t = 5$, assuming that $x(10) = [10°\ 0\ 5\ 2]'$ and $v(t) = 0$ for $5 \le t \le 10$.

(e) Using the state model, compute the state response $x(t)$ for all $t > 0$, when $\theta(0) = 10°, d(0) = 0, \dot{\theta}(0) = 0, \dot{d}(0) = 0$, and $v(t) = 0$ for $t \ge 0$.

(f) Repeat part (e) with $\theta(0) = 0, \dot{\theta}(0) = 1, d(0) = 0, \dot{d}(0) = 0$, and $v(t) = 0$ for $t \ge 0$.

(g) Repeat part (e) with $\theta(0) = 0, \dot{\theta}(0) = 0, d(0) = 0, \dot{d}(0) = 1$; and $v(t) = 0$ for $t \ge 0$.

(h) Verify the results of parts (d)–(g) by simulating the response of the state model.

11.22. A linear time-invariant discrete-time system is given by the state model $x[n + 1] = Ax[n] + Bv[n], y[n] = Cx[n]$, where

$$A = \begin{bmatrix} -1 & 1 \\ -1 & -2 \end{bmatrix}, \quad B = \begin{bmatrix} 0.5 & 1 \\ -1 & -0.5 \end{bmatrix}, \quad C = \begin{bmatrix} 2 & 1 \\ -1 & -2 \end{bmatrix}$$

(a) Compute $x[1], x[2]$, and $x[3]$ when $x[0] = [1\ 1]'$ and $v[n] = [n\ n]'$.

(b) Compute the transfer function matrix $H(z)$.

(c) Suppose that $x[0] = [0\ 0]'$. Find an input $v[n]$ that sets up the state $x[2] = [-1\ 2]'$; that is, the state $x[2]$ of the system at time $n = 2$ resulting from input $v[n]$ is equal to $[-1\ 2]'$.

(d) Now suppose that $x[0] = [1\ -2]'$. Find an input $v[n]$ that drives the system to the zero state at time $n = 2$; that is, $x[2] = [0\ 0]'$.

(e) Verify the results of parts (a)–(d) by simulating the response of the state model.

11.23. The input $x[n] = -2 + 2^n, n = 0, 1, 2, \ldots$, is applied to a linear time-invariant discrete-time system. The resulting response is $y[n] = 3^n - 4(2^n), n = 0, 1, 2, \ldots$, with no initial energy in the system. Find a state model of the system with the smallest possible number of state variables. Verify the model by simulating its response to $x[n] = -2 + 2^n, n \geq 0$.

11.24. Consider the discrete-time system with state model $x[n + 1] = Ax[n] + Bv[n], y[n] = Cx[n]$, where

$$A = \begin{bmatrix} 1 & 0 \\ 0.5 & 1 \end{bmatrix}, \quad B = \begin{bmatrix} 2 \\ 1 \end{bmatrix}, \quad C = \begin{bmatrix} 2 & -1 \\ 1 & 0 \\ 1 & -1 \end{bmatrix}$$

The following parts are independent:

(a) Compute $y[0], y[1]$, and $y[2]$ when $x[0] = [-1\ 2]'$ and the input is $v[n] = \sin(\pi/2)n$.

(b) Suppose that $x[3] = [1\ -1]'$. Compute $x[0]$, assuming that $v[n] = 0$ for $n = 0, 1, 2, \ldots$.

(c) Suppose that $y[3] = [1\ 2\ -1]'$. Compute $x[3]$.

(d) Verify the results of parts (a)–(c) by simulating the response of the state model.

11.25. A discrete-time system has the state model $x[n + 1] = Ax[n] + Bv[n], y[n] = Cx[n]$, where

$$CB = \begin{bmatrix} 6 \\ 3 \end{bmatrix} \quad \text{and} \quad CAB = \begin{bmatrix} 22 \\ 11 \end{bmatrix}$$

When $x[0] = 0$, it is known that $y[1] = [6\ 3]'$ and $y[2] = [4\ 2]'$. Compute $v[0]$ and $v[1]$.

11.26. For the loan/savings system in Example 11.7 do the following:

(a) Determine the number of months it takes to pay off the $300,000 mortgage.

(b) Determine the amount in the savings account when the loan is paid off.

11.27. For the loan/savings system in Example 11.7 do the following:

(a) Determine the constant c so that when the monthly loan payments are $p[n] = c, n = 1, 2, \ldots$, the loan balance remains at $300,000. This is referred to as an "interest only" loan.

(b) For your solution in part (a), determine the smallest constant q so that, when $d[n] = q, n = 1, 2, \ldots$, the amount in the savings account remains at $20,000.

11.28. A continuous-time system has state model $\dot{x}(t) = Ax(t) + bv(t), y(t) = Cx(t)$, where

$$A = \begin{bmatrix} 3 & -2 \\ 9 & -6 \end{bmatrix}, \quad b = \begin{bmatrix} 1 \\ 2 \end{bmatrix}$$

(a) Determine if there is a coordinate transformation $\overline{x}(t) = Px(t)$ such that $\overline{A}$ is in diagonal form. If such a transformation exists, give P and $\overline{A}$.

(b) Verify the results in part (a) by using MATLAB to compute $\overline{A}$ and $\overline{b}$.

11.29. We are given two continuous-time systems with state models

$$\dot{x}(t) = A_1 x(t) + b_1 v(t), y(t) = C_1 x(t)$$
$$\dot{\overline{x}}(t) = A_2 \overline{x}(t) + b_2 v(t), y(t) = C_2 \overline{x}(t)$$

where

$$A_1 = \begin{bmatrix} 1 & 1 \\ 0 & 2 \end{bmatrix}, \quad A_2 = \begin{bmatrix} 4 & 2 \\ -3 & -1 \end{bmatrix}, \quad b_1 = \begin{bmatrix} 1 \\ 1 \end{bmatrix},$$

$$b_2 = \begin{bmatrix} 5 \\ 3 \end{bmatrix}, \quad C_1 = \begin{bmatrix} 1 & 2 \\ 1 & 2 \end{bmatrix}, \quad C_2 = \begin{bmatrix} 0 & 1 \\ 0 & 1 \end{bmatrix}$$

Determine if there is a coordinate transformation $\bar{x}(t) = Px(t)$ between the two systems. Determine P if it exists.

11.30. A linear time-invariant continuous-time system has state model $\dot{x}(t) = Ax(t) + bv(t)$, $y(t) = cx(t)$. It is known that there is a coordinate transformation $\bar{x}(t) = P_1 x(t)$ such that

$$\overline{A} = \begin{bmatrix} -2 & 0 & 0 \\ 0 & -1 & 0 \\ 0 & 0 & 1 \end{bmatrix}, \quad \overline{b} = \begin{bmatrix} 1 \\ 1 \\ 1 \end{bmatrix}, \quad \overline{c} = \begin{bmatrix} -1 & 1 & 1 \end{bmatrix}$$

It is also known that there is a second transformation $\bar{x}(t) = P_2 x(t)$ such that

$$\overline{\overline{A}} = \begin{bmatrix} 0 & 1 & 0 \\ 0 & 0 & 1 \\ -a_0 & -a_1 & -a_2 \end{bmatrix}, \quad \overline{\overline{b}} = \begin{bmatrix} 0 \\ 0 \\ 1 \end{bmatrix}$$

(a) Determine a_0, a_1, and a_2.
(b) Compute the transfer function $H(s)$ of the system.
(c) Compute $\overline{\overline{c}}$.

11.31. A linear time-invariant continuous-time system is given by the state equations

$$\dot{x}_1(t) = x_1(t) + 2v(t)$$

$$\dot{x}_2(t) = 2x_2(t) + v(t)$$

(a) Compute an input control $v(t)$ that drives the system from initial state $x(0) = [1 \ -1]'$ to $x(2) = [0 \ 0]'$.
(b) Verify the results of part (a) by simulating the response of the state model.

11.32. Consider the car on a level surface with mass $M = 1$ and coefficient of friction $k_f = 0.1$. Compute an input $v(t)$ that drives the car from initial position $y(0) = 10$ and initial velocity $\dot{y}(0) = 2$ to $y(10) = 0$ and $\dot{y}(10) = 0$. Sketch $v(t)$.

11.33. A discretized continuous-time system is given by the state model $x(nT + T) = A_d x(nT) + B_d v(nT)$, $y(nT) = Cx(nT)$, where

$$A_d = \begin{bmatrix} 1 & 2 \\ -2 & 4 \end{bmatrix}, \quad B_d = \begin{bmatrix} 1 \\ 1 \end{bmatrix}, \quad C = \begin{bmatrix} 1 & 1 \end{bmatrix}$$

(a) It is known that $y(0) = 1$, $y(T) = -2$ when $v(0) = 2$, $v(T) = 4$. Compute $x(0)$.
(b) It is known that $x(0) = [2 \ -3]'$, $y(T) = -1$. Compute $v(0)$.

11.34. Determine the discrete-time simulation (with $T = 1$) for the continuous-time system $\dot{x}(t) = Ax(t) + bv(t)$, where

$$A = \begin{bmatrix} 1 & 1 \\ -1 & 1 \end{bmatrix}, \quad b = \begin{bmatrix} 0 \\ 1 \end{bmatrix}$$

11.35. Repeat Problem 11.34 for the system with

$$A = \begin{bmatrix} 0 & 1 \\ 0 & -1 \end{bmatrix}, \quad b = \begin{bmatrix} 0 \\ 1 \end{bmatrix}$$

11.36. The state model that interrelates $\theta(t)$ and $v(t)$ in the inverted pendulum (see Problem 11.21) is given by

$$\begin{bmatrix} \dot{\theta}(t) \\ \ddot{\theta}(t) \end{bmatrix} = \begin{bmatrix} 0 & 1 \\ a & 0 \end{bmatrix} \begin{bmatrix} \theta(t) \\ \dot{\theta}(t) \end{bmatrix} + \begin{bmatrix} 0 \\ b \end{bmatrix} v(t)$$

where

$$a = \frac{(M + m)mgL}{(M + m)J + Mm(L^2)} \quad \text{and} \quad b = \frac{-mL}{(M + m)J + Mm(L^2)}$$

Taking $g = 9.8$, $L = 1$, $J = 1$, $M = 1$, $m = 0.1$, and $T = 0.1$, compute the discretized state model.

11.37. Consider the two-car system given by

$$\dot{v}_1(t) + \frac{k_f}{M} v_1(t) = \frac{1}{M} f_1(t)$$

$$\dot{v}_2(t) + \frac{k_f}{M} v_2(t) = \frac{1}{M} f_2(t)$$

$$\dot{w}(t) = v_2(t) - v_1(t)$$

With the state variables $x_1(t) = v_1(t)$, $x_2(t) = v_2(t)$, $x_3(t) = w(t)$ and the outputs $y_1(t) = x_1(t)$, $y_2(t) = x_2(t)$, $y_3(t) = x_3(t)$, and with $M = 1000$, $k_f = 10$, and $T = 1$, compute the discretized state model.

11.38. Consider the dueling pendulums with the state model constructed in Problem 11.12. Take $M = L = 1$, $g = 9.8$, and $k_s = d_0 = 0.5$.

(a) Compute the discretized state model with $T = 0.1$.

(b) Using your result in part (a), write a program that computes the state $x(0.1n)$, starting with initial state $x(0)$ with $f(t) = 0$ for all $t \geq 0$.

(c) Using your program, compute $x(0.1n)$ for $0 \leq n \leq 800$ when $x(0) = [10° \ 0 \ 0 \ 0 \ 0.174]^T$. Sketch $\theta(0.1n)$ and $y(0.1n)$ for $0 \leq n \leq 800$.

(d) Discuss the behavior of the pendulums in response to the initial state $x(0)$ given in part (c).

Index